Antimutagenesis and Anticarcinogenesis Mechanisms III

BASIC LIFE SCIENCES

Ernest H. Y. Chu, Series Editor
The University of Michigan Medical School
Ann Arbor, Michigan

Alexander Hollaender, Founding Editor

A Continuation Order Plan is available for this series. A continuation order will bring delivery of each new volume immediately upon publication. Volumes are billed only upon actual shipment. For further information please contact the publisher.

Antimutagenesis and Anticarcinogenesis Mechanisms III

Edited by

Giorgio Bronzetti

Istituto di Mutagenesi e Differenziamento del CNR
Pisa, Italy

Hikoya Hayatsu

Okayama University
Okayama, Japan

Silvio De Flora

Istituto di Igiene e Medicina Preventia
Genova, Italy

Michael D. Waters

U.S. Environmental Protection Agency
Research Triangle Park, North Carolina

and

Delbert M. Shankel

University of Kansas
Lawrence, Kansas

Technical Editor:

Claire Wilson & Associates
Washington, D.C.

Springer Science+Business Media, LLC

Library of Congress Cataloging-in-Publication Data

Antimutagenesis and anticarcinogenesis mechanisms III / edited by
  Giorgio Bronzetti ... [et al.].
      p.   cm. -- (Basic life sciences , v. 61)
   'Proceedings of the Third International Conference on Mechanisms
 of Antimutagenesis and Anticarcinogenesis, held May 5-10, 1991, at
 Il Ciocco Conference Center, in Lucca, Italy'--T.p. verso.
   Includes bibliographical references and index.
   ISBN 978-0-306-44577-4   ISBN 978-1-4615-2984-2 (eBook)
   DOI 10.1007/978-1-4615-2984-2
   1. Cancer--Chemoprevention--Congresses. 2. Antimutagens-
 -Congresses. 3. Antineoplastic agents--Congresses. I. Bronzetti,
 Giorgio. II. International Conference on Mechanisms of
 Antimutagenesis and Anticarcinogenesis (3rd   1991   Lucca, Italy)
 III. Title Antimutagenesis and anticarcinogenesis mechanisms 3.
 IV. Series.
   [DNLM 1. Antimutagenic Agents--congresses. 2. Anticarcinogenic
 Agents--congresses.   QZ 202 A631 1991]
 RC268.15.A58   1993
 616.99'4071--dc20
 DNLM/DLC
 for Library of Congress   93-27858
                                                                 CIP

Proceedings of the Third International Conference on Mechanisms of Antimutagenesis and
Anticarcinogenesis, held May 5–10, 1991, at Il Ciocco Conference Center in Lucca, Italy

ISBN 978-0-306-44577-4

© 1993 Springer Science+Business Media New York
Originally published by Plenum Press New York in 1993

ACKNOWLEDGMENTS

The success of any major conference, such as this Third International Conference on Mechanisms of Antimutagenesis and Anticarcinogenesis, depends upon the combined, tireless efforts of many individuals and organizations, and upon the financial support of many sponsoring agencies and corporations. The editors of this volume take this opportunity to thank some of the many contributors to the success of the conference.

This series of conferences was initiated in response to a suggestion from Dr. Tsuneo Kada of Japan, who unfortunately passed away prior to the second conference, held in Ohito, Japan, in 1988. The first conference was held in Lawrence, Kansas, in 1985, and Professor Kada was a vigorous participant in it.

The success of this third conference, held near Lucca in the beautiful Tuscany region of Italy, was due largely to the outstanding efforts of Dr. Giorgio Bronzetti, who chaired the overall organizing committee, and Dr. Silvio de Flora, who chaired the Scientific Program Committee. They were ably assisted by a Scientific Organizing Committee which included Dr. H. Hayatsu (Japan), Dr. C. Ramel (Sweden), Dr. Y. Kuroda (Japan), Dr. D. Shankel (USA), Dr. F. Sobels (The Netherlands), and Dr. M. Waters (USA). An outstanding International Advisory Board, with strong representation from 13 different countries, provided welcome advice on the selection of speakers, participants, and topics. And a Local Organizing Committee of distinguished Italian scientists assured outstanding arrangements for the conference and participants. The organizers wish to express their sincere thanks to all those members who helped to assure the success of the conference.

Funding for support of this conference was provided by the following major contributors, sponsors, and companies:

Parmalat (Italy)
Carol-Erba-Farmiltalia
Parmigiano Reggiano
Gentili Pharmaceutical Pisa
Monte dei Paschi di Siena Bank
Banca Nazionale del Lavoro
Commission of the European
   Communities (EEC)
International Agency for Research on
   Cancer

U.S. Environmental Protection Agency,
   Office of Health and Environmental
   Assessment and Office of Health
   Research
U.S. Department of Energy
National Institutes of Health/National
   Cancer Institute (USA)
U.S. Department of Agriculture
Hoffman-La Roche Pharmaceutical
   Company

Smith-Kline Beecham Pharmaceutical Company
The Coca Cola Company (Italy and USA)
Eisai Chemical Company (Japan)
Ajinomoto Company
Fujisawa-Yakuhin Industries
Funakoshi-Yakuhin Company

Kyoua-Hakko Industries
Naris Cosmetics Co. (Japan)
Otsuka-Seiyaku Company
Sumitomo Chemical Industries
Taiho-Yakuhin Industries
Takeda-Yakuhin Industries
Teikokuzoki-Seiyaku Co.
Yamanouchi-Seiyaku Co.

In addition, other contributions were received from a number of industries in Japan and Italy--including welcome contributions from Italian wineries and the Leather Association of S. Croce, which guaranteed traditional Italian hospitality for the participants.

The conference was held (May 5-10, 1991) at "Il Ciocco" Conference Center in the hills above Lucca, and the arrangements were exceptional. The organizers wish to express their appreciation to the management and staff of the Center, and also to the secretarial and administrative staffs at the Istituto di Mutagenesi e Differenziamento in Pisa, and in the Executive Vice Chancellor's and Chancellor's Office at the University of Kansas, for their unfailing and cheerful support work in facilitating a successful conference.

Ultimately, the success of any scientific conference depends upon the quality of the scientific participants and on their presentations, and the organizers are especially grateful for the participation of the superb scientists and students who participated freely and extensively in the shared science that sets the stage for further advances in our knowledge of this growing and important scientific field. We are also deeply appreciative for the work of Claire Wilson, who served as Technical Editor for this volume.

Readers of this volume will be pleased to know that the conference also included a workshop on "Assessment of Antimutagenicity and Anticarcinogenicity: End-points and Systems." The excellent papers presented in this workshop have been published in a special issue of *Mutation Research* (257:153-298, 1992), edited by de Flora, Bronzetti, and Sobels.

Our ability to improve *prevention* of cancer and other genetically based diseases, to delay or prevent the mutations that contribute to the aging process, to deter mutations that lead to drug resistance in pathogenic microorganisms, and to improve the lives and health of all peoples will be the ultimate reward for the participants and for society. To all of those who have helped to make these advances possible, we extend our deep appreciation.

Delbert M. Shankel
for the Editors

CONTENTS

MODULATION OF METABOLISM AND BLOCKING
OF REACTIVE SPECIES

PROSPECTS IN CHEMOPREVENTION OF MUTATION AND CANCER

MECHANISMS OF ANTIMUTAGENESIS AND ANTICARCINOGENESIS:

ROLE IN PRIMARY PREVENTION

Silvio De Flora, Alberto Izzotti, and Carlo Bennicelli

Institute of Hygiene and Preventive Medicine
University of Genoa
I-16132 Genoa, ITALY

INTRODUCTION

Rationale for Chemoprevention as Related to Other Intervention Strategies

Figure 1 reports the possible intervention strategies against cancer, as related to the multistep carcinogenesis process and to growth of the neoplastic mass. Similar concepts may hold true for other mutation-related conditions and, in general, for those chronico-degenerative diseases having a multifactorial origin, a multistep pathogenesis, and a long latency period. Keeping in mind that the neoplastic mass is of monoclonal origin, and assuming a regular doubling of the population of neoplastic cells, 30 cell divisions will be needed to form a mass of 10^9 cells from a single cell undergoing initiation. Very approximately, this mass may weigh 1 g. At this stage, depending on many variability factors, it may be possible to apply secondary prevention, involving early detection and therapy. Otherwise, in 3.25 further divisions the mass will be composed of 10^{10} cells and weigh 10 g. At this stage, which may already involve invasion and spread of metastases, the disease will become clinically manifest and will be treated with the most suitable therapeutic protocol. This intervention will be followed by tertiary prevention, which aims at avoiding relapses, complications, metastases, and second primitive tumors. In the absence of any medical intervention, in 10 cell divisions only (i.e., from the 30th to the 40th division), the neoplastic mass will grow from 1 g to as much as 1 kg (20). Although this computer-drawn growth is a mathematical oversimplification and does not take into account possible natural regressions of the disease, it gives an idea of the paramount importance of early detection and explains the difficulties that are encountered in cancer therapy when the disease becomes clinically manifest. As a primary prevention, which is addressed to healthy individuals, an obvious approach is to minimize exposure to environmental and lifestyle carcinogens. However, as shown in Fig. 1, there is a latency period, lasting years or more often decades in humans before onset of malignancy, during

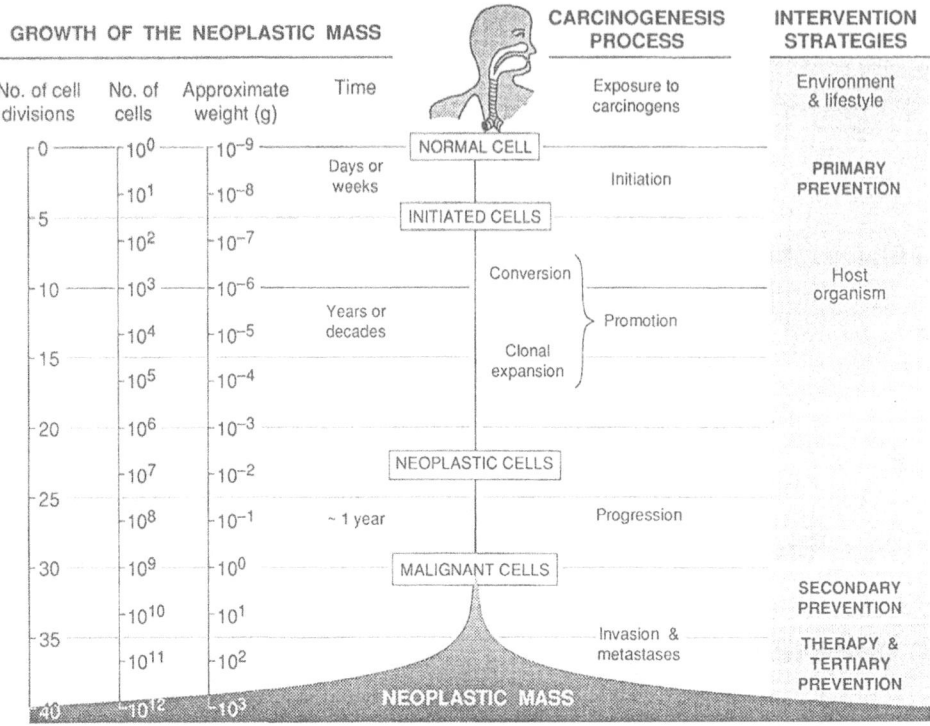

GROWTH OF THE NEOPLASTIC MASS					CARCINOGENESIS PROCESS	INTERVENTION STRATEGIES
No. of cell divisions	No. of cells	Approximate weight (g)	Time		Exposure to carcinogens	Environment & lifestyle
0	10^0	10^{-9}		NORMAL CELL		
	10^1	10^{-8}	Days or weeks		Initiation	PRIMARY PREVENTION
5	10^2	10^{-7}		INITIATED CELLS		
10	10^3	10^{-6}			Conversion	Host organism
	10^4	10^{-5}	Years or decades		Promotion	
15	10^5	10^{-4}			Clonal expansion	
20	10^6	10^{-3}				
	10^7	10^{-2}		NEOPLASTIC CELLS		
25	10^8	10^{-1}	~ 1 year		Progression	
30	10^9	10^0		MALIGNANT CELLS		
	10^{10}	10^1				SECONDARY PREVENTION
35	10^{11}	10^2			Invasion & metastases	THERAPY & TERTIARY PREVENTION
40	10^{12}	10^3		NEOPLASTIC MASS		

Figure 1. Intervention strategies against cancer, as related to the multistep carcinogenesis process and to growth of the neoplastic mass.

which we can try to reinforce the host defense machinery by applying chemopreventive measures. These are intended to minimize the adverse effects of agents of either exogenous or endogenous source to which the human organism is continuously and often inevitably exposed.

Dual Approach to the Primary Prevention of Mutations and Cancer

It is, therefore, evident that the primary prevention of mutations and cancer is based on a dual approach (Fig. 2). The methodological tools include a variety of end-points and systems, from humans to prokaryotes and mathematical models, which are discussed in detail in a special issue of *Mutation Research* (31). Similar techniques are used for assessing genotoxic and/or carcinogenic risks, or intermediate end-points thereof, and chemical interactions evaluating modulation of the genotoxic and/or carcinogenic response. The information provided by risk assessment is exploited for implementing risk management measures, involving appropriate regulations and health education for the control of environmental and lifestyle risk factors, respectively. In parallel, the assessment of protective effects will serve as a base for chemoprevention by means of either pharmacological agents or dietary factors.

2

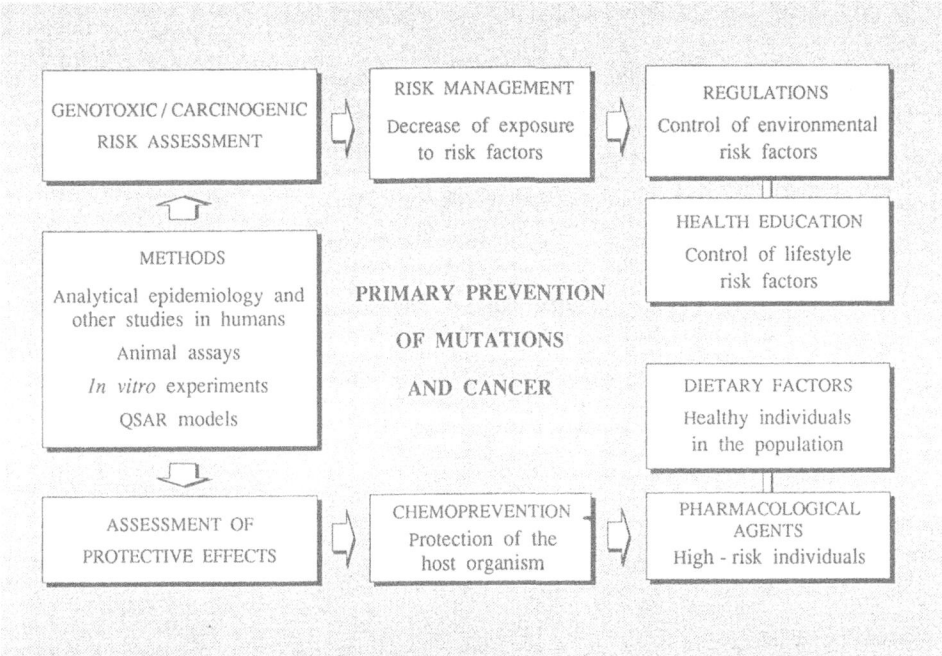

```
┌─────────────────────────┐      ┌─────────────────────────┐      ┌─────────────────────────┐
│ GENOTOXIC / CARCINOGENIC │  ⇨   │    RISK MANAGEMENT       │  ⇨   │      REGULATIONS         │
│    RISK ASSESSMENT       │      │   Decrease of exposure   │      │ Control of environmental │
│                          │      │     to risk factors      │      │      risk factors        │
└─────────────────────────┘      └─────────────────────────┘      └─────────────────────────┘
            ⇧                                                      ┌─────────────────────────┐
┌─────────────────────────┐                                       │    HEALTH EDUCATION      │
│         METHODS          │                                       │    Control of lifestyle  │
│  Analytical epidemiology │        PRIMARY PREVENTION             │      risk factors        │
│  and other studies in    │                                       └─────────────────────────┘
│        humans            │           OF MUTATIONS
│      Animal assays       │                                       ┌─────────────────────────┐
│    In vitro experiments  │           AND CANCER                  │     DIETARY FACTORS      │
│       QSAR models        │                                       │    Healthy individuals   │
└─────────────────────────┘                                       │     in the population    │
            ⇩                                                      └─────────────────────────┘
┌─────────────────────────┐      ┌─────────────────────────┐      ┌─────────────────────────┐
│     ASSESSMENT OF        │  ⇨   │    CHEMOPREVENTION       │  ⇨   │     PHARMACOLOGICAL      │
│   PROTECTIVE EFFECTS     │      │    Protection of the     │      │         AGENTS           │
│                          │      │      host organism       │      │  High - risk individuals │
└─────────────────────────┘      └─────────────────────────┘      └─────────────────────────┘
```

Figure 2. Methodological tools and approaches applicable to the primary prevention of mutations and cancer.

Requisites and Targets for Chemoprevention

In spite of its tremendous prevalence and mortality rates in the population, in relative terms, cancer and other clinically apparent diseases just represent the visible top of a huge "iceberg," which is the target for therapy. The submerged "base" symbolizes preclinical states and the interactions occurring between lifestyle or environmental risk factors and the host defense machinery. In principle, these interactions can be modulated exogenously, which is the premise for chemoprevention in healthy individuals.

Chemopreventive agents are expected to possess some general requisites, such as low cost for long-term administration, practicality of use, efficacy, and tolerability. Indeed, the major problem hampering a wider application of chemoprevention is that most inhibitors of mutagenesis and carcinogenesis, at least under certain conditions, may cause adverse effects (21,22). Therefore, although the highest degree of both efficacy and tolerability would always be desirable, inverse gradients of these requisites can be accepted (Fig. 3). Thus, therapy in cancer patients needs the maximum clinical efficacy, in spite of severe side-effects. On the opposite side, a public health intervention in the general population must be absolutely safe, even when the extent of effectiveness is lower or more uncertain. In fact, even rare side-effects are not acceptable in large-scale

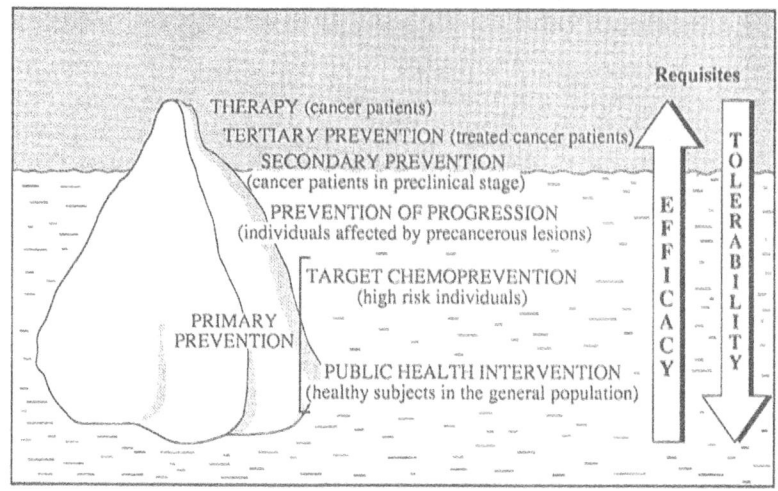

Figure 3. Inverse gradients of efficacy and tolerability in chemotherapy and chemoprevention of cancer and their targets.

treatments addressed to healthy individuals. For this reason, chemoprevention in the general population is so far limited to dietary suggestions, which, incidentally, have already proven to contribute successfully to the primary prevention of other chronico-degenerative diseases, such as cardiovascular diseases. Target chemoprevention, as pursued so far in a number of chemopreventive trials using pharmacological agents, is addressed to high-risk individuals. Hence, under certain limits, unimportant side-effects can be accepted. As shown in Fig. 3, intermediate situations can be envisioned for the other intervention strategies that have been previously defined (Fig. 1).

ANTIMUTAGENESIS AND ANTICARCINOGENESIS MECHANISMS

A mechanistic approach is needed for a safer and more effective implementation of chemoprevention (19), which is the general theme of these proceedings [ICMAA-I (61), ICMAA-II (48), and ICMAA-III (9)]. Understanding of the mechanisms involved in the protection against mutagens and carcinogens is a difficult task and requires appropriate methodological tools (31). Often, it is not easy to differentiate whether an observed end-point represents a real mechanism or an effect due to other mechanisms. For instance, certain end-points observed at the DNA level (e.g., prevention of DNA damage, inhibition of the expression of oncogenes, etc.) may represent the target of chemopreventive agents or, alternatively, be the consequence of earlier protective events occurring in extranuclear sites.

Classification of Individual Mechanisms

Table 1 provides a detailed classification of the mechanisms of inhibitors of mutagenesis and carcinogenesis, which is updated and modified from De Flora

Table 1. Mechanisms of inhibitors of mutagenesis and carcinogenesis [modified from De Flora and Ramel (1988), Ref. 22].

1   Inhibition of mutagens/carcinogens by extracellular mechanisms
    1 1   Inhibition of uptake
        1 1 1   Inhibition of penetration
        1 1 2   Removal from the organism
    1 2   Inhibition of endogenous formation
        1 2 1   Inhibition of the nitrosation reaction
        1 2 2   Modification of the microbial intestinal flora
    1 3   Complexation, dilution and/or deactivation
        1 3 1   By physical or mechanical means
        1 3 2   By chemical reaction
        1 3 3   By enzymatic reaction
2   Inhibition of mutation and cancer initiation by cellular mechanisms
    2 1   Inhibition of cell replication
    2 2   Stimulation of trapping and detoxification in nontarget cells
    2 3   Modification of transmembrane transport
        2 3 1   Inhibition of cellular uptake
        2 3 2   Stimulation of extrusion outside cells
    2 4   Modulation of metabolism
        2 4 1   Inhibition of activation of promutagens/procarcinogens
        2 4 2   Induction of detoxifying mechanisms
        2 4 3   Stimulation of activation, coordinated with detoxification and blocking of reactive metabolites
    2 5   Blocking or competition with reactive molecules
        2 5 1   Reaction of nucleophiles with electrophiles
            2 5 1 1   By chemical reaction
            2 5 1 2   By enzymatic reaction
        2 5 2   Scavenging of reactive oxygen species
        2 5 3   Protection of nucleophilic sites of DNA
    2 6   Modulation of DNA repair and control of gene expression
        2 6 1   Increase of the fidelity of DNA replication
        2 6 2   Stimulation of repair and/or reversion of DNA damage
        2 6 3   Inhibition of error-prone repair pathways
        2 6 4   Inhibition of DNA repair leading to death of damaged cells
        2 6 5   Control of gene expression
            2 6 5 1   Inhibition of oncogene expression
            2 6 5 2   Restoration of deleted suppressor genes
3   Inhibition of tumor promotion, progression, invasion, and metastases
    3 1   Modulation of tumor promotion
        3 1 1   Inhibition of genotoxic effects
        3 1 2   Scavenging of free radicals
        3 1 3   Inhibition of proteases
        3 1 4   Inhibition of cell proliferation
        3 1 5   Protection of intercellular communications
        3 1 6   Induction of cell differentiation
        3 1 7   Modulation of signal transduction
    3 2   Modulation of tumor progression
        3 2 1   Inhibition of genotoxic effects
        3 2 2   Inhibition of proteases
        3 2 3   Modulation of signal transduction
        3 2 4   Effects on growth factors
        3 2 5   Effects on hormones
        3 2 6   Effects on the immune system
        3 2 7   Inhibition of neovascularization
        3 2 8   Physical, chemical, or biological antineoplastic activity
    3 3   Modulation of invasion and metastases
        3 3 1   Inhibition of proteases
        3 3 2   Induction of cell differentiation
        3 2 3   Inhibition of neovascularization
        3 2 4   Effect on cell-adhesion molecules
        3 2 5   Modulation of the interaction with the extracellular matrix

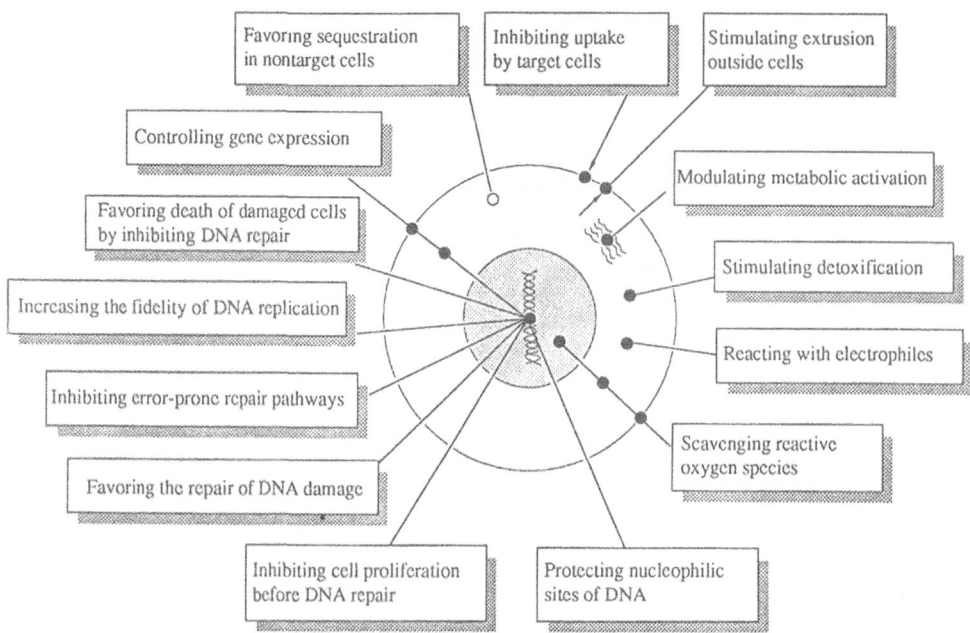

Figure 4. Mechanisms preventing mutations and cancer initiation at the cellular level.

Transport"), or by favoring their removal from the organism. For instance, washing and mechanical removal of exogenous contaminants and endogenous secretions from the skin, oral mucosa, and genital mucosa can prevent adverse local effects or spread to other compartments of the organism. The respiratory tract has formidable aspecific defense mechanisms, including not only the action of sweeping cells (see "Stimulation of Trapping and Detoxification in Nontarget Cells") but also protective devices of the mucosa, such as the muco-ciliatory escalator and shedding with the epithelial lining fluid, having a reducing capacity (53). The optimal efficiency of these mechanisms can be restored, under pathological conditions and/or overexposure to air pollutants or tobacco smoke, by means of suitable pharmacological agents.

A well-recognized role in favoring removal of dietary mutagens and carcinogens from the organisms is played by fibers, including heterogeneous carbohydrate compounds, such as cellulose, hemicellulose, and pectin, as well as noncarbohydrate compounds, such as lignin. A variety of mechanisms can contribute to the protective activity of fibers, also depending on their nature. The increase of fecal bulk results in an accelerated intestinal transit time and excretion, thereby minimizing the possibility of contact of genotoxins and tumor promoters, either introduced with food or of endogenous source (e.g., bile acids), with metabolizing cells (including intestinal bacteria) and target cells of the intestinal mucosa. Additional mechanisms may include altering absorption of bile salts and exogenous mutagens (e.g., food pyrolysis products), lowering of colonic pH, decreased production of fecal mutagens, increased production of short-chain fatty acids, and modification of metabolism during enterohepatic circulation (57,69). Fibers appear to lower the risk not only of colon cancer but also of breast cancer, possibly via endocrine mechanisms (69).

Inhibition of Endogenous Formation

Endogenous formation of mutagens and carcinogens may occur via reactions between inactive precursors, for example, nitrosation of aminocompounds, or metabolic activation by colonizing prokaryotes, such as intestinal bacteria.

A number of compounds and complex mixtures have been shown to inhibit the nitrosation reaction and/or to block the resulting *N*-nitroso compounds (NOC). Examples of chemical compounds include vitamins, such as ascorbic acid and α-tocopherol; phenolic compounds, such as catechol, cinnamic acid, chlorogenic acid, gallic acid, hydroquinones, phenolic acids, pyrogallol, tannic acid and tannins, thymol, BNA, and BHT; sulfur compounds, such as bisulfite, cysteine, GSH, NAC, methionine, sulfamic acid, and sulfur dioxide; a variety of miscellaneous compounds, such as alcohols, azides, caffeine carbohydrates, hydrazine, hydroxylamine, sorbic acid, unsaturated fatty acids, and urea. Complex mixtures include food extracts and beverages, such as alcoholic beverages, betel nut extracts, coffee, tea, fruit juices, milk, radish juice, soya products, cheese casein, pectin, and gelatin (6,28). The chemistry of NOC formation and inhibition, the protective effects produced by inhibitors of the nitrosation reaction in animal models, and their contribution to human cancer prevention are discussed in this volume (6).

The microecological profile of the intestinal flora can be optimized by suppressing putrefactive organisms involved in the local production of putative carcinogens. A protective role has been ascribed to dairy products, especially those containing *Lactobacillus acidophilus*, which is capable of implanting in the intestinal tract and of lowering the activity of enzymes of bacterial source involved in the metabolism of xenobiotics, such as β-glucosidase, nitroreductase, azoreductase, and α-dehydroxylase (56). Indeed, a broad literature is now available on the antitumor properties of fermented milk products (56), which have long been suspected to increase longevity, as reported not only by Metchnikoff at the beginning of this century but also by Old Testament scriptures.

Complexation and Deactivation

Mutagens of either endogenous or exogenous source may undergo complexation or deactivation in extracellular spaces, according to a variety of mechanisms. For instance, entrapping of mutagens by micelles of oleic acid and C16-24 unsaturated fatty acids may contribute to their protective effects in the intestine (38). Porphyrins form a face-to-face complex with multi-ring mutagens (38). A mechanism of this type is probably involved with hemoglobin and myoglobin (5), and chlorophyll and chlorophyllin (28). Formation of insoluble salts with cholic acid and fatty acids may be one of the mechanisms by which calcium prevents colon carcinogenesis (50). Compounds such as bioflavonoids, creatinine, and ergothioneine are capable of chelating metal ions, especially heavy metals (37). A number of nucleophiles can block reactive molecules outside cells. An example is provided by thiols, which eliminate the mutagenicity of several direct-acting compounds (29). GSH exerts this mechanism not only intracellularly, but also in

7

extracellular spaces, for example, in the blood plasma, where GSH levels of about 25 μM can be dynamically maintained by stimulating its synthesis and efflux from the liver (4), or in the intestinal lumen, where enterobacteria contain millimolar ranges of GSH, which can be exported extracellularly (52). There is indirect evidence that GSH concentrations in enterobacteria can be enhanced by adding GSH itself or its precursor NAC (13). Antioxidants, including antioxyenzymes, can afford protective effects also extracellularly. For instance, peroxidase and NADPH-oxidase activities may be responsible for deactivation of food pyrolysis products by vegetable juices (41). Maintenance of a physiological pH in biological fluids is also important. Therefore, administration of dietary or pharmacological protectors of the gastric mucosa may be advisable in case of enhancement of gastric pH. In individuals affected by chronic gastritis, achlorhydria is associated with an increased incidence of gastric cancer (18), and gastric juice becomes mutagenic following enhancement of pH due to treatment with antisecretory drugs (26).

CELLULAR MECHANISMS PREVENTING MUTATIONS AND CANCER INITIATION

Even more intricate is the network of protective mechanisms which can be stimulated at the cellular level in order to prevent the occurrence and fixation of DNA damage. They are schematically reported in Tab. 1 and Fig. 4.

Inhibition of Cell Replication

Antiproliferative effects, for example, those produced by retinoids, anti-inflammatory drugs, food components, and various other agents, provide a crucial defense mechanism not only during tumor promotion but also throughout other stages of the carcinogenesis process. Inhibition of cell replication is very important as an antimutagenesis mechanism. In fact, mitogenesis favors occurrence and fixation of DNA damage, for example, for triggering mitotic recombination, gene conversion, nondisjunction, gene duplication, and expression of oncogenes (3). Moreover, error-free repair of DNA must occur before cell division, and therefore slowly proliferating cells have more time allowed for DNA repair. This explains the lower sensitivity of cells in early G^1 phase, as compared to their particular vulnerability in S phase (45).

Stimulation of Trapping and Detoxification in Nontarget Cells

Pharmacokinetic and metabolic processes are of fundamental importance in discriminating target cells from nontarget cells. Mutagens and carcinogens can be accumulated in nontarget cells, where they can be detoxified via inducible mechanisms. Of particular interest are long-living carrier or sweeping cells, such as erythrocytes and pulmonary alveolar macrophages. For instance, sulfhydryl groups in erythrocytes, which have been shown to be increased by in vivo treatment with N-acetylcysteine (24), can reduce hexavalent chromium, which is accumulated in these cells during transport in the blood (23), and can denitrosate nitroso compounds, such as nitrosocimetidine (42). Alveolar macrophages provide a formidable bulwark of terminal airways, not only by phagocytizing particulate material containing harmful substances, but also by metabolizing certain mutagens. The ability to reduce hexavalent chromium and the activity of phase II enzymes, such as GSH S-transferase, were enhanced in these cells by the in vivo administration of N-acetylcysteine (25).

Modification of Transmembrane Transport

A two-way modulation of transmembrane transport may provide an interesting defense mechanism of target cells. Inhibition of cellular uptake may occur in several ways. For instance, the damage of colon epithelium by fat and bile acids can be reduced by dietary calcium, protecting against an increased permeability of the epithelium and a loss in integrity of the mucosal barrier (11). Acylglucosylsterols, isolated from green fruits, have been suspected to alter the cell permeability to mitomycin C (35). Oral vaccines against carcinogens, aimed at stimulating formation of secretory immunoglobulin A (IgA) protecting the intestinal mucosa, have even been proposed (62), but their practical applicability remains to be established. It is also possible to prevent the generation of mutagens by competitive cellular uptake. For instance, iodide is administered in case of nuclear accidents because it competes with the uptake of ^{131}I by thyroid cells. Other examples are putrescin, competing with the cellular uptake of the oxidative mutagen paraquat (10), and aromatic aminoacids, competing with the transport of azaserine in bacterial cells (2).

On the other hand, an attractive hypothesis is that mechanisms responsible for the elimination of xenobiotics outside cells may be stimulated. Studies are now in progress in our laboratory, in collaboration with Dr. B. Kurelec (Zagreb, Yugoslavia), in order to assess whether the multidrug resistance factor may play a role in defending the cell against mutagens, and whether this mechanism may be modulated by thiols. This factor, coded by the *mdr* gene, acts as a transmembrane pump, determining the elimination of cytostatic drugs outside tumor cells (47). Expression of the *mdr* gene is elevated in rat liver preneoplastic and neoplastic lesions, as well as in regenerating liver cells (66).

Modulation of Metabolism

Regulation of metabolic pathways involved in the intracellular bio-transformation of mutagens/carcinogens is a very difficult and delicate task because several pathways have a common genetic regulation and many inhibitors possess pleiotropic properties interfering with the intricate network of biochemical processes governing the balance between metabolic activation and detoxification. We refer to our previous reviews for more details on these problems (22, 28).

As shown in Tab. 1, an intuitive approach is to inhibit activation of promutagens/procarcinogens and, at the same time, induce detoxifying mechanisms. Some monofunctional inducers, such as synthetic phenols (e.g., BHA and BHT) and dithiolethiones (e.g., oltipraz), are capable of selectively stimulating phase II detoxification enzymes, such as GSH S-transferase and DT diaphorase (46,55,65). Other agents, such as flat planary molecules, at the same time induce detoxifying cytosolic enzymes and cytochrome PI-450, by binding a specific *Ah* receptor (65). Therefore, depending on the experimental conditions and on the monitored parameters, a positive or negative modulation may be observed with these agents, which are also contained in vegetables known for their protective properties, such as cruciferous plants (64). However, it should be noted that stimulation of metabolic activation, when coordinated with detoxification and blocking of reactive metabolism, has the advantage of preventing accumulation of unmetabolized hydrophobic procarcinogens in the organism and of favoring the excretion of their water-soluble conjugated metabolites. A mechanism of this type has been postulated with indole derivatives (43) and with thiols, such as NAC (29).

A crucial role in the cellular defense against toxic, mutagenic, and carcinogenic agents is played by GSH. This tripeptide and its precursors and analogs possess a variety of antimutagenic and anticarcinogenic properties, due to multiple mechanisms (29). Therefore, maintenance of optimal intracellular stores and modulation of the enzymes involved in the synthesis and metabolism of GSH are very important. This is particularly true when GSH stores are depleted, which occurs not only following exposure to toxic substances but also in viral diseases that are associated with cancer such as viral hepatitis B (27) and AIDS (12).

Blocking or Competition with Reactive Molecules

GSH itself and other intracellular nucleophiles can block electrophilic molecules, either by direct chemical reaction or by enzymatic conjugation, leading to formation of inactive products, such as GSH conjugates, glucuronides, and sulfates. Phase II enzymes, such as GSH S-transferase isoenzymes, uridine diphosphate glucuronyl transferases, sulfotransferases, and acteyltransferases can be stimulated by a variety of inhibitors acting as either monofunctional or bifunctional inducers, as discussed in the previous section. Blocking agents have been proposed as first priority chemopreventive tools applicable to the general population (67).

Scavenging of reactive oxygen species and stimulation of antioxidant enzymes are often associated with blocking of electrophiles and induction of phase II enzymes. Reactive oxygen species are involved in genotoxicity and various stages of carcinogenesis, including the oxidative metabolism of certain xenobiotics as well as cancer initiation, promotion, and progression (54). These mechanisms have been extensively reviewed (14, 40,63). Antioxidant mechanisms contribute to the protective effects of a large variety of inhibitors of mutagenesis and carcinogenesis (22,28).

Although most blocking agents work by trapping reactive molecules, thereby hampering their reaction with DNA, the possibility has been also envisaged that certain inhibitors may bind nucleophilic sites of DNA and, therefore, protect them from the attack of electrophiles. A typical example is provided by ellagic acid, a naturally occurring phenol present in plants, which interferes with metabolic activation, traps reactive epoxides, and, in addition, is capable of binding DNA and preventing its methylation (32).

Modulation of DNA Repair and Control of Gene Expression

As already mentioned, antiproliferative agents favor error-free repair of DNA damage by prolonging the G_2 phase of cell replication. As shown in Tab. 1 and Fig. 4, several other strategies are available for modulating DNA repair. For instance, according to Kada et al. (44), protective effects can be obtained by (a) increase of the fidelity of DNA replication, (b) stimulation of the repair of DNA damage, and (c) inhibition of errorprone repair systems. A variety of compounds, such as cobaltous chloride, sodium arsenite, tannic acid, protease inhibitors, *p*-aminobenzoic acid, and α-tocopherol, have been shown to interfere with these processes in several systems, and especially in bacteria (28). However, it is evident that regulation of DNA repair mechanisms is extremely delicate. Certain anticarcinogens, such as NAC, protect nuclear enzymes, for example, poly(ADP-ribose)polymerase (ADPRP), from depletion by carcinogens (15). A different strategy, which may be pursued, for instance, in chemotherapeutic and radiotherapeutic regimens, is to use DNA repair modulators in such a way that damaged cells are preferentially killed rather than being repaired defectively and potentially subjected to neoplastic development. This is mainly achieved by inhibiting early DNA repair processes, which ensure cell survival at the risk of increased mutations (8).

In principle, there is the possibility of modulating genotoxicity by correcting wrong cellular messages at the level of genes and gene products, forming a chain of products from the outer cell membrane via the cytoplasm to the nucleus (68). It is nowadays possible to correct genetic changes associated with cancer, both in in vitro and in vivo experimental systems (21,34), but clearly application of this approach to cancer prevention encounters both technical and ethical problems. Inhibition of oncogene expression has been obtained in a variety of experimental systems by using known anticarcinogens, such as protease inhibitors (16), retinoic acid (71), a metabolite of vitamin D^3 (58), and *S*-adenosyl-Lmethionine (33). A novel pharmacological approach to controlling oncogene-mediated malignant transformation has been offered by the discovery that a connection exists between the cholesterol biosynthetic pathway and transformation by the *ras* oncogene (60). In additon, human immunoglobulin genes have been shown to have anticarcinogenic effects in c-*myc* bearing transgenic mice (51). Inhibitors of ADPRP induce a loss of exogenous transforming genes transfected into cultured mammalian cells (49). Synthetic oligonucleotides are also being investigated in order to control oncogene expression by inhibiting either translation (antisense strategy) or transcription (antigene strategy) (39). In any case, the possibility that chemopreventive agents may inhibit oncogene expression or function is complicated by the fact that many proto-oncogenes play essential roles in normal cell physiology, which stimulates the development of strategies to selectively target oncogene proteins (17). As to suppressor genes, it is well documented that their deletion can be successfully repaired in vitro by inserting normal human chromosomes, such as by microcell hybridization (70), and somatic cell genetic analyses of hybrids between normal cells and neoplastic cells have generally shown that the normal cell has a dominant effect in suppressing the tumorigenicity of the malignant cell (36).

MECHANISMS PREVENTING TUMOR PROMOTION, PROGRESSION, INVASION, AND METASTASES

All mechanisms that have been discussed so far bear relevance to prevention of genotoxic effects and of cancer initiation. However, it should be borne in mind that the operative distinction of the carcinogenesis process into a classical initiation-promotion-progression-invasionmetastasis scheme or in further fractionations thereof, although scholastically useful, is not so sharp. It has been argued that the two-step (initiation-promotion) paradigm may be inverted, for instance, in hormonal carcinogenesis, in which cellular proliferation precedes genetic damage (59). In any case, there is no doubt that events of a similar nature may occur along all stages of the carcinogenesis process, and also in very advanced stages, such as invasion and metastases.

For this reason, at lesst some of the mechanisms of modulation of tumor promotion, tumor progression, invasion, and metastases that are reported in Tab. 1 are the same as those already discussed in previous pages. We refer to our previous reviews (22,28) and to specifically addressed articles in this volume (e.g., Ref. 1) for more details on mechanisms of anticarcinogenesis and for examples of inhibitors affecting advanced stages of the carcinogenesis process.

REFERENCES

1. Albini, A., and A. Colacci (1992) Inhibition of malignant cell invasion: An approach to anti-progression. This volume.
2. Ames, G.F. (1964) Uptake of amino acids by *Salmonella typhimurium*. *Arch. Biochem. Biophys.* 18:1-18.
3. Ames, B.N., and L.S. Gold (1990) Chemical carcinogenesis: Too many rodent carcinogens. *Proc. Natl. Acad. Sci.*, USA 87:7772-7776.
4. Anderson, M.E., and A. Meister (1980) Dynamic state of glutathione in blood plasma. *J. Biol. Chem.* 255:9530-9533.
5. Arimoto, S., and H. Hayatsu (1989) Role of hemin in the inhibition of mutagenic activity of 3-amino-1-methyl-5H-pyrido[4,3b]indole (Trp-P-2) and other aminoazaarenes. *Mutat. Res.* 213:217-226.
6. Bartsch, H., B. Pignatelli, S. Calmels, and H. Ohshima (1992) Inhibition of nitrosation. This volume.
7. Bertram, J.S., L.N. Kolonel, and F.L. Meyskens, Jr. (1987) Rationale and strategies for chemoprevention of cancer in humans. *Cancer Res.* 47:3012-3031.
8. Boothman, D.A., R. Schlegel, and M.A. Pardee (1988) Anticarcinogenic potential of DNA-repair modulators. *Mutat. Res.* 202:393-411.
9. Bronzetti, G., H. Hayatsu, S. De Flora, M.D. Waters, and D.M. Shankel, eds. (1992). This volume.
10. Brooke-Taylor, S., L.L. Smith, and G.M. Cohen (1983) The accumulation of polyamines and paraquat by human peripheral lung. *Biochem. Pharmacol.* 32:717-720.
11. Bruce, W.R., R.P. Bird, and J.J. Rafter (1986) The effect of calcium on the pathogenicity of high fat diets to the colon. In *Diet, Nutrition and Cancer*, Y. Hayashi, M. Nagao, T. Sugimura, S. Takayama, L. Tomatis, L.W. Wattenberg, and J.N. Wogan, eds. Japan Scientific Societies Press, Tokyo, Japan/VNU Science Press BV, Utrecht, The Netherlands, pp. 291-294.
12. Buhl, R., K.J. Holroyd, A. Mastrangeli, A.M. Cantin, H.A. Jaffe, F.B. Wells, C. Saltini, and R.G. Crystal (1989) Systemic glutathione deficiency in symptom-free HIV-seropositive individuals. *Lancet* i:1294-1297.
13. Camoirano, A., G.S. Badolati, P. Zanacchi, M. Bagnasco, and S. De Flora (1988) Dual role of thiols in *N*-methyl-*N*-nitro-*N*-nitrosoguanidine genotoxicity. *Life Science Adv. Exp. Oncol.* 7:21-25.

14. Cerutti, P., G. Shah, A. Peskin, and P. Amstad (1992) Oxidant carcinogenesis and antioxidant defense. Presented at 3rd International Conference on Mechanisms of Antimutagenesis and Anticarcinogenesis, Italy, 1991.

15. Cesarone, C.F., A.I. Scovassi, L. Scarabelli, R. Izzo, M. Orunesu, and U. Bertazzoni (1988) Depletion of adenosine diphosphate-ribosyl transferase activity in rat liver during exposure to *N*-2-acetylaminofluorene: Effect of thiols. *Cancer Res.* 48:3581-3585.

16. Chang, J.D., P.C. Billings, and A.R. Kennedy (1985) c-*myc* expression is reduced in antipain-treated proliferating C3H 10T1/2 cells. *Biochem. Biophys. Res. Comm.* 133:830-835.

17. Cooper, G.M. (1992) Oncogenes and chemoprevention. In *Cancer Chemoprevention*, L.W. Wattenberg, ed. CRC Press, Boca Raton, Florida (in press).

18. Correa, P., W. Haenszel, C. Cuello, M. Archer, and S. Tannenbaum (1975) A model for gastric cancer epidemiology. *Lancet* ii:58-60.

19. De Flora, S., ed. (1988) Role and mechanisms of inhibitor in prevention of mutation and cancer. *Mutat. Res.* 202:277-446.

20. De Flora, S. (1988) Editorial. Problems and prospects in antimutagenesis and anticarcinogenesis research. *Mutat. Res.* 202:279-283.

21. De Flora, S. (1990) Mechanisms of inhibitors of genotoxicity: Relevance in preventive medicine. In *Mutation and the Environment*, Part E, M.L. Mendelsohn and R.J. Albertini, eds. Wiley-Liss, Inc., New York, pp. 307-318.

22. De Flora, S., and C. Ramel (1988) Mechanisms of inhibitors of mutagenesis and carcinogenesis. *Mutat. Res.* 202:285-306.

23. De Flora, S., and K.E. Wetterhahn (1989) Mechanisms of chromium metabolism and genotoxicity. *Life Chem. Rep.* 7:169-244.

24. De Flora, S., C. Bennicelli, A. Camoirano, D. Serra, M. Romano, G.A. Rossi, A. Morelli, and A. De Flora (1985) In vivo effects of *N*-acetylcysteine on glutathione metabolism and on the biotransformation of carcinogenic and/or mutagenic compounds. *Carcinogenesis* 6:1735-1745.

25. De Flora, S., M. Romano, C. Basso, M. Bagnasco, C.F. Cesarone, G.A. Rossi, and A. Morelli (1986) Detoxifying activities in alveolar macrophages of rats treated with acetylcysteine, diethyl maleate and/or Aroclor. *Anticancer Res.* 6:1009-1012.

26. De Flora, S., A. Picciotto, V. Savarino, C. Bennicelli, A. Camoirano, G. Garibotto, and G. Celle (1987) Circadian monitoring of gastric juice mutagenicity. *Mutagenesis* 2:115-119.

27. De Flora, S., E. Hietanen, H. Bartsch, A. Camoirano, A. Izzotti, M. Bagnasco, and I. Millman (1989) Enhanced metabolic activation of chemical hepatocarcinogens in woodchucks infected with hepatitis B virus. *Carcinogenesis* 10:1099-1106.

28. De Flora, S., P. Zanacchi, A. Izzotti, and H. Hayatsu (1991) Mechanisms of food-borne inhibitors of genotoxicity relevant to cancer prevention. In *Mutagens in Food*. Detection and Prevention, H. Hayatsu, ed. CRC Press, Boca Raton, Florida, pp. 157-180.

29. De Flora, S., A. Izzotti, F. D'Agostini, and C.F. Cesarone (1991) Antioxidant activity and other mechanisms of thiols in chemoprevention of mutation and cancer. *Am. J. Med.* 91(suppl. 3C):122-130.

30. De Flora, S., M. Bagnasco, and P. Zanacchi (1992) Classification and mechanism of action of chemopreventive compounds. In *Progress and Perspectives in Chemoprevention of Cancer*, G. De Palo, U. Veronesi, and M. Sporn, eds. Raven Press, New York (in press).

31. De Flora, S., G. Bronzetti, and F.H. Sobels, eds. (1992) Assessment of antimutagenicity and anticarcinogenicity. End points and systems. *Mutat. Res.* Special issue (in press).

32. Dixit, R., and B. Gold (1986) Inhibition of *N*-methyl-*N*-nitrosourea-induced mutagenicity and DNA methylation by ellagic acid. *Proc. Natl. Acad. Sci., USA* 83:8039-8043.

33. Feo, F., R. Garcea, L. Daino, R. Pascale, S. Frassetto, P. Cozzolino, M.G. Vannini, M.E. Ruggiu, M.M. Simile, and M. Puddu (1988) S-Adenosyl methionine antipromotion and antiprogression effect in hepatocarcinogenesis. Its association with inhibition of gene expression. In *Chemical Carcinogenesis: Models and Mechanisms*, F. Feo, P. Pani, A. Columbano, and R. Garcea, eds. Plenum Press, New York, pp. 407.

34. Friedman, T. (1989) Progress toward human gene therapy. *Science* 244:1275-1281.

35. Guevara, A.P., C. Lim-Sylianco, F. Dayrit, and P. Finch (1990) Anti-mutagens from *Marmodica carantia. Mutat. Res.* 230:121-126.

36. Harris, H. (1988) The analysis of malignancy by cell fusion: The position in 1988. *Cancer Res.* 48:3302-3306.

37. Hartman, P.E., and D.M. Shankel (1990) Antimutagens and anticarcinogens: A survey of putative interceptor molecules. *Env. Molec. Mutagen.* 15:145-182.

38. Hayatsu, H., S. Arimoto, and T. Negishi (1988) Dietary inhibitors of mutagenesis and carcinogenesis. *Mutat. Res.* 202:429-446.

39. Helene, C. (1991) Rational design of sequence-specific oncogene inhibitors based on antisense and antigene nucleotides. *Eur. J.* Cancer 27:1466-1471.

40. Hochstein, P., and A. Atallah (1988) The nature of oxidants and antioxidant systems in the inhibition of mutation and cancer. *Mutat. Res.* 202:363-375.

41. Inoue, T., K. Morita, and T. Kada (1981) Purification and properties of a plant desmutagenic factor for the mutagenic principle of tryptophan pyrolysates. *Agr. Biol. Chem.* 45:345-353.

42. Jensen, D.E., G.J. Stelman, and A. Spiegel (1987) Species differences in blood-mediated nitrosocimetidine denitrosation. *Cancer Res.* 47:353-359.

43. Jongen, W.M.F., R.J. Topp, H.G.M. Tiedink, and E.J. Brink (1987) A co-cultivation system as model for in vitro studies of modulating effects of naturally occurring indoles on the genotoxicity of model compounds. *Toxicol. In Vitro* 1:105-110.

44. Kada, T., T. Inoue, and N. Namiki (1982) Environmental desmutagens and antimutagens. In *Environmental Mutagenesis and Plant Biology*, E.J. Klekowski, ed. Praeger, New York, pp. 137-151.

45. Kaufmann, W.K. (1989) Pathways of human cell post-replication repair. *Carcinogenesis* 10:1-11.

46. Kensler, T.W., P.A. Egner, M.A. Trush, E. Bueding, and M.D. Groopman (1985) Modification of aflatoxin B^{1} binding to DNA in vivo in rats fed phenolic antioxidants and dithiothione. *Carcinogenesis* 6:759-763.

47. Klohs, W.D., and R.W. Steinkampf (1988) Possible link between the intrinsic drug resistance of colon tumors and a detoxification mechanism of intestinal cells. *Cancer Res.* 48:3025-3030.

48. Kuroda, Y., D.M. Shankel, and M.D. Waters, eds. (1990) *Antimutagenesis and Anticarcinogenesis Mechanisms, II.* Plenum Press, New York and London, 485 pp.

49. Nakayasu, M., H. Shima, S. Aonuma, H. Nakagama, M. Nagao, and T. Sugimura (1988) Deletion of transfected oncogenes from NIH 3T3 transformants by inhibitors of poly(ADP-ribose)polymerase. *Proc. Natl. Acad. Sci., USA* 85:9066-9070.

50. Newmark, H.L., M.J. Wargovich, and W.R. Bruce (1984) Colon cancer and dietary fat, phosphate and calcium: An hypothesis. *J. Natl. Cancer Inst.* 72:1323-1325.

51. Nussenzweig, M.C., E.V. Schmidt, A.C. Shaw, E. Sinn, J. CamposTorres, B. Mathey-Prevot, P.K. Pattengale, and P. Leder (1988) A human immunoglobulin gene reduces the incidence of lymphomas in c*myc*-bearing transgenic mice. *Nature* 336:446-450.

52. Owens, R.A., and P.E. Hartman (1986) Glutathione: A protective agent in *Salmonella typhimurium* and *Escherichia coli* as measured by mutagenicity and by growth delay assay. *Env. Mutagen.* 8:659-673.

53. Petrilli, F.L., G.A. Rossi, A. Camoirano, M. Romano, D. Serra, C. Bennicelli, A. De Flora, and S. De Flora (1986) Metabolic reduction of chromium by alveolar macrophages and its relationships to cigarette smoke. *J. Clin. Invest.* 77:1917-1924.

54. Pryor, W.A. (1986) Cancer and free radicals. In *Antimutagenesis and Anticarcinogenesis Mechanisms,* D.M. Shankel, P.E. Hartman, T. Kada, and A. Hollaender, eds. Plenum Press, New York, pp. 4559.

55. Ramel, C., U.K. Alekperov, B.N. Ames, T. Kada, and L.W. Wattenberg (1986) Inhibitors of mutagenesis and their relevance to carcinogenesis. *Mutat. Res.* 168:47-65.

56. Rao, D.R., S.R. Pulusani, and C.B. Chawan (1986) Natural inhibitors of carcinogenesis: Fermented milk products. In *Diet, Nutrition and Cancer: A Critical Evaluation, Vol. II. Micronutrients, Nonnutritive Dietary Factors, and Cancer,* B.S. Reddy and L.A. Cohen, eds. CRC Press, Boca Raton, Florida, pp. 63-75.

57. Reddy, B.S. (1986) Diet and colon cancer: Evidence from human and animal model studies. In *Diet, Nutrition and Cancer: A Critical Evaluation, Vol. I, Macronutrients and Cancer,* B.S. Reddy and L.A. Cohen, eds. CRC Press, Boca Raton, Florida, pp. 47-65.

58. Reitsma, P.H., P.G. Rothberg, S.M. Astrin, J. Trial, Z. BarShavit, A. Hall, S.L. Teitelbaum, and A.J. Kahn (1983) Regulation of *myc* gene expression in HL-60 leukaemia cells by a vitamin D metabolite. *Nature* 306:492-494.

59. Roe, F.J.C. (1989) Non-genotoxic carcinogenesis: Implications for testing and extrapolation to man. *Mutagenesis* 4:407-411.

60. Schaffer, W.R., R. Kim, T. Sterne, J. Thorner, S.-H. Kim, and J. Rine (1989) Genetic and pharmacological suppression of oncogenic mutations in *ras* genes of yeast and humans. *Science* 245:379-385.

61. Shankel, D.M., P.E. Hartman, T. Kada, and A. Hollaender, eds. (1986) *Antimutagenesis and Anticarcinogenesis Mechanisms,* Plenum Press, New York and London, 605 pp.

62. Silbart, I.K., and D.F. Keren (1989) Reduction of intestinal absorption by carcinogen-specific secretory immunity. *Science* 243:14621464.

63. Simic, M.G. (1988) Mechanisms of inhibition of free radical processes in mutagenesis and carcinogenesis. *Mutat. Res.* 202:377-386.

64. Sparnins, V.L., P.L. Venegas, and L.W. Wattenberg (1982) Glutathione *S*-transferase activity: Enhancement by compounds inhibiting chemical carcinogenesis and by dietary constituents. *J. Natl. Cancer Inst.* 68:493-496.

65. Talalay, P., and H.J. Prochaska (1987) Mechanisms of induction of NAD(P)H:quinone reductase. *Chem. Scripta* 27A:61-66.

66. Thorgeirsson, S.S., B.E. Huber, S. Sorrell, A. Fojo, I. Pastan, and M.M. Gottesman (1987) Expression of the multidrug-resistant gene in hepatocarcinogenesis and regenerating rat liver. *Science* 236:1120-1122.

67. Wattenberg, L.W. (1992) Chemoprevention of cancer by naturally occurring and synthetic compounds. In *Cancer Chemoprevention*, L.W. Wattenberg, M. Lipkin, C. Boone, and G. Kelloff, eds. CRC Press, Boca Raton, Florida (in press).

68. Weinstein, I.B. (1988) Strategies for inhibiting multistage carcinogenesis based on signal transduction pathways. *Mutat. Res.* 202:413-420.

69. Weisburger, J.H., B.S. Reddy, D.P. Rose, L.A. Cohen, M.E. Kendall, and E.L. Wynder (1992) Protective mechanisms of dietary fibers in nutritional carcinogenesis. This volume.

70. Weissman, B.E., P.J. Saxon, S.R. Pasquale, G.R. Jones, A.G. Geiser, and E.J. Stanbridge (1987) Introduction of a normal human chromosome 11 into a Wilm's tumor cell line controls its tumorigenic expression. *Science* 236:175-180.

71. Westin, E.H., F. Wong-Staal, E.P. Gelmann, R. Dalla Favera, T.S. Papas, J.A. Lautenberger, A. Eva, E.P. Reddy, S.R. Tronick, S.A. Aaronson, and R.C. Gallo (1982) Expression of cellular homologues of retroviral *onc* genes human hemopoietic cells. *Proc. Natl. Acad. Sci., USA* 79:2490-2494.

MULTIPLE MECHANISMS: THE EXAMPLE OF VITAMIN A

Luigi M. De Luca

National Cancer Center Institute
Bethesda, Maryland 20892 USA

ABSTRACT

It has become evident that retinoids control differentiation, embryonal development, and tumorigenesis. In animal models, skin tumorigenesis has been shown to be prevented by retinoids, which in this organ function as antitumor promoters in the two-stage system using 7,12-dimethylbenz(a)anthracene (DMBA) as the initiator, and 12-tetradecanoyl-phorbol13-acetate (TPA) as tumor promoter. Even though pharmacological doses applied topically appear to inhibit tumor formation, we found that papilloma and keratoacanthoma growth required physiological concentrations of retinoic acid and that vitamin A deficiency was even more effective than excess retinoid in inhibiting SENCAR mouse skin tumorigenesis. In human beings, oral administration of retinoic acid after tumor resection was effective in inhibiting the appearance of new tumors on the skin of four patients with Xeroderma Pigmentosum, and was effective in preventing new primary tumor formations in patients treated for head and neck cancer. The newly-discovered nuclear receptors for retinoic acid function as transcriptional activators for several genes. In patients with acute promyelocytic leukemia presenting with a reciprocal translocation of chromosome 17 to chromosome 15, the breakpoint has been identified in the retinoic acid receptor alpha gene, which forms a fusion gene with a new gene termed *myl*, on chromosome 15. Treatment of the patients with retinoic acid causes complete remission of the APL. It also appears to generate cells that do not bear the translocation. Therefore, retinoids may well function as modulators of carcinogenesis both at the promotion level as well as by causing differentiation of neoplastically transformed cells.

INTRODUCTION

Retinoids are a family of compounds derived from dietary retinol and β-carotene. An intestinal enzyme is responsible for the cleavage of β-carotene to yield retinal (Fig. 1) (1). Retinol is esterified and is found in liver tissue as

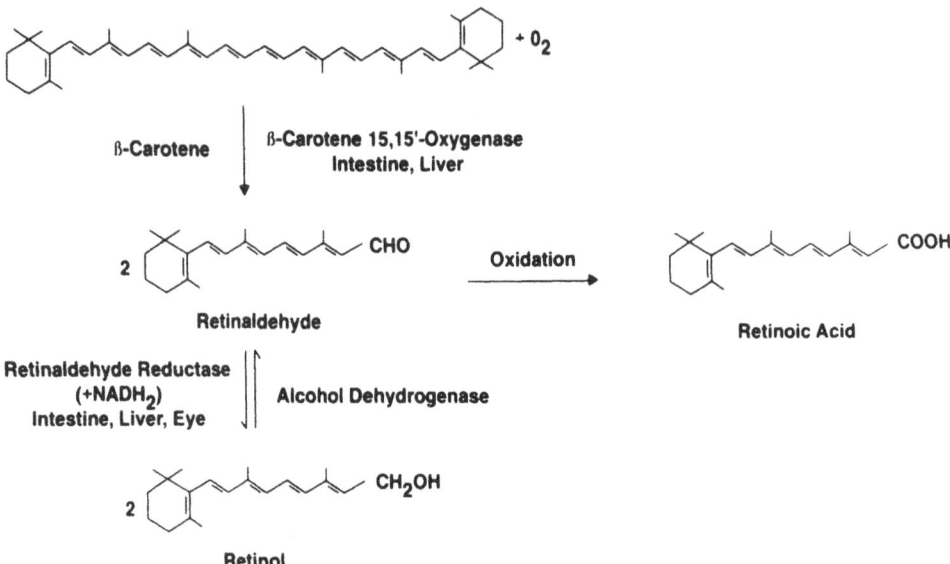

Figure 1. Schematic representation of the central cleavage of β-carotene to yield retinal and conversion of retinal to retinol and retinoic acid.

retinylpalmitate. An esterase enzyme cleaves retinylpalmitate to yield retinol, which is complexed with the retinol binding protein (RBP) and secreted into the bloodstream (2). Formation of retinoic acid takes place at the target tissue.

Retinoid delivery and function are tightly regulated by the availability of various binding proteins specific for various retinoids (3). In fact, under normal conditions, retinoids are bound to these specific proteins, some of which have a transport function (e.g., RBP); others may regulate free retinoid concentrations (e.g., CRABP), and others yet may function as transcriptional activators (e.g., the nuclear receptors, RARα, β, and γ; see Tab. 1 for a summary).

Retinoids are essential for embryogenesis, growth, and maintenance of normal epithelial differentiation. Dietary deprivation of vitamin A causes the replacement of normally mucociliary epithelia by a squamous metaplastic and eventually keratinizing epithelium (4). This is a gradual process that initially takes place by stratification of the normally pseudostratified epithelium. For instance, in the hamster trachea, all cells are in contact with the basement membrane. When the hamsters consume a vitamin A-depleted diet for appropriate times, the basement membrane becomes totally occupied by basal cells, which can be stained by immunohistochemistry using an antibody to the keratin K5 (Lancillotti et al., manuscript in preparation). More severe depletion causes the formation of squamous metaplastic lesions which are K5 positive. These lesions appear to be a necessary transition, marking the change toward the formation of bronchogenic squamous cell carcinoma.

Retinoids in vivo and in vitro totally reverse the squamous metaplastic phenotype (5). These data are consistent with in vivo data by Saffiotti, who

Table 1. Retinoid-binding proteins and receptors.

Protein	Approximate mass (kD)	Main ligand	Suggested function
RBP	21	Retinol	Blood plasma transport
IRBP	140	Retinol, retinal	Intercellular transport in visual cycle
Four proteins secreted from pig uterus	22	Retinol	Transport to the fetus
Two luminal proteins in rat epididymis	20	RA	Intercellular transport
CRBP (I)	16	Retinol	Donor for esterification, intracellular transport
CRBP (II)	16	Retinol	Donor for esterification
CRBP (III) from fish eye	15	Retinol	
CRABP (I)	16	RA	Intracellular transport, regulate free RA
CRABP (II) from neonatal rat	15	RA	Intracellular transport, regulate free RA
CRABP (II) from embryonal chick	16	RA	Intracellular transport, regulate free RA
CRALBP	36	Retinal	Enzymatic reactions in the visual cycle
RARα (7 isoforms)	48	RA	Ligand-dependent transcription factors
RARβ (3 isoforms)	48	RA	Ligand-dependent transcription factors
RARγ (7 isoforms)	48	RA	Ligand-dependent transcription factors
RXRα β γ	48	RA metabolite	Ligand-dependent transcription factors

demonstrated that retinylpalmitate in the diet of hamsters prevents the formation of squamoid tumors (6). We suggest that squamous carcinoma cells may derive from squamous metaplastic cells upon the action of initiating agents (7). Therefore, the action of retinoids as preventive agents of bronchopulmonary tumor formation may be explained on the basis of their action on differentiation of the bronchial epithelium.

SKIN TUMORIGENESIS

Do retinoids affect tumorigenesis of normally squamoid cells such as the mouse epidermal keratinocytes? The classic work of Verma, Boutwell, and collaborators was done in the two-stage system of mouse skin tumorigenesis. In this system DMBA is used as the tumor initiator, and TPA as the promoting agent (8). Retinoic acid (RA) was applied topically two hours before the application of the tumor promoting agent, and was effective in inhibiting tumorigenesis by 60% to 80% (8). Equally effective were 5,6-epoxyretinoic acid and 5,6-dehydroretinoic acid (9). These compounds do not have vitamin A activity in growth and differentiation and, therefore, their preventive activity must be independent of the physiological role of retinoids.

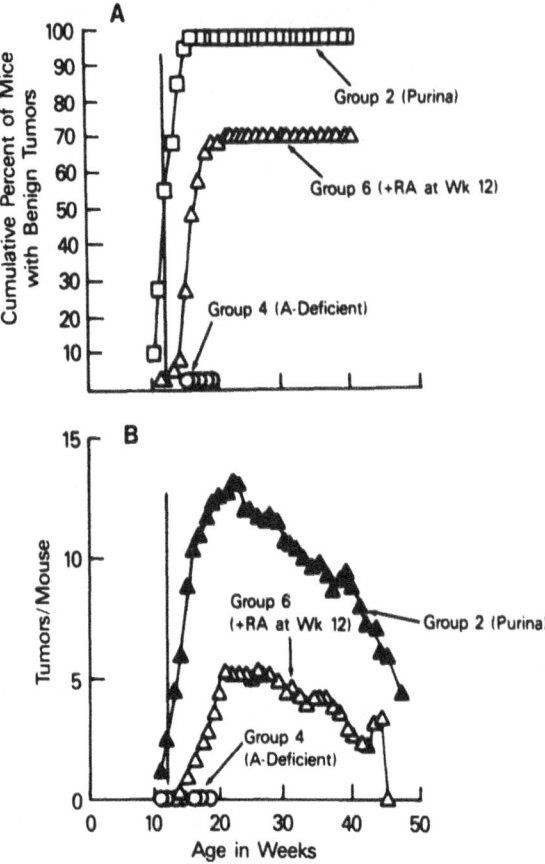

Figure 2. Vitamin A depletion inhibits and retinoic acid repletion permits two-stage tumorigenesis in female SENCAR mouse skin. The switching of the diet from vitamin A-deficient to RA-repleted in group 6 was gradual: 1 μg RA/g between weeks 13 and 15.5 to 2 μg/g between weeks 15.5 and 18.8 to 3 μg/g after 18.8 weeks. The vertical line at 12 weeks indicates the time at which group 6 mice were switched from the vitamin A-deficient to the RA-containing diet. Tumorigenic response was the same as in group 2 (Purina) in another experiment (10) where the purified diet supplemented with 3 μg/g A/g was fed to mice.

We asked whether the condition of dietary depletion of vitamin A might function as a tumor-promoting agent in the skin of female SENCAR mice (10). Our results showed a 99% protection of skin tumor formation in the two-stage system (10) in depleted mice and a relatively fast tumorigenic response to RA administration (Fig. 2). This effect of deficiency could not be due to caloric differences, because mice showed similar food consumption and body weight at 12 weeks of age. Moreover, bladder tumor formation under similar conditions is enhanced (11). Therefore, retinoid status modulates tumor formation in different ways, depending on the tissue type.

In very recent experiments, we have expanded these observations to include the effect of excess (30 μg/g diet) of dietary retinoic acid. As in previous work, retinoid depletion greatly reduced skin tumorigenesis and retinylpalmitate (6 μg/g diet) permitted skin tumor formation in mice that were tumor-free when fed the depleted diet. Interestingly, the excess retinoic acid did not influence papilloma formation. However, excess RA decreased carcinoma formation (De Luca et al., manuscript in preparation), supporting a possible effect of RA on the conversion of papillomas to carcinomas.

IN VIVO STUDIES IN HUMANS

Clinical trials have been conducted to study the effect of retinoids in human cancer. Kraemer et al. (12) have used oral administration of 13cis retinoic acid (isotretinoin) to study the rate of skin tumor formation in patients with Xeroderma Pigmentosum, an autosomal recessive disease that manifests itself by a multiplicity of skin tumors. In this trial the patients were treated surgically to remove visible skin tumors. The patients were then followed for two years during the isotretinoin treatment, then for two years after removal of the drug. The patients were selected for high production of tumors. Oral consumption of isotretinoin greatly reduced the yearly rate of tumor formation (Tab. 2) in 4 patients. Patient 5 did not show a decrease in yearly tumor rate, and showed an even higher rate of tumor formation during the two years following isotretinoin.

This study raises the following questions:

1.   Why was the tumor rate in some patients higher after RA treatment than before?
2.   Does tumor resection affect the following tumorigenic response?

In an approach to study the effect of ingestion of isotretinoin on head and neck cancer, Hong and collaborators (13) have accrued 100 patients, 49 of whom took the retinoid and 51 took the placebo for a period of two years after resection of the primary tumors. The formation of second primaries was followed using only those tumors that developed at about 2 cm from the first primaries. A highly significant decrease in the development of second primaries was observed (13) (see Fig. 3) in patients taking the isotretinoin.

Table 2. Number of skin cancers in patients with Xeroderma Pigmentosum before, during, and after therapy with oral isotretinoin (2 mg/kg/day). (Reproduced with permission from Kenneth H. Kraemer.)

Patient	Age/Sex	Before Treatment (2 Yr)	During Treatment (2 Yr)	After Treatment (12-14 Mo)
		number (number per year)		
1	19/F	43 (21.50)	3 (1.5)	18 (18.0)
2	12/F	37 (18.5)	4 (2.0)	29 (38.7)
3	17/M	23 (11.5)	6 (3.0)	20 (20.0)
4	39/M	10 (5.0)	3 (1.5)	4 (3.4)
5	10/M	8 (4.0)	9 (4.5)	10 (10.0)

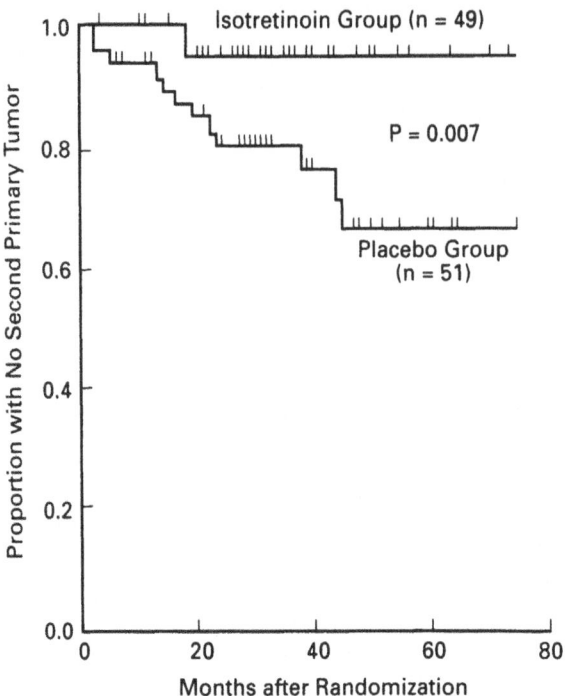

Figure 3. Kaplan-Meier estimate of the patients free of a second primary tumor, according to study group. The treatment period lasted from month 0 to month 12. The follow-up period began on the date of randomization. Reproduced with permission from *New England Journal of Medicine*.

DIFFERENTIATION THERAPY

Recent excitement in retinoid research has ensued following the discovery that several cell lines from tumors can be induced to differentiate into normal cells in an irreversible manner. The mouse embryonal carcinoma cell line F-9 was shown to respond to retinoic acid by differentiating into parietal endoderm (14) in an irreversible manner. These cells, when treated with retinoic acid, are stimulated to synthesize laminin (15). Also, recent studies have shown that RA-treated cells display an increased adhesiveness to laminin, collagen, and fibronectin (16).

Retinoids also induce differentiation of the human leukemia cell line HL-60. These cells become granulocytic in an irreversible process (17). Following these observations, Huang et al. (18) tested the ability of RA to influence acute promyelocytic leukemia in man. In the Huang study, 24 patients were given RA (45 to 100 mg/m^2/day). Twenty-one patients responded with complete remission. These data were confirmed in the laboratory of Castaigne and Chomienne (19), who showed that the leukemic cells were induced to differentiate into granulocytes (normal) by the retinoid. Thus, RA holds great potential for the therapy of acute promyelocytic leukemia.

The observation by de The et al. (20) that RARα maps to chromosome 17q 21, close to the t(15;17) (q21-q11-22) translocation specifically associated with APL, paved the way to an exciting recent development in this field: the demonstration that in leukemic cells of patients with acute promyelocytic leukemia, the RARα gene was translocated from chromosome 17 to a locus, *myl*, on chromosome 15, eventually resulting in the formation of a fusion transcript mRNA for *myl*/RARα (20,21). These rearrangements and breaks were found in almost all patients with APL. Therefore, the RARα gene appears to be somehow involved in leukemogenesis (22).

CONCLUSION

Retinoids should be regarded as highly potent regulators of embryogenesis, growth, differentiation, and carcinogenesis.

Retinoids, when applied topically to the backskin of mice, strongly inhibit tumorigenesis by interfering with the tumor-promoting step (8,23). This pharmacological effect of the retinoids has been used in the clinic to prevent tumor formation in patients with Xeroderma Pigmentosum (12) and in head and neck cancer (13).

We found that the condition of vitamin A deficiency is also very effective in inhibiting skin tumorigenesis by both the two-stage (10) and the complete tumorigenesis system (24). This would suggest a nutritional requirement for retinoids in growth of benign tumors.

The literature on vitamin A spans nearly eight decades with two major highlights: the discovery in the 1960s that retinal functions as the chromophore for rhodopsin (25) and the relatively recent discovery of the nuclear receptors for retinoic acid (26,27). These receptors function as transcriptional activators, and are obviously involved in normal differentiation processes. In certain cases they may be directly involved in the neoplastic process.

The relationship, if any, between the differentiation effects of RA on leukemic cells and the translocation of the RARα gene remains to be characterized, but this latter phenomenon puts on a solid molecular footing previous expectations that retinoids are somehow involved in neoplastic transformation. Although the biological consequences of the translocation of the RARα gene to the new *myl* locus are not clear, it is evident that the generation of new *myl*/RARα transcripts may well influence differentiation and growth potential of the cells.

This area of research will certainly be expanded in future years to investigate whether other tumor cells also contain anomalous RAR genes. These findings may or may not be related to the reported prophylactic affects of RA in tumor development, or to the effects of RA on differentiation of various tumor cells, including HL-60, neuroblastoma, embryonal carcinoma, and others; but both phenomena, the translocation as well as the differentiation effects, should be the subject of intense investigations involving clinicians, molecular biologists, and other basic scientists in coming years.

REFERENCES

1. Goodman, D.S., and J.A. Olson (1969) The conversion of all-trans-β-carotene into retinol. *Methods Enzymol.* 15:463.

2. Goodman, D.S. (1984) Plasma retinol binding protein. In *The Retinoids,* M.B. Sporn, A.B. Roberts, and D.S. Goodman, eds. Academic Press, Orlando, Florida, pp. 41-88.

3. Blomhoff, R., M.H. Green, T. Berg, and K.R. Norum (1990) Transport and storage of vitamin A. *Science* 250:399-404.

4. Wolbach, S.B., and P.R. Howe (1925) Tissue changes following deprivation of fat-soluble A vitamin. *J. Exp. Med.* 42:753-777.

5. De Luca, L.M., F.L. Huang, and D.R. Roop (1990) Retinoids and control of epithelial differentiation and keratin biosynthesis in hamster trachea. *Methods Enzymol.* 190:91-100.

6. Saffiotti, V., R. Montesano, A.R. Sellakumar, and S.A. Borg (1967) Experimental cancer of the lung: Inhibition by vitamin A of the induction of tracheobronchial squamous metaplasia and squamous cell tumors. *Cancer* 20:857-864.

7. De Luca, L.M. (1989) The concept of nutritional exotrophism in carcinogenesis. *Clin. Nutr.* 8:187-191.

8. Verma, A.K., E.A. Conrad, and R.K. Boutwell (1982) An overview of cutaneous carcinogenesis. *Cancer Res.* 42:3519.

9. Verma, A.K., T.J. Slaga, P.W. Wertz, G.C. Mueller, and R.K. Boutwell (1980) Inhibition of skin tumor promotion by retinoic acid and its metabolite 5,6-epoxyretinoic acid. *Cancer Res.* 40:2367-2371.

10. De Luca, L.M., R.L. Shores, E.F. Spangler, and M.L. Wenk (1989) Inhibition of initiator-promoter-induced skin tumorigenesis in female SENCAR mice fed a vitamin A-deficient diet and reappearance of tumors in mice fed a diet adequate in retinoid or beta-carotene. *Cancer Res.* 49:5400-5406.

11. Cohen, S.M., J.F. Wittenberg, and G.T. Brian (1976) Effects of avitaminosis A and hypervitaminosis A on urinary bladder carcinogenesis of N-(4-(nitro-2-furyl)-2-thiasolylformanide. *Cancer Res.* 36:2334-2339.

12. Kraemer, K.H., J.J. DiGiovanna, A.N. Moshell, R.E. Tarone, and G.L. Peck (1988) Prevention of skin cancer in Xeroderma Pigmentosum with the use of oral isotretinoin. *N. Engl. J. Med.* 318:1633-1637.

13. Hong, W.K., S.M. Lippman, L.M. Itri, D.D. Karp, J.S. Lee, R.M. Byers, S.P. Schantz, A.M. Kramer, R. Lotan, L.J. Peters et al. (1990) Prevention of second primary tumors with isotretinoin in squamous-cell carcinoma of the head and neck (see comments). *N. Engl. J. Med.* 323:795-801.

14. Strickland, S., and V. Mahdavi (1978) The induction of differentiation in teratocarcinoma stem cells by retinoic acid. *Cell* 15:393-403.

15. Vasios, G.W., D.G. Gold, M. Petkovich, P. Chambon, and L.J. Gudas (1989) A retinoic acid-responsive element is present in the 5' flanking region of the laminin B_1 gene. *Proc. Natl. Acad. Sci., USA* 86:9099-9103.

16. Ross, S.A., R.A. Ahrens, and L.M. De Luca (1991) Retinoic acid enhances attachment of F9 teratocarcinoma cells to laminin. *Proc. Am. Assoc. Cancer Res.* 32(Abs.):128.

17. Breitman, T.R. (1990) Growth and differentiation of human myeloid leukemia cell line HL60. *Methods Enzymol.* 190:118-130.

18. Huang, M.E., Y.C. Ye, S.R. Chen, J.R. Chai, J.X. Lu, L. Zhao, L.J. Gu, and Z.Y. Wang (1989) Use of all-trans retinoic acid in the treatment of acute promyelocytic leukemia. *Haematol. Bluttransfus.* 32:88-96.

19. Castaigne, S., C. Chomienne, M.T. Daniel, P. Ballerini, R. Berger, P. Fenaux, and L. Degos (1990) All-trans retinoic acid as a differentiation therapy for acute promyelocytic leukemia. I. Clinical results. *Blood* 76:1704-1709.

20. De The, H., C. Chomienne, M. Lanotte, L. Degos, and A. Dejean (1990) The t(15;17) translocation of acute promyelocytic leukaemia fuses the retinoic acid receptor alpha gene to a novel transcribed locus. *Nature* 347:558-561.

21. Borrow, J., A.D. Goddard, D. Sheer, and E. Solomon (1990) Molecular analysis of acute promyelocytic leukemia breakpoint cluster region on chromosome 17. *Science* 249:1577-1580.

22. Chang, K.S., J.M. Trujillo, T. Ogura, C.M. Castiglione, K.K. Kidd, S.R. Zhao, E.J. Freireich, and S.A. Stass (1991) Rearrangement of the retinoic acid receptor gene in acute promyelocytic leukemia. *Leukemia* 5:200-204.

23. Verma, A.K., B.G. Shapas, H.M. Rice, and R.K, Boutwell (1979) Correlation of the inhibition by retinoids of tumor promoter-induced ornithine decarboxylase activity and of skin tumor promotion. *Cancer Res.* 39:419-425.

24. De Luca, L.M., L.M. Sly, and C.S. Jones (1991) Retinoic acid is required for tumor formation in the skin of female SENCAR mice. *Proc. Am. Assoc. Cancer Res.* 32(Abs.):871.

25. Wald, G. (1968) The molecular basis of visual excitation. *Nature* 219:800-807.

26. Petkovich, M., N.J. Brand, A. Krust, and P. Chambon (1987) A human retinoic acid receptor which belongs to the family of nuclear receptors. *Nature* 330:444-450.

27. Giguere, V., E.S. Ong., P. Segui, and R.M. Evans (1987) Identification of a receptor for the morphogen retinoic acid. *Nature* 330:624-629.

INHIBITION OF NITROSATION

H. Bartsch, B. Pignatelli, S. Calmels, and H. Ohshima

Unit of Environmental Carcinogens and Host Factors
International Agency for Research on Cancer
150 cours Albert-Thomas
69372 Lyon, Cedex 08, FRANCE

ABSTRACT

Humans are exposed through ingestion or inhalation to preformed N-nitroso compounds (NOC) in the environment and through the endogenous nitrosation of amino precursors in the body. Activated macrophages and bacterial strains isolated from human infections can enzymatically produce nitrosating agents and NOC from precursors at neutral pH. As a consequence, endogenous nitrosation may occur at various sites of the body, such as the oral cavity, stomach, urinary bladder, and at other sites of infection or inflammation. Numerous substances to which humans are exposed have been identified and shown to inhibit formation of NOC. Such inhibitors include vitamins C and E, certain phenolic compounds, and complex mixtures such as fruit and vegetable juices or other plant extracts. Nitrosation inhibitors normally destroy the nitrosating agents and, thus, act as competitors for the amino compound that serves as substrate for the nitrosating species. Independently, epidemiological studies have already established that fresh fruits and vegetables that are sources of vitamin C, other vitamins, and polyphenols have a protective effect against cancers at various sites and in particular gastric cancer. This article briefly reviews (a) the chemistry of NOC formation and inhibition; (b) the studies in experimental animals that showed that inhibition of endogenous NOC synthesis leads to a reduction of toxic, mutagenic, and carcinogenic effects; (c) recent studies in humans where the degree of inhibition of endogenous NOC synthesis was directly quantified; and (d) the possible contribution of nitrosation inhibitors to human cancer prevention.

INTRODUCTION

Increasing attention has been focused on estimating the extent of human exposure to N-nitroso compounds (NOC) formed in vivo, from both exogenous

Antimutagenesis and Anticarcinogenesis Mechanisms III,
Edited by G. Bronzetti *et al.*, Plenum Press, New York, 1993

and endogenous precursors (17). Humans are indeed exposed to a wide range of nitrogen-containing compounds that can react with nitrosating agents to form NOC, a versatile class of carcinogens that produce tumors in 40 animal species. In addition, nitrosation of certain polyaromatic hydrocarbons and primary amines or phenolic compounds also results in the formation of C-nitroso, nitro, and reactive diazo compounds, several of which have been reported to be mutagenic and/or carcinogenic (4,26,58,73). There is also documented human exposure to various types of nitrosating agents, such as nitrite, nitrate, and nitrogen oxides (NOx), through diet, tobacco smoke, and drinking water. Nitrite, nitrate, and nitrosating agents can also be synthesized endogenously by reactions mediated by bacteria, activated macrophages, and hepatocytes (5,8,43,66). As a consequence, endogenous nitrosation may occur at many sites in the body, including nitrosation at organ sites where there is infection or inflammation.

Mirvish et al. (40) were the first to show that ascorbate reduces tumor formation in animals following feeding of nitrite and amine by inhibiting the formation of NOC in vivo. Since then, numerous substances to which humans are exposed have been shown to inhibit the formation of NOC in vitro and in animal models. In 1981, a method (the N-nitrosoproline, NPRO, test) was described for quantifying nitrosation in humans in vivo (3,48). This test made it possible to demonstrate for the first time the blocking effect of nitrosation inhibitors in man. Independently, epidemiological studies had already clearly established that fresh fruits and vegetables, which are a source of vitamin C, other vitamins, and poly-phenols, have a protective effect against cancers at various sites, in particular gastric cancer (review by Forman, Ref. 18). Although the evidence that endogenously formed NOC are an important class of carcinogenic agents involved in human cancer is far from conclusive, it is at least cohesive, and it justifies the reduction of exposure to NOC as a preventive measure.

The Chemistry of NOC Formation and Its Inhibition

The formation of NOC from amines and nitrosating species (Y-NO) can be described in general by equation (A):

$$\geq\text{N-H} + \text{Y-NO} \longrightarrow \geq\text{N-NO} + \text{HY} \qquad \text{(A)}$$

Several of these nitrosating agents are known to be formed under physiological conditions, for instance, in the stomach. Under mildly acidic conditions, for example, the nitrosating agent N_2O_3 is formed from two molecules of nitrous acid, and reacts with unprotonated amine to form nitrosamines. Therefore, the rate of nitrosation is pH-dependent, proportional to the square of the nitrite concentration and to the amine concentration (equation B), and is inversely related to the basicity of the amine.

$$\text{Rate (NOC)} = k \, [\text{amine}] \, [\text{nitrite}]_2 \qquad \text{(B)}$$

Because of the counteracting effects of decreasing pH on the concentration of unprotonated amine and the concentration of nitrous acid, there is an optimal pH (2.5-3.4) for nitrosation of a given amine. Certain amides, carbamates, and alkylureas are too unreactive to be nitrosated by N_2O_3. However, at pH < 2, they react readily with nitrosonium ion (NO^+), and the rate of nitrosamide formation is proportional to the concentration of nitrite, amide, and hydrogen ions (equation C):

$$\text{Rate (NOC)} = k \ [\text{amide}] \ [\text{nitrite}] \ [\text{H}^+] \qquad (C)$$

N-Nitrosamine formation by reaction of gaseous nitrogen oxides (N_2O_3, N_2O_4, NO_2) at neutral/alkaline pH is usually faster and more extensive than with nitrites in acidic aqueous solution (12). In the presence of certain anions (X^-), powerful nitrosating agents (NOX) are formed. Secondary amino compounds are major substrates from NOC. Nitrosation of primary and tertiary amino compounds also leads to NOC but occurs at slower rates, except for a few compounds such as the drug aminopyrine, which can rapidly form N-nitrosodimethylamine through more complex mechanisms. Certain primary aromatic amines or phenols have been reported to undergo reaction with nitrite to yield diazo compounds, which have been shown to be mutagenic and carcinogenic (50,73).

Free and protein-bound tyrosine residues easily react with nitrating/nitrosating agents to yield 3-nitrotyrosine (NTYR). NTYR formation in vivo showed a dose-dependent increase in NTYR in both plasma proteins and hemoglobin obtained from rats 24 hrs after ip injection of various doses of tetranitromethane (51). Thus, NTYR in proteins or its metabolites in urine can be readily analyzed as a new/additional marker for endogenous nitrosation and nitration.

Formation of Nitrosating Agents and NOC by Bacterial and Mammalian Enzymes

The first observation of bacterially mediated nitrosation was made by Sander in 1968 (57). It was only recently shown unequivocally that various bacterial strains isolated from urinary tract infections are capable of catalyzing nitrosation of secondary amines enzymically at neutrality in vitro (7-9,33). In *Escherichia coli*, this nitrosation capacity induced only by nitrate during culture is closely related to nitrate-reductase activity, and was shown to be linked to the expression of the nitrate-reductase gene narGHI (8,10,55). In contrast, in *Pseudomonas aeruginosa* (a denitrifying bacterial strain that shows extremely high nitrosation activity compared to *E. coli*), the nitrosating activity was induced by nitrite and nitrate, revealing an additional pathway operating in this strain (10,33). This enzyme catalyzing nitrosamine formation was isolated from two denitrifying microorganisms, *P. aeruginosa* and *Neisseria mucosae*. The soluble enzyme has a molecular weight of 66 kDa and a pH optimum for *P. aeruginosa* of 7.25 (11). Nine out of thirty strains isolated from human achlorhydric gastric juice and 12 out of 14 of those isolated from urinary tract infections possessed nitrosating enzymes, as determined by assays containing morpholine and nitrite (9). The mechanism by which *E. coli* and probably other nitrosation-proficient strains can catalyze the nitritedependent nitrosation, given the requirement for air, probably proceeds via formation of the strong nitrosating agents N_2O_3 and N_2O_4 formed from nitric oxide (NO) in the presence of oxygen (75).

Macrophages, when stimulated with lipopolysaccharide (LPS) and/or interferon-γ (INF-γ), can produce nitrite, nitrate, and nitrosating agents (21,28,31,43,44,66,67). These are derived from NO which is produced from the terminal guanidine nitrogen atom(s) of L-arginine. The process is catalyzed via an enzyme called NO-synthase (22,38), and is present in many other cells and organs such as endothelial cells, neutrophils, the Kupffer cells, and hepatocytes (14,20,39,45).

Nitrosamines are formed in macrophage cultures on addition of appropriate amines and incubation with macrophage stimulators for 24-72 hrs (28,31,43,44). The mechanism of macrophage-mediated nitrosamine formation was studied using a cell-free reaction system (52). The cytosol fraction of macrophages activated with LPS and INF-γ could nitrosate a wide range of secondary and tertiary amines without exogenously added nitrite or nitrate. Nitrosation occurred rapidly in a reaction mixture containing only L-arginine, NADPH (or $NADP^+$), and macrophage cytosol at about pH 7-7.5. Formation of nitrosamines also required oxygen and was inversely correlated with the basicity of nitrosatable amines, following a three-step mechanism as proposed for macrophage-mediated nitrosamine formation: (i) generation of NO from L-arginine by NO-synthase; (ii) chemical or enzymatic oxidation of NO to NO_2, which exists in equilibrium with the potent nitrosating agents N_2O_3 and N_2O_4; and (iii) the reaction with amines to form nitrosamines.

Inhibitors of Nitrosation

A great variety of synthetic or naturally occurring compounds or mixtures has been shown to inhibit the N-nitrosation reaction (reviewed in Bartsch et al., Ref. 2).

The formation of NOC can be reduced, minimized, or completely prevented by the presence of reagents that rapidly destroy nitrosating agents or reduce them to unreactive products. Such reagents usually act as competitors for the amine that serves as substrate for the nitrosating species; therefore, the degree of inhibition of N-nitrosation depends on several factors: (i) both the absolute and relative concentrations of the nitrosating species, the inhibitor, and the amine, and (ii) the relative rates of reactions of the nitrosating agent with the inhibitor and with the amine. N-nitrosation is also modified by the pH and the presence of catalysts. Generally, inhibitors of NOC formation involve compounds that either reduce Y-NO to N_2, N_2O, or NO_2 or bind the NO^+ group irreversibly. N_2 and N_2O are devoid of any nitrosating activity, but NO is converted to nitrosating species in the presence of oxygen, anions, and certain metal salts. NO is readily oxidized to NO_2, leading to N_2O_3, N_2O_4, and then the nitrosating capability is quickly restored.

Ascorbic Acid and L-Ascorbate

Since the original discovery by Mirvish et al. (40), the blocking effects of L-ascorbic acid and ascorbate on NOC formation have been demonstrated in many in vitro and in vivo systems (reviewed by Mirvish, Ref. 42; Bartsch et al., Ref. 2). Ascorbic acid and the ascorbate anion inhibit NOC formation over a wide pH range of 2-5 in aqueous solution by a rapid reduction of nitrous acid to NO and production of dehydroascorbic acid.

In the presence of certain catalysts or air, an excess of ascorbic acid must be used in order to prevent N-nitrosation. Although in most studies in vivo ascorbic acid inhibited the intragastric formation of NOC (see sections below), increased endogenous formation of N-nitrosoproline (NPRO) in the oral cavity of betal-quid and tobacco chewers has been observed in the presence of ascorbic acid (46).

L-Ascorbic acid also effectively inhibits nitrosamine formation catalyzed by bacterial and macrophage enzymes. The formation of N-nitrosomorpholine from morpholine and nitrite, when mediated by a nitrosation-proficient denitrifying bacterium (P. aeruginosa, BM1030) at neutral pH was strongly inhibited by less than 1 mM concentrations of ascorbic acid (37), despite the fact that it ordinarily shows little reactivity towards nitrite at neutral pH. This study provides some justification for the use of ascorbate as an inhibitor of endogenous N-nitrosation regardless of gastric pH (see section on gastric cancer below). L-Arginine-dependent nitrosation of morpholine by macrophage cytosol was inhibited by ascorbate at a wide range of concentrations of 0.01 to 1,000 µM, 50% inhibition being observed with about 2 µM ascorbate (52). In contrast, in similar assays using stimulated macrophages, inhibition of nitrosomorpholine formation was seen only at $\geq$ 50 µM ascorbate in the medium; at a lower concentration, an increase in nitrosation was found (31).

Tocopherols, Phenols, and Polyphenols

α-Tocopherol, the principal component of vitamin E, reduces nitrite to NO and is, therefore, an excellent inhibitor of N-nitrosation in lipids and emulsions in water. α-Tocopherol esters (particularly the acetate) are only active as inhibitors of N-nitrosation after in vivo hydrolysis of the ester by lipase.

Some simple phenols and phenolic compounds can reduce the rate of NOC formation depending on their structure and reaction conditions (53). Under acidic conditions, phenolics usually react with nitrite more rapidly than most amino compounds. Ortho and para-substituted phenols and many naturally occurring polyphenols that are oxidized to quinoid derivatives inhibit NOC formation by reducing the nitrosating agent to NO. Certain phenolics (e.g., phenol; 1,3-dihydroxyphenols such as resorcinol, catechin) can form C-nitroso derivatives that can act as powerful nitrosating agents. At high nitrite concentrations, phenols can also form reactive diazonium compounds (50,73).

Inhibition of Endogenous Nitrosation in Animals

The endogenous formation of NOC and the effect of inhibitors of Nnitrosation have been studied in experimental animals, mostly mice, rats, and dogs (reviewed in Bartsch et al., Ref. 2). Four main end-points have been used to monitor the extent of in vivo nitrosation and its inhibition: (i) the levels of NOC and their metabolites in isolated organs, the whole body, or urine and feces; (ii) mutagenicity in host-mediated assays; and (iii) acute toxic effects (histopathology or marker enzymes in serum) and carcinogenic effects in selected organs, mostly liver.

Ascorbic acid has been shown to be one of the most effective inhibitors of intragastric nitrosation. When it is administered together with nitrosating agents and a variety of secondary amines, amino acids, alkyl ureas, and nitrosatable drugs, the formation of NOC and the toxic, mutagenic, and carcinogenic effects of

NOC were significantly reduced. Other inhibiting substances such as α-tocopherol, phenolics, and plant and fruit extracts have also been studied in these systems. As many of these substances occur in the human diet, their levels and frequency of intake were expected to influence nitrosation reactions in the human body. However, recently it has been confirmed that, qualitatively, the inhibitory effects of many of these substances in humans are similar to those in experimental animals (see section below).

Inhibition of Endogenous Nitrosation in Humans

Monitoring of urinary levels of N-nitrosamino acids, such as NPRO, could be a useful procedure for the quantitative estimation of nitrosation in humans in vivo (48). Thus, the difference between the amount of NPRO excreted in 24-hr urine and that ingested in foods can be used as an indicator of daily endogenous nitrosation (the NPRO test). This noninvasive test was used to study the effects of various agents on endogenous nitrosation in humans for the first time. Results from ingested nitrate and proline, with or without dietary nitrosation modifiers, or consuming a standard diet consisting of a high-nitrate meal with and without foods rich in vitamin C, demonstrated unequivocally that in vivo nitrosation can occur in humans; it can be inhibited by dietary constituents like vitamins C and E and plant phenols, such as caffeic and ferulic acid, coffee, tea, and areca-nut extracts (reviewed in Bartsch et al., Ref. 2).

Ingestion of ascorbic acid (2 g/day), together with proline, inhibited the incorporation of [^{15}N]-nitrate into NPRO by 81%. However, the urinary level of [^{14}N]-NPRO excreted by the subjects was not reduced by this vitamin, suggesting that endogenous formation of NPRO could take place at sites other than the stomach, which are not accessible to this blocking agent (72). Similarly, Garland et al. (19) reported no effect of vitamin C and E supplements on the basal urinary excretion of NPRO and N-nitrosodimethylamine by healthy volunteers.

Ten times molar excess of ascorbic acid over the available nitrite concentration in the stomach is required to achieve complete inhibition of endogenous nitrosation of proline (34). Furthermore, vitamin C from dietary sources was also effective in inhibiting endogenous NPRO formation from dietary nitrate plus proline (30). The NPRO test has also been utilized to show the inhibitory effects of betel-nut extracts, caffeic and ferulic acids, and coffee and tea on the endogenous nitrosation of proline in human subjects (64,65).

Complex mixtures, including numerous foods and beverages, have been tested for their effect on NOC formation (summarized in Bartsch et al., Ref. 2). Inhibition of N-nitrosation has been observed with tea, coffee, vegetable and fruit juices, alcoholic beverages (54), milk and milk products, soya products, and betel-nut extracts. The inhibitory action could be due in part to the presence of ascorbic acid and/or phenolics in these items.

N-Nitrosation Inhibitors and Prevention of Human Cancer

A number of epidemiological studies, mostly of the case-control type, have shown that regular consumption of fruits and vegetables is associated with lower risks for cancers at a number of sites, in particular gastric cancer. Whether this protective effect is due primarily to ascorbate, other vitamins, and polyphenols as protective agents has yet to be determined.

(a) Gastric Cancer. Despite this uncertainty, the hypothesis that intragastric, acid-catalyzed formation of nitrosamides and other nitritederived carcinogens early in life is mainly responsible for at least the initiation of gastric cancer is consistent with many observations on its etiology (18,25,41,74).

One of the reasons that no convincing epidemiological evidence has so far been available has been the lack of reliable data for assessing endogenous nitrosation in humans. The development of the noninvasive dosimetry method (the NPRO test; Ref. 48) made possible measurements in high- and low-risk areas for gastric cancer in Northern Japan, Poland, and Costa Rica (Tab. 1, Nos. 1-3). Endogenous nitrosation of proline was found to be greater in subjects living in the three high-risk areas, and this could be efficiently inhibited by intake (100-mg doses) of ascorbic acid after each meal. In contrast, subjects living in low-risk areas showed no increased nitrosation (after proline intake), and ascorbic acid had no inhibitory effect, indicating that subjects in low-risk areas, for instance in Japan, are protected against endogenous nitrosation by sufficient intake of inhibitory substances in the diet (29). Even though their nitrate intake was positively correlated with the consumption of vegetables, the protective effect of certain vegetables appears to outweigh the contribution to dietary nitrate. The protective effect of vegetables against gastric cancer probably results from the high molar ascorbate:nitrate ratio found in certain vegetables (up to 3.5).

Present knowledge on the etiology of gastric cancer (intestinal type) suggests that it is initiated early in life, most likely by (nitrosamidetype) carcinogens derived from diets rich in nitrite and nitrate and dried, smoked or cured fish, and meat products or fava beans, or when the gastric mucosa is damaged by a high-salt diet and infection by *Helicobacter pylori* (16).

Thus, in order to minimize gastric cancer risk, it would appear prudent to eliminate bacterial infection, to limit the intake of these items, and to raise the regular consumption of fresh fruits and vegetables from early childhood onwards, so as to maximize the levels of ascorbate and polyphenols in the gastric contents where nitrosamides are most likely to be formed. It is not known how dietary intake of vitamins and antioxidants at various stages affects the gastric cancer process, which proceeds via precancerous conditions such as chronic atrophic gastritis, intestinal metaplasia, and dysplasia. In this context, recent reports by Sobala et al. (61) and Reed et al. (56) are of interest. Significantly lower ascorbic acid and vitamin C (defined as ascorbic acid plus dehydroascorbic acid) levels were observed in gastric juice of chronic gastritis patients (CAG), in particular with intestinal metaplasia, that were *not* associated with higher concentrations of nitrite and NOC (Tab. 1). However, plasma levels of vitamin C did not differ in patients with and without precancerous conditions of the stomach. These new results (60,61) draw attention to a lowered antioxidant defense and lack of nitrosation inhibitors in the stomach of CAG patients.

In another study, fasting gastric juice samples were obtained from 62 high-risk patients (atrophic gastritis, pernicious anemia, partial gastrectomy, vagotomy and drainage, and highly selective vagotomy) before, during four weeks' treatment with

Table 1. Ecological and case-control studies on human exposure to endogenously formed *N*-nitroso compounds: principal findings in subjects who are at higher cancer risk due to lifestyle, disease state, or geographical location.

No.	Characteristics of study subjects	Principal findings	References
STOMACH:			
1	Inhabitants from high- and low-risk areas for stomach cancer in Japan	Significantly increased ($p < 0.001$) nitrosation in the high-risk population, and reduction by protective factors in the diet of a low-risk population ($p < 0.001$).	Kamiyama et al. 1987 (29)
2	Inhabitants from high- (rural) and low- (urban) risk areas for stomach cancer in Poland	Slightly increased nitrosation in subjects from the rural area ($p < 0.02$). Moderate reduction by ascorbate.	Zatonski et al. 1989 (76)
3	Children from high- and low-risk areas for stomach cancer in Costa Rica	Higher NPRO output in the urine of children from the high risk area ($p < 0.04$); reduction by ascorbate ($p < 0.05$).	Sierra et al., 1991 (59)
4	Clinical studies on intragastric nitrosation in patients with precancerous conditions of the stomach	Patients with CAG and IM had lower gastric but not plasma ascorbate levels ($p < 0.001$) but no elevated nitrite and total N–nitroso compounds	Sobala et al., 1989, 1991 (60,61)
OESOPHAGUS:			
5	Inhabitants from high- (Lin-xian) and low-risk (Fan-xian) areas for oesophageal cancer in northern PR China	Higher levels of endogenous nitrosation in the high-risk population, ($p < 0.001$), and its inhibition by vitamin C ($p < 0.001$).	Lu et al., 1986 (35)
6	Inhabitants from eight counties with different cancer mortality in PR China	Nitrosamino acid levels in urine correlate positively with mortality rate from oesophageal cancer ($r = 0.95$).	Lu et al., 1987 (36)
7	Ecological study on inhabitants in 26 counties in PR China	Moderately positive correlation between oesophageal cancer mortality rates and endogenous NPRO formation ($r = 0.30$; ns); negative correlation of plasma ascorbate levels with cancer mortality ($r = -0.42$; $2p < 0.03$).	Chen et al., 1987 (13)

34

No.	Characteristics of study subjects	Principal findings	References
8	Subjects with liver fluke (Opisthorchis viverrini) infestation in Thailand at high-risk for cholangio-carcinoma	a) Higher urinary excretion of NPRO and nitrate in subjects with liver fluke.	Srianujata et al., 1987 (62)
		b) Several-fold increased urinary NPRO formation in infested subjects ($p < 0.01$) and reduction by ascorbate ($p < 0.01$).	Srivanatakul et al., 1991 (63)
9	Cigarette smokers	Increased formation of urinary NPRO and other N-nitrosamino acids; partial inhibition by ascorbate; effect of diet on their urinary excretion.	Hoffmann & Brunnemann, 1983 (23); Ladd et al.; 1984 (32); Bartsch et al., 1984 (1); Ohshima et al., 1984 (49); Tsuda et al., 1986 (68); Tsuda et al. 1987 (69)
10	Chewers of betel quid with or without tobacco	a) Increased urinary NPRO and other nitrosamino acids.	Nair et al., 1986, 1987 (46,47)
		b) Increased nitrosation of some alkaloids in the oral cavity; failure to inhibit nitrosation by ascorbic acid (added to quid) at salivary pH.	Nair et al., 1986; 1987 (46,47)

1 g ascorbic acid four times daily, and four weeks after treatment. Treatment with ascorbic acid lowered the median pH only in the vagotomized patients ($p < 0.001$), resulted in a reduction in nitrate-reducing bacterial counts and in nitrite and total N-nitroso compound concentrations in all groups, except for an increase in the nitrate-reducing bacterial count in atrophic gastritis patients and in nitrite in those with pernicious anemia. These data suggest that treatment with a high dose of ascorbic acid reduces the intragastric formation of nitrite and N-nitroso compounds, irrespective of the gastric pH (56). Intervention studies, some of which are in progress, are clearly warranted to investigate whether vitamins or antioxidant supplements could reduce the risk for progression of gastric precursor lesions to malignancies in such patients.

(b) Esophageal Cancer. Results of epidemiological and clinical investigations suggested that environmental factors such as NOC and their precursors, together with micronutrient deficiencies, were important (15). That endogenously formed NOC play a role in the causation of esophageal cancer is now supported by three correlation studies conducted in inhabitants of high- and low-risk areas for esophageal cancer in China, using urinary N-nitrosamino acids as an index of individual exposure to NOC and their precursors (Tab. 1, No. 5-7). In the first study, the amounts of NPRO, total nitrosamino acids, and nitrate excreted in 24-hr urine of subjects in Lin-xian (a high-risk area) were significantly higher ($p < 0.001$) than those in Fan-xian (a low-risk area). Intake of ascorbic acid by Lin-xian subjects effectively reduced the urinary levels of NAA to those found in undosed subjects in the low-risk area ($p < 0.001$). In the second study, the nitrosation potential in eight populations was highly positively correlated (r = 0.95, N = 1,500 subjects) with the mortality rates for esophageal cancer. Similarly, inhibition of increased nitrosation by ascorbate in inhabitants of 5 other high-risk areas in northern China was demonstrated following doses of proline and of proline plus vitamin C.

In the third study, samples of 12-hr urine from 1,035 subjects were collected (and pooled to give one sample for each commune) representing approximately 40 male adults in each of the 26 counties with a wide range of mortality rates (1.4 to 435 per 100,000) for esophageal cancer. One sample with proline alone and another after a dose of proline and ascorbic acid were collected from each subject. There was a moderate tendency (r = 0.3; n.s.) for esophageal cancer mortality rates to be associated positively with nitrosation potential and negatively with background ascorbate levels in plasma (r = - 0.42; $2 p < 0.03$). A dietary supplement of ascorbic acid significantly reduced the levels of N-nitrosamino acids excreted in the urine, indicating inhibition of endogenous nitrosation. These data offer a rational basis for long-term intervention studies in high-risk areas for esophageal cancer in China that are currently under way (6).

(c) Cholangiocarcinoma in Thailand and Parasitic Infestation. Cholangiocarcinoma in northeast Thailand has been associated with infestation by liver fluke, *Opisthorchis viverrini* (OV). Inhibitants in five areas of contrasting incidence for cholangiocarcinoma were dosed with proline and ascorbic acid or proline alone, and overnight urine was collected and analyzed for nitrosamino acids, nitrate, and creatinine (63). Blood samples from the same subjects were examined serologically for hepatitis B virus infection and OV infestation. Endogenous nitrosation was severalfold increased ($p < 0.01$) in the subjects positive for *OV* antibody, that could be blocked by vitamin C ($p < 0.01$); no such difference was observed between subjects who were negative and positive for HBsAg.

Taken together, the results suggest that interaction between chemical carcinogens, especially N-nitrosamines, and OV infestation may play an important role in the development of cholangiocarcinoma (CCA) in Thailand. This process could involve an increased nitrite/nitrate and NOC synthesis in liver cells or activated macrophages as a result of chronic inflammation caused by parasitic infestation. Recently, in a rodent model of induced hepatic damage (by injection of *Corynebacterium parvum* followed by lipopolysaccharide), enzymic production of nitric oxide from L-arginine was dramatically increased in the injured liver (5). It is, therefore, possible that the increased nitrosation potential in OV-infested subjects results from the increased enzymic production of nitric oxide in the liver damaged by OV infestation.

(d) NOC in Tobacco-associated Cancers. Tobacco users are all exposed to high levels of nitrosamine precursors, such as nicotine, nitrate, and nitrite, and to nitrosation modifiers, such as thiocyanate (24). Endogenous NOC synthesis occurs at a higher rate in tobacco users than in persons without such habits (Tab. 1, No. 9), thus adding to the body burden of carcinogens ingested or inhaled from exogenous sources. Endogenous NPRO formation in smokers could be inhibited partially by daily addition of 1 g ascorbic acid to the diet (23). These and other results (46) imply that a substantial fraction (as yet undetermined) of tobacco-related and other nitrosamines may be synthesized in vivo. There is now good evidence that tobacco-specific nitrosamines are major factors in cancers associated with tobacco use--both smoking and, to an even greater degree, use of smokeless tobacco products (24). The finding that ascorbic acid also inhibits endogenous NOC synthesis in smokers could explain in part why regular consumption of fresh vegetables and fruits, sources of ascorbic acid, has some protective effect against tobacco associated malignancies in the upper digestive tract (70,71).

Chewing of betel quid-containing tobacco is causally associated with cancer of the oral cavity in humans (27), and it has been suggested that tobacco-specific and areca nut-specific nitrosamines play a major role in the etiology (27). Nair et al. (46) demonstrated that a substantial fraction of tobacco-specific nitrosamines and other NOC are synthesized in vivo in the oral cavity of betel quid chewers, in addition to the preformed NOC present in betel quid-tobacco mixtures (Tab. 1, No. 10). In such subjects, endogenous nitrosation at salivary pH could not be inhibited by adding ascorbic acid to the betel quid mixture.

CONCLUSION

Although endogenous formation of NOC from ingested precursors has been suspected for some time of being the largest single source of exposure to these compounds for the general population, in the past only limited data existed to quantify human exposure. In addition, activated macrophages have been shown to synthesize nitrate, nitrite, and nitrosating agents that can form nitrosamines, and a number of bacterial strains isolated from infected humans produce nitrosamines from precursors at neutral pH. Thus, nitrosation could occur at sites remote from the stomach and could lead to exposure to NOC in persons with infections or inflammatory diseases. Recently, sensitive procedures have been developed and applied to estimate exposure of humans to exogenous and endogenous NOC (e.g., the NPRO test). The results indicate higher exposure to endogenous NOC of high-risk subjects for cancer of the stomach in northern Japan, Poland, and Costa Rica, and for cancer of the esophagus in northern China. Tobacco users and betel quid chewers have high levels of exposure to NOC formed, in part, endogenously. Parasite-infested subjects, at high risk for cholangiocarcinoma, had elevated nitrosation rates.

Most importantly, moderate doses of vitamin C taken after each meal lowered the body burden of intragastrically formed NOC in subjects living in high-risk areas for cancer of the stomach and of the esophagus, as well as in parasite-infested subjects. These results point to the role of endogenous NOC in human cancer etiology, and indicate preventive measures for reducing exposures. The results also provide plausible interpretation of epidemiological studies that have shown the protective effects of fruits and vegetables--sources of vitamin C, other vitamins, and polyphenols--against various malignancies and, particularly, stomach cancer.

Current knowledge and the consistency of observations that vitamin C and polyphenols can inhibit NOC formation in vitro in animals and in humans, however, justify application of this knowledge in preventive measures for high-risk populations. Until information concerning at what age and what point in the multistage process of carcinogenesis dietary nitrosation inhibitors are operating becomes available, it would be prudent to instigate intervention studies using nitrosation-blocking agents and adequate diets in early childhood in populations at high risk for gastric and esophageal cancer. The importance of regular consumption over time should be stressed in such studies to maximize ascorbate and polyphenol levels in the gastric environment at periods when NOC or other nitritederived carcinogens are formed.

ACKNOWLEDGEMENTS

A part of the work was supported by U.S. NIH Grant CA 47591. The authors gratefully acknowledge the scientific contributions and collaborative studies of numerous colleagues from IARC and the national cancer research institutes who have made this work possible, and which are referred to in the text.

REFERENCES

1. Bartsch, H., H. Ohshima, N. Muñoz, M. Crespi, V. Cassale, V. Ramazotti, R. Lambert, Y. Minaire, J. Forichon, and C.L. Walters (1984) In-vivo nitrosation, precancerous lesions and cancers of the gastrointestinal tract. On-going studies and preliminary results. In *N-Nitroso Compounds, Occurrence, Biological Effects and Relevance to Human Cancer,* I.K. O'Neill, R.C. von Borstel, C.T. Miller, L.E. Long, and H. Bartsch, eds. IARC Scientific Publications No. 57, International Agency for Research on Cancer, Lyon, France, pp. 955-962.
2. Bartsch, H., H. Ohshima, and B. Pignatelli (1988) Inhibitors of endogenous nitrosation. Mechanisms and implications in human cancer prevention. *Mutat. Res.* 202:307-324.
3. Bartsch, H., H. Ohshima, B. Pignatelli, and S. Calmels (1989) Human exposure to endogenous *N*-nitroso compounds: Quantitative estimates in subjects at high risk for cancer of the oral cavity, oesophagus, stomach and urinary bladder. *Cancer Surveys* 88:335-362.
4. Bartsch, H., H. Ohshima, B. Pignatelli, C. Malaveille, and M. Friesen (1990) Nitrite-reactive phenols present in smoked foods and amino-sugars formed by the Maillard reaction as precursors of genotoxic arenediazonium ions or nitroso compounds. In *Mutagens in Food: Detection and Prevention,* H. Hayatsu, ed. CRC Press, Boca Raton, Florida, and Ann Arbor, Michigan, pp. 87-100.

5. Billiar, T.R. R.D. Curran, D.J. Stuehr, J. Stadler, R.L. Simmons, and S.A. Murray (1990) Inducible cytosolic enzyme activity for the production of nitrogen oxides from L-arginine in hepatocytes. *Biochem. Biophys. Res. Comm.* 168:1034-1040.

6. Blot, W.J., and J.Y. Li (1985) Some considerations in the design of a nutrition intervention trial in Lin-Xian. *China, Natl. Cancer Inst. Monogr.* 69:29-34.

7. Calmels, S., H. Ohshima, P. Vincent, A.-M. Gounot, and H. Bartsch (1985) Screening of microorganisms for nitrosation catalysis at pH 7 and kinetic studies on nitrosamine formation from secondary amines by *E. coli* strains. *Carcinogenesis* 6:911-915.

8. Calmels, S., H. Ohshima, H. Rosenkranz, E. McCoy, and H. Bartsch (1987) Biochemical studies on the catalysis of nitrosation by bacteria. *Carcinogenesis* 8:1085-1088.

9. Calmels, S., H. Ohshima, M. Crespi, H. Leclerc, C. Cattoen, and H. Bartsch (1987) Nitrosamine formation by microorganisms isolated from human gastric juice and urine; biochemical studies on the bacterial catalysed nitrosation. In *The Relevance of N-Nitroso Compounds in Human Cancer: Exposures and Mechanisms,* H. Bartsch, I.K. O'Neill, and R. Schulte-Hermann, eds. IARC Scientific Publications No. 84, International Agency for Research on Cancer, Lyon, France, pp. 391-395.

10. Calmels, S., H. Ohshima, and H. Bartsch (1988) Nitrosamine formation by denitrifying and non-denitrifying bacteria: Implication of nitrite reductase and nitrate reductase in nitrosation catalysis. *J. Gen. Microbiol.* 134:221-226.

11. Calmels, S., N. Dalla Venezia, and H. Bartsch (1990) Isolation of an enzyme catalysing nitrosamine formation in *Pseudomonas aeruginosa* and *Neisseria mucosae. Biochem. Biophys. Res. Comm.* 171:655-660.

12. Challis, B.C., A. Edwards, R.R. Hunma, S.A. Kyrtopoulos, and J.R. Outram (1978) Rapid formation of *N*-nitrosamines from nitrogen oxides under neutral and alkaline conditions. In *Environmental Aspects of N-Nitroso Compounds,* E.A. Walker, L. Griciute, M. Castegnaro, and R.E. Lyle, eds. IARC Scientific Publications No. 19, International Agency for Research on Cancer, Lyon, France, pp. 127-142.

13. Chen, J., H. Ohshima, J. Yang, J. Li, T.C. Campbell, R. Peto, and H. Bartsch (1987) A correlation study on urinary excretion of *N*-nitroso compounds and cancer mortality in China: Interim results. In *The Relevance of N-Nitroso Compounds in Human Cancer: Exposures and Mechanisms,* H. Bartsch, I.K. O'Neill, and R. SchulteHermann, eds. IARC Scientific Publications No. 84, International Agency for Research on Cancer, Lyon, France, pp. 503-506.

14. Collier, J., and P. Vallance (1989) Second messenger role for NO widens to nervous and immune systems. *Trends Pharmacol. Sci.* 10:428-431.

15. Coordinating Group for Research on the Etiology of Esophageal Cancer of North China (1975) The epidemiology of esophageal cancer in north China and preliminary results in the investigation of its etiological factors. *Sci. Sin.* 18:131-148.

16. Correa, P. (1988) A human model of gastric carcinogenesis. *Cancer Res.* 48:3354-3560.

17. Forman, D., and D. Shuker (1989) Nitrate, nitrite and nitroso compounds in human cancer. *Cancer Surveys* 8:205-487.

18. Forman, D. (1991) The etiology of gastric cancer. In *Relevance to Human Cancer of N-Nitroso Compounds, Tobacco Smoke and Mycotoxins,* I.K. O'Neill, J. Chen, and H. Bartsch, eds. IARC Scientific Publications No. 105, International Agency for Research on Cancer, Lyon, France, pp. 22-32.

19. Garland, W.A., W. Kuenzig, F. Rubio, H. Kornychuk, E.P. Norkus, and A.H. Conney (1986) Studies on the urinary excretion of nitroso-dimethylamine and nitrosoproline in humans: Interindividual and intraindividual differences and the effect of administered ascorbic acid and α-tocopherol. *Cancer Res.* 46:5392-5400.

20. Griffith, T., and M. Randall (1989) Nitric oxide comes of age. *Lancet* ii:875-876.

21. Hibbs, Jr., J.B., R.R. Taintor, and Z. Vavrin (1987) Macrophage cytotoxicity: Role for L-arginine deiminase and imino nitrogen oxidation to nitrite. *Science* 235:473-476.

22. Hibbs, Jr., J.B., R.R. Taintor, Z. Vavrin, and E.M. Rachlin (1988) Nitric oxide: A cytotoxic activated macrophage effector molecule. *Biochem. Biophys. Res. Comm.* 157:87-94.

23. Hoffmann, D., and K. Brunnemann (1983) Endogenous formation of *N*-nitrosoproline in cigarette smokers. *Cancer Res.* 43:5570-5574.

24. Hoffmann, D., and S. Hecht (1985) Nicotine-derived *N*-nitrosamines and tobacco related cancer: Current status and future direction. *Cancer Res.* 45:935-944.

25. Howson, C.P., T. Hiyama, and E.L. Wynder (1986) The decline in gastric cancer: Epidemiology of an unplanned triumph. *Epidemiol. Rev.* 8:1-27.

26. IARC (1984) *IARC Monographs on the Evaluation of the Carcinogenic Risk of Chemicals to Humans, Vol. 33, Polynuclear Aromatic Compounds, Part 2, Carbon Blacks, Mineral Oils and Some Nitroarenes.* IARC, Lyon, France, pp. 167-222.

27. IARC (1985) *IARC Monographs on the Evaluation of the Carcinogenic Risk of Chemicals to Humans, Vol. 37, Tobacco Habits Other than Smoking: Betel-Quid and Areca-Nut Chewing; and Some Related Nitrosamines.* IARC, Lyon, France, pp. 7-269.

28. Iyengar, R., D.J. Stuehr, and M.A. Marletta (1987) Macrophage synthesis of nitrite, nitrate, and *N*-nitrosamines: Precursors and role of the respiratory burst. *Proc. Natl. Acad. Sci., USA* 84:6369-6373.

29. Kamiyama, S., H. Ohshima, A. Shimada, N. Saito, M.-C. Bourgade, P. Ziegler, and H. Bartsch (1987) Urinary excretion of *N*-nitrosamino acids and nitrate by inhabitants in high- and low-risk areas for stomach cancer in Northern Japan. In *The Relevance of N-Nitroso Compounds in Human Cancer: Exposures and Mechanisms,* H. Bartsch, I.K. O'Neill, and R. Schulte-Hermann, eds. IARC Scientific Publications No. 84, International Agency for Research on Cancer, Lyon, France, pp. 497-502.

30. Knight, T.M., and D. Forman (1987) The availability of dietary nitrate for the endogenous nitrosation of L-proline. In *The Relevance of N-Nitroso Compounds in Human Cancer: Exposures and Mechanisms,* H. Bartsch, I.K. O'Neill, and R. Schulte-Hermann, eds. IARC Scientific Publications No. 84, International Agency for Research on Cancer, Lyon, France, pp. 518-523.

31. Kosaka, H.A., J.S. Wishnok, M. Miwa, C.D. Leaf, and S.R. Tannenbaum (1989) Nitrosation by stimulated macrophages. Inhibitors, enhancers and substrates. *Carcinogenesis* 10:563-566.

32. Ladd, K.F., H.L. Newmark, and M.C. Archer (1984) *N*-Nitrosation of proline in smokers and nonsmokers. *J. Natl. Cancer Inst.* 73:83-97.

33. Leach, S.A., A.R. Cook, B.C. Challis, M.J. Hill, and M.H. Thompson (1987) Bacterially mediated *N*-nitrosation reactions and endogenous formation of *N*-nitroso compounds. In *The Relevance of N-Nitroso Compounds in Human Cancer: Exposures and Mechanisms,* H. Bartsch, I.K. O'Neill, and R. Schulte-Hermann, eds. IARC Scientific Publications No. 84, International Agency for Research on Cancer, Lyon, France, pp. 396-399.

34. Leaf, C.D., A.J. Vecchio, D.A. Roe, and J.H. Hotchkiss (1987) Influence of ascorbic acid dose on *N*-nitrosoproline formation in humans. *Carcinogenesis* 8:791-795.

35. Lu, S.H., H. Ohshima, H.-M. Fu, Y. Tian, F.-M. Li, M. Blettner, J. Wahrendorf, and H. Bartsch (1986) Urinary excretion of *N*-nitrosamino acids and nitrate by inhabitants of high- and low-risk areas for esophageal cancer in northern China: Endogenous formation of nitrosoproline and its inhibition by vitamin C. *Cancer Res.* 46:1485-1491.

36. Lu, S.H., W.X. Yang, L.P. Guo, F.M. Li, G.J. Wang, J.S. Zhang, and P.Z. Li (1987) Determination of *N*-nitrosamines in gastric juice and urine and a comparison of endogenous formation of *N*-nitrosoproline and its inhibition in subjects from high- and low-risk areas for oesophageal cancer. In *The Relevance of N-Nitroso Compounds in Human Cancer: Exposures and Mechanisms,* H. Bartsch, I.K. O'Neill, and R. Schulte-Hermann, eds. IARC Scientific Publications No. 84, International Agency for Research on Cancer, Lyon, France, pp. 538-543.

37. Mackerness, C.W., S.A. Leach, M.H. Thompson, and M.J. Hill (1989) The inhibition of bacterially-mediated *N*-nitrosation by vitamin C: Relevance to the inhibition of endogenous *N*-nitrosation in the achlorhydric stomach. *Carcinogenesis* 10:397-399.

38. Marletta, M.A., P.S. Yoon, R. Iyengar, C.D. Leaf, and J.S. Wishnok (1988) Macrophage oxidation of L-arginine to nitrite and nitrate: Nitric oxide is an intermediate. *Biochemistry* 27:8706-8711.

39. Marletta, M. (1989) Nitric oxide: Biosynthesis and biological significance, trends. *Biochem. Sci.* 14:488-492.

40. Mirvish, S.S., L. Wellcave, M. Eagen, and P. Shubik (1972) Ascorbate-nitrite reaction: Possible means of blocking the formation of carcinogen *N*-nitroso compounds. *Science* 177:65-68.

41. Mirvish, S.S. (1983) The etiology of gastric cancer, intragastric nitrosamide formation and other theories. *J. Natl. Cancer Inst.* 71:629-647.

42. Mirvish, S.S. (1986) Effects of vitamins C and E on *N*-nitroso compound formation, carcinogenesis and cancer. *Cancer* 58:18421850.

43. Miwa, M., D.J. Stuehr, M.A. Marletta, J.S. Wishnok, and S.R. Tannenbaum (1987) Nitrosation of amines by stimulated macrophages. *Carcinogenesis* 8:955-958.

44. Miwa, M., M. Tsuda, Y. Kurashima, H. Hara, Y. Tanaka, and K. Shinohara (1989) Macrophages-mediated *N*-nitrosation of thioproline and proline. *Biochem. Biophys. Res.* Comm. 159:373-378.

45. Moncada, S., R.M.J. Palmer, and E.A. Higgs (1989) Biosynthesis of nitric oxide from L-arginine. A pathway for the regulation of cell function and communication. *Biochem. Pharmacol.* 38:1709-1715.

46. Nair, J., H. Ohshima, B. Pignatelli, M. Friesen, C. Malaveille, S. Calmels, and H. Bartsch (1986) Modifiers of endogenous carcinogen formation:

Studies on in vivo nitrosation in tobacco users. In *New Aspects of Tobacco Carcinogenesis,* D. Hoffmann and C. Harris, eds. Banbury Report No. 23, Cold Spring Harbor Laboratory, New York, pp. 45-61.

47. Nair, J., U.J. Nair, H. Ohshima, S.V. Bhide, and H. Bartsch (1987) Endogenous nitrosation in the oral cavity of chewers while chewing betel quid with or without tobacco. In *The Relevance of N-Nitroso Compounds in Human Cancer: Exposures and Mechanisms,* H. Bartsch, I.K. O'Neill, and R. Schulte-Hermann, eds. IARC Scientific Publications No. 84, International Agency for Research on Cancer, Lyon, France, pp. 465-469.

48. Ohshima, H., and H. Bartsch (1981) Quantitative estimation of endogenous nitrosation in humans by monitoring *N*-nitrosoproline excreted in the urine. *Cancer Res.* 41:3658-3662.

49. Ohshima, H., I.K. O'Neill, M. Friesen, J.-C. Béréziat, and H. Bartsch (1984) Occurrence in human urine of new sulphur-containing *N*-nitrosamino acids *N*-nitrosothiazolidine 4-carboxylic acid and its 2-methyl derivative and their formation. *J. Cancer Res. Clin. Oncol.* 108:121-128.

50. Ohshima, H., M. Friesen, C. Malaveille, I. Brouet, A. Hautefeuille, and H. Bartsch (1989) Formation of direct-acting genotoxic substances in nitrosated smoked fish and meat products: Identification of simple phenolic precursors and phenyldiazonium ions as reactive products. *Food Chem. Tox.* 27:193-203.

51. Ohshima, H., M. Friesen, I. Brouet, and H. Bartsch (1990) Nitrotyrosine as a new marker for endogenous nitrosation and nitration of proteins. *Food Chem. Tox.* 28:647-652.

52. Ohshima, H., M. Tsuda, H. Adachi, T. Ogura, T. Sugimura, and H. Esumi (1992) L-Arginine-dependent formation of *N*-nitrosamines by the cytosol of macrophages activated with lipopolysaccharide and interferon-γ. *Carcinogenesis* 12:1217-1220..

53. Pignatelli, B., J.-C. Béréziat, G. Descotes, and H. Bartsch (1982) Catalysis of nitrosation in vitro and in vivo in rats by catechin and resorcinol and inhibition by chlorogenic acid. *Carcinogenesis* 3:1045-1049.

54. Pignatelli, B., R. Scriban, G. Descotes, and H. Bartsch (1984) Modifying effects of polyphenols and other constituents of beer on the formation of *N*-nitroso compounds. *Am. Soc. of Brewing Chemists* 42:18-23.

55. Ralt, D., J.S. Wishnok, R. Fitts, and S.R. Tannenbaum (1988) Bacterial catalysis of nitrosation: Involvement of the *nar* operon of *Escherichia coli*. *J. Bacteriol.* 170:359-364.

56. Reed, P.I., B.J. Johnston, C.L. Walters, and M.J. Hill (1991) Effect of ascorbic acid on the intragastric environment in patients at increased risk of developing gastric cancer. In *Relevance to Human Cancer of N-Nitroso Compounds, Tobacco Smoke and Mycotoxins,* I.K. O'Neill, J. Chen, and H. Bartsch, eds. IARC Scientific Publications No. 105, International Agency for Research on Cancer, Lyon, France, pp. 139-142.

57. Sander, J. (1968) Nitrosaminosynthese durch Bakterien. *Hoppe Seyler's Z. Physiol. Chem.* 349:429-432.

58. Shephard, S.E., C. Schlatter, and W.K. Lutz (1987) Assessment of the risk of formation of carcinogenic *N*-nitroso compounds from dietary precursors in the stomach. *Food Chem. Toxicol.* 25:91-108.

59. Sierra, R., H. Ohshima, N. Muñoz, A.S. Peña, S. Teuchmann, C. Malaveille, B. Pignatelli, A. Chinnock, F. El Ghissassi, C.S. Chen, A. Hautefeuille, C. Gamboa, and H. Bartsch (1991) Exposure to *N*-nitrosamines and other risk

factors for gastric cancer in Costa Rican children. In *Relevance to Human Cancer of N-Nitroso Compounds, Tobacco Smoke and Mycotoxins,* I.K. O'Neill, J. Chen, and H. Bartsch, eds. IARC Scientific Publications No. 105, International Agency for Research on Cancer, Lyon, France, pp. 162-167.

60. Sobala, G.M., C.J. Schorah, M. Sanderson, M.F. Dixon, D.S. Tompkins, P. Godwin, and A.T.R. Axon (1989) Ascorbic acid in the human stomach. *Gastroenterology* 97:357-363.

61. Sobala, G.M., B. Pignatelli, C.J. Schorah, H. Bartsch, M. Sanderson, M.F. Dixon, S. Shires, R.F.G. King, and A.T.R. Axon (1991) Levels of nitrite, nitrate, *N*-nitrosocompounds, ascorbic acid and total bile acids in gastric juice of patients with and without precancerous conditions of the stomach. *Carcinogenesis* 12:193-198.

62. Srianujata, S., S. Tonbuth, S. Bunyaatvej, A. Valyseva, N. Promvanit, and W. Chiavatsagul (1987) High urinary excretion of nitrate and *N*-nitrosoproline in *Opisthorchiasis* subjects. In *The Relevance of N-Nitroso Compounds in Human Cancer: Exposures and Mechanisms,* H. Bartsch, I.K. O'Neill, and R. Schulte-Hermann, eds. IARC Scientific Publications No. 84, International Agency for Research on Cancer, Lyon, France, pp. 544-546.

63. Srivatanakul, P., H. Ohshima, M. Khlat, M. Parkin, S. Sukaryodhin, I. Brouet, and H. Bartsch (1991) Endogenous nitrosamines and liver fluke as risk factors for cholangiocarcinoma in Thailand. In *Relevance to Human Cancer of N-Nitroso Compounds, Tobacco Smoke and Mycotoxins,* I.K. O'Neill, J. Chen, and H. Bartsch, eds. IARC Scientific Publications No. 105, International Agency for Research on Cancer, Lyon, France, pp. 88-95.

64. Stich, H.F., H. Ohshima, B. Pignatelli, J. Michelon, and H. Bartsch (1983) Inhibitory effect of betel nut extracts on endogenous nitrosation in humans. *J. Natl. Cancer Inst.* 70:1047-1050.

65. Stich, H.F., B.P. Dunn, B. Pignatelli, H. Ohshima, and H. Bartsch (1984) Dietary phenolics and betel nut extracts as modifiers of *N*-nitrosation in rat and man. In *N-Nitroso Compounds: Occurrence, Biological Effects and Relevance to Human Cancer,* I.K. O'Neill, R.C. von Borstel, R.C. Miller, C.T. Long, and H. Bartsch, eds. IARC Scientific Publications No. 57, International Agency for Research on Cancer, Lyon, France, pp. 213-222.

66. Stuehr, D.J., and M.A. Marletta (1985) Mammalian nitrate biosynthesis: Mouse macrophages induce nitrite and nitrate in response to *Escherichia coli* lipopolysaccharide. *Proc. Natl. Acad. Sci., USA* 82:7738-7742.

67. Stuehr, D.J., and M.A. Marletta (1987) Synthesis of nitrite and nitrate in murine macrophage cell lines. *Cancer Res.* 47:5590-5594.

68. Tsuda, M., J. Niitsuma, S. Sato, T. Hirayama, T. Kakizoe, and T. Sugimura (1986) Increase in levels of *N*-nitrosoproline, *N*-nitrosothioproline and *N*-nitroso-2-methylthioproline in human urine by cigarette smoking. *Cancer Lett.* 30:117-124.

69. Tsuda, M., A. Nagai, H. Suzuki, T. Hayashi, M. Ikeda, M. Kuratasune, S. Sato, and T. Sugimura (1987) Effect of cigarette smoking and dietary factors on the amounts of *N*-nitrosothiazolidine 4-carboxylic acid and *N*-nitroso-2-methylthiazolidine 4-carboxylic acid in human urine. In *The Relevance of N-Nitroso Compounds in Human Cancer: Exposures and Mechanisms,* H. Bartsch, I.K. O'Neill, and R. Schulte-Hermann, eds. IARC Scientific Publications No. 84, International Agency for Research on Cancer, Lyon, France, pp. 446-450.

70. Tuyns, A.J. (1983) Protective effect of citrus fruit on esophageal cancer. *Nutr. Cancer* 5:195-200.

71. Tuyns, A.J., E. Riboli, G. Doornbos, and G. Péguignot (1987) Diet and esophageal cancer in Calvados. *Nutr. Cancer* 9:81-92.

72. Wagner, D.A., D.E.G. Shuker, C. Bilmazes, M. Obiedzinki, I. Baker, V.R. Young, and S.R. Tannenbaum (1985) Effect of vitamins C and E on endogenous synthesis of *N*-nitrosamino acids in humans: Precursor-product studies with [15]*N*-nitrate. *Cancer Res.* 45:65196522.

73. Wakabayashi, K., M. Nagao, M. Ochiai, Y. Fijita, T. Tahira, M. Nakaysu, H. Ohgagki, S. Takayama, and T. Sugimura (1987) Recently-identified nitrite-reactive compounds in food: Occurrence and biological properties of the nitrosated products. In *The Relevance of N-Nitroso Compounds in Human Cancer: Exposures and Mechanisms*, H. Bartsch, I.K. O'Neill, and R. Schulte-Hermann, eds. IARC Scientific Publications No. 84, International Agency for Research on Cancer, Lyon, France, pp. 287-291.

74. Weisburger, J.H., H. Marquardt, H. Hirota, N. Mori, and G.M. Williams (1980) Induction of cancer of the glandular stomach in rats of nitrite-treated fish. *J. Natl. Cancer Inst.* 64:163-167.

75. Xiao-Bing, J.I., and T.C. Hollocher (1988) Mechanism for nitrosation of 2,3-diaminonaphthalene by *Escherichia coli*: Enzymatic production of NO followed by O_2-dependent chemical nitrosation. *Appl. Env. Microbiol.* 54:1791-1794.

76. Zatonski, W., H. Ohshima, K. Przewozniak, K. Drosnik, J. Mierzwimska, M. Krygier, W. Chmielarczyk, and H. Bartsch (1989) Urinary excretion of *N*-nitrosamino acids and nitrate by inhabitants of high- and low-risk areas for stomach cancer in Poland. *Int. J. Cancer* 44:823-827.

PROTECTIVE MECHANISMS OF DIETARY FIBERS IN

NUTRITIONAL CARCINOGENESIS

John H. Weisburger, Bandaru S. Reddy, David P. Rose,
Leonard A. Cohen, Marcia E. Kendall, and Ernst L. Wynder

American Health Foundation
Valhalla, New York 10595-1599 USA

ABSTRACT

Fibers in foods are complex carbohydrates. There are several types of fiber, but, for the purpose of mechanistic insight into their mode of protective action in carcinogenesis, classification into two broad types, soluble and insoluble fibers, is warranted. Soluble fibers are present in fruits, vegetables, and certain grains like oats. This type of fiber undergoes metabolism in the small intestine and especially in the large intestine through bacterial enzymes, converting it to products that increase stool size only moderately. But, they have appreciable effects in modifying the metabolism of colon carcinogens like azoxymethane to yield detoxified products and, thus, reducing colon carcinogenesis. In contrast, insoluble fibers present in sizeable amounts in bran cereals, like wheat or rice, are not significantly metabolized by enzymes in the intestinal flora. Such fibers increase stool size substantially through several mechanisms, including higher water retention. The larger bulk dilutes carcinogens, especially tumor promoters such as secondary bile acids, resulting in lower risk of colon cancer in animals and in humans. Evidence in animal models and in humans also indicates that fiber may lower the risk of breast cancer, possibly via an endocrine mechanism. Based on these concepts, increased intake of total fiber, but especially of wheat bran cereal fiber, to yield a daily stool in adults of about 200 grams can significantly reduce the risk of colon cancer and, to a lesser but definite extent, of breast cancer. Thus, adequate fiber intake from cereals, fruits, and vegetables can help prevent important types of human cancer.

INTRODUCTION

Lifestyle, in any part of the world, is associated with the occurrence of major chronic diseases, including coronary heart disease, hypertension, and stroke; many distinct types of cancer; and adult onset diabetes (11,22,70,75). One important element of lifestyle is tobacco use, which appreciably increases the risk of a number of chronic diseases (21). Other lifestyle factors relate to nutritional traditions, alcohol use, and exercise (11,30,58,60,70,71). The underlying evidence is, in part, based on the evidence of specific diseases in relation to dietary habits in a given geographic region, as well as changes in such disease occurrences in migrant populations from a low risk to a high risk area, and vice versa (25,42).

Populations that customarily consume fruits and vegetables and also whole grain cereal products generally have a lower risk of cardiovascular diseases and specific types of cancer than populations with a lower intake of such foods, everything else being equal (63). In the last few years, research has provided mechanistic explanations as to the reasons for the protective effects of specific foods. This paper will emphasize the mechanisms whereby cancer risks are reduced by dietary modification.

MECHANISMS OF CARCINOGENESIS

Cancer causation involves a complex series of steps (80). An early essential point is the modification of normal cellular DNA by specific genotoxic carcinogens, by radiation, or by certain viruses. Cells with such an altered DNA are subject to distinct growth and development involving regulatory mechanisms that are functions of the type of cell and are controlled through endogenous or exogenous factors.

Virtually all human cancers, whether those caused by heavy exposure to specific chemical carcinogens at the workplace or those due to lifestyle elements such as cigarette smoke, tobacco, or nutritional factors, involve genotoxic carcinogens (80). One exception is the transplacental action of the powerful estrogenic hormone, diethylstilbestrol, that has caused clear cell cancer of the vagina at puberty in female offspring of treated mothers through elements of faulty differentiation and pronounced hormonal imbalances. In this instance, the modification of DNA may depend on the generation of hydroxy radicals (41). Usually, however, carcinogens that are proven genotoxins are likely human cancer risks under appropriate conditions of dosage and chronic exposure. Excellent methods are available to determine the attribute of genotoxicity in chemicals (80). One scheme is a battery of the bacterial mutagenicity tests developed by Ames (80) and the mammalian cell DNA repair test in hepatocytes of Williams (80). Every type of human cancer is caused by one or more specific genotoxic carcinogen(s). Cancer prevention, in part, depends on effectively decreasing the extent of exposure of man to such carcinogens.

Furthermore, many types of human cancers involve the operation of nongenotoxic carcinogens or tumor promoters (80). The important principle underlying the operation of promoters is that their action is highly dependent on dose and is reversible. Therefore, a sharply lower risk for a given type of cancer can be achieved by decreasing the concentration of promoters. Diseases associated with smoking (i.e., cancer of the respiratory tract), including lung, pancreas, kidney, and bladder, are due to the presence in tobacco smoke of relatively small amounts of genotoxic carcinogens, but of appreciable amounts of promoters. Thus, total elimination of promoters is the situation prevailing in smoking cessation. For this reason, the risk of these cancers is progressively reduced in individuals who stop smoking (21,76).

Similarly, it may be, but is not yet proven, that adjustment of dietary conditions with a view to lowering the promoting phenomenon would likewise yield a lower risk for the nutritionally-linked cancers, such as those in the colon, postmenopausal breast, pancreas, prostate, ovary, and endometrium (3,8,22,39). The type and amount of dietary fat in relation to calories play an important role in promoting these types of cancer. In contrast, fruits, vegetables, and several types of fiber have an inhibiting or antipromoting effect, the mechanisms of which will be discussed in detail (3,60,63,74). The field of nutrition and risk for breast and colorectal or large bowel cancer is sometimes controversial. In part, problems arise because the mechanisms are complex and, therefore, need to be analyzed in detail. In the case of breast cancer, it is probable that premenopausal breast cancer has somewhat distinct risk factors from those associated with postmenopausal breast cancer. The type and amount of fats clearly play a major role in postmenopausal breast cancer, but perhaps not in premenopausal disease. Likewise, colorectal cancer represents three distinct diseases with their own risk factors. The etiology of proximal colon cancer is poorly understood, although fibers, selenium salts, and antioxidants may be protective. As in postmenopausal breast cancer, fat, through control of intestinal fatty acids and bile acids, plays a key role (3,46). The situation for rectal cancer is similar, yet clearly distinct; alcohol intake increases the risk of this type of cancer (58,77).

FRUITS AND VEGETABLES

In many instances, the health benefits of the regular intake of fruits and vegetables have been attributed to the presence in such foods of essential vitamins (63). One important means of meeting the recommended daily allowances (RDA) for vitamins is to consume four to six servings of varied mixed fruits and vegetables a day. In fact, the recommended daily allowances for vitamins are designed to provide a sufficient amount of these micronutrients to avoid specific deficiency diseases (36). In North America, as in most of the Western world, adoption of the RDA has, indeed, been effective in eliminating almost completely the occurrence of specific vitamin deficiency diseases. Unfortunately, this finding is not universal, and in certain parts of the world, there are still groups of individuals, especially children, with symptoms of acute vitamin-deficiency diseases.

In the Western world, however, we need to go beyond the successful eradication of deficiency diseases and ask whether there are optimal amounts of certain vitamins to decrease the risk of chronic diseases including coronary heart diseases and cancer. For example, β-carotene, vitamin E, and vitamin C are excellent antioxidants and free radical traps. It can be surmised that levels higher than the RDA would prove beneficial, since, in part, the molecular mechanisms of action of genotoxic carcinogens and of promoting substances involve the intracellular generation of hydroxy radicals and of hydrogen peroxide (26,80,81). The antioxidant vitamins would be expected to decrease these reactive intermediates that are also associated with premature aging (43,63).

Vegetables and fruits also contain other "chemopreventive" substances. For example, glucaric acid is present in certain vegetables. In animal models, glutaric acid has lowered the incidence of colon cancer (72). There are additional protective agents in fruits and vegetables, such as phytate, phenethyl isothiocyanate, ellagic acid, and glycyrrhizinic acid (11,70,74). This is currently an active field of research, especially regarding relevant mechanisms. Green tea and black tea contain polyphenols that can serve as radical traps and, thus, may serve to lower specific chronic disease risk (78).

FIBERS

Dietary fiber is defined as the endogenous components of plant materials in the diet that are resistant to digestion by enzymes produced by human intestinal flora (6,12,29,34,45,55,59,69). Dietary fiber includes predominantly nonstarch polysaccharides, namely, cellulose, hemicellulose, pectins, gums, mucilages, and lignin. The chemical and physical nature of the fiber and the prevailing bacterial flora determine whether a dietary fiber can be degraded in the human colon. Based on properties of fermentability, polysaccharides have been classified as soluble and insoluble. Soluble fibers have structures such that they can be partially digested in the intestinal tract, mainly as a result of microbial enzyme action. They include pectins, gums, mucilages, and some hemicelluloses. Insoluble fibers include cellulose, some hemicelluloses, and lignin. In describing the physiological effects of dietary fiber, distinctions should be made among fiber-containing foods (whole grains and grain products, legumes, beans, fruits, and vegetables), fiber-rich supplements (wheat, oat, corn and rice brans, psyllium, and soy isolates), and isolated, purified fibers (cellulose, pectin, guar gum, arabic or locust-bean gum, and lignin). Also, starch that is not completely digested and absorbed may increase water retention and modify bulk intestinal and fecal contents (5,7,57).

Insoluble fibers are so named because they are poorly converted by mammalian or bacterial enzymes in the intestinal tract. Such fibers include lignin, cellulose, and some hemicelluloses. This type of fiber with a property of retaining water does increase stool bulk and effectively dilutes mutagens, carcinogens, and promoters in the intestinal tract, leading to their elimination. While fruits and vegetables contain some insoluble fibers, the major source is whole grain cereals, particularly in the seed coat. Regular intake of whole grain cereals with bran and phytate is thought to be instrumental in lowering the risk of intestinal diseases, including diverticulosis, appendicitis, and distal colon cancer, and may have a similar role in breast cancer (1,3,28,33,44-48,50,54,69). The relevant mechanisms will now be discussed.

ADSORPTION OF GENOTOXIC CARCINOGENS

Several studies have demonstrated that synthetic carcinogens like 3,2'-dimethyl-4-aminobiphenyl or the food carcinogen 2-amino-3-methylimidazo[4,5-f] quinoline are irreversibly adsorbed to insoluble fiber (4,64). These chemicals induce, among others, cancer in the mammary gland, prostate colon, and pancreas (13,40,79). The finding that the heterocyclic amine carcinogens are adsorbed to insoluble fiber may be important, since current concepts suggest that these types of agents are the genotoxic carcinogens in human nutritional carcinogenesis. Ingestion of foods such as fried meat, which contains these carcinogens, might optimally be accompanied by the consumption of whole grain cereal bread or other sources of insoluble fiber to lower the carcinogen available in vivo.

MECHANISM OF COLON AND BREAST CANCER INHIBITION BY FIBERS

Soluble and especially insoluble fibers increase water retention and, thus, increase intestinal bulk. A major consequence is the dilution of potentially harmful substances in the large bowel. This can include genotoxic carcinogens for which a more dilute solution in the intestinal tract results in a lower rate of absorption. Thus, less is available in the liver for the metabolic activation of those carcinogens for which key steps occur mainly in this organ (breast or colon) or the proximate carcinogen produced in the liver (80,81). Therefore, a lower overall carcinogenic effect is expected, when there is adequate total fiber intake, with an increased intestinal content and stool bulk.

The breast and endometrium are target organs of estrogen action. Overstimulation by estrogenic steroids is involved in the carcinogenic process (8,10), Androgens probably have a similar role in prostate cancer (53). In turn, circulating levels of estrogens and androgens influence the production and secretion of pituitary hormones that also participate in the hormonal action on endocrine target organs. The principal circulating estrogens--estradiol and estrone--undergo conjugation by type 2 hepatic enzymes and conversion to glucuronides and sulfate esters that are secreted in the bile and reach the intestinal tract. In the presence of fiber and, thus, higher intestinal bulk, the concentration of these conjugates is reduced, as is microbial enzyme activity, and, therefore, lower amounts are hydrolyzed by bacterial glucuronidase and sulfatase. Hence, fewer unconjugated steroids are available for reabsorption and enterohepatic cycling and, in consequence, there are lower amounts of circulating steroids to exert their specific effects on target organs such as the breast (10,17,20,52,54,82). Similar fiber-mediated effects probably apply to testosterone metabolism and the prostate, but this mechanism requires documentation.

Also, fibers, particularly insoluble fibers, modify the enzyme activities of the intestinal bacterial flora, usually yielding a lower level of such enzymes (48,54,55,82). Some of the fibers also generate a lower intestinal pH so that the hydrolytic enzymes are not in a medium at their optimal pH. The overall effect is a lesser degree of hydrolysis of the estrogen conjugates. The lower pH, in the presence of fiber, arises from a release of lower molecular weight fatty acids, as a consequence of the partial hydrolysis of fiber components (12,14,24,33,67). There is a greater extent of fermentation that can be assessed directly by measuring exhaled methane and other gases.

An important aspect of intestinal biochemistry stemming from fiber, particularly insoluble fibers, is the lower rate of conversion of primary bile acids produced in the liver and secreted in the bile. The total amount of bile acids found is a function of the type and amount of total fat consumed. Individuals on a high fat Western diet (about 40% of calories) display higher levels of intestinal bile acids than individuals on a low total fat intake (10-20% of calories), typical of Far Eastern dietary traditions. Primary bile acids are converted to a series of secondary bile acids through the action of enzymes from the bacterial flora, the activity of which is a function of intestinal fiber content. The overall effect of fat and fiber is to control the concentration of intestinal bile acids. Secondary bile acids display cytotoxicity to intestinal cells as a function of concentration. The promoting effect of bile acids in colon carcinogenesis is, in great part, a reflection of cytotoxicity. Other parameters involved in promotion are the enzyme protein kinase C, and elements associated with hyperproliferation (17-19,24,31,41,45,49,51,56,66). Thus, people on a Western high-fat, low-fiber regimen have a higher risk of colon and breast cancer compared to populations on a high-fat, high-fiber diet as in Finland, or in certain lacto-ovo-vegetarian populations (1,3,22,35,46,47). On the other hand, individuals on the traditional Japanese diet have a low-fat intake and display a low risk for colon cancer, even though their fiber intake is comparable to that of North American people. These same populations, as a result of the influence of fat and fiber on hormonal action described above, exhibit pronounced differences in their breast cancer risks.

Populations in India, where tradition and customary nutrition are vegetarian, with relatively low fat intake and appreciable total fiber intake, have a low risk of the nutritionally linked cancers. In the Western world, vegetarians, through religious precepts such as Seventh Day Adventists, are not necessarily on low-fat diets, since such individuals may consume appreciable amounts of fats of dairy origin or as vegetable oils and seeds. Nonetheless, their overall fruit, vegetable, and fiber intake is considerably higher than that of nonvegetarian U.S. populations, and their risk of several types of cancer, including cancer of the colon and breast, and of heart disease is lower than that of nonvegetarians.

LABORATORY ANIMAL MODEL STUDIES IN COLON CANCER

A variety of compounds, namely, 1,2-dimethylhydrazine (DMH), azoxymethane (AOM), 3,2'-dimethyl-4-amiobiphenyl (DMAB), and nitrosomethylurea (NMU), that are carcinogenic in the colon have been used in a number of rodent models to study the effect of types and amounts of dietary fiber on tumorigenesis at this site (3,15,23,24,26,39,50,61,66). Administration of these compounds to rodents induces both benign and malignant carcinomas of the colon. With the carcinogens requiring metabolic activation, namely, DMH, AOM, and DMAB, 14 of 17 studies demonstrated evidence of protection by dietary wheat bran against colon tumor development, whereas three studies showed no effect. Dietary soybean bran and rice bran had no effect, whereas oat bran had an enhancing effect on DMH-induced colon carcinogenesis in rats. Dietary corn bran either had an enhancing effect or no effect on colon carcinogenesis. Results of dietary cellulose indicate that 6 of 9 experiments showed a protective effect and 3 studies showed no effect. Dietary pectin and lignin had a protective effect. The results thus far generated on the effect of dietary fiber in colon carcinogenesis suggest strongly

Table 1. Colon tumor incidence in F344 rats fed various fibers

Amount of dietary fiber	fat	Sex of animals	Percentage of animals with colon tumors	
			Azoxymethane treated	3,2'-dimethyl-4-aminobiphenyl treated
Experiment 1				
Control (5% alphacel)	20% corn oil	Female	57	--
+15% wheat bran	"	"	33[a]	--
+15% pectin	"	"	10[a]	--
Experiment 2				
Control (5% alphacel)	5%	Male	90	--
+15% wheat bran	"	"	71[a]	--
+15% citrus pulp	"	"	63[a]	--
Experiment 3				
Control (5% alphacel)	"	"	--	46
+15% wheat bran	"	"	--	26[a]

[a] significantly different from the control group, $P < 0.05$.

that (a) the inhibitory effect depends on the type of fiber, and (b) wheat bran appears to inhibit colon tumor development more consistently in animal models. The varied conclusions on the other types of fiber appear to be due to differences in the experimental protocols, the amount and degree of milling of the fiber involved, the amount and type of carcinogen used to induce colon tumors, and, perhaps even more importantly, elements of the diet other than the fiber. Several of the diets used to study the effect of fiber were deficient in certain micronutrients that are involved in carcinogen metabolism or detoxification.

The effect of semipurified diet containing 15% pectin, or wheat bran and 20%, fat on colon carcinogenesis induced by AOM was studied in F344 rats (see Reddy, Ref. 3). The control and experimental diets were fed to rats before, during, and after carcinogen treatment and then were continued until termination of the experiment. The control diet contained about 5% alphacel as fiber source. The addition of pectin or wheat bran to the diet greatly inhibited colon tumor incidence induced by AOM, a carcinogen requiring host-mediated metabolic activation (Tab. 1). In another study, the effect of a semipurified diet containing 15% wheat bran or dehydrated citrus fiber and 5% dietary fat on intestinal carcinogenesis induced by AOM and DMAB was studied in male F344 rats. The composition of diets was adjusted so that all animals in different experimental groups consumed approximately the same amount of protein, fat, minerals, and vitamins. Experimental diets were begun before carcinogen administration and continued until the termination of the experiment. The animals fed the wheat bran or citrus fiber and treated with AOM had a lower incidence (number of animals with tumors) and multiplicity (number of tumors/tumor-bearing rats) of colon tumor and tumors of the small intestine than did those fed the control diet and treated with AOM (Tab. 1). The animals fed the wheat bran and treated with DMAB had a lower incidence and multiplicity of colon tumors compared with those fed the control diet and given DMAB (Tab. 1).

Recent studies suggested the possibility that dietary fiber may protect also against breast cancer (1,20,28,52,54,82). In order to test the fiber hypothesis with regard to breast cancer, Cohen et al. (10) used the NMU-induced mammary tumor model. Groups of rats were fed a high fat (HF 23% wt/wt) diet; a high-fat diet supplemented with 10% soft white wheat bran (SWWB); a low-fat (LF 5% wt/wt) diet; and a LF diet supplemented with 10% SWWB. The diets were administered 3 days after administration of NMU and fed for 15 weeks. It was found that SWWB supplementation reduced the incidence of tumors from 90% to 67% in the HF groups and from 63% to 47% in the LF groups. A similar pattern was seen when total tumor number was assessed (Figs. 1-4). These results suggest that doubling fiber intake and lowering fat intake can result in a 50% reduction in mammary cancer incidence in this laboratory animal model.

A few experiments in animal models for colon cancer have yielded an enhancing effect of bran fiber (23,24). The underlying mechanism rests on the fact that a high amount of fiber, 20% of the diet, was fed. In turn, this large amount of insoluble material in the intestinal tract caused an increased cell duplication rate that, in the presence of a genotoxic carcinogen, yielded a higher carcinogenic effect, in part because of the lower possibility of DNA repair. Furthermore, increased cell duplication rates enhanced the rate of early tumor

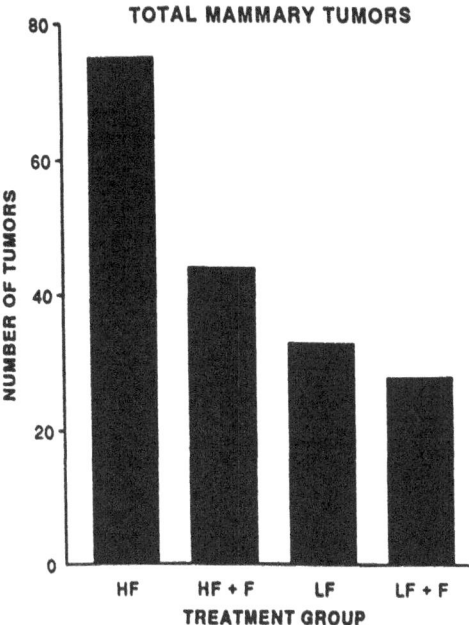

TOTAL MAMMARY TUMORS

NUMBER OF TUMORS

80
60
40
20
0

HF   HF + F   LF   LF + F

TREATMENT GROUP

Figure 1. Effect of dietary wheat bran supplementation on number of mammary tumors. Thirty rats were in each dietary group. Data points represent mean ± SE. The four dietary groups were: HF = high fat (23.5% corn oil); HF + F = high fat plus fiber (10% soft white wheat bran); LF = low fat (5% corn oil); and LF + F = low fat plus fiber.

development (31,49). Indeed, this is how the effect of promoters in general can be visualized (80). Clearly, humans, even on a high total fiber intake, never reach intestinal fiber concentrations that would prove to be irritating or increase rates of cell duplication. Rather, fibers act by increasing the retained water content in the intestinal tract, thus diluting promoters and effectively lowering the rate of cell duplication. In addition, certain mineral salts like magnesium and especially calcium, together with vitamin D, have a documented protective effect in colon and breast cancer, mainly because of a favorable action on lowering cell duplication rates (2,9,16,27,31,37,38,44,49,56,62,65,73). The molecular intracellular and intercellular events on the controlling elements in cell duplication rates remain to be defined. Kinases and phosphatases are likely key elements.

The distinct action of carcinogens in humans and in rodents in yielding cancer more frequently in the distal colon may rest on specific conditions of enzyme-generated reactive metabolites at that site (80,81). It was found that different types of fiber also had a select action in distinct parts of the bowel. Sprague-Dawley male rats were fed diets with 1) low 2% fiber, 2) 10% guar, 3) 10% oat bran, and 4) 1% wheat bran (32). With diets 1 to 3, there was a decreasing gradient of the pH and of short-chain fatty acids, acetate, propionate, and butyrate. On the other hand, with the wheat bran diet 4, the butyrate and pH were unaltered from the cecum to the feces, maintaining the higher level of

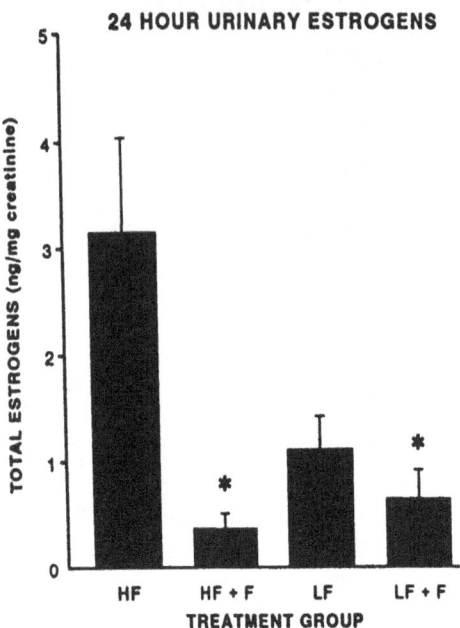

24 HOUR URINARY ESTROGENS

Figure 2. Effect of dietary wheat bran supplementation on 24 hr urinary total (E^1 + E^2) estrogen levels. Urine was obtained from 4-5 rats/group while the rats were in metabolism cages. Assays were performed using a kit obtained from ICN/RSL. Data points represent mean ± SE. * Indicates signficance at $p < 0.05$ compared to the unsupplemented controls. The four dietary groups were: HF = high fat (23.5% corn oil); HF + F = high fat plus fiber (10% soft white wheat bran); LF = low fat (5% corn oil); and LF + F = low fat plus fiber.

butyrate in the distal colon. It was suggested that this exerted an antitumor action, accounting for the protective effect of wheat bran (32).

METABOLIC EPIDEMIOLOGICAL STUDIES IN HUMANS

In view of an inverse relationship between dietary fiber and colon cancer risk, and the potential significance of fecal secondary bile acids and bacterial enzymes in the pathogenesis of colon cancer, a series of studies on the effect of types of fiber on these fecal constituents was studied in healthy subjects.

One investigation dealt with the effect of types of dietary fiber and fiber fraction, namely, wheat bran, oat fiber, and cellulose, on bile acids in healthy subjects from New York consuming high-fat/moderately lowfiber diets (47). A 4-day dietary record and 2-day stool samples were collected from each volunteer before the fiber intervention study (control diet period). The subjects were randomly divided into various dietary groups and were given about 10 g per day of one of the dietary fibers (micrograde roasted hard red wheat bran, Williamson's

24 HOUR FECAL LIPID

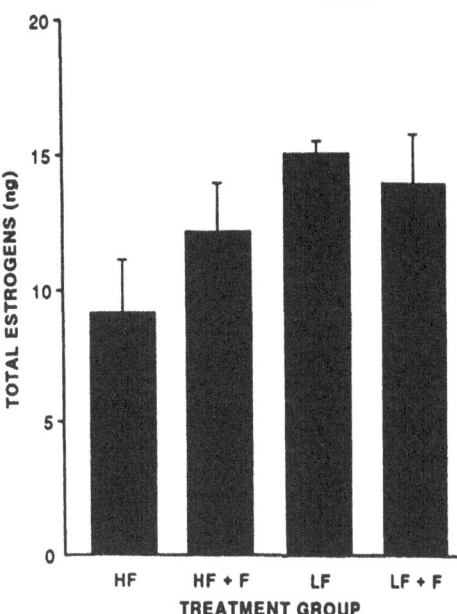

Figure 3. Effect of dietary wheat bran supplementation on 24 hr fecal total (E^1 + E^2) estrogen levels. Fecal samples (4/group) were obtained while the rats were in metabolism cages. Assays were performed using a kit obtained from ICN/RSL. Data points represent mean ± SE. The four dietary groups were: HF = high fat (23.5% corn oil); HF + F = high fat plus fiber (10% soft white wheat bran); LF = low fat (5% corn oil); and LF + F = low fat plus fiber.

Better Basic oat fiber, or Solka-Floc BW-40FCC or BW-200FCC cellulose) in the form of bread, pasta, or muffins. At the end of 5 weeks on each high-fiber dietary regimen, 24-hr stool samples for 2 days and 4-day dietary records were obtained. A 4-week period of control diet without supplemental fiber followed each fiber supplementation period. At the end of those 4 weeks, 24-hr stool samples for 2 days and a 4-day dietary record were obtained. Stool samples were analyzed for bile acids.

Table 2 summarizes the fecal secondary bile acids and total bile acids excretion. The excretion of secondary bile acids--deoxycholic acid, lithocholic acid, and 12-ketolithocholic acid--and total bile acids was lower during the cellulose and wheat bran periods, compared with their respective control diet periods. On the other hand, the concentrations of secondary bile acids were unaffected during the oat fiber period. Changing the diets from cellulose and wheat bran to their respective follow-up control diets significantly increased the secondary bile acid concentration.

The results generated from these studies indicate that certain dietary fibers may decrease the concentration of secondary bile acids that seemingly play a role

24 HOUR FECAL ESTROGENS

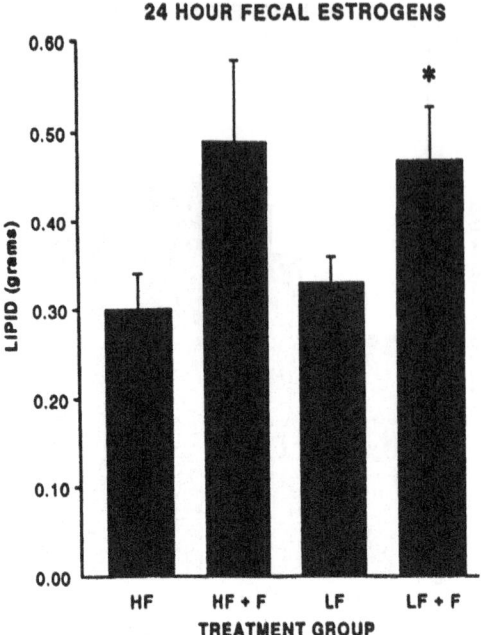

Figure 4. Effect of dietary wheat bran supplementation on 24 hr fecal lipid content. Fecal samples (4/group) were obtained while the rats were in metabolism cages. Assays were performed using standard Soxhlet extraction procedures. Data points represent mean ± SE. * Indicates significance at p < 0.05 compared to the unsupplemented control. The four dietary groups were: HF = high fat (23.5% corn oil); HF + F = high fat plus fiber (10% soft white wheat bran); LF = low fat (5% corn oil); and LF + F = low fat plus fiber.

in colon carcinogenesis. The effect of bile acid excretion may depend on the type of fiber consumed. These controlled experiments extend the early findings demonstrating that people in Kupio, Finland, had a lower concentration of fecal bile acid than controls in New York, paralleling the risk. People in areas with intermediate risk displayed a corresponding intermediate bile acid concentration. In part, however, risk may also depend on the intestinal calcium and magnesium levels (45-48,68,73). In a comparative study of Chinese residing in China and in San Francisco, fat and protein and low carbohydrate intakes were enhancing factors, but fiber was not significantly protective, being similar and low in the reduced risk population in China and the higher risk group in San Francisco (83). Such findings were also made in Japan, where fiber intake is fairly low, and where the protection stems from a low fat intake (77). These conditions were mimicked in rat models for colon cancer (3,61). Thus, fiber displays preventive ability for distal colon cancer in the face of a high risk due to elevated fat intake. Calcium and magnesium from dairy products, as in Finland, enhance the protective effect through distinct mechanisms (31,73).

Table 2. Fecal bile acids in healthy subjects during the control and supplemental fiber periods

mg/g of dry feces

Bile acids	Control diet	Cellulose diet	Control diet	Oat fiber diet	Control diet	Wheat bran diet	Control diet
Cholic	0.12[a]	0.12	0.27	0.41	0.39	0.40	0.10
Chenodeoxycholic	0.42	0.31	0.51	0.29[b]	0.29	0.33	0.30
Deoxycholic	3.98	2.64[b,c]	3.42	3.24	3.80	1.60[b,c]	3.41
Lithocholic	3.28	2.16[b,c]	2.96	2.46	2.98	1.77[b,c]	3.01
12-Ketolithocholic	0.46	0.30[b]	0.34	0.29	0.40	0.13[b,c]	0.40
Other	2.38	1.66	1.99	1.69	2.48	1.80	2.40
Total	10.74	7.19[b,c]	9.49	8.38[c]	10.34	6.03[b,c]	9.62

[a] Mean (n = 19).

[b] significantly different from its control diet period diet, $P < 0.05$.

[c] significantly different from its follow-up control diet period, $P < 0.05$.

CONCLUSIONS

Fiber is an important part of the human nutritional environment. There are distinct types of fiber as a function of the specific foods. Basically, fruits and vegetables contain more soluble fibers that are partially hydrolyzed through bacterial enzymes in the intestinal tract, yielding products with specific metabolic attributes. The hulls of grains contain bran, mainly insoluble fibers, so-called because they are not highly susceptible to metabolic change in the intestinal tract. Depending on the degree of milling, such fibers lead to water retention and, thus, increase in stool bulk. Coarse fibers have this property to a much greater extent than finely milled cereal fibers. Some of the discrepancies in the literature dealing with fiber can be accounted for, in part, by a failure to take into account and describe the degree of milling of the bran fiber.

The mechanism of action of soluble and insoluble fibers in modifying cancer risks is reasonably well understood, based on observations in human populations and deliberate studies in animal models. Data in humans who, by tradition, consume more fruits and vegetables, for example, in the Mediterranean countries, or more bran cereal fiber, in Finland and in certain religious groups, provide evidence that these habits lead to a lower incidence of cancer in the colon and a somewhat lower risk of breast cancer in post-menopausal women. These human findings have their parallel in animal experimentation. One component of insoluble fiber, phytic acid, has proven to be particularly effective in preventing colon cancer induction in models. In this connection, concern has been expressed that phytic acid intake, and insoluble fiber intake in general, might have adverse effects owing to the alleged lower availability of essential micronutrients and, in particular, minerals like magnesium, zinc, and calcium. This fear is based on historic observations in areas like Iran, where the traditional daily nutritional regime involved chiefly the consumption of high phytate fiber pita bread without many other foods serving as sources of minerals, vitamins, and like essential micro-nutrients. This kind of unbalanced nutrition has led to poor development and growth in children and resulted in dwarfism. However, in other parts of the world, particularly in the Western world, with the general availability of a great variety of many foods, including fruits and vegetables as well as meats and fish, as sources of essential minerals and vitamins, there is no fear of mineral and vitamin deficiencies, even in the presence of an adequate intake of cereal fiber and even phytate. The Finnish people have a low colon cancer rate and a lower breast cancer rate than Europeans and Americans, generally because of a high cereal in-soluble fiber intake, and these populations do not display any signs of mineral and vitamin deficiencies leading to abnormal growth and development.

What, then, is an adequate fiber intake? The formal recommendations of experts in nutrition suggest that we consume four to six portions of fruits and vegetables a day as a source of essential vitamins, minerals, and certain fibers, mostly soluble fibers. Maintenance of intestinal health in particular, but proper functioning of other organs as well, requires optimal amounts of insoluble fibers. The specific amount varies from individual to individual, not only as a result of metabolic considerations but also as a function of the amount and type of exercise performed regularly. Nutrition authorities recommend a total daily fiber intake of 25-35 grams. Based on the dramatic reduction in intestinal diseases and breast cancer in Finland, where the measured stool weight per day was 250 grams, and the observations of Burkitt in Africa, who recorded a virtual absence of intestinal diseases, including colon cancer, in African populations, it would seem that an adequate total fiber intake is one that avoids constipation and yields a daily stool of 200-250 grams in one or two passes a day (77).

ACKNOWLEDGEMENTS

As in our previous paper at ICMAA II, we dedicate this report to the memory of one of the originators of ICMAA, the late Dr. Tsuneo Kada, Genetics Institute, Mishima, Japan, a distinguished scientist and wonderful musician. Also recognized with admiration is a pioneer in fiber and health, Denis Burkitt, CMG, MD, FRS, FRCS.

Research in our laboratory is supported by USPHS grants CA-42381, CA-45720, CA-40839, the Kellogg Company, and a grant, CN-29, from the American Cancer Society. Ms. Beth-Alayne McKinney provided excellent editorial support.

REFERENCES

1. Adlercreutz, H. (1990) Diet, breast cancer, and sex hormone metabolism. *Ann. N.Y. Acad. Sci.* 595:281-290.
2. Appleton, G.V.N., P.W. Davies, J.B. Bristol, and R.C.N. Williamson (1987) Inhibition of intestinal carcinogenesis by dietary supplementation with calcium. *Br. J. Surg.* 74:523-525.
3. American Health Foundation (1987) Proceedings of a workshop on new developments on dietary fat and fiber in carcinogenesis (optimal types and amounts of fat or fiber). *Prev. Med.* 16:449-595.
4. Barnes, W.S., J. Maiello, and J.H. Weisburger (1983) In vitro binding of the food mutagen 2-amino-3-methylimidazo-[4,5-*f*]quinoline to dietary fibers. *JNCI* 70:757.
5. Bartram, H.P., W. Scheppach, C. Heid, C. Fabian, and H. Kasper (1991) Effect of starch malabsorption on fecal bile acids and neutral sterols in humans: Possible implications for colonic carcinogenesis. *Cancer Res.* 51:4238-4242.
6. Bingham, S.A. (1988) Meat, starch, and nonstarch polysaccharides and large bowel cancer. *Am. J. Clin. Nutr.* 48:762-767.
7. Caderni, G., F. Bianchini, P. Dolara, and D. Kriebel (1991) Starchy foods and colon proliferation in mice. *Nutr. Cancer* 15:33-40.
8. Carroll, K.K. (1991) Dietary fats and cancer. *Am. J. Clin. Nutr.* 53:1064S-1067S.
9. Carroll, K.K., E.A. Jacobson, L.A. Eckel, and H.L. Newmark (1991) Calcium and carcinogenesis of the mammary gland. *Am. J. Clin. Nutr.* 54:206S-208S.
10. Cohen, L.A., M.E. Kendall, E. Zang, C. Meschter, and D.P. Rose (1991) Modulation of NMU-induced mammary tumor promotion by dietary fiber and fat. *J. Natl. Cancer Inst.* 83:496-501.
11. Committee on Diet and Health, Food and Nutrition Board (1989) *Diet and Health: Implications for Reducing Chronic Disease Risk.* Washington, D.C., National Academy Press.

12. Cummings, J.H., and S.A. Bingham (1987) Dietary fiber, fermentation and large bowel cancer. *Cancer Res.* 6:601-621.

13. Felton, J.S., and M.G. Knize (1991) Occurrence, identification and bacterial mutagenicity of heterocyclic amines in cooked foods. *Mutat. Res.* 259:205-218.

14. Fleming, S.E., M.D. Fitch, and M.W. Chansler (1989) High-fiber diets: Influence on characteristics of cecal digesta including short-chain acid concentrations and pH. *Am. J. Clin. Nutr.* 50:93-97.

15. Freeman, H.J. (1986) Effects of differing purified cellulose, pectin and hemicellulose fiber diets on fecal enzymes in 1,2-dimethylhydrazine-induced rat colon carcinogenesis. *Cancer Res.* 46:5529-5532.

16. Garland, C.F., F.C. Garland, and E.D. Gorham (1991) Can colon cancer incidence and death rates be reduced with calcium and vitamin D? *Am. J. Clin. Nutr.* 54:193S-201S.

17. Goettler, D., A.V. Rao, and R.P. Bird (1987) The effects of a "low-risk" diet on cell proliferation and enzymatic parameters of preneoplastic rat colon. *Nutr. Cancer* 10:149-162.

18. Gregoire, R., K.S. Yeung, J. Stadler, H.S. Stern, H. Kashtan, G. Neil, and W.R. Bruce (1991) Effect of high-fat and low-fiber meals on the cell proliferation activity of colorectal mucosa. *Nutr. Cancer* 15:21-26.

19. Hashiba, H., M. Fukushima, K. Chida, and T. Kuroki (1987) Systemic inhibition of tumor promoter-induced ornithine decarboxylase in 1α-hydroxy-vitamin D^3-treated animals. *Cancer Res.* 47:5031-5035.

20. Heber, D., J.M. Ashley, D.A. Leaf, and R.J. Barnard (1991) Reduction of serum estradiol in postmenopausal women given free access to low-fat high-carbohydrate diet. *Nutrition* 7:137-141.

21. Hoffmann, D., and S.S. Hecht (1990) Advances in tobacco carcinogenesis. In *Chemical Carcinogenesis and Mutagenesis I*, C.S. Cooper and P.L. Grover, eds. Springer-Verlag, New York, pp. 61102.

22. Ip, C., D.F. Birt, A.E. Roger, and C. Mettlin, eds. (1986) *Dietary Fat and Cancer.* Alan R. Liss, New York.

23. Jacobs, L.R. (1986) Modification of experimental colon carcinogenesis by dietary fibers. *Adv. Exp. Med. Biol.* 206:105-118.

24. Jacobs, L.R., and J.R. Lupton (1986) Relationship between colonic luminal pH, cell proliferation, and colon carcinogenesis in 1,2dimethylhydrazine treated rats fed high fiber diets. *Cancer Res.* 46:1727-1734.

25. Jensen, O.M., J. Estève, H. Møller, and H. Renard (1990) Cancer in Europe community and its member states. *Eur. J. Cancer* 26:1167-1256.

26. Liehr, J.G. (1990) Genotoxic effects of estrogens. *Mutat. Res.* 238:269-276.

27. Kawaura, A., N. Tanida, K. Sawada, M. Oda, and T. Shimoyama (1989) Supplemental administration of 1α-hydroxyvitamin D^3 inhibits promotion by intrarectal instillation of lithocholic acid in Nmethyl-N-nitrosurea-induced colonic tumorigenesis in rats. *Carcinogenesis* 10:647-649.

28. Kohlmeier, L., J. Rehm, and H. Hoffmeister (1991) Lifestyle and trends in worldwide breast cancer rates. *Ann. N.Y. Acad. Sci.* 609:259-268.

29. Kritchevsky, D. (1988) Dietary fiber. *Ann. Rev. Nutr.* 8:301-328.

30. Lee, I.M., R.S. Paffenbarger, and C.C. Hsieh (1991) Physical activity and risk of developing colorectal cancer among college alumni. *J. Natl. Cancer Inst.* 83:1324-1329.

31. Lipkin, M. (1991) Application of intermediate biomarkers to studies of cancer prevention in the gastrointestinal tract: Introduction and perspective. *Am. J. Clin. Nutr.* 54:188S-192S.

32. McIntyre, A., G.P. Young, T. Taranto, P.R. Gibson, and P.B. Ward (1991) Different fibers have different regional effects on luminal contents of rat colon. *Gasterenterology* 101:1274-1281.

33. McKeown-Eyssen, G.E. (1987) Fiber intake in different populations and colon cancer risk. *Prev. Med.* 16:532-539.

34. Messina, M., and S. Barnes (1991) The role of soy products in reducing the risk of cancer. *J. Natl. Cancer Inst.* 83:541-546.

35. Mills, P.K., W.L. Deeson, R.L. Phillips, and G.E. Fraser (1989) Dietary habits and breast cancer incidence among Seventh-Day Adventists. *Cancer* 64:582-590.

36. National Research Council (1989) *Recommended Dietary Allowances, 10th ed.* National Academy Press, Washington, D.C.

37. Newmark, H., M. Lipkin, and N. Maheshwari (1990) Colonic hyperplasia and hyperproliferation induced by a nutritional stress diet with four components of Western-style diet. *J. Natl. Cancer Inst.* 82:491-496.

38. Newmark, H., M. Lipkin, and N. Maheshwari (1991) Colonic hyperproliferation induced in rats and mice by nutritional-stress diets containing four components of a human Western-style diet (series 2). *Am. J. Clin. Nutr.* 54:209S-214S.

39. Nigro, N.D., and A.W. Bull (1987) The impact of dietary fat and fiber on intestinal carcinogenesis. *Prev. Med.* 16:451-459.

40. Ohgaki, H., S. Takayama, and T. Sugimura (1991) Carcinogenicities of heterocyclic amines in cooked food. *Mutat. Res.* 259:399-410.

41. Otsuka, M., S. Satchithanandam, and R.J. Calvert (1989) Influence of meal distribution of wheat bran on fecal bulk, gastrointestinal transit time and colonic thymidine kinase activity in the rat. *J. Nutr.* 119:566-572.

42. Parkin, D.M., E. Läärä, and C.S. Muir (1988) Estimates of the worldwide frequency of sixteen major cancers in 1980. *Int. J. Cancer* 41:184-197.

43. Patterson, B., and G. Block (1988) Food choices and the cancer guidelines. *Am. J. Public Health* 78:282-286.

44. Pence, B., and F. Buddingh (1988) Inhibition of dietary fat-promoted colon carcinogenesis in rats by supplemental calcium or vitamin D_3. *Carcinogenesis* 9:187-190.

45. Reddy, B.S. (1990) Effect of types of dietary fiber on fecal mutagens and bacterial enzymes in relation to colon cancer. In *New Developments in Dietary Fiber*, I Fuda and C.J. Brine, eds. Plenum Press, New York, pp. 159-167.

46. Reddy, B.S. (1991) Metabolic epidemiology of colon cancer. In *Large Bowel Cancer: Policy, Prevention, Research and Treatment*, P. Rozen, C.G. Reich, and S.J. Winawer, eds. Frontiers in Gastrointestinal Research, Vol 18, Krager, Basel, Switzerland, pp. 88-98.

47. Reddy, B.S., A. Engle, S. Katsifis, B. Simi, H. Bartram, P. Perrino, and C. Mahan (1989) Biochemical epidemiology of colon cancer: Effect of types of dietary fiber on fecal mutagens, and acid and neutral sterols in healthy subjects. *Cancer Res.* 49:4629-4635.

48. Reddy, B.S., A. Engle, B. Simi, and M. Goldman (1992) Effect of types of dietary fiber on colonic bacterial enzymes and bile acids in relation to cancer in women. *Gastroenterology* (in press).

49. Risio, M., M. Lipkin, G. Candelaresi, C.S. Berone, F.P., Rossini, et al. (199) Correlations between rectal mucosa cell proliferation and the clinical and intestine. *Cancer Res.* 51:1917-1921.

50. Roberfroid, R.B. (1991) Dietary modulation of experimental neoplastic development: Role of fat and fiber content and calorie intake. *Mutat. Res.* 259:351-362.

51. Roncucci, L., A. Medline, and W.R. Bruce (1991) Classification of aberrant crypt foci and microadenomas in human colon. *Cancer Epid., Biomarkers & Prevention* 1:57-60.

52. Rose, D.P. (1992) Dietary fiber, phytoestrogens and breast cancer. *Nutrition* 7:139-140.

53. Rose, D.P., and J.M. Connolly (1992) Dietary fat, fatty acids and prostate cancer. *Lipids* (in press).

54. Rose, D.P., M. Goldman, J.M. Connolly, and L.E. Strong (1991) A high-fiber-supplemented diet reduces serum estrogens in premenopausal women. *Am. J. Clin. Nutr.* 54:520-525.

55. Rowland, I.R., ed. (1988) *Role of the Gut Flora in Toxicity and Cancer.* Academic Press, London.

56. Rozen, P., Z. Fireman, N. Fine, Y. Wax, and E. Ron (1989) Oral calcium suppresses increased rectal epithelial proliferation of persons at risk of colorectal cancer. *Gut* 30:650-655.

57. Scheppach, W., C. Fabian, F. Ahrens, M. Spengler, and H. Kasper (1988) Effect of starch malabsorption on colonic function and metabolism in humans. *Gastroenterology* 95:1549-1555.

58. Seitz, H.K., and U.A. Simanowski (1991) Alcohol and cancer: A critical review. In *Alcoholism: A Molecular Perspective*, T.N. Palmer, ed. Plenum Press, New York, pp. 265-296.

59. Shankar, S., and E. Lanza (1991) Dietary fiber and cancer prevention. *Hematol./Oncol. Clinics North America* 5:25-41.

60. Simopoulos, A.P., ed. (1987) Diet and health: Scientific concepts and principles. *Am. J. Clin. Nutr.* 45:1015-1407.

61. Sinkeldam, E.J., C.F. Kuper, and M.C. Bosland (1990) Interactive effects of dietary wheat bran and lard on N-methyl-N-nitro-Nnitrosoguanidine-induced colon carcinogenesis in rats. *Cancer Res.* 50:1092-1096.

62. Sitrin, M.D., A.G. Halline, C. Abrahams, and T.A. Brasitus (1991) Dietary calcium and vitamin D modulate 1,2-dimethylhydrazineinduced colonic carcinogenesis in the rat. *Cancer Res.* 51:56085613.

63. Slater, T.F., and G. Block, eds. (1991) Antioxidant vitamins and β-carotene in disease prevention. *Am. J. Clin. Nutr.* 53:189S-396S.

64. Smith-Barbaro, P., D. Hanson, and B.S. Reddy (1981) Carcinogen binding to various types of dietary fiber. *JNCI* 67:495-497.

65. Sorenson, A., M. Slattery, and M. Ford (1988) Calcium and colon cancer: A review. *Nutr. Cancer* 11:135-145.

66. Summerton, J., N. Goeting, G.A. Trotter, and I. Taylor (1985) Effect of deoxycholic acid on the tumour incidence, distribution, and receptor status of colorectal cancer in the rat model. *Digestion* 31:77-81.

67. Spiller, G.A., M.C. Chernoff, R.A. Hill, J.E. Gates, J.J. Nassar, and E.A. Shipley (1980) Effect of purified cellulose, pectin, and a low-residue diet on fecal volatile fatty acids, transit time, and fecal weight in humans. *Am. J. Clin. Nutr.* 33:754-759.

68. Trock, B., E. Lanza, and P. Greenwald (1990) Dietary fiber, vegetables, and colon cancer: Critical review and meta-analyses of the epidemiologic evidence. *JNCI* 82:650-661.

69. Trowell, H., D. Burkitt, and K. Heaton (1985) *Dietary Fibre, Fibredepleted Foods and Disease.* Academic Press, San Diego.

70. U.S. Public Health Service (1988) *The Surgeon General's Report on Nutrition and Health.* DHHS (PHS) Publ. No. 88-50210. U.S. Government Printing Office, Washington, D.C.

71. Vogel, V.G., and R.S. McPherson (1989) Dietary epidemiology of colon cancer. *Hematol. Oncol. Clin. North Am.* 3:35-63.

72. Walaszek, Z. (1990) Potential use of D-glucaric acid derivatives in cancer prevention. *Cancer Lett.* 54:1-8.

73. Wargovich, M.J., P.M. Lynch, and B. Levin (1991) Modulating effects of calcium in animal models of colon carcinogenesis and shortterm studies in subjects at increased risk for colon cancer. *Am. J. Clin. Nutr.* 54:202S-205S.

74. Wattenberg, L.W. (1990) Inhibition of carcinogenesis by naturallyoccurring and synthetic compounds. *Basic Life Sci.* 52:155-166.

75. Weed, D.L., P. Greenwald, and J.W. Cullen (1990) The future of cancer prevention and control. *Semin. Oncol.* 17:504-509.

76. Weisburger, J.H. (1990) The mechanism of lung carcinogenesis and smoking cessation. *Epidemiology* 1:314-317.

77. Weisburger, J.H. (1991) Causes, relevant mechanisms, and prevention of large bowel cancer. *Semin. Oncol.* 18:316-336.

78. Weisburger, J.H. (1992) On the role of tea in modifying causes of major human cancers. *Prev. Med.* Vol. 21 (3,4).

79. Weisburger, J.H., and R.C. Jones (1990) Prevention of formation of important mutagens/carcinogens in the human food chain. In *Antimutagenesis and Anticarcinogenesis Mechanisms II,* Y. Kuroda, D.M. Shankel, and M.D. Waters, eds. Plenum Press, New York.

80. Williams, G.M., and J.H. Weisburger (1991) Chemical carcinogenesis. In *Casarett and Doull's Toxicology, 4th Edition,* M.O. Amdur, J. Doull, and C.D. Klaassen, eds. Pergamon Press, New York.

81. Witmer, M.C., R. Snyder, D.J. Jollow et al., eds. (1991) Molecular and cellular effects and their impact on human health. In *4th International Symposium on Biological Reactive Intermediates.* Plenum Press, New York.

82. Woods, M.N., S.L. Gorbach, C. Longcope, et al. (1989) Low-fat, high-fiber and serum estrone sulfate in premenopausal women. *Am. J. Clin. Nutr.* 49:1179-1183.

83. Yeung, K.S., G.E. McKeown-Eyssen, G.F. Li, E. Glazer, et al. (1991) Comparisons of diet and biochemical characteristics of stool and urine between Chinese populations with low and high colorectal cancer rates. *Natl. Cancer Inst.* 83:46-50.

EXTRACELLULAR INTERCEPTION OF MUTAGENS

D.M. Shankel, S. Kuo, C. Haines, and L.A. Mitscher*

Department of Microbiology and
*Medicinal Chemistry
The University of Kansas
Lawrence, Kansas 66045 USA

ABSTRACT

Extracellular interception of mutagens by excreted enzymes or by chemical agents that react with or bind to formed mutagens provides an important means of defense against chemical mutagens/carcinogens. Kada and Shimoi (25) have classified molecules that function in this manner as "desmutagens," and many of them are natural cellular metabolites (23). Among the specific mechanisms that such agents may employ are: prevention of the activation of "promutagens" to mutagens; stimulation of enzymes (e.g., glutathione-S-transferase) that catalyze the binding/inactivation of damaging electrophiles; direct binding and concomitant inactivation of promutagens or mutagens; interference with uptake of mutagens into cells; etc. De Flora and Ramel (11) have provided an excellent discussion of the mechanisms of these agents and a proposed classification scheme. Drawing on work from our own laboratories and other recent examples in the literature, several examples of mechanistic approaches to these studies using natural plant-derived materials, e.g., humic acid, *Glycyrrhiza glabra* extract, glutathione, and bioflavonoids, are also described. Antioxidants and agents that conjugate electrophiles will be among the modes of action described for obtaining the goal of intercepting mutagens/carcinogens.

INTRODUCTION

In spite of the great expansion of knowledge about antimutagens that has occurred during the past twenty years, our understanding of the *mechanisms* of antimutagenesis is still quite limited. Perhaps the best understood mechanism, at the molecular level, is that of the mutant DNA polymerases of phage T4 originally described by Drake and his coworkers (14,15). In this case, the alterations in the DNA polymerase gene result in an enzyme that copies DNA with greater fidelity than normal, resulting in an antimutagenic effect.

It is a well-established fact that some "promutagens" require activation (as by S9 in the Ames test) in order to exert their mutagenic or genotoxic effects. If the action of an enzyme can alter a promutagen to a mutagen, then it is logical to assume that enzymes can also convert active mutagens to inactive compounds. Indeed, this is undoubtedly a major defensive mechanism against the numerous insults to which cells are constantly subjected.

Glutathione-S-transferase is a major detoxifying enzyme system that catalyzes the binding of damaging electrophiles to GSH and their subsequent inactivation. This suggests that agents that stimulate the production or activity of this enzyme would be probable antimutagens. Lam et al. (27) have shown that the coffee constituents kahweol and cafestol are inducers of this activity in mice and have identified the key functional chemical groups.

The role of enzymes such as the "Cytochrome P-450" type has been confusing, with an apparent potential for inactivation or activation of promutagens and procarcinogens. Wood et al. (41) demonstrated that a chemical model of cytochrome P-450 could oxidize aflatoxin B_1 to $AFB_1 8,9$-epoxide-the compound generally believed to be the ultimate carcinogenic and mutagenic derivative of AFB_1. This suggests that agents that stimulate or inhibit these cytochromes could have a major effect, positively or negatively, on mutagenesis, depending on the mutagen involved and the cellular environments. Similarly, any chemical or physical agent that results in the breakdown of a mutagen or promutagen would exert a "desmutagenic" effect, and an agent that has the capability of binding to, or absorbing, a damaging compound would produce the same result.

De Flora and Ramel (11) have summarized the general mechanisms of antimutagenesis. Table 1 describes the general mechanisms of *extracellular* interception which will be the focus of the remainder of this chapter. These can be summarized as follows:

1. Agents that act by inhibiting the uptake of mutagens or other precursors;

2. Agents that act by inhibiting the endogenous formation of mutagens; and

3. Agents that act by deactivating mutagens.

Representative examples of each class are also shown in Tab. 1.

In our recent review article (23) we focused on compounds that act by nonenzymatic interception, such as compounds that act by chemical reactivity with already-formed mutagens or by binding to such agents. These agents have generally been classified as "desmutagens" by Kada and his colleagues (25). The terms "desmutagens" or "interceptors" can probably be used interchangeably.

Table 1. Mechanisms of inhibitors of mutagenesis and carcinogenesis inhibitors acting extracellularly. (From De Flora and Ramel, *Mutat. Res.* 202:285-306.)

CLASSIFICATION	EXAMPLES
1. INHIBIT THE UPTAKE OF MUTAGENS OR PRECURSORS	
A. BY HINDERING THEIR PENETRATION	BODY SHIELDING DEVICES, WASHING
A. INTO THE ORGANISM	FATTY ACIDS, PULRESCINE, AROMATIC
B. INTO CELLS	AMINO ACIDS, IODIDE
B. BY FAVORING THEIR REMOVAL	FIBERS
2. INHIBIT THE ENDOGENOUS FORMATION OF MUTAGENS	
A. BY INHIBITING NITROSATION REACTIONS	ASCORBIC ACID, TOCOPHEROLS, PHENOLS
B. BY MODIFYING MICROBIAL INTESTINAL FLORA	FERMENTED DAIRY PRODUCTS
3. DEACTIVATE MUTAGENS	
A. BY PHYSICAL REACTION	MAINTENANCE OF PHYSIOLOGICAL pH
B. BY CHEMICAL REACTION	THIOLS, ANTIOXIDANTS
C. BY ENZYMATIC REACTION	VEGETABLES WITH PEROXIDASE ACTIVITY

Many of the agents that appear to be of most practical use as antimutagens and anticarcinogens are derivatives from plant or microbial agents that are active against a variety of mutagens/carcinogens. In our review mentioned earlier, we described the current state of knowledge about a number of these agents. Table 2 provides a brief summary (and a key reference) for many of those agents that are derived from plant or microbial systems. The list is not totally inclusive, since it continues to grow rapidly, but these data illustrate the broad variety of natural sources that have yielded active antimutagens/anticarcinogens. It should also be noted that there is a wide variety of structures that are able to produce this type of activity.

SOME SIMPLE POSSIBLE MECHANISMS AND EXAMPLES

A paper by Sato et al. (37) on the desmutagenic effect of humic acid illustrates one of the simpler mechanisms by which such agents may function--the physical binding of the mutagen. Figure 1 presents the effects of increasing concentrations of humic acid in a standard Ames test assay, utilizing strains TA100 and TA98, respectively. The data indicate a dramatic dose-dependent decrease in the frequency of revertants. Figure 2 presents the results of an experiment designed to demonstrate whether the humic acid was functioning as a typical "biomutagen" or as a typical "desmutagen." The bioantimutagenicity test utilized standard Ames test procedures in which the bacterial suspension was mixed with the mutagen, and the mixture was incubated at 30° C for one hour with slow shaking, the bacteria were washed twice by centrifugation and resuspended in cold phosphate buffer, and a standard Ames test was performed with or without added humic acid. In the desmutagenicity test, the mutagen was mixed with humic acid and incubated at 37° C for one hour with full shaking, following which the mutagen was extracted, dissolved in DMSO, and plated with the bacteria. The data (Fig. 2) show that the desmutagen test reduced the frequency of mutations dramatically, while the standard bioantimutagen test did not. This demonstrates clearly that the mechanism of action of humic acid is not by inhibition of metabolic activation, but the humic acid is instead binding to or inactivating the mutagen.

Table 2. Selected antimutagens/anticarcinogens which function as putative interceptor molecules. (From Hartman and Shankel, *Env. Molec. Mutagen.* 15:145-182.)

ANTIMUTAGEN/ANTICARCINOGEN	KEY REFERENCE
1. N-ACETYL-L-CYSTEINE	DeFLORA, ET AL, 1989
2. ACYLGLUCOSYLSTEROLS	GUEVARA, ET AL, 1988
3. p-AMINOBENZOIC ACID	GICHNER & VELEMINSKY, 1988
4. AROMATIC ISOTHIOCYANATES	WATTENBERG, 1987
5. ASCORBATE (VITAMIN C)	GHASKADBI & VAIDYA, 1988
6. BIOFLAVONOIDS	FRANCIS, ET AL, 1989
7. β-CAROTENE	KRINSKY, 1987
8. CHLOROPHYLLS & CHLOROPHYLLIN	HAYATSU, ET AL, 1988
9. CINNAMALDEHYDE	DeSILVA & SHANKEL, 1987
10. CURCUMINS	NAGABHUSHAN, ET AL, 1987
11. DIALLYL SULFIDE	WARGOVICH, 1988
12. DITHIOLTHIONES	ANSHER, ET AL, 1986
13. FIBER	DeFLORA & RAMEL, 1988
14. GLUTATHIONE	DeFLORA, ET AL, 1988
15. HYDROXYCHAVICOL	NAGABHUSHAN, ET AL, 1989
16. POLYAMINES	DeLANGE & GLAZER, 1989
17. POLYPHENOLS	BARTSCH, ET AL, 1988
18. RETINOIDS	BUDROE, ET AL, 1987
19. TOCOPHEROLS	CHEN, ET AL, 1988
20. THIOSULFATE	WATTENBERG, ET AL, 1987

Another simple and nontoxic antimutagen is chlorophyllin. Ong and colleagues (34,39) have utilized a Salmonella arabinose-resistant assay system to examine the antimutagenic activity of chlorophyllin against a number of different significant mutagens/carcinogens. Table 3 presents some of the results of their study. Various concentrations of each chemical and a number of compounds or complex mixture extracts were assayed for mutagenic activity with and/or without S9 in a preincubation test. The results demonstrate that chlorophyllin at concentrations of 2.5 mg per plate or less completely or almost completely inhibits the mutagenicity of a range of mutagens. Chlorophyllin has been shown to be an antioxidant, but the exact mechanism by which it inhibits mutagenicity is unknown at this time. It has been suggested that it may exert its inhibitory effect by suppressing metabolic activation and/or by carrying out radical scavenging activity. Studies by other authors (32) indicate that it may also interact with mutagens to form a complex and that this complex formation inactivates the mutagens and decreases their ability to attack DNA. It should be noted that it is important not to utilize **only** reverse mutation assays, since these may tend to pick up only specific base changes; it is also important to utilize forward mutation assays in these tests, since these make it possible to pick up a much wider array of base changes.

SOME ADDITIONAL RESULTS

In our laboratory, we have been studying effects of extracts of *Glycyrrhiza glabra* that contain a variety of phenolic compounds (29). Many plant-derived phenolics have been implicated as chemoprotective agents against mutagenesis. Possible mechanisms of *G. glabra*-induced protection against growth inhibition and mutation include induction of the adaptive response, induction of other DNA repair systems, scavenging of genotoxins by the compounds present in cells, and/or the induction of nonenzymatic scavenging systems for potential genotoxins.

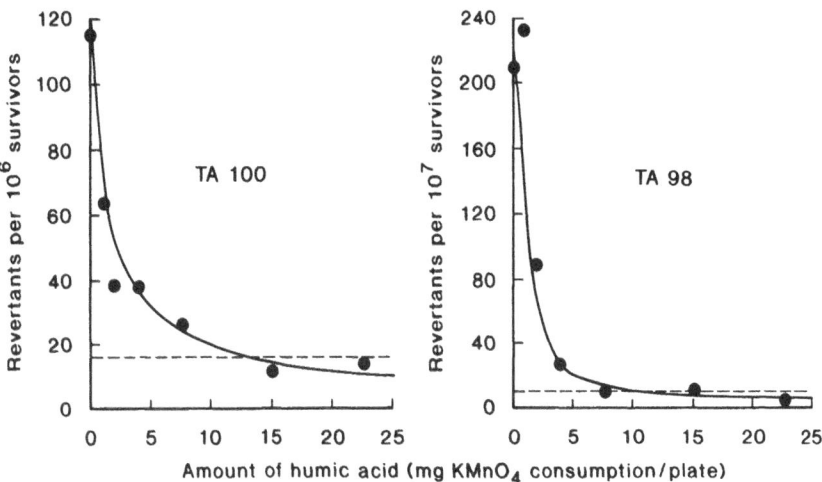

Figure 1. The effects of increasing concentrations of humic acid in a standard Ames test assay utilizing strains TA100 and TA98, respectively. Reproduced with permission from Elsevier Science Publishers.

In our laboratory tests, the scavenging ability of *G. glabra* polar lipid fraction has been demonstrated using a cellular cytotoxicity assay and the phenosafranin/N, N/-dimethyl-4-nitrosoaniline assay.

In a cellular cytotoxicity assay, *Escherichia coli* K12 was incubated in the presence of the photosensitizer rose bengal, then illuminated to generate singlet oxygen (9). Addition of *G. glabra* extract resulted in an increase in survival of cells in a dose-dependent fashion.

We also discovered that cells grown in the presence of 100 µg/ml *G. glabra* exhibited higher levels of reduced glutathione that could also act to scavenge reactive species (Fig. 3). The total amount of reduced glutathione was measured using the modifications of Owens and Hartman (35) of the glutathione reductase recycling assay and employing *E. coli* K12.

FLAVONOIDS

It would probably be appropriate to conclude this review with some discussion of a very large heterogeneous group of phenolic compounds that show interesting activities.

Flavonoids are a large heterogeneous group of phenolic compounds from vascular plants and are reported by many authors to have a range of biological activities that include: anti-inflammation, antiallergic, and antiviral activities; inhibition of DNA synthesis in tumor cells; inhibition of tumor promotion; and scavenging of oxygen free radicals (6,8,16,33). A number of flavonoids, both synthetic and naturally occurring, have been shown to markedly antagonize the formation of chemically induced tumors in experimental animals.

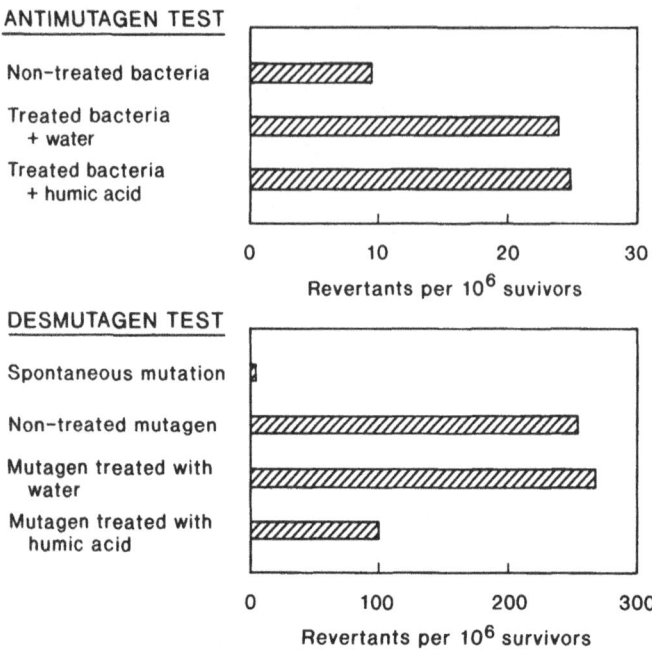

ANTIMUTAGEN TEST

Non-treated bacteria

Treated bacteria
+ water

Treated bacteria
+ humic acid

Revertants per 10^6 suvivors

DESMUTAGEN TEST

Spontaneous mutation

Non-treated mutagen

Mutagen treated with
water

Mutagen treated with
humic acid

Revertants per 10^6 survivors

Figure 2. Results of an experiment designed to demonstrate whether the humic acid was functioning as a typical "bioantimutagen" or as a typical "desmutagen." Reproduced with permission from Elsevier Science Publishers.

A number of studies have shown that flavonoids exert a range of effects, inhibitory and stimulatory, related to the production of arachidonate metabolites via cyclooxygenase and lipoxygenase (18,28).

Middleton (28) has shown that a number of flavonoids are active as inhibitors of protein kinase C and that certain flavonoids interact with calmodulin, which is involved in numerous Ca++-dependent cell processes and enzyme activities. Recent studies by Geahlen (19) have revealed a very complex structure-activity relationship of flavonoids that inhibits protein-tyrosine kinase activity. For example, flavonols and flavones with hydroxy groups at C5 and C7 or with three hydroxy groups in the phenyl ring were potent p40 kinase inhibitors. Quercetin inhibited all three kinases.

Flavonoids have been proposed to act as antioxidants based on their radical-scavenging capabilities. For example, several studies have demonstrated that certain flavonoids (i.e., quercetin and rutin) are scavengers of superoxide (36,42). Bors et al. (3,4) have elucidated, by scavenging assays, a structure-activity relationship defining the structural requirement, a 3',4'-catechol structure, as the structure that gives optimum antioxidant potential. The 3',4-catechol structure apparently increases aroxyl radical stability, one of the critical features in determining whether a substance (a flavonoid) is an antioxidant. Thus, a flavonoid such as kaempferol may be a highly efficient radical scavenger; however, it is not a potent antioxidant (e.g., quercetin) because it lacks the 3',4'-catechol structure and, therefore, has a high aroxyl radical decay rate.

Table 3. Antimutagenic activity of chlorophyllin against selected mutagenic chemicals and complex mixtures in the Ara[R] assay system with tester strain SV50[A].

SAMPLE (DOSE/PLATE)	S9	NUMBER OF MUTANTS/PLATE (PERCENT INHIBITION)					
		0[B]	0.31[B]	0.63[B]	1.25[B]	2.5[B]	5.0[B]
CHLOROPHYLLIN, CONTROL	–	96[D]	103	106	93	95	72
CHLOROPYLLIN, CONTROL	+	170[D]	206	200	185	163	148
AFLATOXIN B$_1$ (0.005 UG)	+	453	302(53)[E]	218(92)	196(95)	201(81)	171(89)
2-AMINOANTHRACENE (2.0 UG)	+	1003	1068(0)	767(29)	197(98)	162(100)	136(100)
BENZO[α]PYRENE (2.0 UG)	+	480	352(37)	253(77)	211(89)	178(93)	159(95)
N-METHYL-N'-NITRO- N-NITROSOGUANIDINE (0.3 UG)	–	1483	204(92)	170(95)	152(95.6)	156(95.5)	126(96)
RED WINE[C] (50 ML)	+	287	NT[F]	132(100)	123(100)	128(100)	95(100)

[A]RESULTS ARE AVERAGED FROM TWO INDEPENDENT EXPERIMENTS IN DUPLICATE.
[B]CONCENTRATION OF CHLOROPHYLLIN (MG/PLATE).
[C]AMOUNT OF ORIGINAL MATERIAL PER PLATE.
[D]NUMBER OF SPONTANEOUS MUTANTS PER PLATE.
[E] THE NUMBERS IN PARENTHESES REPRESENT THE PERCENT INHIBITION.
[F]NT, NOT TESTED.

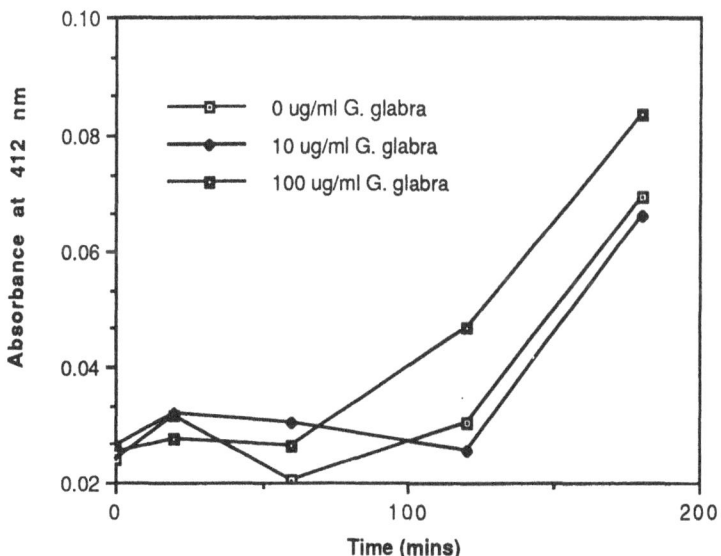

Figure 3. Reduced glutathione levels in *E. coli* K12 grown in the presence of *G. glabra* polar lipid fraction.

CONCLUSIONS

1.   It is clear that various antimutagenic agents, especially naturally derived mixed extracts, may act by a variety of different mechanisms.

2.   It should be noted that a single screening system may miss important activity profiles. In particular, it is useful to include a forward mutation assay as well as a reverse mutation assay.

3.   The mechanisms by which antimutagens/anticarcinogens may act are clearly complex and involve numerous biochemical and repair pathways that require elucidation.

ACKNOWLEDGEMENT

This research and the preparation of this paper have been supported by Grant #CA 43731 from the National Cancer Institute.

REFERENCES

1.   Ansher, S.S., P. Dolan, and E. Bueding (1986) Biochemical effects of dithiolthiones. *Fd. Chem. Toxicol.* 24:405-415.
2.   Bartsch, H., H. Oshima, and B. Pignatelli (1988) Inhibitors of endogenous nitrosation. Mechanisms and implications in human cancer prevention. *Mutat. Res.* 202:307-324.
3.   Bors, W., W. Heller, C. Michel, and M. Saran (1989) Flavonoids as antioxidants: Determining of radical-scavenging efficiencies. *Methods in Enzymol.* 186:343-355.
4.   Bors, W., W. Heller, C. Michel, and M. Saran (1990) Radical chemistry of flavonoid antioxidants. In *Antioxidants in Therapy and Preventive Medicine,* I. Emerit et al., eds. Plenum Press, New York.
5.   Budroe, J.D., J.G. Shaddock, and D.A. Casciano (1987) Modulation of ultraviolet light-, ethylmethane-sulfonate-, and 7,12-dimethylbenzanthracene-induced unscheduled DNA synthesis by retinol and retinoic acid in the primary rat hepatocyte. *Env. Molec. Mut.* 10:129-139.
6.   Cassady, J.M., W.M. Baird, and C. Chang (1990) Natural products as a source of potential chemotherapeutic and chemopreventive agents. *J. Med. Chem.* 53:23-41.
7.   Chen, L.H., G.A. Boissonneault, and H.P. Glauert (1988) Vitamin C, vitamin E and cancer. *Anticanc. Res.* 8:739-748.
8.   Cody, V., E. Middleton, J.B. Harborne, and A. Beretz, eds. (1988) *Plant Flavonoids in Biology and Medicine II.* Alan R. Liss, Inc., New York.
9.   Dahl, T.A., W.R. Midden, and P.E. Hartman (1987) Pure singlet oxygen cytotoxicity for bacteria. *Photochem. Photobiol.* 46:345-352.
10.   De Flora, S., C. Bennicelli, D. Serra, A. Izzotti, and C.F. Cesarone (1989) Role of glutathione and N-acetyl cysteine as inhibitors of mutagenesis and carcinogenesis. In *Absorption and Utilization of Amino Acids,* M. Friedman, ed. CRC Press, Boca Raton, Florida, pp. 19-55.

11. De Flora, S., and C. Ramel (1988) Mechanisms of inhibitors of mutagenesis and carcinogenesis. Classification and overview. *Mutat. Res.* 202:285-306.

12. De Lange, R.J., and A.N. Glazer (1989) Phycoerythrin fluorescencebased assay for peroxy radicals: A screen for biologically relevant protective agents. *Analyt. Biochem.* 177:300-306.

13. De Silva, H.V., and D.M. Shankel (1987) Effects of the antimutagen cinnamaldehyde on reversion and survival of selected *Salmonella* tester strains. *Mutat. Res.* 187:11-19.

14. Drake, J.W. (1973) The genetic control of spontaneous and induced mutation rates in bacteriophage T4. *Genetics* 73(Suppl.):45-64.

15. Drake, J.W., and E.O. Greening (1974) Suppression of chemical mutagenesis in bacteriophage T4 by genetically modified DNA polymerases. *Proc. Natl. Acad. Sci., USA* 66:823-829.

16. Di Giovanni, J. (1990) Inhibition of chemical carcinogenesis. In *Chemical Carcinogenesis and Mutagenesis II.* Springer-Verlag, Berlin, Heidelberg, New York.

17. Francis, A.R., T.K. Shetty, and R.K. Bhattacharya (1989) Modifying role of dietary factors on the mutagenicity of aflatoxin B^1: In vitro effects of plant flavonoids. *Mutat. Res.* 22:393-401.

18. Ferrandiz, M.L., A.G.R. Nair, and M.J. Alcaraz (1990) Inhibition of sheep platelet arachidonate metabolism by flavonoids from Spanish and Indian medicinal herbs. *Pharmazie* 45:206-208.

19. Geahlen, R.L., N.M. Koonchanok, and J.L. McLaughlin (1989) Inhibition of protein-tyrosine kinase activity by flavonoids and related compounds. *J. Natl. Prod.* 52:982-986.

20. Ghaskadbi, S., and V.G. Vaidya (1988) In vitro antimutagenic effect of L-cysteine and vitamin C against the mutagenicity of the common antiamoebic drug diiodohydroxy-quinoline. In *Antimutagenesis and Anticarcinogenesis Mechanisms II,* Y. Kuroda et al., eds. Plenum Press, New York, p. 87.

21. Gichner, T., and J. Veleminsky (1988) Mechanisms of inhibition of N-nitroso compounds-induced mutagenicity. *Mutat. Res.* 202:325-334.

22. Guevara, A., C.-L. Sylianco, F. Dayrit, and P. Finch (1988) Acylglucosyl-sterols: Antimutagens from *Momordica charantia* L. In *Antimutagenesis Mechanisms II,* Y. Kuroda et al., eds. Plenum Press, New York, p. 58.

23. Hartman, P.E., and D.M. Shankel (1990) Antimutagens and anticarcinogens: A survey of putative interceptor molecules. *Env. Molec. Mutagen.* 15:145-182.

24. Hayatsu, H., S. Arimoto, and T. Negishi (1988) Dietary inhibitors of mutagenesis and carcinogenesis. *Mutat. Res.* 202:429-446.

25. Kada, T., and K. Shimoi (1987) Desmutagens and bio-antimutagens: Their modes of action. *Bioessays* 7:113-116.

26. Krinsky, N.I. (1989) Carotenoids and cancer in animal models. *J. Nutr.* 199:123-126.

27. Lam, L.K.T., V.L. Sparnins, and L.W. Wattenberg (1987) Effects of derivatives of kahweol and cafestol on the activity of glutathione S-transferase in mice. *J. Med. Chem.* 30:1399-1403.

28. Middleton, E. (1988) Some biological properties of plant flavonoids. *Annals of Allergy* 61:53-57.

29. Mitscher, L.A., S. Drake, S.R. Gollapudi, J. Harris, and D.M. Shankel (1986) Isolation and identification of higher plant agents active in antimutagenic assay systems: *Glycyrrhiza glabra.* In *Antimutagenesis and*

Anticarcinogenesis Mechanisms I, D.M. Shankel et al., eds. Plenum Press, New York, pp. 153-165.

30.   Nagabhushan, M., A.J. Amoncar, A.V. D'Souza, and S.V. Bhide (1987) Antimutagenicity of curcumins and related compounds: The structural requirement for the antimutagenicity of curcumins. *Ind. Drugs* 25:91-95.

31.   Nagabhushan, M., A.J. Amoncar, U.J. Nair, A.V. D'Souza, and S.V. Bhide (1989) Hydroxychavicol: A new antinitrosating phenolic compound from betel leaf. *Mutagenesis* 4:200-204.

32.   Negishi, T., S. Arimoto, C. Nishizaki, and H. Hayatsu (1988) Inhibitory effect of chlorophyll on the genotoxicity of Trp P2. In *Antimutagenesis and Anticarcinogenesis Mechanisms II,* Y. Kuroda et al., eds. Plenum Press, New York, p. 59.

33.   Newmark, H.L. (1987) Plant phenolics as inhibitors of mutational and precarcinogenic events. *Can. J. Physiol. Pharmacol.* 65:461-466.

34.   Ong, T.M., W.Z. Whong, J. Stewart, and H.E. Brockman (1986) Chloro-phyllin: A potent antimutagen against environmental and dietary complex mixtures. *Mutat. Res.* 173:111-115.

35.   Owens, R.A., and P.E. Hartman (1986) Glutathione: A protective agent in *Salmonella typhimurium* and *Escherichia coli* as measured by mutagenicity and by growth delay assays. *Env. Muta.* 8:659-673.

36.   Robac, J., and R. Gryglewski (1988) Flavonoids are scavengers of superoxide anions. *Biochem. Pharmacol.* 37:837-841.

37.   Sato, T., Y. Ose, H. Nagase, and K. Hayase (1987) Mechanism of the desmutagenic effect of humic acid. *Mutat. Res.* 176:199-204.

38.   Wargovich, M.J., C. Woods, V.W.S. Eng, L.C. Stephens, and K. Gray (1988) Chemoprevention of n-nitrosomethylbenzyl-amine-induced esophageal cancer in rats by the naturally-occurring thioether, diallyl sulfide. *Cancer Res.* 48:6872-6875.

39.   Warner, J.R., J. Nath, and T.M. Ong (1991) Antimutagenicity studies of chlorophyllin using the Salmonella arabinose-resistant assay system. *Mutat. Res.* 262:25-30.

40.   Wattenberg, L.W., J.B. Hochalter, and A.R. Galbraith (1987) Inhibition of B-propiolactone-induced mutagenesis and neoplasia by sodium thiosulfate. *Cancer Res.* 47:4351-4354.

41.   Wood, M.L., J.R. Lindsay Smith, and R.C. Garner (1987) Aflatoxin B[1] activation to a plasmid mutagen by a chemical model of cytochrome P-450. *Mutat. Res.* 176:11-20.

42.   Yuting, C., Z. Rongliang, J. Zhongjian, and J. Yong (1990) Flavonoids as superoxide scavengers and antioxidants. *Free Radical Biol. and Med.* 9:19-21.

PROTEINS AS SCAVENGERS OF NITRITE: ANTIMUTAGENIC IMPLICATIONS

Kiyomi Kikugawa and Tetsuta Kato

Tokyo College of Pharmacy
1432-1 Horinouchi
Hachioji, Tokyo 192-03, JAPAN

INTRODUCTION

Nitrate taken from the diet and secreted into saliva is converted into nitrite by bacterial reduction in the oral cavity, and nitrate and nitrite are produced from arginine in mammalian cells (9). Several epidemiological studies demonstrated that there is a good correlation between nitrate intake and gastric cancer mortality (4). It has been shown that various mutagenic or carcinogenic compounds are produced from food components by reactions with nitrite under simulated gastric conditions. For example, nitrosamines, nitrosoindoles, and diazoquinones are produced from their precursors (9). Nitrosamines are produced from secondary amines in fish, meats and other food\stuffs. Nitrosoindoles are produced from indole compounds contained in fava beans and chinese cabbage. Diazoquinones are produced from phenolics in smoked foods, soy sauce, and medical drugs. It is well known that nitrosamines are carcinogenic after absorption and metabolic activation. Nitrosoindoles and diazoquinones have been found to be direct-acting for Salmonella bacteria.

There are several natural food components that scavenge or destroy nitrite available for production of nitrosamines. Mirvish et al. (13,14) found that ascorbate scavenges nitrite effectively, and that it inhibited nitrosamine formation in vitro and prevented tumor development due to coadministration of nitrite and several amines. Unsaturated fatty acids, tocopherols, polyphenolics, and vegetable juice have been shown to inhibit nitrosamine formation by scavenging nitrite (8). Amounts of these components in normal foods are, however, relatively low.

Proteins are abundant in most foodstuffs, and protein digests or amino acids are produced by enzymatic digestion of proteins in the stomach. The daily intake of proteins for man may amount to 60-80 g, and the concentrations of proteins, protein

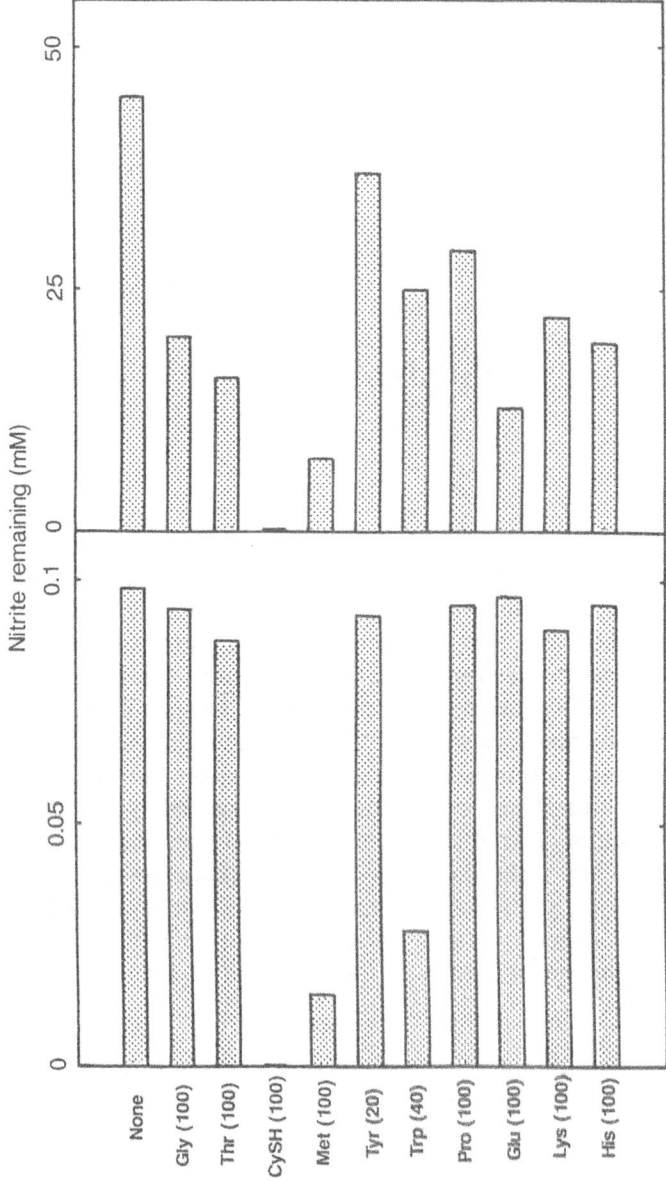

Figure 1. Loss of 0.1 mM nitrite in 0.05 mM citrate buffer (pH 3.0) and 50 mM nitrite in 0.20 M citrate buffer (pH 3.0) by treatment with amino acids at 30°C for 1 hr. Nitrite remaining was determined by Griess reagent.

digest, or amino acids in gastric juice after one meal may be 50 mg/ml or 100 mM amino acid equivalents. This paper describes the effects of proteins, protein digests, and amino acids at these high levels of on the loss of nitrite and the inhibition of mutagenic and/or carcinogenic nitrosamine and diazoquinone formation.

LOSS OF NITRITE BY AMINO ACIDS

Loss of nitrite by various amino acids including Gly, Thr, CySH, met, Tyr, Trp, pro, Blu, Lys, and His was investigated (Fig. 1). Two concentrations of nitrite, 0.1 mN\M and 50 mM, were selected for the study. The concentrations of the amino acids ranged from 20 mM to 100 mM. The mixture was incubated at pH 3 and 37 C for 1 hr. Each amino acid lost nitrite, especially at the higher nitrite concentration. Among the amino acids CySH was found to be most effective.

Nitrite may be lost mainly by the following five reactions. First, the reaction of nitrite with the alpha-amino group of the amino acid may produce nitrogen by the well-known van Slyke reaction (1) and thus may decrease the nitrite level. All the amino acids except pro may undergo this reaction to lose available nitrite. Second, the loss of nitrite by Pro may be due to the formation of N-nitrosproline, which is not mutagenic and usually found in human urine (6,17). Third, the reaction of nitrite with CySH may produce S-nitrosocysteine (12). Under the conditions of the present investigation, the reaction mixture of nitrite and CySH revealed a characteristic absorption maximum of S-nitrosocysteine at 335 nm (12). The amount of S-nitrosocysteine can be determined by its molar extinction coefficient (12). Time course studies indicated that more than 80% of the nitrite was converted into S-nitrosocysteine at pH 3 within 5 min. While CySH was an effective scavenger of nitrite, it has been pointed out that S-nitroso-cysteine can act as a nitrosating agent (2). Fourth, Trp may yield nitrosated tryptophan or 2-hydroxy-(1-N-nitrosoindole) propionic acid, which may be mutagenic (15,16), and can act as a nitrosating agent (11).

Fifth, we found new products in the reaction of nitrite with Tyr under the present reaciton conditions. While it has been reported that 3-nitrotyrosine was produced (10), formation of diazotyrosine was suggested (3). Figure 2 shows the HPLC pattern of the reaction mixture of nitrite and Tyr at pH 3. Three products were formed. Product II was identified as a diazoquinone-type compounds, and product I was identified as a deaminated product of I. Product III was not been identified yet. In the mutagenicity test using *Salmonella typhimurium* TA98 and TA 100 strains in the presence of S9 mix, products I and II were not mutagenic. This reaction may be effective in the loss of available nitrite.

LOSS OF NITRITE BY BOVINE SERUM ALBUMIN AND TRYPSINIZED CASEIN

Figure 3 shows the loss of nitrite by bovine serum albumin (BSA) and trypsinized casein at pH 3. BSA lost 0.1 mM and 50 mM nitrite dependent on the dose. Trypsinized casein lost nitrite more effectively, and more than 80% of 0.1 mM and 50 mM nitrite was lost by 50 mg/ml trypsinized casein.

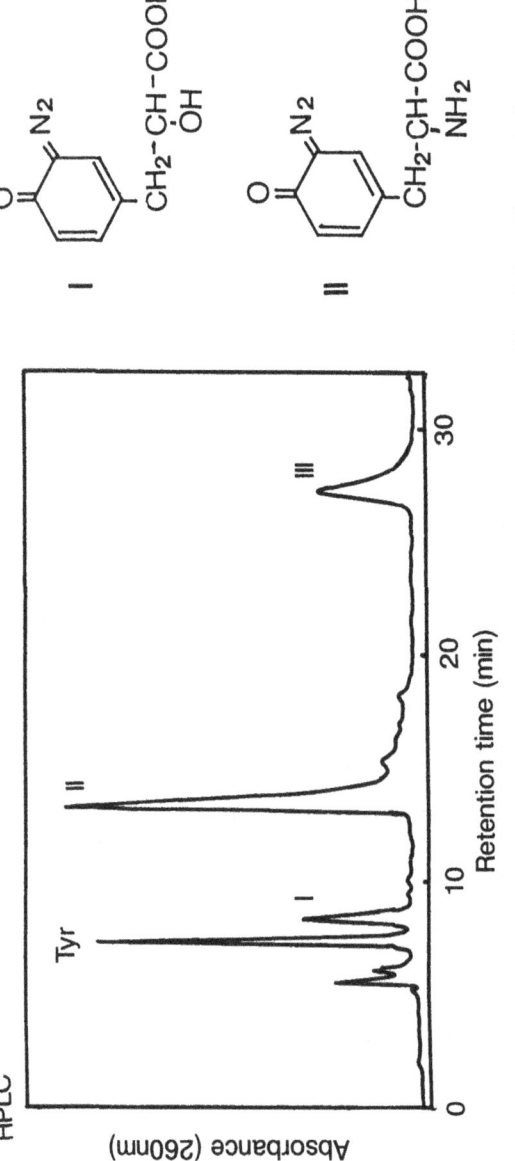

Figure 2. HPLC of the reaction mixture of 25 mM Tyr and 160 mM sodium nitrite in water (pH 3.0) and 37°C for 4 hr. The reaction was stopped by addition of 160 mM ammonium sulfamate. HPLC: YMC A-303 ODS (4.6 mm X 25 cm), methyl alcohol-0.1% acetic acid (3:17, v/v), flow rate at 0.5 ml/min.

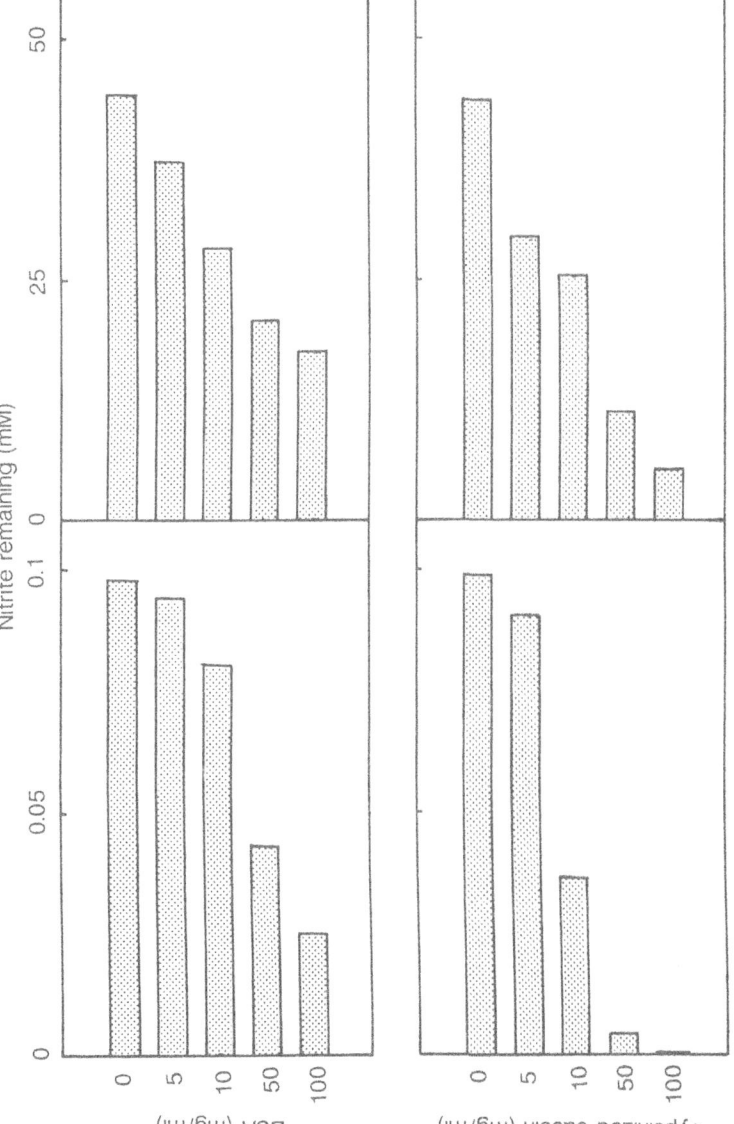

Figure 3. Loss of 0.1 mM and 50 mM nitrite by treatment with BSA and trypsinized casein in citrate buffer (pH 3.0) and 37°C for 1 hr.

79

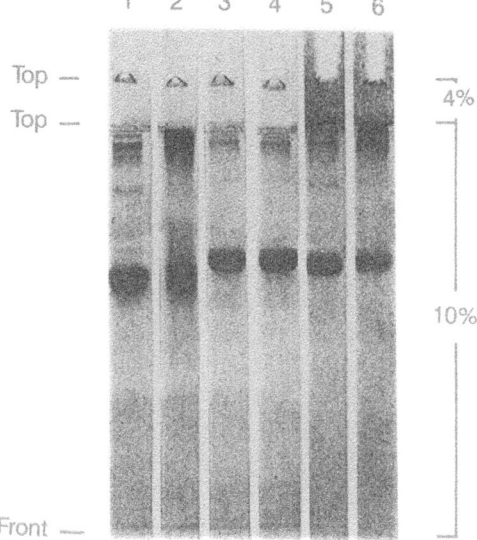

Figure 4. SDS-PAGE (reduced) of BSA and acetylated BSA treated with nitrite. A mixture of BSA or acetylated BSA (10 mg/ml) and sodium nitrite (2.5 mg/ml) in water (pH 3.0) was incubated at 37°C for 3 hr. The reaction was stopped by addition of ammonium sulfamate (5 mg/ml), and the mixture was diluted to make the concentration of protein at 1 mg/ml for SDS-PAGE. Lane 1, BSA; 2, BSA+NaNO$_2$; 3, BSA acetylated by N-acetylimidazole [55.1% amino and 11.2% hydroxyl groups modified]; 4, BSA acetylated by N-acetylimidazole+NaNO$_2$; 5, BSA acetylated by aced\tic anhydride [92% amino and 27.4% hydroxyl groups modified]; 6, BSA acetylated by acetic anhydride+NaNO$_2$.

Treatment of BSA with nitrite produced extensive cross-links (Fig. 4). When BSA was treated with nitrite at pH 3, high-molecular-weight polymers were detected at the top of the separating gel of SDS-PAGE. In contrast, BSA acetylated with N-acetylimidazole or acetic anhydride did not produce cross-links by nitrite treatment. Because extensive amounts of Tyr and Lys residues were modified by the acetylation, formation of cross-links may be due to the derivatization of Tyr and Lys residues in the protein.

INHIBITION OF NITROSAMINE FORMATION BY AMINO ACIDS, PROTEIN, AND PROTEIN DIGEST

The second investigation was the effect of amino acids, protein, and protein digest on nitrosamine formation from dimethylamine and nitrite. Figure 5A (left panel) shows the inhibitory effect of amino acids on the nitrosodimethylamine (NDMA) formation at pH 3. Most amino acids at 20-100 mM effectively inhibited NDMA formation from 50 mM nitrite. It should be noted that although 100 mM CySH lost most of 50 mM nitrite (Fig. 1), it inhibited the nitrosamine formation by only 70%. Figure 5A (right panel) shows the inhibitory effect of amino acids on NDMA formation at pH 5. Most amino acids were inhibitory.

Table 1. Formation of nitrosodimethylamine (NDMA) and *p*-diazoquinone (*p*-DQ) in the mixture of nitrite pretreated with CySH

Pretreatment (pH 3, 37°C, 10 min)			Reaction (pH 3, 37°C)	
Nitrite added (mM)	CySH added (mM)	Nitrite determined (mM)	Dimethylamine [or phenol] added (mM)	NDMA [or *p*-DQ] formed (mM)
50	0	49	200	17.0
50	100	0	200	4.7
100	0	96	[25]	[2.05]
100	100	6	[25]	[1.06]

A mixture of nitrite and CySH in 0.1 M citrate buffer (pH 3.0) was pretreated at 37°C for 10 min. To the mixture was added dimethylamine or phenol, and the mixture was adjusted at pH 3.0 and incubated at 37°C for 5 (NDMA) or 4 (*p*-DQ) hr.

Figure 5B shows the effect of BSA and trypsinized casein on NDMA formation. At both pH 3 and 5, they were inhibitory at 100 mg/ml concentration.

INHIBITION OF DIAZOQUINONE FORMATION BY AMINO ACIDS, PROTEIN, AND PROTEIN DIGEST

The third investigation was the effect of amino acids, protein, and protein digest on diazoquinone formation from phenol and nitrite. Figure 6A shows the effect of amino acids on the p-diazoquinone (p-DQ) formation from phenol and nitrite at pH 3 (7). Most amino acids (except for CySH) effectively inhibited the formation of p-DQ at 10 to 100 mM concentrations. It is interesting to note that CySH was stimulatory at 10 mM but inhibitory at 100 mM.

Figure 6B shows the effect of BSA and trypsinized casein on p-DQ formation. Both were inhibitory, the inhibitory effect of trypsinized casein being extensive.

While CySH was the most effective scavenger of nitrite among the amino acids tested (Fig. 1), its inhibitory effect n NDMA and p-DQ formation was smaller than that expected or rather stimulatory under certain conditions (Fig. 5A, Fig. 6A). Table 1 shows that nitrosocysteine was an effective nitrosating or diazotizing agent for NDMA and p-DQ formation. When nitrite was pretreated with CySH to lose most of the nitrite, the mixture was still effective in producing NDMA and p-DQ. CySH could be regarded as a strong scavenger of nitrite, but not as an inhibitor of the NDMA and p-DQ formation.

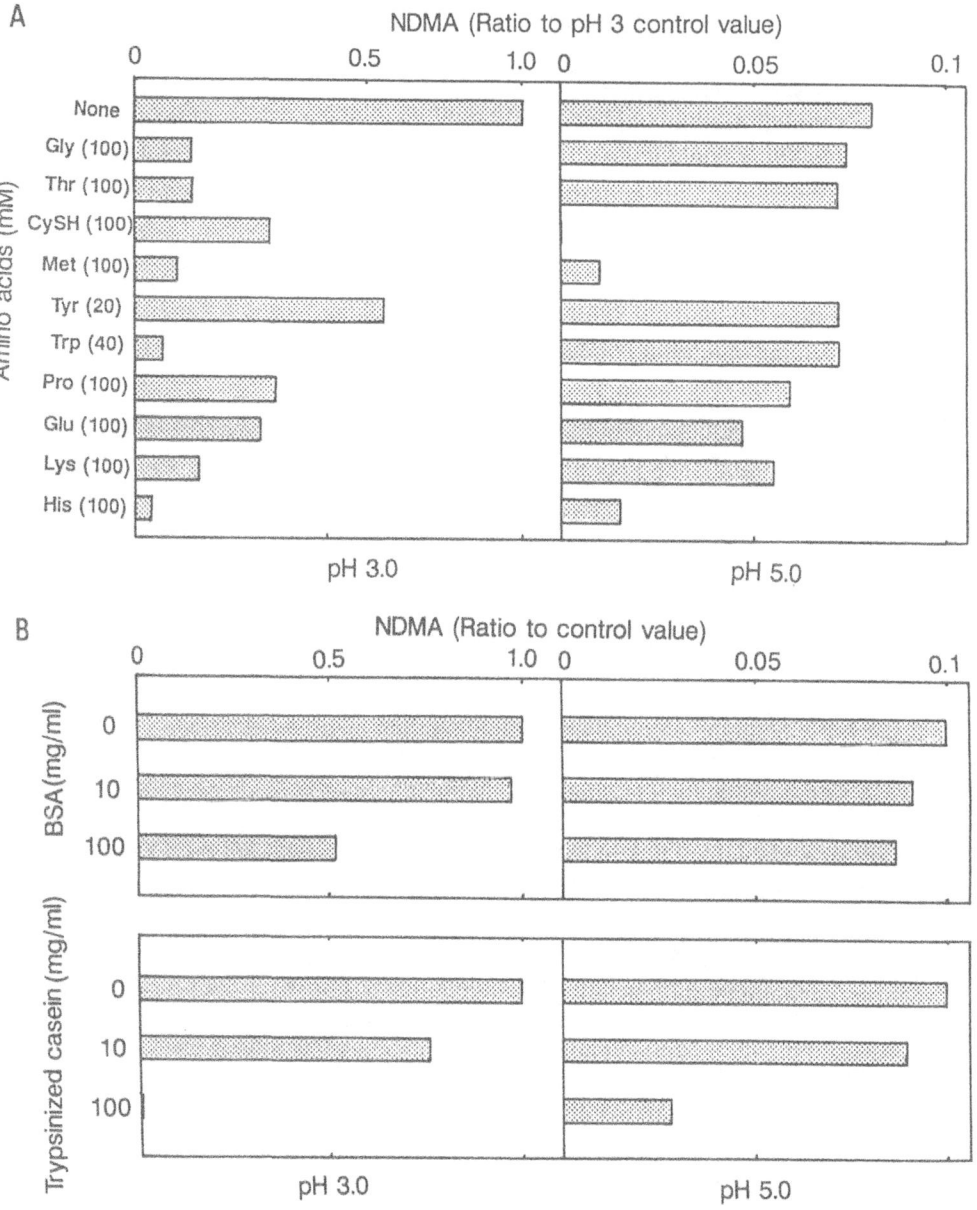

Figure 5. Effect of amino acids [A], BSA and trypsinized casein [B] on nitrosodimethylamine (NDMA) formation in the reaction of 50 mM sodium nitrite and 200 mM dimethylamine in 0.1 M citrate buffer at 37°C for 5 hr. The reaction was stopped by addition of 50 mM ammonium sulfamate (for pH 3) or sulfanylic acid (for pH 5). Control without test samples at pH 3.0 produced 18-23 mM NDMA. GLC: polyethylene glycol 6,000 (25%) on Chromosorb W AW (3 mm X 3 m), at 100°C, nitrogen gas flow at 30 ml/min.

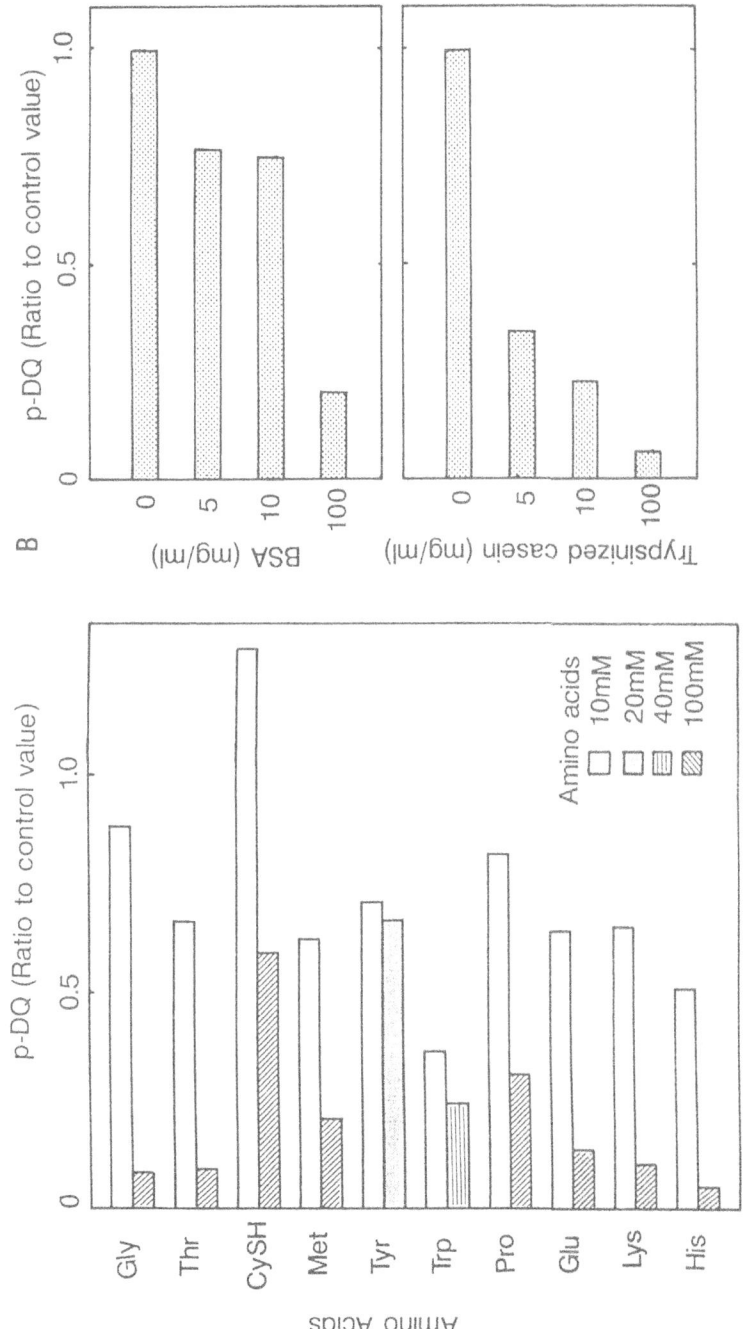

Figure 6. Effect of amino acids [A], BSA and trypsinized casein [B] on *p*-diazoquinone (*p*-DQ) formation in the reaction of 100 mM sodium nitrite and 25 mM phenol in 0.2 M citrate buffer (pH 3.0) at 37°C for 4 hr. The reaction was stopped by addition of 100 mM ammonium sulfamate. Control without test samples at pH 3.0 produced 2.5 mM *p*-DQ. HPLC:L YMC A-303 ODS column, acetonitrile-0.05 M citrate buffer (pH 3.0) (1:19, v/v) for CySH and methyl alcohol-0.1% acetic acid (3:17, v/v) for others, flow rate at 0.5 ml/min, detection of *p*-DQ at 350 nm.

DISCUSSION

Previously, Gray and Dugan (5) observed the inhibitory effect of CySH and Met on NDMA formation. In the present investigation we observed that most amino acids, BSA, and trypsinized casein at the concentrations in gastric juice effectively scavenged nitrite, and thus inhibited nitrosamine and diazoquinone formation. It is known that CySH (2) and Trp (16) are able to transnitrosate. It was found in the present investigation that CySH yielded NDMA and p-DQ after nitrite was completely lost by the amino acid, and thus CySH could not be regarded as an inhibitor of nitrosamine and diazoquinone formation. Nevertheless, proteins in foods are composed of various amino acids and they may be digested into a mixture of various amino acids. Hence, proteins, protein digests, and a mixture of various amino acids may serve as inhibitors of nitrosamine and diazoquinone formation.

Form a toxicological point of view, it should be pointed out that the inhibitory activities of proteins, protein digest, and a mixture of amino acids on nitrosamine and diazoquinone formation do not always result in the disappearance of mutagenic and/or carcinogenic activity. It has been reported that treatment of Gly, Tyr, CySH, and Met with nitrite produced significant but weak mutagenicity in *S. typhimurium* TA98 and TA100 strains in the presence of S9 mix (19). Treatment of Trp with nitrite produces nitrosated mutagenic products (15,16,18), but the mutagenicity of these products is much lower than that of the reaction mixtures of several other indole derivatives with nitrite (15). Hence, we concluded that proteins, protein digests, and a mixture of amino acids may effectively scavenge nitrite in the stomach and thus may inhibit the formation of more harmful mutagenic and/or carcinogenic compounds.

REFERENCES

1. Austin, A.T. (1950) the deamination of amino-acids by nitrous acid with particular reference to glycine. The chemistry underlying the Van Slyke determination of alpha-amino acids. *J. Chem. Soc.*, p. 149.
2. Dennis, M.J., R. Davies, and D.J. McWeeny (1979) The transnitrosation of secondary amines S-nitrosocysteine in relation to N-nitrosamine formation in cured meats. *J. Sci. Food Agric.* 30:639.
3. Dorie, J.P., P. Mellet, and B.G. Mechin (1990) Spectroscopic evidence for an aryldiazonium salt formed via the action of sodium nitrite on N-acetyltyrosine and tyrosyl groups of bovine serum albumin. *J. Agric. Food Chem.* 38:262.
4. Fine, D.H., B.C. Challis, P. Hartman, and J. Van Ryzin (1982) *N-Nitroso Compounds: Occurrence and Biological Effects,* H. Bartsch, I.K. O'Neill, M. Castegnaro, M. Okada, and W.
    Davis, eds. IARC Scientific Publ. No. 41, International Agency for Research on Cancer, Lyon, France.
5. Gray, J.I., and L.R. Dugan, Jr. (1975) Inhibition of N-nitrosamine formation in model food systems. *J. Food Sci.* 40:981.
6. Greenblatt, M., V.R.C. Kommineni, and W. Lijinsky (1973) Null effect of concurrent feeding of sodium nitrite and amino acids to MRC rats. *J. Natl. Cancer Inst.* 50:799.
7. Kikugawa, K., and T. Kato (1988) Formation of a mutagenic diazoquinone by interaction of phenol with nitrite. *Food Chem. Toxicol.* 26:209.

8. Kikugawa, K., and T. Kato (1991) *Mutagens in Food: Detection and Prevention,* H. Hayatsu, ed. CRC Press, Boca Raton, Florida, pp. 205-217.

9. Kikugawa, K., and M. Nagao (1991) *Mutagens in Food: Detection and Prevention,* H. Hayatsu, ed. CRC Press, Boca Raton, Florida, pp. 67-85.

10. Knowles, M.E., D.J. McWeeny, L. Couchman, and M. Thorogood (1974) Interaction of nitrite with proteins at gastric pH. *Nature* 247:288.

11. Mellet, P.O., P.R. Noel, and R. Goutefongea (1986) Nitrite-tryptophan reaction: Evidence for an equilibrium between tryptophan and its nitrosated form. *J. Agric. Food Chem.* 34:892.

12. Mirna, A., and K. Hoffmann (1969) Uber den Verbleib von Nitrit in Fleischwaren I. Umsetzung von Nitrit mit Sulfhydryl-Verbindungen. *Fleischwirtschaft* 49:1361.

13. Mirvish, S.S., A. Cardesa, L. Wallcave, and P. Shubik (1975) Induction of mouse lung adenomas by amines or ureas plus nitrite and by N-nitroso compounds: Effect of ascorbate, gallic acid, thiocyanate, and caffeine. *J. Natl. Cancer Inst.* 55:633.

14. Mirvish, S.S., L. Wallcave, M. Eagen, and P. Shubik (1972) Ascorbate-nitrite reaction: Possible means of blocking the formation of carcinogenic N-nitroso compounds. *Science* 177:65.

15. Ochiai, M., K. Wakabayashi, T. Sugimura, and M. Nagao (1986) Mutagenicities of indole and 30 derivatives after nitrite treatment. *Mutat. Res.* 172:189.

16. Ohara, A., M. Mizuno, G. Danno, K. Kanazawa, Y. Yoshioka, and M. Natake (1988) Mutagen formed from tryptophan reacted with sodium nitrite in acidic solution. *Mutat. Res.* 206:65.

17. Ohshima, H., I.K. O'Neill, M. Friesen, J.-C. Bereziat, and H. Bartsch (1984) Occurrence in human urine of new sulfur-containing N-nitrosamino acids, N-nitrosothiazolidine 4-carboxylic acid and its 2-methyl derivative, and their formation. *J. Cancer Res. Clin. Oncol.* 108:121.

18. Ohta, T., M. Isa, Y. Suzuki, N. Yamahata, S. Suzuki, and T. Kurechi (1981) Formation of mutagens from tryptophan by the reaction with nitrite. *Biochem. Biophys. Res. Comm.* 100:52.

19. Ohta, T., and S. Suzuki (1983) Formation of mutagens by the reaction of amino acids with nitrite in acidic solution. *Agric. Biol. Chem.* 47:405.

ANTIMUTAGENIC AGENTS FROM NATURAL PRODUCTS OF

TERRESTRIAL AND MARINE ORIGIN

Monroe E. Wall and Mansukh C. Wani

Research Triangle Institute
Research Triangle Park, North Carolina 27709 USA

INTRODUCTION

Both chemists and biologists have been intrigued for many years by the role of secondary metabolites in plants of terrestrial and marine origins. Of equal interest is whether such compounds have antitumor, antiviral, or antimutagenic properties. For example, the highly active antitumor agents, camptothecin and taxol, have been isolated from the bark and wood of trees (13,21), and another active agent, bryostatin I, from a marine animal (11). Dudley Williams has proposed that "secondary metabolites are a measure of the fitness of the organism to survive by repelling or entrapping other organisms" (23). Mitscher has stated that "as some constituents of higher plants are mutagens, it seems reasonable that substances capable of antimutagenicity also be produced by such plants" (10).

Initially influenced by the first conference on Antimutagenesis and Anticarcinogenesis Mechanisms (12), we have been working for several years on isolation and structure determination of potential antimutagens from terrestrial and marine sources.

GENERAL METHODS

Antimutagenic Screening

Our procedure has been described in detail (14-16). It is based on the inhibition of the mutagenicity of 2-aminoanthracene (2AN) by crude or semipurified plant extracts (14). The procedure incorporates methodology first described by Birt et al. (1) and utilizes an enzyme metabolic activator for the mutagen (8). In brief, 300-600 μg aliquots of crude extract in DMSO solution are

1. <u>Salmonella</u> <u>typhimurium</u> (TA-98)

   +

2. Ames S-9 preparation

   +

3. 600 µg/plate of test substance

   +

4. 2-Aminoanthracene (2AN) (2.5 µg/plate)

5. Mix with top agar and incubate 48-72 hour

TOXICITY PROCEDURE

1. Steps 1-3 as above

2. Omit 2AN in step 4 and replace by histidine

3. Step 5 as above

Figure 1. Inhibition procedure.

added to the nutrient broth containing *Salmonella typhimurium* TA98, the Ames S9 metabolic activation preparation, and the mutagen, 2AN. The procedure is outlined in Fig. 1.

Extraction and Isolation of Antimutagenic Compounds

The extraction and isolation of antimutagenic agents which, as defined by Kada, may include both desmutagenic and bioantimutagenic agents (6), is guided by the bioassay at every step. The general extraction and isolation procedure has been given in detail in previous papers from our laboratory (7,14-18). In brief, antimutagenic ethanolic extracts are concentrated, partitioned between methylene chloride or chloroform-water, the organic solvent fraction concentrated, and the residue partitioned between 90% methanol-10% water and hexane. The methanolic fraction is concentrated and chromatographed on silica gel or Sephadex LH-20. Chromatographic fractions are tested for inhibition of 2AN and toxicity. Active, nontoxic fractions are usually rechromatographed and then subjected to preparative HPLC for final purification.

Determination of Physical Properties

Isolated compounds are initially crystallized, and melting points and optical rotations are determined. If the compounds have ultraviolet absorption (UV), purity can then be carefully checked by a technique involving HPLC and diode array UV. In addition to UV, infrared (IR), nuclear magnetic resonance (both ^{1}H and ^{13}C-NMR), and high resolution mass spectra (HRMS), which give the molecular formula are obtained. By means of the molecular formula, the compounds can be rapidly checked via Chemical Abstracts and other computer search compendiums to determine if the compounds are novel, or, if previously known, have ever received antimutagenic testing. In certain cases, when it is evident we are dealing with a complex structure with several asymmetric centers, the precise structure is determined by X-ray crystallography, which, in spite of all the improvements in NMR techniques, is probably still the best single structural technique. It is limited, however, to compounds or their derivatives that are available in appropriate crystal form.

ACTIVE PLANTS

Edible Plants

As described previously, we have had little or no success in finding antimutagenic agents in edible plants by our techniques (19). A number of edible vegetables in the Brassicaceae found to have anticarcinogenic constituents (22) were inactive when screened for antimutagenicity toward 2AN (19).

Wild Plants

We have had considerably more success in studying wild plants of both terrestrial and marine origin. As described previously, out of approximately 2,750 plants, 3.5% were found to be active and nontoxic (19). In the case of marine samples, 50 algae samples gave 20% active, nontoxic extracts; from almost 100 samples of corals and sponges, only 3.1% were active, and from various marine animals, only one out of approximately 185 samples showed activity. Recently, plants with reputed therapeutic activity were examined, and these gave a much higher success rate. From 100 plants obtained from Sri Lanka, accurately identified botanically and reported to have various therapeutic activities, the success rate was almost 20%. We plan to investigate this specialized area and to obtain additional plants utilized in other cultures and areas for medicinal activity.

ANTIMUTAGENIC COMPOUNDS ISOLATED

Flavonoids

Figure 2 shows a number of putative antimutagenic agents, all of which were isolated in our laboratory and which received antimutagenicity testing for the first time. In addition, some of these compounds also have completely novel structures. Flavonoids have received considerable attention, both in our laboratory and by other groups. Previously, we have described the antimutagenic

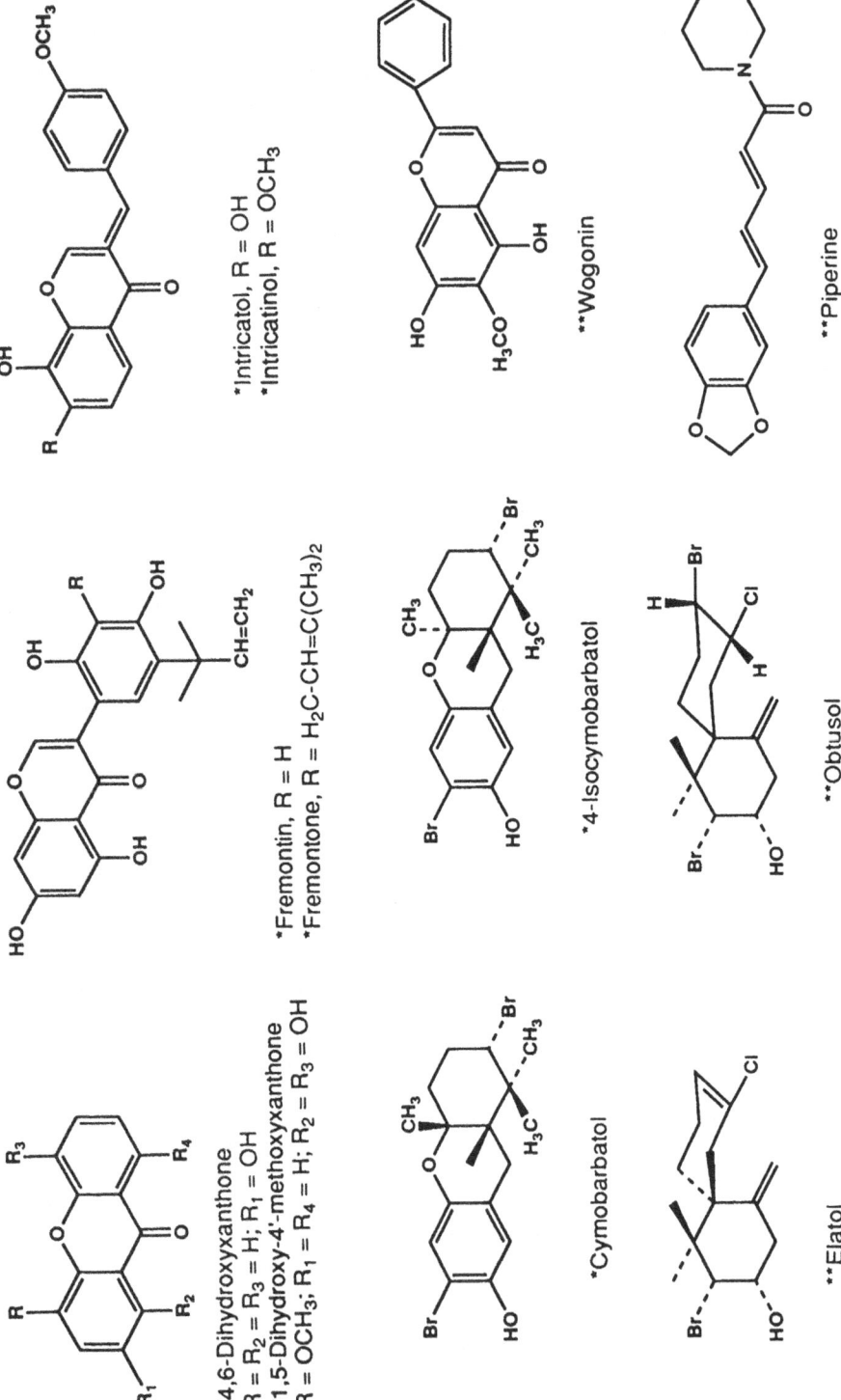

Figure 2. New* and known** putative antimutagens.

90

activity of a number of known flavonoids, including, in some cases, flavonoid glycosides (15). The latter category have never shown any activity. Recently, we have isolated from the roots of *Psorothamnus fremontii* Fabiaceae two new isoflavonoids, fremontin and fremontone (7). The latter is considerably more potent as an antimutagen than the former compound. Two homoisoflavonoids that have novel structures are the compounds intracatol and intracatinol, isolated from *Hoffmanosseggia intricata* (17). Of these, we have found that intracatinol is much more potent against a number of mutagens, including 2AN, acetylaminofluorine (AAF), and ethyl methanesulfonate (EMS) (17). We have developed synthetic schemes for the synthesis both of the fremontin and the intracatinol systems, and, hence, these compounds can be produced in quantity if further study shows good therapeutic activity.

Recently, we isolated a known flavonoid, Wogonin, from *Oroxylum indicum* (Bignoniaceae) (20). The antimutagenic activity of this compound is of a low order at concentrations that are not toxic.

Flavonoids isolated by other groups include the compound glabrene, a known isoflavone isolated from *Glycyrrhiza glabra* (10). Glabrene was highly active against EMS-induced mutations of *S. typhimurium* (TA100). Recently, the well-known flavonoid, Biochanin A, has been reported to inhibit B(a)P activation in hamster embryo cells in a tissue culture test system (2). In general, the flavonoid group will require careful study in a variety of antimutagenesis assays and should also be tested for mutagenic activity. Flavonoids have been reported to have a wide variety of biological activities (3). Certain flavonoids have been considered to display mutagenic and carcinogenic effects (3) whereas other, and sometimes the same, flavonoids have been reported to exert inhibition of tumor promotion (22). An excellent discussion of the multitudinous biological activities attributed to flavonoids has been well summarized in a symposium report (3).

Coumarins

Coumarins are a group of well-known compounds that have received attention both as anticarcinogens (22) and antimutagens (6). Coumarin, umbelliferone, and 8-methoxypsoralen have been previously described by Kada et al. (6). Recently, we have made a thorough investigation of antimutagenic coumarins isolated from various sources (16). We found, in particular, that the known coumarins, imperatorin and osthol, isolated from *Selinum monniere*, have particularly high activity in the inhibition of 2AN and B(a)P. Because of the well-known photochemical activity of these compounds, we have not investigated this group further.

Xanthones

The two xanthones shown in Fig. 2 were isolated from *Vismia amazonica* (Clusiaceae). Both xanthones shown in Fig. 2 have novel structures and display considerable antimutagenicity at a number of dose levels against 2AN and EMS (20).

Acetophenone Analog

A simple acetophenone analog, 2,3-dihydroxy-5-ethoxy-acetophenone, was isolated from a rather unusual source, a lichen, *Lasallia papulosa* (Umbellicariaceae). This acetophenone analog was active and nontoxic in the 2AN assay.

Piperine

The structure of piperine is shown in Fig. 2. This is a known compound, but it had never received antimutagenicity testing. Piperine was isolated from *Scindapsus officianallis* (Araceae) (20). Of interest is the fact that it is used as a medicinal herb in Sri Lanka. The compound showed considerable activity in the 2AN antimutagenicity assay and was nontoxic. A number of synthetic analogs of piperine were also prepared and are also active. Further studies of the activity in this series and in other in vitro test systems is underway.

MARINE ANIMALS AND PLANTS

Elatol and Obtusol

The structures of these compounds are shown in Fig. 2. Of interest is the fact that they are halogenated, both containing bromine and chlorine (4). These compounds were isolated from extracts from a marine animal known as the "sea hare," *Aplysia dactylomela* (20). It is almost certain that these and a large number of similar compounds are derived originally from marine algae belonging to the Laurencia (4). Although both elatol and obtusol show significant activity in the inhibition of 2AN mutagenicity, the latter compound is far more potent (20).

ISOLATED AND STRUCTURE OF CYMOBARBATOL AND RELATED COMPOUNDS FROM GREEN ALGAE

Background

The following sections, which exemplify our methodology, describe the isolation, structure, and antimutagenic activity of several new secondary metabolites from a green algae, *Cymopolia barbata*. Marine algae have been the subject of many investigations dealing with the structure of various secondary metabolites. Numerous halogenated metabolites have been found in red and brown algae. Green algae received less attention until a group of prenylated bromohydroquinones, collectively termed cymopols, were isolated from *Cymopolia barbata* (L.), Lamouroux (*Dasycladaceae*), by Hogberg et al. (5). Of particular interest were two compounds, cymopol (#1) and cyclocymopol (#2), the structures of which are shown below. The structure of one of the diastereomers of (#2) as the monomethyl ether was confirmed in an X-ray crystallographic study (9). We have found two new cymopols in *C. barbata* extracts guided by antimutagenicity assay.

Cymopol (1)   Cyclocymopol (2)

Isolation of New Cymopols from *Cymopolia barbata*

Crude methanol chloroform extracts of *C. barbata* that had been collected by Professor William Gerwick, Oregon State University, Corvallis, Oregon, were tested for both antimutagenicity and toxicity and were found to be active in the inhibition of 2AN mutagenesis and to be relatively nontoxic. Guided by antimutagenic bioassay, two new compounds were obtained, cymobarbatol (#3) and 4-isocymobarbatol (#4), isolated by the procedure presented in Fig. 3 (18). In addition, a somewhat more polar compound was isolated from the same preparation in a fraction that was obtained from a more polar gradient. A pure known compound, cymopol (#1) (5), was isolated utilizing techniques involving preparative HPLC similar to those shown in Fig. 3.

Physical Properties of Compounds (#3) and (#4)

Some of the important physical properties of compounds (#3) and (#4) are shown in Fig 4. Both compounds show similar ultraviolet spectra and HRMS in high resolution mass spectra. However, the optical rotations of (#3) and (#4) differ considerably.

Structure of Cymobarbatols

The structure of the cymobarbatols was deduced primarily from NMR data using both ^{1}H and ^{13}C analyses. From the ^{1}H spectra of (#3) and (#4) shown in Fig. 5, it was immediately evident that (#3) and (#4) had identical structures in the aromatic ring, but there was an obvious differentiation in the ring fusion including carbons 3 and 4. X-ray spectrometry gave complete structural proof as shown in Fig. 6 (18).

Antimutagenic Activity of Cymopols

Cymobarbatol and 4-isocymobarbatol were nontoxic at doses from 75300 μg/plate in both *S. typhimurium* TA98 and TA100 strains, whereas cymopol was very toxic in both these strains at doses from 75-600 μg/plate. Cymobarbatol and 4-isocymobarbatol were highly active in the inhibition of 2AN mutagenicity toward *S. typhimurium* TA98 strain at doses of 75, 150, and 300 μg/plate. These two compounds showed even greater activity (80-95%) in the inhibition of EMS mutagenicity toward *S. typhimurium* (TA100) at doses from 32-300 μg/plate. On the other hand, cymopol was toxic even at low concentrations.

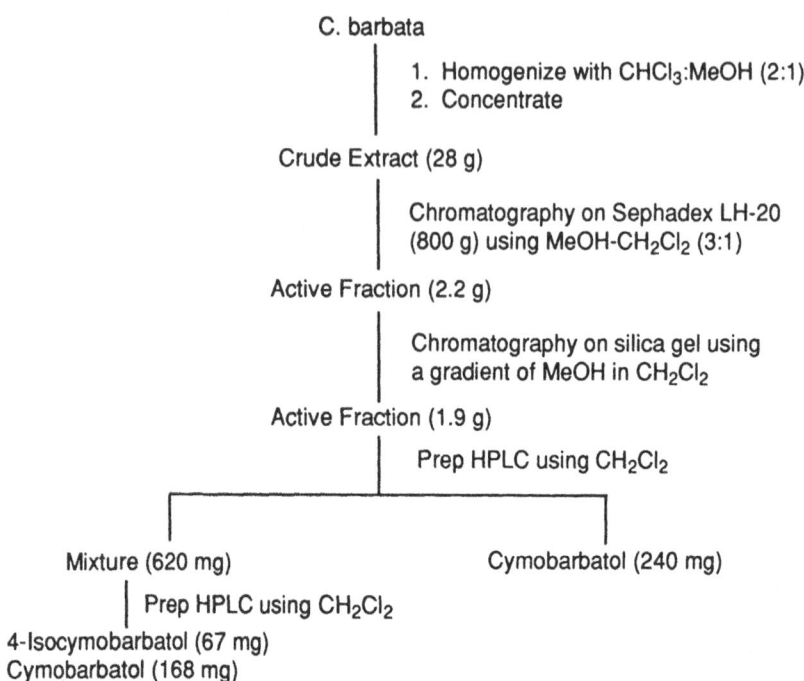

C. barbata

    1. Homogenize with $CHCl_3$:MeOH (2:1)
    2. Concentrate

Crude Extract (28 g)

    Chromatography on Sephadex LH-20
    (800 g) using MeOH-CH_2Cl_2 (3:1)

Active Fraction (2.2 g)

    Chromatography on silica gel using
    a gradient of MeOH in CH_2Cl_2

Active Fraction (1.9 g)

    Prep HPLC using CH_2Cl_2

Mixture (620 mg)   Cymobarbatol (240 mg)

    Prep HPLC using CH_2Cl_2

4-Isocymobarbatol (67 mg)
Cymobarbatol (168 mg)

Figure 3. Isolation of cymobarbatol and 4-isocymobarbatol from *Cymopolia barbata*.

<u>Cymobarbatol</u>	<u>4-Isocymobarbatol</u>
Mp 166°C (Aq MeOH)	Mp 147°C (Aq MeOH)
$[\alpha]^{23}$ - 15.4°	$[\alpha]^{23}$ - 51.4°
UV (MeOH), nm (log Σ) , 306 (3.72), 225 Sh (4.00); (meOH + NaOH), 327 (3.84)	UV (MeOH), nm (log Σ) , 305 (3.76), 225 Sh (3.76); (MeOH + NaOH), 326 (3.79)
HRMS m/z 401.9831 ($C_{16}H_{22}Br_2O_2$ = 401.9831	HRMS m/z 401.9835 ($C_{16}H_{22}Br_2O_2$ = 401.9831

Figure 4. Physical properties of cymobarbatol and 4-isocymobarbatol.

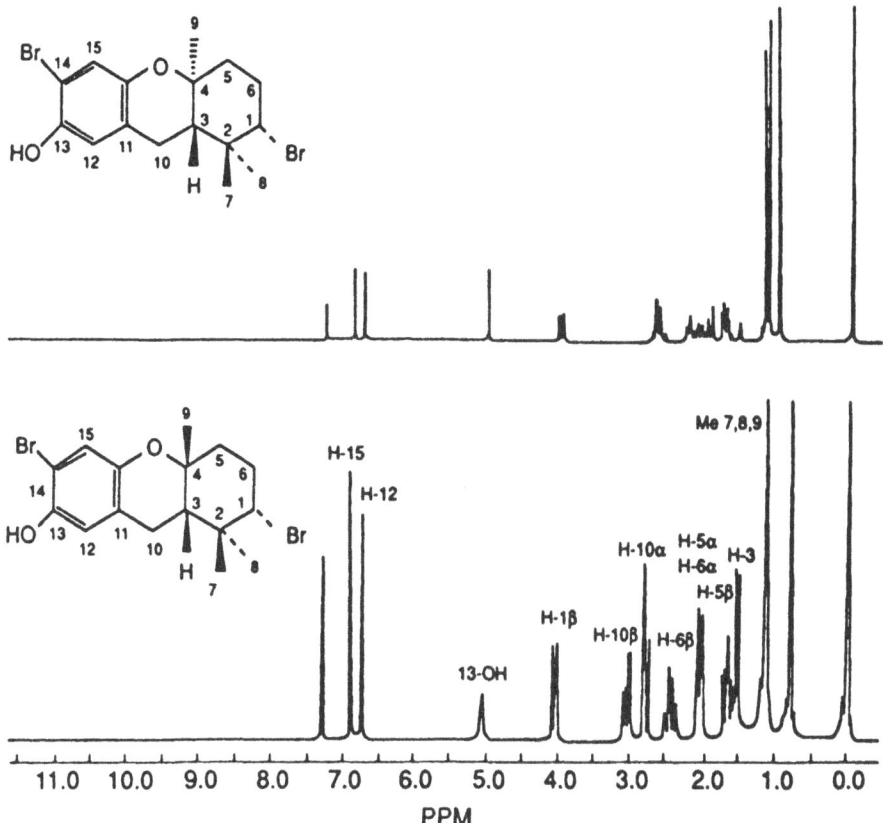

Figure 5. ^{1}H NMR (250 MHz) spectrum of 4-isocymobarbatol in CDCl$_3$ (top) and ^{1}H NMR (250 MHz) spectrum of cymobarbatol in CDCl$_3$ (bottom).

1. HRMS shows that cymobarbatol and cyclocymopol are isomeric with a molecular formula of C$_{16}$H$_{20}$Br$_2$O$_2$.
2. Acetylation of cymobarbatol gives a monoacetate. Cyclocymopol gives diacetate.
3. H-NMR of cymobarbatol shows the presence of 3 Me groups whereas that of cyclocymopol shows 2 Me groups and olifinic CH$_2$ group.
4. Therefore cymobarbitol appears to be formed by the cyclization of cyclocymopol.
5. The stereochemistry of ring fusion at C-3 and C-4 and the configuration of Br at C-1 in both cymobarbatol and 4-isocymobarbatol were determined by single crystal X-ray crystallography.

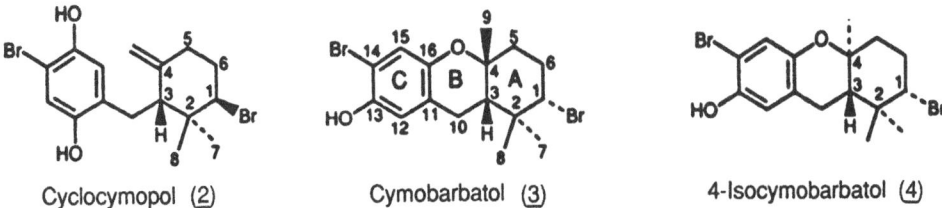

Cyclocymopol (2)   Cymobarbatol (3)   4-Isocymobarbatol (4)

Figure 6. Structures of cymobarbatol.

CONCLUSIONS

Secondary metabolites found in terrestrial and marine plants and organisms are capable of inhibiting the mutagenicity of a number of mutagens toward *S. typhimurium* TA98 and TA100. The sensitivity of the antimutagenicity assay is such that crude extracts can be evaluated and purification of the extracts by various analytical methods readily followed. Major classes of antimutagenic compounds that have been isolated include flavonoids, coumarins, xanthones, and cymopols.

REFERENCES

1. Birt, D.F., B. Walker, M.G. Tibbels, and E. Bresnick (1986) Antimutagenesis and antipromotion by apigenin, robinetin and indole-3 carbinol. *Carcinogen* 7:959-963.
2. Cassady, J.M. (1990) Natural products as a source of potential cancer chemotherapeutic and chemopreventive agents. *J. Nat. Prod.* 53:23-41.
3. Cody, V., E. Middleton, Jr., and J.B. Harborn, eds. (1986) *Plant Flavonoids in Biology and Medicine.* Allen R. Liss, Inc., New York, New York.
4. Gonzalez, A.G., J.D. Martin, V.S. Martin, M. Martinez-Ripoli, and J. Fayos (1979) X-ray study of sesquiterpene constituents of the alga *L. obtusa* leads to structure revision. *Tetrahedron Lett.* pp. 2717-2718.
5. Hogberg, H.E., and R.H. Thomson (1976) The cymopols, a group of prenylated bromohydroquinones from the green calcareous alga *Cymopolea barbata. J. Chem. Soc.* pp. 1696-1701.
6. Kada, T., I. Tadashi, O. Toshihiro, and Y. Shirasu (1986) Antimutagens and their mode of action. In *Antimutagenesis and Anticarcinogenesis Mechanisms,* D.M. Shankel, P.E. Hartman, T. Kada, and A. Hollaender, eds. Plenum Press, New York, New York, pp. 181-196.
7. Manikumar, G., K. Gaetano, M.C. Wani, H. Taylor, T.J. Hughes, J. Warner, R. McGivney, and M.E. Wall (1989) Plant antimutagenic agents 5. Isolation and structure of two new isoflavones, Fremontin and fremontone from *Psorothamnus fremontii. J. Nat. Prod.* 52:769773.
8. Maron, D.M., and B.N. Ames (1983) Revised methods for the Salmonella mutagenicity tests. *Mutat. Res.* 113:173-215.
9. McConnell, O.J., P.A. Hughes, and A.M. Targett (1982) Diastereoisomers of cyclocymopol and cyclocymopol mono methyl ether from *Cymopolia barbata. Phytochem.* 21:2139-2141.
10. Mitscher, L.A., S. Drake, S.R. Gollapudi, J.A. Harris, and D.M. Shankel (1986) Isolation and identification of higher plant agents active in antimutagenic assay systems: *Glycyrrhiza glabra.* In *Antimutagenesis and Anticarcinogenesis Mechanisms,* D.M. Shankel, P.E. Hartman, T. Kada, and A. Hollaender, eds. Plenum Press, New York, New York, pp. 153-165.
11. Pettit, G.R., C.L. Herald, D.L. Doubek, and D.L. Herald (1982) Isolation and structure of bryostatin 1. *J. Am. Chem. Soc.* 104:6846-6848.
12. Shankel, D.M., P.E. Hartman, T. Kada, and A. Hollaender, eds. (1986) *Antimutagenesis and Anticarcinogenesis Mechanisms.* Plenum Press, New York, New York.
13. Wall, M.E., M.C. Wani, C.E. Cook, Keith H. Palmer, A.T. McPhail, and G.A. Sim (1966) Plant antitumor agents. I. The isolation and structure of

camptothecin, a novel alkaloidal leukemia and tumor inhibitor from *Camptotheca acuminata*. *J. Am. Chem. Soc.* 88:38883890.

14. Wall, M.E., M.C. Wani, T.J. Hughes, and H. Taylor (1988) Plant antimutagenic agents 1. General bioassay and isolation procedures. *J. Nat. Prod.* 51:866-873.

15. Wall, M.E., M.C. Wani, G. Manikumar, P. Abraham, H. Taylor, T.J. Hughes, J. Warner, and R. McGivney (1988) Plant antimutagenic agents 2. Flavonoids. *J. Nat. Prod.* 51:1084-1091.

16. Wall, M.E., M.C. Wani, G. Manikumar, T.J. Hughes, H. Taylor, R. McGivney, and J. Warner (1988) Plant antimutagenic agents 3. Coumarins. *J. Nat. Prod.* 51:1148-1152.

17. Wall, M.E., M.C. Wani, G. Manikumar, H. Taylor, and R. McGivney (1989) Plant antimutagens 6. Intricatin and intricatol, new antimutagenic homoisoflavonoids from *Hoffmanosseggia intricata*. *J. Nat. Prod.* 52:774-778.

18. Wall, M.E., M.C. Wani, G. Manikumar, H. Taylor, T.J. Hughes, and K. Gaetano (1989) Plant antimutagenic agents 7. Structure and antimutagenic properties of barbatol and 4-isobarbatol, new cymopols from green algae (*Cymopolia barbata*). *J. Nat. Prod.* 52:1092-1099.

19. Wall, M.E., M.C. Wani, T.J. Hughes, and H. Taylor (1989) Plant anti-mutagens. In *Antimutagenesis and Anticarcinogenesis Mechanisms II*, Y. Kuroda, D.M. Shankel, and M.D. Waters, eds. Plenum Press, New York, New York, pp. 61-78.

20. Wall, M.E., M.C. Wani, G. Manikumar, H. Taylor, and R. McGivney (1992) Plant antimutagenic agents 9. Isolation, structure elucidation and antimutagenic properties of various phenolic compounds. (Submitted for publication) *J. Nat. Prod.*

21. Wani, M.C., H.L. Taylor, M.E. Wall, P. Coggin, and A.T. McPhail (1971) Plant antitumor agents. VI. The isolation and structure of taxol, a novel antileukemic and antitumor agent from *Taxus brevifolia*. *J. Am. Chem. Soc.* 93:2325-2327.

22. Wattenberg, L.W. (1983) Inhibition of neoplasia by minor dietary constituents. *Cancer Res.* 43:2448-2453.

23. Williams, D.H., M.J. Stone, P.R. Hawk, and S.K. Rahman (1989) Why are secondary metabolites (natural products) synthesized? *J. Nat. Prod.* 52:1189-1208.

INHIBITORY EFFECTS OF DIETARY LEAFY VEGETABLES

ON MUTAGENS AND ON ACTIVE OXYGENS

Junko Ebata,[1] Kazuaki Kawai,[2] and Hideyuki Furukawa[2]

[1]Faculty of Science of Living
Osaka City University
Sugimoto 3-3-138, Sumiyoshi-ku, Osaka 558, JAPAN

[2]Faculty of Pharmaceutical Sciences
Meijyo University
Nagoya 468, JAPAN

INTRODUCTION

Epidemiological studies have made it clear that cancer is closely related to dietary factors, and the incidence of cancer of various organs is inversely related to estimated amounts of vegetable consumption. Initially, we screened approximately eighty different edible plants and found that antimutagenicity against furyl-furamide (AF2) was widely distributed in vegetables.

In this study, the effects of the growing conditions of the leafy vegetables *Brassica rapa* L. "Komatsuna" and *Spinacia oleraceae* L. on their antimutagenic and active oxygen-scavenging activities were examined.

MATERIALS AND METHODS

Assay of the Antimutagenic Activity

The mutagenicity and antimutagenicity were determined by the Ames Salmonella microsomal protocol (4) with *Salmonella typhimurium* strain TA100. The streptomycin-dependent strain SD100 of *S. typhimurium* developed by Kada et al. (3) was also employed for the assay of antimutagenicity against AF2.

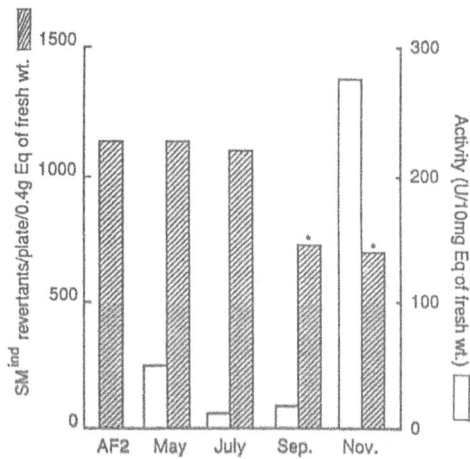

Figure 1. Metabolic activation/deactivation of IQ. Antimutagenic and hydroxyl radical scavenging activities of "Komatsuna" harvested in different growing seasons. "Komatsuna" plants were grown 4 times in a year at the Agricultural Experimental Station in Nara Prefecture applying organic fertilizer. The fresh leaves were thoroughly washed with water and freeze-dried followed by extraction with water. The filtrates were subjected to each analysis. Mutagenic activities (▨) were significantly different from the AF2-only control at 1% level with the Sep. and Nov. samples (*). Hydroxyl radical scavenging activities (□) were estimated by the method of spin-trapping with DMPO. One hundred units of the activity correspond to the scavenging ability of 50 nmoles of L-ascorbic acid.

Assay of Active Oxygen-Scavenging Activity

The superoxide dismutase (SOD)-like activity of vegetable extracts was measured spectrophotometrically by the nitroblue tetrazolium (NBT) method coupled with xanthine-xanthine oxidase (X-XOD) system (1). The hydroxyl radicals formed by the Fenton reaction were measured by the ESR method using the spin-trapping agent, 5,5'-dimethyl-1-pyrroline-N-oxide (DMPO), and the scavenging activity of vegetable extracts was estimated by the lowering of the signal height of the spin adduct. Catalase was assayed by the method of Beers et al. (2).

RESULTS AND DISCUSSION

The Effect of the Growing Season of the "Komatsuna" Plant

"Komatsuna" plants were cultured in different seasons using organic or chemical fertilizer in an open field or in a greenhouse using chemical fertilizer. The antimutagenicity of "Komatsuna" extracts against AF2 was different depending on the harvest time, but not on the fertilizer, and the activity increased in September and November (Fig. 1). The hydroxyl radical scavenging activity was also significantly high in November (Fig. 1), but SOD-like activity was not affected by the growing season. Similar results were obtained with spinach, and catalase

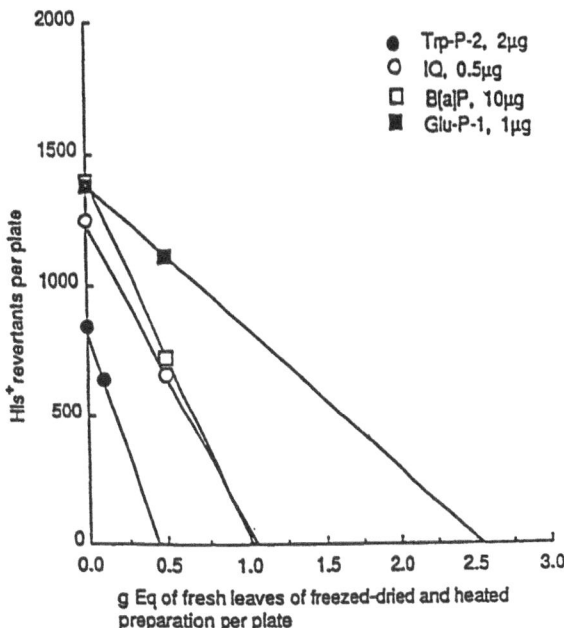

Figure 2. Antimutagenic activity of "Komatsuna" leaves on *Salmonella typhimurium* TA100.

activity was also the highest in November (data not shown). From these results, it was clear that vegetables grown in their optimum growing season have the greatest antimutagenicity and scavenging activity for active oxygen radicals.

The Effect of Open and House Cultures

"Komatsuna" plants were harvested from open and greenhouse plants in the same field and during the same season. The activities of both inhibition of AF2 and scavenging for active oxygens such as superoxide anion radical, hydrogen peroxide, and hydroxyl radical were significantly higher in the extract from "Komatsuna" plants grown in open fields than from those in the greenhouses (data not shown). The amounts of dry matter and protein in the fresh leaves were also richer in the former than in the latter case. It was considered that the low yield of dry matter with greenhouse cultures might be affected by the lower sunshine, higher temperature, and moisture in the greenhouse.

Antimutagenic Activity of "Komatsuna" Leaves Against Different Mutagens

Inhibitory effects of "Komatsuna" extracts on different mutagens were determined as shown in Fig. 2. By linear approximation, it was estimated that the mutagenicities of 90 x 10⁻3 μg AF2, 2 μg Trp-P-2, 1 μg GluP-1, 10 μg B(a)P, 0.5 μg IQ, and 2.5 mg NDMA might be completely suppressed by 0.8. 0.49, 2.5, 1.1, 0.69, and 0.14 g of fresh "Komatsuna" leaves, respectively. Similar results were obtained with spinach leaves.

CONCLUSION

The leafy vegetable "Komatsuna" and spinach, grown in their normal growing seasons, had the greatest antimutagenicity and scavenging activity for active oxygen radicals resulting from the metabolic activation of mutagens. The fresh weight of each vegetable required to completely suppress a specific amount of various mutagenic substances in the diet is defined.

ACKNOWLEDGEMENT

The author is very much indebted to Dr. Tsuneo Taimatsu for his cooperative cultivation of vegetables.

REFERENCES

1. Beauchamp, C., and I. Fridovich (1971) Superoxide dismutase: Improved assays and an assay applicable to acrylamin gels. *Analyt. Biochem.* 44:276-287.
2. Beers, Jr., R.F., and I.W. Sizer (1952) A spectrophotometric method for measuring the breakdown of hydrogen peroxide by catalase. *J. Chem.* 195:133-140.
3. Kada, T., K. Aoki, and T. Sugimura (1983) Isolation of streptomycin-dependent strains from *Salmonella typhimurium* TA98 and TA100 and their use in mutagenicity tests. *Env. Muta.* 5:9-15.
4. Maron, D.M., and B.N. Ames (1983) Revised methods for the Salmonella mutagenicity test. *Mutat. Res.* 113:173-215.

EXTRAPOLATION OF IN VITRO ANTIMUTAGENICITY TO THE

IN VIVO SITUATION: THE CASE FOR ANTHRAFLAVIC ACID

C. Ioannides, A.D. Ayrton, D.F.V. Lewis, and R. Walker

Molecular Toxicology and Food Safety Research Groups
Division of Toxicology
School of Biological Sciences
University of Surrey
Guildford, Surrey, GU2 5XH, UNITED KINGDOM

INTRODUCTION

The current approach commonly employed by "mutageneticists" to test chemicals for possible antimutagenic/anticarcinogenic potential involves using the Ames Salmonella/microsome test as the biological endpoint. Various plant compounds or extracts are incorporated into the postmitochondrial (S9) activation system and their ability to inhibit the mutagenicity of model precarcinogens is determined. However, such a protocol suffers from a number of major disadvantages which make any extrapolation to the in vivo situation virtually impossible.

(i) The hepatic S9 preparation used to activate carcinogens is derived from animals pretreated with Aroclor 1254, a mixture of polychlorinated biphenyls, which induces two families of cytochrome P450, namely, P450 I and IIB, which comprise only 8% and 4%, respectively, of total cytochrome P450 in the uninduced animal (1). Since the P450 I activity plays a major role in the activation of many major groups of chemical carcinogens (2), high activity is required to facilitate the activation of carcinogens, especially those which are activated only with difficulty and/or are weak mutagens. However, it must be emphasized that this is a totally "artificial" system, as in the untreated animals these two families of cytochrome P450 are only minor components (3), and so any conclusions reached on the basis of the Ames test may not reflect the in vivo situation where the cytochrome P450 composition is completely different and may comprise additional proteins that also activate the carcinogen, albeit less effectively. Moreover, it is now being recognized that other families of cytochrome P450, such as the P450 IIE, are responsible for the metabolic activation of some carcinogens, such as certain nitrosamines. These families are not only not inducible by Aroclor 1254 but are actually suppressed by this treatment (1).

(ii)   In the Ames test the only enzyme system monitored is the one that catalyzes the activation of the model carcinogen. For example, it is not ascertained whether the inhibitory effect of the potential anticarcinogen is selective for the activating enzyme, or is a general one showing no discrimination for the activating cytochrome P450 protein. If cytochrome P450 proteins which catalyze the deactivation of the chemical carcinogen are more susceptible to the inhibitory effects of the potential anticarcinogen, then no anticarcinogenic effect will occur in vivo, and even an exacerbation of the carcinogenic response may be evident. Moreover, many cytochrome P450 proteins have important functions in the metabolism of endogenous substrates so that, if modulated by the potential anticarcinogen, they may give rise to unacceptable adverse effects. For example, the imidazole-containing antifungal agents, in addition to inhibiting the cytochrome P450 proteins involved in xenobiotic metabolism (3), also inhibit testicular cytochromes P450 involved in steroid metabolism (4), giving rise to decreased circulating levels of testosterone in man (5).

(iii)   The Ames test favors activation reactions. The activation system is not supplemented with the necessary cofactors for phase II metabolism. Although it is not unusual for a phase II reaction to catalyze activation, in most cases the outcome is deactivation. A potential anticarcinogen may also inhibit a critical phase II deactivation reaction, for instance, conjugation with glutathione, and this attenuates any benefit resulting from the impairment of the activation process. On the other hand, stimulation of the glutathione S-transferase activity would probably potentiate the anticarcinogenic effect of a chemical.

(iv)   The indicator bacterial strains are frequently considered as metabolically inert because of their inability to oxidize xenobiotics. However, they contain azo- and nitro-reductases, acetylases, and peroxidases that profoundly influence the mutagenic response in the Ames test. Lack of appreciation of the contribution of these pathways may lead to incorrect evaluation of the anticarcinogenic potential of a chemical, if the contribution of these pathways to the mammalian activation of the carcinogen is not significant.

(v)   In in vivo systems, high levels of the potential anticarcinogen may be achieved that in vivo may be impossible to reach either because of poor absorption and/or presystemic metabolism (first pass effect). Moreover, the anticarcinogen may display toxicity that cannot be detected in in vitro tests. For these reasons it is imperative that preliminary pharmacokinetic studies are undertaken at an early stage during the evaluation of anticarcinogenic potential.

(vi)   It is frequently assumed that an antimutagenic effect in the Ames test, being the consequence of inhibition of the activation pathway(s), will also occur following in vivo administration, as long as the necessary tissue concentrations are achieved. However, many in vitro inhibitors of cytochrome P450 activity, when administered in vivo, exhibit a biphasic effect. An initial transient inhibitory phase is followed by a more sustained induction of the proteins that were initially inhibited. Examples include the ellipticines (6), safrole (7), the imidazole-containing antifungal agents (8), and cimetidine (9), etc. Under such circumstances the activity of the cytochrome P450 catalyzing the activation of a carcinogen will be enhanced, leading to an increased production of the reactive intermediates that, through their covalent interaction to DNA, will initiate the process of carcinogenesis. As a result, the putative anticarcinogen will not only not antagonize the action of the carcinogen but may actually potentiate the carcinogenic response. This sequence of events is very clearly illustrated in our studies with anthraflavic acid (2,6-dihydroxyanthraquinone), whose ability to act as an antimutagen/anticarcinogen was evaluated using IQ (2-amino-3-methylimidazo[4,5-f]quinoline) as the model carcinogen.

Metabolic Activation/Deactivation of IQ

IQ belongs to the group of amino-imidazoazaarenes that are generated during the broiling of fish and meat (10) and are believed to be formed from the interaction of creatin(in)e with Maillard reaction products (11). It is a very potent mutagen in the Ames test, in the presence of an activation system (12), and is carcinogenic to rodents, such as rats and mice (13,14), giving rise to tumors in a number of tissues, including the liver and intestinal tract.

IQ expresses its mutagenicity/carcinogenicity only following metabolic activation that proceeds through N-hydroxylation, the resulting hydroxylamine serving as the proximate carcinogen (Fig. 1). The most likely ultimate carcinogen is the electrophilic nitrenium ion that may also be formed from the sulphate and acetoxy esters of the hydroxylamine. The major pathway of deactivation is hydroxylation on the 5-position to produce the phenol that is excreted as a conjugate. Other pathways of deactivation include the formation of an N-sulphamate and to a lesser extent of an N-glucuronide.

The activation of IQ, such as N-hydroxylation, is catalyzed by the cytochrome P450-dependent mixed-function oxidase system. As IQ is essentially a planar compound having an area/depth^2 ratio of 5.1 (15), it interacts selectively with the P450 I family, whose substrates are essentially planar compounds (16). For this reason, the activation of IQ appears to be catalyzed exclusively by the P450 I family (17). This family contains only one subfamily comprising two proteins, namely, A1 and A2 (2), and it is the latter in particular that mediates the activation of IQ. Agents that induce P450 I activity also enhance the activation of IQ (18). IQ itself selectively induces the P450 I family (15) and enhances its own activation (19).

Effect of Anthraflavic Acid on the In Vitro Mutagenicity of IQ

The plant phenol anthraflavic acid was chosen as a potential anti-mutagen/anticarcinogen, on the basis of a computergraphic analysis (COMPACT: Computer-Optimized Molecular Parametric Analysis of Chemical Toxicity) which showed that it is a planar compound, having an area/depth^2 of 10.9, and as such it would be expected to interact with the P450 I family and consequently inhibit the metabolic activation of IQ (20). Indeed, subsequent studies (21) demonstrated that anthraflavic acid was a very potent inhibitor of the mutagenicity of IQ in the Ames test ($I^{50} \approx 5 \times 10^{-6}$ M) even at concentrations as low as 5×10^{-7} M. Such tissue concentrations may be achievable through dietary means, and clearly

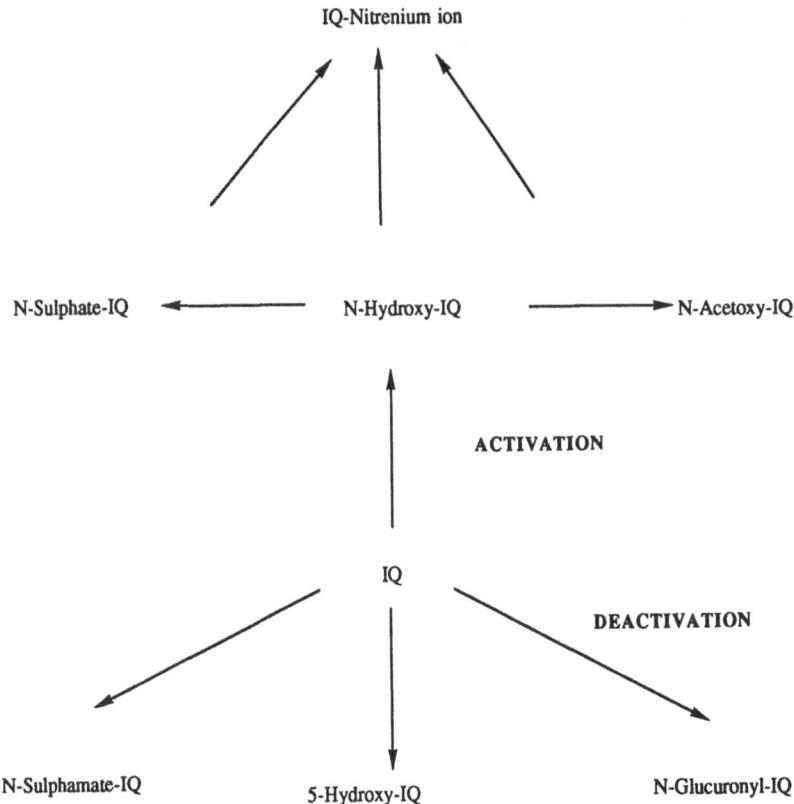

Figure 1. Metabolic activation/deactivation of IQ.

anthraflavic acid has a high potential for anticarcinogenic action. Its antimutagenic action could be accounted for by one or both of the following mechanisms: (a) A direct interaction between anthraflavic acid and the mutagenic intermediate(s) of IQ, rendering them inactive; or (b) inhibition of the P450 I activation of IQ as predicted by the computergraphic COMPACT analysis. The inhibitory effect of anthraflavic acid was abolished when the phenol was incorporated into the activation system subsequent to completion of the microsomal metabolism of IQ, indicating that a direct interaction between IQ-derived mutagens and anthraflavic acid was unlikely (21). Moreover, anthraflavic acid was a potent inhibitor of the O-deethylation of ethoxyresorufin (probe for P450 IA1) and the bioactivation of Glu-P-1 (probe for P450 IA2) but had no effect on the dealkylation of pentoxyresorufin (probe for P450 IIB). The I^{50} value for the inhibition of the two reactions was about 10^{-6} M, for example, similar to that of the inhibition of IQ mutagenicity by the plant phenol.

So, clearly, these in vitro studies established that anthraflavic acid is a potent inhibitor of IQ-mutagenicity by virtue of its ability to selectively inhibit its metabolic activation.

Table 1. Hepatic mixed-function oxidase activity following treatment of rats with anthraflavic acid

PARAMETER	CONTROL	ANTHRAFLAVIC ACID
Ethoxyresorufin O-deethylase (pmol/min per mg protein)	92 ± 5	314 ± 50*
Pentoxyresorufin O-depentylase (pmol/min per mg protein)	15.0 ± 2.4	30.9 ± 7.8
Bioactivation of Glu-P-1 (2.0 µg) (Histidine revertants/mg protein)	711 ± 56	2225 ± 88*
NADPH-cytochrome c reductase (nmol/min per mg protein)	16.7 ± 2.3	11.6 ± 0.6
Cytochromes P-450 (nmol/mg protein)	0.47 ± 0.04	0.52 ± 0.04
Glutathione S-transferase (µmol/min per mg protein)	0.39 ± 0.02	0.46 ± 0.03
Microsomal protein (mg/g liver)	17.8 ± 1.8	15.7 ± 2

Results are presented as Mean $\pm$ SEM for 5 animals except in the case of Glu-P-1 where triplicate plates were used.

* $P < 0.05$

Interactions of Anthraflavic Acid with Cytochrome P450

Since the anticarcinogenic potential of anthraflavic acid relies on its ability to inhibit the P450 I activity, it is pertinent to investigate the possibility of such a mechanism of action also operating in vivo. Treatment of rats with anthraflavic acid resulted in a 2.5-fold increase in the O-deethylation of ethoxyresorufin with a similar increase in the metabolic activation of Glu-P-1, indicating that anthraflavic acid actually induced rather than inhibited P450 I activity in rat hepatic microsomes (Tab. 1). This was confirmed by immunoblot analysis of microsomal proteins, resolved by electrophoresis, using monospecific polyclonal proteins to the P450 I family, where anthraflavic acid caused an increse in the apoprotein levels of both A1 and A2 proteins (22). In contrast, the O-dealkylation of pentoxyresorufin was unaffected by the same treatment with anthraflavic acid. Since glutathione S-transferase activity, which may have a role in the detoxification of the reactive intermediates of IQ, was unaffected by the same treatment, it appears that the detoxification pathway is not similarly stimulated to compensate for the enhanced activation. The consequence of these changes in enzyme activity is that hepatic microsomes from anthraflavic acid-treated animals were markedly more effective (2.5-fold) than similar preparations from control animals in converting IQ to mutagenic intermediates in the Ames test (22).

Clearly, under such circumstances anthraflavic acid treatment would be expected to exacerbate the carcinogenicity of IQ and of any other chemical carcinogen whose activation is mediated by the P450 I family of cytochromes. In previous studies, P450 I-inducing agents, such as flavonoids and indoles, have been shown to inhibit the carcinogenicity of chemicals, particularly polycyclic aromatic hydrocarbons, in a variety of models (23,24), and this was attributed, but not experimentally demonstrated, to the P450 I-inducing properties of these chemicals. However, in these studies the carcinogen in question was administered to animals in which the intracellular levels of these inducing agents were likely to be very high, so that the protection against the chemical carcinogen most probably represents competitive inhibition of P450 I activity, resulting in decreased carcinogen activation. The time of administration of the anticarcinogen in relation to that of the carcinogen appears to be of paramount importance. If they are both simultaneously administered, or the carcinogen is administered before the anticarcinogenic-inducing agent is eliminated, protection against the carcinogenic effect may be evident. If, however, the inducing agent is given prior to the carcinogen, so that P450 I activity is induced but carcinogen exposure is effected following elimination of the inducing agent, then an increase in carcinogenic response cannot be ruled out. Moreover, if the inducing agent is given orally, increased P450 I activity in the gastrointestinal tract and/or in the liver may limit the amount of carcinogen reaching target organs, such as the lung (25), giving a false impression of an anticarcinogenic effect when only this tissue is evaluated for tumorigenicity.

CONCLUSION

The procedures adopted by many "mutageneticists" to evaluate anti-carcinogenic potential of chemicals by incorporating them into the activation systems of in vitro tests in the presence of model precarcinogens is grossly inadequate and misleading. Such studies: use an artificial activating system, totally different from the cytochrome P450 composition seen in animals and man; ignore deactivating phase I and phase II enzyme systems; do not take into account any pharmacokinetic considerations; and assume that the potential anticarcinogen is devoid of any other biological effect. Moreover, chemicals that inhibit the metabolic activation of carcinogens, and consequently their mutagenic response in the Ames test, when administered in vivo may actually elicit precisely the opposite effect, such as acting as co-carcinogens by potentiating the metabolic activation pathway. Extrapolation of in vitro test data to the in vivo situation without due consideration to all the above factors is unlikely to prove productive.

ACKNOWLEDGEMENT

Part of this work was supported by the Food Science Division, Ministry of Agriculture, Fisheries and Food (Grant No. 530), United Kingdom.

REFERENCES

1. Ryan, D., and W. Levin (1990) Purification and characterization of hepatic microsomal cytochrome P-450. *Pharmac. Ther.* 45:153-239.
2. Ioannides, C., and D.V. Parke (1990) The cytochrome P450 I gene family of microsomal hemoproteins and their role in the metabolic activation of chemicals. *Drug Met. Rev.* 22:1-85.

3. Rodrigues, A.D., G.G. Gibson, C. Ioannides, and D.V. Parke (1987) Interactions of imidazole antifungal agents with purified cytochrome P-450 proteins. *Biochem. Pharmacol.* 36:4277-4281.

4. Rajfer, J., S.C. Sikka, F. Rivera, and D.J. Handelsman (1986) Mechanism of inhibition of human testicular steroidogenesis by oral ketoconazole. *J. Clin. Endocr. Met.* 63:1193-1198.

5. Schurmeyer, T., and E. Nieschlag (1982) Ketoconazole-induced drop in serum and saliva testosterone. *Lancet* 2:1098.

6. Phillipson, C.E., P.M.M. Godden, C. Ioannides, and D.V. Parke (1982) The mutagenicity of 9-hydroxyellipticine and its induction of cytochrome P-448 in rat liver microsomes. *Carcinogenesis* 3:11791182.

7. Ioannides, C., M. Delaforge, and D.V. Parke (1981) Safrole: Its metabolism, carcinogenicity and interactions with cytochrome P-450. *Food Cosmet. Toxicol.* 19:657-666.

8. Rodrigues, A.D., P.R. Waddell, E. Ah-Sing, B.A. Morris, C.R. Wolf, and C. Ioannides (1988) Induction of the rat hepatic mixed-function oxidases by 3 imidazole-containing antifungal agents: Selectivity for the cytochrome P450 IIB and P450 III families of cytochromes P-450. *Toxicology* 50:283-301.

9. Ioannides, C. A.D. Rodrigues, A.D. Ayrton, C.R. Barnett, J. Chown, and D.V. Parke (1989) Induction of the rat hepatic microsomal mixed-function oxidases by cimetidine. *Toxicol. Lett.* 49:6168.

10. Kasai, H., Z. Yamaizumi, K. Wakabayashi, M. Nagao, T. Sugimura, S. Yokoyama, T. Miyazawa, N.E. Springarn, J.H. Weisburger, and S. Nishimura (1980) Potent novel mutagens produced by broiling fish under normal conditions. *Proc. Japan Acad.* 56(b):278-283.

11. Jagerstad, M., S. Grivas, K. Olsson, A. Laser-Reutersward, C. Negishi, and S. Sato (1986) Formation of food mutagens via Maillard reactions. In *Genetic Toxicology of the Diet*, I. Knudsen, ed. Alan R. Liss, Inc., New York, New York, pp. 155-167.

12. Sugimura, T. (1982) Mutagens, carcinogens and tumor promoters in our daily food. *Cancer* 49:1970-1984.

13. Ohgaki, H., K. Kusama, N. Matsukara, K. Marino, H. Hasegawa, S. Sato, S. Takayama, and T. Sugimura (1984) Carcinogenicity in mice of a mutagenic compound, 2-amino-3-methylimidazo[4,5-f]quinoline, from broiled sardine, cooked beef and beef extract. *Carcinogenesis* 5:921-924.

14. Takayama, S., Y. Nakatsuru, M. Masuda, H. Ohgaki, S. Sato, and T. Sugimura (1984) Demonstration of carcinogenicity in F344 rats of 2-amino-3-methylimidazo[4,5-f]quinoline from broiled sardine, fried beef and beef extracts. *Gann* 75:467-470.

15. Rodrigues, A.D., A.D. Ayrton, E.J. Williams, D.F.V. Lewis, R. Walker, and C. Ioannides (1989) Preferential induction of the rat hepatic P450 I proteins by the food carcinogen 2-amino-3-methylimidazo[4,5-f]quinoline. *Eur. J. Biochem.* 181:627-631.

16. Lewis, D.F.V., C. Ioannides, D.V. Parke (1986) Molecular dimensions of the substrate binding site of cytochrome P-488. *Biochem. Pharmacol.* 35:2179-2185.

17. Kato, R., T. Kamataki, and Y. Yamazoe (1983) N-Hydroxylation of carcinogenic and mutagenic aromatic amines. *Env. Health Perspect.* 49:21-25.

18. Abu-Shakra, A., C. Ioannides, and R. Walker (1986) Metabolic activation of 2-amino-3-methylimidazo[4,5-f]quinoline by hepatic preparations: Contribution of cytosolic fraction and its significance to strain differences.

Mutagenesis 1:367-370.

19. Ayrton, A.D., E.J. Williams, A.D. Rodrigues, C. Ioannides, and R. Walker (1989) The food pyrolysis product IQ enhances its own activation. *Mutagenesis* 4:205-207.

20. Ayrton, A.D., D.F.V. Lewis, C. Ioannides, and R. Walker (1987) Anthraflavic acid is a potent and specific inhibitor of cytochrome P-448 activity. *Biochim. Biophys. Acta* 916:328-331.

21. Ayrton, A.D., C. Ioannides, and R. Walker (1988) Anthraflavic acid inhibits the mutagenicity of the food mutagen IQ: Mechanism of action. *Mutat. Res.* 207:121-125.

22. Ayrton, A.D., C. Ioannides, and R. Walker (1988) Induction of rat hepatic cytochrome P450 I proteins by the antimutagen anthraflavic acid. *Food Chem. Tox.* 26:909-915.

23. Wattenberg, L.W. (1978) Inhibition of chemical carcinogenesis. *J. Natl. Cancer Inst.* 60:11-18.

24. Wattenberg, L.W. (1983) Inhibition of neoplasia by minor dietary constituents. *Cancer Res.* 43:2448S-2453S.

25. Legraverend, C., T.M. Guenthner, and D.W. Nebert (1984) Importance of the route of administration for the genetic differences in benzo(a)pyrene in utero toxicity and teratogenicity. *Teratology* 29:35-47.

PREVENTIVE EFFECT OF MAGNESIUM HYDROXIDE ON

CARCINOGEN-INDUCED LARGE BOWEL CARCINOGENESIS IN RATS

Hideki Mori, Yukio Morishita, Tukuro Shinoda, and Takuji Tanaka

Department of Pathology
Gifu University School of Medicine
40 Tsukasa-machi, Gifu 500, JAPAN

ABSTRACT

Preventive effect of magnesium hydroxide on carcinogen-induced, large bowel carcinogenesis was examined in three experiments using F344 rats. Experiment I: Rats received dietary administration of magnesium hydroxide at concentrations of 500 or 1,000 ppm after treatment with methylazoxymethanol (MAM) acetate (25 mg/kg, 3 times). These rats had a lower incidence of large bowel neoplasms than animals given MAM acetate alone. Reduction of the tumor incidence was especially significant at a dose of 500 ppm. Experiment II: Rats given magnesium hydroxide (250, 500, or 1,000 ppm) together with 1,2-dimethylhydrazine (DMH) (20 mg/kg, 10 times) showed a lower multiplicity of large bowel tumors than those given DMH alone. Experiment III: The labeling indices of the cryptal cells of the large bowel (cecum or proximal colon or distal colon) or rates given magnesium hydroxide for 4, 6, or 8 weeks after treatment with MAM acetate (25 mg/kg, 3 times) were smaller than those of animals given MAM acetate alone, indicating that magnesium hydroxide suppressed, carcinogen-induced epithelial cell (large bowel) proliferation. The results of the three experiments suggest that magnesium, one of the essential metals, is a promising chemopreventive agent in humans.

INTRODUCTION

Nutritional factors are known to play a significant role in the development of neoplasms. Certain chemopreventive agents proved effective against chemical carcinogenesis in different organs (16). Some trace elements and metals such as selenium have been reported to have some inhibitory effects on tumor induction and development (8,9). There is limited information on the possibility that magnesium, one of the essential metals, may be a chemopreventive agent. Some epidemiological data imply that high concentrations of magnesium in water or food are related to a

Table 1. Incidence of intestinal tumors in each group

Group (treatment)	No. of effective rats[a]	No. of rats with tumors (no. of tumors) at: Entire intestine			Small intestine			Large intestine		
		Total	AD[b]	ADC	Total	AD	ADC	Total	AD	ADC
1 (MAM acetate alone)	32	18(26)	8(8)	15(18)	2(2)	0	2(2)	17(24)	8(8)	14(16)
2 (MAM acetate/ 500 ppm Mg(OH)2)	30	5(5)[c]	0[d]	5(5)[e]	2(2)	0	2(2)	3(3)[f]	0[g]	3(3)[h]
3 (MAM acetate/ 1000 ppm Mg(OH)2)	30	11(13)	3(3)	9(10)	2(2)	0	2(2)	9(11)	3(3)	7(8)
4 (500 ppm Mg(OH)2)	20	0	0	0	0	0	0	0	0	0
5 (1000 ppm Mg(OH)2)	19	0	0	0	0	0	0	0	0	0
6 (No treatment)	20	0	0	0	0	0	0	0	0	0

[a]Rats survived more than 126 days. [b]AD, adenoma; ADC, adenocarcinoma. [c-h]Significantly different from Group 1 ([c]$p < 0.001$, [d]$p < 0.003$, [e]$p < 0.008$, [f]$p < 0.0002$, [g]$p < 0.003$ and [h]$p < 0.003$)

low cancer mortality (2,14). In experimental studies, an increased incidence of lymphoma or thymoma was observed in rats kept on a magnesium-deficient diet (4), and administration of magnesium was shown to suppress skin tumorigenesis (3,13). Magnesium as well as calcium are also known to be fundamental regulators of the cell cycle (6,15). However, systematic studies of the anticarcinogenic action of magnesium have not yet been performed. We have examined effects of magnesium hydroxide in the development of colon neoplasms in rat models using MAM acetate and DMH. Furthermore, effects of this agent on the proliferative activity of cryptal cells of the large bowel of rats given MAM acetate were also studied in a rat model.

EXPERIMENT I: EFFECT OF MAGNESIUM HYDROXIDE ON MAM ACETATE-INITIATED INTESTINAL CARCINOGENESIS

A total of 155 male F344 rats were divided into six groups. At 6 weeks of age, rats in Groups 1, 2, and 3 were given ip injections of MAM acetate (25 mg/kg body weight) once a week for 3 weeks. Two weeks after the final MAM acetate exposure, rats in Groups 2 and 3 were transferred to a diet containing 500 or 1,000 ppm magnesium hydroxide, respectively, and kept on the diet until the end of the experiment. Rats in Groups 4 and 5 were given an equal volume of normal saline and received the diets with magnesium hydroxide at concentrations of 500 or 1,000 ppm, respectively. Animals in Groups 1 and 6 were continued on the basal diet, CE-2 (CLEA Japan, Tokyo) throughout the experiment. The experiment was terminated after 255 days following the first MAM acetate treatment. At necropsy, all organs, especially the intestine, were carefully inspected grossly and all abnormal lesions were examined histologically.

In this experiment, one rat from Group 2, which was moribund and necropsied on the 126th day, had a colon tumor. The rats that were alive on that day were regarded as effective animals. Intestinal tumors developed in rats in Groups 1, 2, and 3. Most of them were present in the large intestine. They were sessile or pedunculated polyps macroscopically, and were histologically adenomas or adenocarcinomas with a higher incidence of adenocarcinomas. The dietary administration of magnesium hydroxide at a level of 500 ppm resulted in a significant inhibition of MAM acetate-induced colon carcinogenesis as revealed by a low incidence of both adenomas and adenocarcinomas of the colon (Tab. 1). The incidence of colon tumors in rats treated with MAM acetate and 1,000 ppm of magnesium hydroxide was lower than that in rats given MAM acetate alone, but the difference was not significant between the two groups. In rats of Groups 4-6, no intestinal neoplasms were found. The results of the present study indicate that the induction of colon neoplasms by MAM acetate was inhibited by magnesium hydroxide when it was given after the carcinogen exposure.

EXPERIMENT II: EFFECT OF MAGNESIUM HYDROXIDE ON DMH-INDUCED INTESTINAL CARCINOGENESIS

A total of 218 male F344 rats, 6 weeks of age, were divided into 8 groups. Rats of Groups 1-5 received sc injection of DMH (20 mg/kg) 10 times (once every two weeks) starting at the commencement of the experiment. Animals of Group 1 were kept on the basal diet (CE-2) after the discontinuation of carcinogen exposure.

Table 2. Multiplicity (number of tumors/rat) of intestinal tumors in each group

Group (treatment)	No. of effective rats	Entire intestine			Small intestine			Large intestine		
		Total	AD	ADC	Total	AD	ADC	Total	AD	ADC
1 (DMH alone)	30	1.94	1.00	1.88	1.17	–	1.17	1.86	1.00	2.09
2 (DMH + 250 ppm Mg(OH)$_2$)	32	1.29[a]	1.00	1.14[b]	1.20	–	1.20	1.09[c]	1.00	1.00[e]
3 (DMH + 500 ppm Mg(OH)$_2$)	32	1.53	1.00	1.50	1.33	–	1.33	1.38	1.00	1.31[f]
4 (DMH + 1000 ppm Mg(OH)$_2$)	32	1.36	1.00	1.31	1.17	–	1.17	1.09[d]	1.00	1.19[g]
5 (DMH + 2000 ppm Mg(OH)$_2$)	31	1.75	1.00	1.69	1.09	1.00	1.10	1.45	–	1.45
6 (250 ppm Mg(OH)$_2$)	20	–	–	–	–	–	–	–	–	–
7 (2000 ppm Mg(OH)$_2$)	20	–	–	–	–	–	–	–	–	–

Significantly different from Group 1 ([a,f] $p < 0.05$, [b] $p < 0.02$, [c,d,g] $p < 0.01$, [e] $p < 0.001$). AD: adenoma, ADC: adenocarcinoma

Rats of Groups 2-5 were fed the diet mixed with magnesium hydroxide at concentrations of 250, 500, 1,000, or 2,000 ppm, respectively, throughout the experiment. Animals of Groups 6 or 7 were not given the carcinogen. They were, however, exposed to magnesium hydroxide (250 or 2,000 ppm) like those of Groups 2-5. The experiment was terminated one year after the start. At the end of the experiment, animals were sacrificed and complete autopsies were done.

Intestinal tumors were seen in the animals of Groups 1-5. In Group 1, 57% of effective rats (those that survived for more than 200 days after the start f the experiment) developed tumors (20%, small intestinal tumors; 47%, large intestinal tumors). In Group 2, 44% of the effective animals had tumors (15 ^, small intestinal tumors; 34%, large intestinal tumors). In Group 3, the tumors were seen in 53% of the animals (9%, small intestinal tumors; 50%, large bowel tumors). Similarly, the incidence of tumors of Groups 4 or 5 were 44% (19%, small intestinal tumors; 34%, large intestinal tumors) or 52% (35%, small intestinal tumors; 35%, large bowel tumors). These tumors were, histologically, adenomas or adenocarcinomas. No intestinal tumors were seen in the rats of Groups 6 and 7. Statistically, no significant difference of the incidence of intestinal tumors was obtained between Group 1 and Group 2 or Groups 3, 4, and 5. Multiplicity of intestinal tumors (number of tumors per rat) of rats in each group in shown in Tab. 2. In Group 1, the mean number of intestinal tumors was 1.94 (1.17, small intestinal tumors; 1.86, large intestinal tumors). In Group 2, the number of intestinal tumors was 1.29 (1.20, small intestinal tumors; 1.09, large intestinal tumors). Similarly, the numbers of tumors in Group 3, 4, and 5 were 1.53 (1.33, small intestinal tumors; 1.38, large intestinal tumors); 1.36 (1.17, small intestinal tumors; 1.09, large intestinal tumors); and 1.75 (1.09, small intestinal tumors; 1.45, large intestinal tumors); respectively. Statistically, the mean number of intestinal tumors, that of adenocarcinoma of the entire intestine, the number of large intestinal tumors, and that of adenocarcinoma of Group 2 were smaller than the corresponding numbers of Group 1 (P $\langle$ 0.05, $\langle$ 0.02, $\langle$ 0.01, and $\langle$ 0.001, respectively). Multiplicity of adenocarcinomas of the large bowel of Group 3 was significantly smaller than that of Group 1 (P $\langle$ 0.05). Furthermore, those of the large intestinal tumor and of the adenocarcinoma of Group 4 were, respectively, smaller that those of Group 1 (P $\langle$ 0.01). The results indicate an inhibitory effect of magnesium hydroxide on DMH-induced large bowel carcinogenesis.

EXPERIMENT III: EFFECT OF MAGNESIUM HYDROXIDE ON MAM ACETATE-INDUCED EPITHELIAL PROLIFERATION

A total of 110 male, 6-week old F344 rats were divided into 10 groups. Rats of Groups 1-5 were given ip injections of MAM acetate (25 mg/kg) once a week for 3 weeks. Animals of Groups 2-5 were transferred to the basal diet mixed with magnesium hydroxide with doses of 250, 500, 1,000, or 2,000 ppm one week after the final MAM acetate treatment. Rats of Groups 6-9 also received dietary exposure of magnesium hydroxide as in Groups 2-5 (250, 500, 1,000 or 2,000 ppm). However, they were not given MAM acetate. Animals in Group 10 were maintained on the basal diet throughout the experiment and treated as controls. Animals of all groups were sequentially sacrificed at 2-week intervals, starting 4 weeks after the start of the experiment. The experiment was terminated 8 weeks after the start. Prior to sacrifice, rats were given ip injections of 50 mg/kg bromodeoxyuridine (BrdUrd). At necropsy, the large bowel was divided into 3 segments (cecum, proximal colon, and

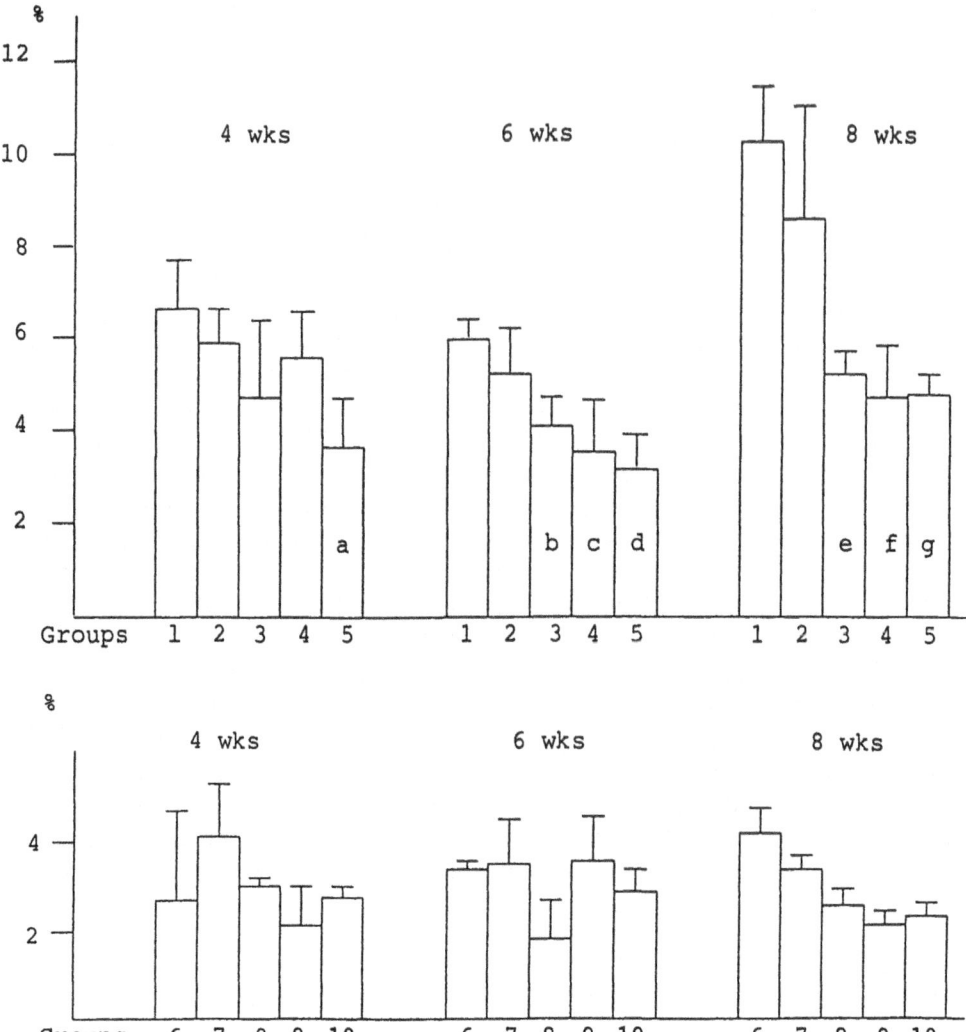

Figure 1. Labeling indices of the cryptal cells of proximal colons of rats of each group. The indices of animals given MAM acetate and magnesium hydroxide (Groups 2-5) were in general smaller than those given MAM acetate alone (Group 1) [a]P $\langle$ 0.04); [b]P $\langle$ 0.02; [c]P $\langle$ 0.05; [d]P $\langle$ 0.005; [e]P $\langle$ 0.002; [f]P $\langle$ 0.003; [g]P $\langle$ 0.001).

distal colon) and fixed in 10% buffered formalin. Two serial sections were made from each segment; one section was stained with H.E. and the other section was used for immunohistological examination to detect BrdUrd incorporation (7). Measurement of labeling indices was done using 12 crypts having more than 35 clearly identified cells.

The labeling indices of the cells from three segments of the large bowels of rats given magnesium hydroxide alone were similar to those of control animals at any sacrificing point of 4, 6, or 8 weeks (Groups 6-9; Group 10). The indices of rats given MAM acetate alone (Group 1) were significantly higher than those of controls at any segment of the large intestine and at any sacrificing point. In general, the labeling indices of the cryptal cells of the large bowel of rats given magnesium hydroxide after the exposure of MAM acetate (Groups 2-5) were smaller than those of animals that were not given magnesium hydroxide after the carcinogen treatment (Group 1). This tendency of the labeling indices was more obvious in the cells of animals given higher doses of the trace element (only the result of the proximal colon is shown in Fig. 1).

DISCUSSION

Some experimental investigations have been done suggesting the ability of magnesium, one of the essential metals, to provide chemoprevention against carcinogenesis. These studies basically report the ability of magnesium (as a salt) to prevent the induction of local tumors (i.e., skin tumors) by its local application (3,13). The effect of a magnesium salt on chemically-induced large bowel carcinogenesis was examined by a systematic application. The results of Experiment I indicate that the induction of colon neoplasms by MAM acetate was inhibited by magnesium hydroxide when it was given during the promotional phase of carcinogenesis. The inhibitory effect of magnesium hydroxide on carcinogen-induced large bowel carcinogenesis was also demonstrated in Experiment II with a different protocol using DMH. Results of both experiments suggest that magnesium is a promising chemopreventive agent against large bowel neoplasms in humans. In both experiments, however, a dose-relation was not confirmed in the inhibitory effect of the agent. We do not have any clear reason for this. Nevertheless, it may be true that there is some optimal dose for the chemopreventive action of these agents. In fact, we have demonstrated a chemopreventive action of a phenolic compound on carcinogen-induced intestinal and liver carcinogenesis in a very low dose (12). Magnesium is known to relate to the control of cell cycle (6,15). In the present study, magnesium hydroxide suppressed MAM acetate-induced cell proliferation in the large bowels of rats. this suppression appears to be important. Epithelial cell proliferation has been known to increase in gastrointestinal tracts, including the colon, with increased susceptibility to cancer before the appearance of tumors (10,11). It has been reported that dietary calcium decreases hyperproliferation of colonic epithelial cells in human subjects at high risk for colon cancer (5) and that the calcium decreases incidence of carcinogen-induced colonic tumors (1). The inhibitory effect of magnesium hydroxide on MAM acetate-initiated large bowel carcinogenesis may occur through a suppressive action on the carcinogen-induced cell proliferation during the post-initiation phase of carcinogenesis. Besides these cell-biological functions of magnesium, some other important biological effects may be related to the protective effect of the agent against carcinogenesis.

REFERENCES

1   Appleton, G.V.N., P.W. Davies, J.B. Bristol, and R.C.N. Williamson (1987) Inhibition of intestinal carcinogenesis by dietary supplementation with calcium. *Br. J. Surg.* 74:523-525.

2.   Armstrong, R.W. (1972) Cancer and infectious diseases related to geochemical environment. Is there a particular kind of soil or geologic environment that predisposes to cancer? *Ann. N. Y. Acad. Sci.* 199:239-248.

3.   Bazikan, K.A., and A.A. Akimov (1968) Anticarcinogenic effect of magnesium. *Vop. Onkol.* 14:57-61.

4.   Bois, P. (1964) Tumors of the thymus in magnesium deficient rats. *Nature* 204:1316.

5.   Buset, M., M. Lipkin, S. Winawer, S. Swaroop, and E. Friendman (1986) Inhibition of human colonic epithelial cell proliferation in vivo and in vitro by calcium. *Cancer Res.*46:5426-5430.

6.   Duffus, J.H., and G.M. Walker (1987) Magnesium in mitosis and the cell cycle. In *Magnesium in Cellular Processes and Medicine,* B.M. Altura, J. Durlach, and M.S. Seeling, eds. Karger, Basal, Switzerland, pp. 131-141.

7.   Fujimoto, Y., M. Oyamada, A. Hattori, H. Takahashi, M. Sawaki, K. Dempo, M. Mori, and M. Nagao (1989) Accumulation of abnormally high ploid nuclei in the liver of LEC rats developing spontaneous hepatitis. *Jan. J. Cancer Res.* 80:45-50.

8.   Griffin, A.C. (1979) Role of selenium in the chemoprevention of cancer. *Adv. Cancer Res.* 29:419-442.

9.   Jacob, M.M. and A.C. Griffin (1981) Trace elements and metals as anticarcinogens. In *Inhibition of Tumours Induction and Development,* M.S. Zedeck and M. Lipkin, eds. Plenum Press, New York, pp. 169-188.

10.   Lipkin, M. (1987) Biomarkers of increased susceptibility to gastrointestinal cancer. Their development and application to studies of cancer prevention. *Gastroenterology* 92:1083-1086.

11.   Lipkins, M. (1988) Biomarkers of increased susceptibility to gastrointestinal cancer: New application to studies of cancer prevention in human subjects. *Cancer Res.* 48:235-245.

12.   Mori, H., T. Tanaka, H. Shima, T. Kuniyasu, and M. Takahashi (1986) Inhibitory effect of chlorogenic acid on methylazoxymethanol acetate-induced carcinogenesis in large intestine and liver of hamsters. *Cancer Lett.* 30:49-54.

13.   Poirier, L.A., K.S. Kasprzak, K.L. Hoover, and M.L. Wenk (1983) Effects of calcium and magnesium acetate on the carcinogenicity of cadmium chloride in Wistar rats. *Cancer Res.* 43:4575-4581.

14.   Schrauzer, G.N. (1976) Cancer mortality correlation studies. II. Regional association of mortalities with the consumption of foods and other commodities. *Med. Hypotheses* 2:39-49.

15.   Walker, G.M. (1986) Magnesium and cell cycle control: An update. *Magnesium* 5:9-23.

16.   Wattenberg, L.W. (1985) Chemoprevention of cancer. *Cancer Res.* 45:1-8.

ANTIMUTAGENIC INVESTIGATIONS ON COMMERCIAL YOGURT

C. Della Croce, E. Morichetti, G. Bronzetti, C. Salvadori,*
and E. Macri*

Istituto di Mutagenesi e Differenziamento
CNR via Svezia 10
56124 Pisa, Italy

*Direzione Scientifica Parmalat SpA
Castellaro di Sala Baganza
43039 Parma, Italy

INTRODUCTION

The role of diet in the etiology of cancer has received increasing attention in the last ten years (1-5). Consumption of milk reduced the incidence of human stomach cancer induced by alkylating agents (6). Human subjects (7) and experimental animals (8) receiving dietary supplements of *Lactobacillus acidophilus* exhibited significantly lower levels of fecal enzymes associated with colon carcinogenesis. In addition, extracts prepared from *L. bulgaricus, L. casei*, and *L. acidophilus* spent media were shown to inhibit the growth of Sarcoma and Ehrlich tumors, suggesting that lactobacillic cultures have an antitumorigenic effect (9,10).

Other reports indicate that dietary supplementation with viable *L. acidophilus* significantly lowered the level of fecal bacterial azoreductase, β-glucuronidase, and nitroreductase in rats given meat diets (8). These enzymes have been postulated to play an important role in the conversion of chemical procarcinogens to carcinogens in the bowel and, thus, have been used to monitor colon carcinogenesis (11). Recent studies have shown a similar effect on β-glucuronidase, β-glucosidase, and nitroreductase activities in human subjects receiving *L. acidophilus* cells (7,12).

In this present study, we tested the antimutagenic effects of commercial fermented milk (Kyr from Parmalat), containing four different strains of bacilli: *L. bulgaricus and Streptococcus thermophilus*, present in all common yogurts, and *L. acidophilus* and *Bifidobacterium bifidum*, contained in this product.

We used cells of D_7 strain of *Saccharomyces cerevisiae* as the test organism, and methylmethane-sulfonate (MMS) and ethylmethanesulfonate (EMS) as standard mutagens.

MATERIALS AND METHODS

Microorganism Strain

D_7 strain of yeast *S. cerevisiae* was obtained from F.K. Zimmermann (13). With this strain we can detect, simultaneously, mitotic gene conversion, point reverse mutations at trp5 and ilv1 loci, respectively, and mitotic recombination between the centromere and the ade2 locus. Mitotic crossing over can be detected visually as pink colonies and red twin-sectored colonies, which are due to the formation of homozygous cells of the genotype ade2-40/ade2-40 (deep red) and ade2-119/ade2-119 (pink) from the originally heteroallelic cells, which form white colonies. Mitotic gene conversion can be detected by the appearance of tryptophan-nonrequiring colonies on selective media.

Mutation can be followed by the appearance of isoleucine-requiring colonies on selective media.

Preparation of Samples

A commercial fermented milk (Kyr) was prepared from a stock yogurt culture of *L. bulgaricus* and *S. thermophilus* by adding fermented milk obtained using *L. acidophilus* and *B. bifidum* (Sample 1). This product was treated with a flow of N_2 for 30 min (Sample 2).

N_2 treatment allowed us to separate the product in two phases: a more solid portion containing a larger amount of casein (Sample 5) and a more liquid one containing mainly whey proteins (Sample 4). The commercial product was heated for 30 min at 85° C in order to decrease total bacterial count (Sample 3). Samples 4 and 5 were submitted to the same heat treatment.

Experimental Procedure

Yogurt samples were assayed in a suspension test: 2, 5 x 10^6 cells of the D_7 strain of *S. cerevisiae* from stationary growth phase were suspended in 4 ml of H_2O and incubated for two hours with alkylating agents (7 mM MMS and 110 mM EMS) with several concentrations of different yogurt samples.

Suitable dilutions of samples were plated on complete medium for survival determination and on selective minimal media for genetic effect evaluation.

Determination of bacterial numbers in different Kyr samples was performed in research laboratories of Parmalat using culture media specific for each bacterial strain.

Table 1. Induction of gene conversion and point reverse mutation in the D^7 strain of *S. cerevisiae* by different doses of Kyr samples (S). Data represent mean of 3 independent experiments ± S.D.

DOSE	SURVIVAL %	TRP CONVERTANTS/ 10^5 SURVIVORS	ILV REVERTANTS/ 10^6 SURVIVORS
CONTROL		0.72±0,11	0,22±0,07
S.1 0,2 ml	46.0±8.53	0.71±0.20	0.51±0.09
S.2 0,2 ml	70.0±11.53	1.02±0.31	0.37±0.09
S.3 0,2 ml	71.0±10.54	1.19±0.22	0.10±0.02
S.4 0,2 ml	94.7±6.53	0.57±0.21	0.59±0.10
S.5 0,2 ml	68.5±17.51	1.02±0.30	0.30±0.07

Statistical Analysis

The results were analyzed using the Student's "T" test.

RESULTS AND DISCUSSION

Results of suspension test with yogurt showed no significant influence on genetic activity and survival of yeast cells (Tab. 1).

In Tab. 2 we can see that the samples did not significantly influence genetic activity induced by MMS in yeast cells, nor did they have any effect on survival percentage detected following the treatment with MMS alone.

In Tab. 3 the results with EMS showed that Samples 2 and 4 significantly decreased gene conversion and induced point reverse mutation frequencies. Sample 3 decreased gene conversion at the highest dose and Sample 5 decreased point mutation frequency at the highest dose.

Tab. 4 shows the determination of bacteria numbers in every sample of Kyr: the number of bacteria (*L. bulgaricus, S. thermophilus, L. acidophilus,* and *B. bifidum*) decreased in Samples 2, 4, and 5.

In this study, two known alkylating agents have been employed: methylmethanesulfonate (MMS) as a methylating agent that reacts by the SN_1 mechanism, and ethylmethanesulfonate (EMS) as an ethylating agent that shows an SN_1-SN_2 mechanism (14). Alkylating agents are mutagenic and carcinogenic compounds that have DNA as their biological target; they alkylate it by two different mechanisms.

Table 2. Kyr effect on survival, point reverse mutation, and gene conversion in D_7 strain in *S. cerevisiae* induced by MMS (7 mM). Data represent mean of three independent experiments ± S.D.

DOSE	SURVIVAL %	TRP CONVERTANTS/ 10^5 SURVIVORS	ILV REVERTANTS/ 10^6 SURVIVORS
CONTROL		0.76±0.05	0.26±0.01
MMS (7mM)	59.6±6.9	130.65±40.02	50.85±10.31
MMS+S.1 (50µl)	66.8±15.9	119.60±32.40	42.01±7.48
MMS+S.1(100µl)	61.7±10.8	139.22±59.04	40.68±12.96
MMS+S.2 (50µl)	55.3±12.4	124.27±30.41	51.81±19.41
MMS+S.2(100µl)	52.8±13.5	133.30±35.49	49.60±11.56
MMS+S.3 (50µl)	63.1±11.2	117.90±49.74	42.30±9.10
MMS+S.3(100µl)	48.2±13.1	144.50±55.41	57.20±7.29
MMS+S.4 (50µl)	58.4±11.2	130.45±40.91	40.74±10.41
MMS+S.4(100µl)	47.3±16.8	188.95±61.14	52.14±13.41
MMS+S.5 (50µl)	77.5±10.4	118.50±50.41	34.09±9.64
MMS+S.5(100µl)	45.9±9.6	137.70±41.10	53.79±11.39

SN_1: monomolecular nucleophilic substitution, with the formation of a stable carbocation that rapidly reacts with DNA.

SN_2: bimolecular nucleophilic substitution, with the formation of an intermediate compound that is a transition complex between electrophilic and nucleophilic (DNA) groups (15-17).

The above results showed that yogurt samples do not react generally against genetic effects induced by different alkylating agents, indicating that they could possess specificity of action.

The selective protection against EMS by yogurt suggests that the effect is not due to some general protection against alkylating agents but to some form of specific interaction.

To better understand the mechanism of action of these samples, other experiments with different mutagens and different experimental procedures are

Table 3. Kyr effect on survival, point reverse mutation, and gene conversion in D^7 strain in *S. cerevisiae* induced by EMS (110 mM). Data represent mean of three independent experiments ± S.D.

DOSE	SURVIVAL %	TRP CONVERTANTS/ 10^5 SURVIVORS	ILV REVERTANTS/ 10^6 SURVIVORS
CONTROL		1.01±0.08	0.16±0.06
EMS (110 mM)	48.5±6.5	189.26±50.11	1285.41±99.72
EMS+S.1 (50 μl)	58.5±10.5	106.40±32.21	1069.90±100.03
EMS+S.2(100 μl)	35.0±4.5	126.49±21.10	1220.20±79.45
EMS+S.2 (50 μl)	48.5±9.5	101.90±10.50	1043.00±87.48
EMS+S.2(100 μl)	77.5±22.5	80.68±21.90	1059.19±85.30
EMS+S.3 (50 μl)	55.5±18.5	119.59±26.50	1292.95±284.55
EMS+S.3(100 μl)	56.0±0.5	65.29±11.10	1167.99±159.41
EMS+S.4 (50 μl)	97.0±1.5	54.44±11.17	387.50±75.28
EMS+S.4(100 μl)	85.5±14.5	92.32±15.70	938.80±100.32
EMS+S.5 (50 μl)	55.5±10.5	155.63±35.80	1249.31±196.24
EMS+S.5(100 μl)	85.5±11.5	116.35±21.01	752.95±134.84

necessary. Another consideration is that the more marked antimutagenic effect was produced by Sample 4, a component of the common yogurt containing mainly whey proteins.

It would be very interesting to analyze why Sample 1 (that is, the commercial preparation of yogurt also containing Sample 4) does not have the same effect. The measurement of protein concentration should give further information. These results indicate that short-term tests are very useful in evaluating the antimutagenic potentiality of many substances in common use.

ACKNOWLEDGEMENTS

This work was supported by grants from CNR, Comitato Ambiente ed Habitat No. 88.03.600.13, and by Progetto Finalizzato Fatma. The authors wish to thank A. Galli, R. Del Carratore, R. Vellosi, and R. Fiorio for scientific assistance; D. Rosellini for technical assistance; and M. Minks for revising the manuscript.

Table 4. Determination of bacteria number (*L. bulgaricus, S. thermophilus, L. ascidophilus, B. bifidum*) in different Kyr samples.

	L.bifidum	L.acidophilus	S.thermophilus	L.bulgaricus
S.1	4×10^8	1.2×10^8	5.4×10^8	1.8×10^8
S.2	6×10^2	1×10^4	$1,5 \times 10^4$	1×10^2
S.3	4×10^7	1×10^8	$2,4 \times 10^8$	2×10^7
S.4	3×10^2	1.2×10^4	4×10^3	1×10^2
S.5	3.5×10^2	4×10^3	2×10^3	9×10^3

REFERENCES

1. Armstrong, B., and R. Doll (1975) Environmental factors and cancer incidence and mortality in different countries, with special reference to dietary factors. *J. Cancer* 15:617-631.
2. Correa, P. (1981) Nutrition and cancer: Epidemiological correlations. In *Nutrition and Cancer Etiology and Treatment,* G.R. Nerwell and N.M. Ellison, eds. Raven Press, New York, pp. 1-10.
3. Brown, R.R. (1983) The role of diet in cancer causation. *Food Technol.* 37:49-57.
4. Gori, G.B. (1979) Diet and nutrition in cancer causation. *Nutr. Cancer* 1:5-8.
5. National Dairy Council (1980) An update on nutrition, diet and cancer. *Dairy Council Digest* 51:5.
6. Yano, K. (1979) Effect of vegetable juice and milk on alkylating activity of N-methyl-N-nitrosourea. *J. Agric. Food Chem.* 27:456-458.
7. Ayebo, A.D., I.A. Angelo, and K.M. Shahani (1980) Effect of ingesting *Lactobacillus acidophilus* milk upon fecal flora and enzyme activity in humans. *Milchwissenschaft* 35:730-733.
8. Goldin, B.R., and S.L. Gorbach (1977) Alterations in fecal microflora enzymes related to diet, age, lactobacillus supplements and dimethylhydrazine. *Cancer* 40:2421-2426.
9. Bogdanov, I., P. Popkhristov, and L. Marinov (1962) Anticancer effect of *Antibioticum bulgaricum* on Sarcoma 180 and on the solid form of Ehrlich carcinoma. Abstract of *VIII International Cancer Congress,* Moscow, pp. 364-365.
10. Friend, B.A., R.E. Farmer, and K.M. Shahani (1982) Effect of feeding and intraperitoneal implantation of yogurt culture cells on Ehrlich ascites tumor. *Milchwissenschaft* 37 12:708,710.
11. Goldin, B.R., and S.L. Gorbach (1976) The relationship between diet and rat fecal bacterial enzymes implicated in colon cancer. *J. Natl. Cancer Inst.* 57:371-375.
12. Goldin, B.R., L. Swenson, J. Dwyer, and S.L. Gorbach (1980) Effect of diet and *L. acidophilus* supplements on human fecal bacterial enzymes. *J. Natl. Cancer Inst.* 64:255-261.

13. Zimmermann, F.K., R. Kern, and H. Rasenberger (1975) A yeast strain for simultaneous detection of induced mitotic crossing-over, mitotic gene conversion and reverse mutation. *Mutat. Res.* 28:381-388.

14. Babudri, N., and M.G. Politi (1989) Different action of MMS and EMS in UV sensitive strains. *Mutat. Res.* 217:211-217.

15. Singer, B., and D. Grunberg (1983) In *Molecular Biology of Mutagens and Carcinogens.* Plenum Press, New York, pp. 31-34.

16. Lindahl, T., and B. Sedgwick (1988) Regulation and expression of the adaptative response to alkylating agents. *Ann. Rev. Biochem.* 57:133-157.

17. Horsfall, M.J., J.E.A. Gordon, and P.A. Burns (1990) Mutational specificity of alkylating agents and the influence on DNA repair. *Envir. Molec. Mutag.* 15:107-122.

CHEMOPROTECTION BY INDUCERS OF

ELECTROPHILE DETOXICATION ENZYMES

Thomas W. Kensler,[1,2] Nancy E. Davidson,[3]
John D. Groopman,[1] Bill D. Roebuck,[4] Hans J. Prochaska,[2,5]
and Paul Talalay[2]

[1]Department of Environmental Health Sciences
School of Hygiene & Public Health
Johns Hopkins University

[2]Department of Pharmacology and Molecular Sciences
Johns Hopkins University

[3]Oncology Center, School of Medicine
Johns Hopkins University
Baltimore, Maryland 21205 USA

[4]Department of Pharmacology and Toxicology
Dartmouth Medical School
Hanover, New Hampshire 03756 USA

[5]Department of Molecular Pharmacology
Memorial Sloan-Kettering Cancer Center
New York, New York 10021 USA

INTRODUCTION

Many potential strategies exist for chemical protection against carcinogenesis. Multiple stages of carcinogenesis, including initiation, promotion, and progression, can serve as targets for different types of interventions [see reviews by Wattenberg (32); De Flora and Ramel (10)]. However, in the majority of experimental systems, protection has been achieved by administering the chemoprotective agent prior to and/or concurrent with the exposure to carcinogen. Given this temporal relationship between administration of anticarcinogen and carcinogen, it seems likely that these agents act principally by affecting the metabolism and disposition of carcinogens, thereby altering events critical to the initiation of carcinogenesis. Using this experimental approach, it has been possible to document protection against a diverse array of chemical carcinogens acting at many different target organ sites. Important classes of chemoprotectors that modulate the metabolic processing of carcinogens include phenolic antioxidants, indoles, organic isothiocyanates, coumarins, flavones, allyl sulfides, dithiocarbamates, and dithiolethiones.

A key component in understanding the initial events of carcinogenesis was the recognition by the Millers that many chemical carcinogens are not chemically reactive per se, but must undergo metabolic activation to form electrophilic reactants (19). These reactive species can interact with nucleophilic groups in DNA to induce point mutations and other genetic lesions, leading to activation of proto-oncogenes and inactivation of tumor suppressor genes. The importance of metabolic activation in carcinogenesis is highlighted by the fact that target organ specificities and even species susceptibilities can be determined through the presence or absence of metabolic activation pathways. The metabolism of chemicals to proximate carcinogens often involves an initial two-electron oxidation to a hydroxylated or epoxidated product and is typically catalyzed by the cytochrome P-450 system. Collectively, the enzymes that catalyze the formation of these reactive intermediates are termed Phase I enzymes.

Cells also have a variety of enzymatic and nonenzymatic mechanisms that protect against damage by electrophilic metabolites. A number of enzymes transfer or conjugate various endogenous substrates, such as glutathione, glucuronide, and sulfate, to the products of Phase I metabolism. These Phase II reactions, which often add large polar molecules to the primary metabolite, generally limit further biotransformation by enhancing elimination, thereby leading to detoxication. Thus, the amount of ultimate carcinogen available for interaction with its target represents, in part, a balance between competing activating and detoxifying reactions. While this balance is under genetic control, it is easily modulated by a variety of factors including nutritional status, age, hormones, and exposure to drugs or other xenobiotics (8). In this setting, chemoprotective agents can profoundly modulate the constitutive metabolic balance between activation and inactivation of carcinogens through their actions on both Phase I and II enzymes. This chapter considers the general role for inducers of electrophile detoxication enzymes, principally the Phase II enzymes, as anticarcinogens. To illustrate the effectiveness of this strategy, a discussion is also presented on the actions of one class of selective inducers of Phase II enzymes, the dithiolethiones, against aflatoxin-induced hepatocarcinogenesis.

MECHANISMS OF PHASE II ENZYME INDUCTION

Although it has been known for many years that antioxidants such as butylated hydroxytoluene (BHT), butylated hydroxyanisole (BHA), and ethoxyquin exert an anticarcinogenic effect when given simultaneously (or prior or both) with a carcinogen, there have been few experiments designed to investigate the mechanisms of such protective actions. Possible mechanisms to explain the protective effects of these compounds might include (a) trapping of either electrophilic or free radical intermediates; (b) induction of Phase I enzymes (i.e., cytochrome P-450s) to enhance carcinogen detoxication; (c) inhibition of Phase I enzymes to retard metabolic activation; and (d) induction of Phase II xenobiotic-metabolizing enzymes [e.g., NAD(P)H:quinone reductase, glutathione S-transferases] to enhance carcinogen detoxication and elimination. Although there is experimental evidence to support each of these mechanisms for selected agents, only the last--induction of Phase II enzymes--accounts for the protective actions of the many chemically diverse classes of chemoprotectors that block tumor initiation.

One of the earliest studies to implicate a role for the induction of Phase II enzymes, particularly glutathione S-transferases, in the protective actions of antioxidants was that of Benson and coworkers (1). They showed that liver cytosols from BHA- or ethoxyquin-fed rats or mice exhibited much higher glutathione S-transferase activities than controls and that cytosols prepared from the livers of these rodents eliminated the mutagenic activity in urine from mice treated with benzo(a)pyrene. Subsequent studies demonstrated that dietary administration of antioxidants increased glutathione S-transferase activity in extrahepatic tissues such as lung, stomach, small intestine, and kidney (2). As summarized in Tab. 1, there is now substantial evidence to support the view that induction of Phase II enzymes is a critical and sufficient mechanism to engender protection against the toxic and carcinogenic actions of reactive intermediates. Perhaps most direct are the observations by Fahl and colleagues (20,26) that overexpression of glutathione S-transferases protects against the cytotoxic actions of benzo(a)pyrene and other toxins. Further, the studies of Seidegård et al. (29) suggest that deficiencies in the levels of expression of glutathione S-transferases in humans may be important determinants for susceptibility to lung cancer. It is also notable that several laboratories now effectively use the induction of Phase II enzymes like glutathione S-transferases and quinone reductase to guide the isolation and purification of new classes of naturally occurring anticarcinogens.

Initial studies on the possible molecular mechanisms of induction of glutathione S-transferases by antioxidants were conducted by Pearson et al. (21), who observed a 20-fold increase in mRNA for the major glutathione S-transferase in the livers of mice several days after feeding 0.75% BHA. Benson and colleagues subsequently reported that significant increases in mRNA levels could be observed as early as 24 hours after placing mice on the BHA diet (3). More recently, Pearson et al. (22) have studied the mechanisms of tissue-specific induction of murine glutathione S-transferase mRNAs by BHA. In these studies, measurements of transcription rates in isolated nuclei indicated that increased mRNA levels were due to increased rates of transcription.

The molecular mechanisms regulating the transcriptional activation of Phase II enzymes by antioxidants and other inducers have also been investigated. As originally proposed by Wattenberg (32), two families of Phase II enzyme inducers exist, based upon their ability to elevate Phase I enzymatic profiles. Prochaska and Talalay (25) have coined the terms "bifunctional" and "monofunctional" inducers to describe these compounds.

Bifunctional inducers (e.g., polycyclic hydrocarbons, dioxins, azo dyes, and flavones) can all be characterized as large planar polycyclic aromatics and elevate Phase II as well as selected Phase I enzymatic activities, such as aryl hydrocarbon hydroxylase. These compounds are potent ligands for the Ah (aryl hydrocarbon) receptor, and the direct participation of the Ah receptor in the activation of aryl hydrocarbon hydroxylase gene transcription has been demonstrated (13). Moreover, since Phase II enzyme inducibility by bifunctional inducers segregates in mice that possess functional Ah receptors, it had been presumed that these enzymes were under the direct control of the Ah receptor. Monofunctional inducers (phenols, lactones, isothiocyanates, dithiocarbamates, and 1,2-dithiole-3-thiones) elevate Phase

Table 1. Evidence for major role of induction of Phase II enzymes in chemo protection

1. Many chemoprotectors are most effective if administered prior to carcinogens

2. Treatment with chemoprotectors profoundly alters carcinogen metabolism

3. Induced Phase II enzymes inactivate electropniles (ultimate carcinogens)

4. Chemoprotection is achieved against a wide variety of carcinogens, suggesting a low specificity mechanism

5. Enzyme induction and chemoprotection are produced by the same compounds (of many chemical classes), occur at similar doses, and have similar tissue specificities

6. Overexpression of glutathione *S*-transferase by cDNA transfection protects cells against carcinogen toxicity

7. Deficiencies in the levels of expression of glutathione *S*-transferases may be important determinants for susceptibility to cancer

8. Monitoring of enzyme induction has led to the recognition or isolation of novel chemoprotectors

---

II enzymatic activities without significantly elevating the afore-mentioned Phase I activities, and do not possess an obvious defining structural characteristic. These distinctions in enzyme induction are illustrated in Fig. 1, which indicates that the bifunctional inducer, β-naphthoflavone, induces NAD(P)H:quinone reductase activity six-fold in the parental Hepa 1c1c7 cell line, but is without effect in the Ah- or aryl hydrocarbon hydroxylase gene transcript-impaired mutants. By contrast, the monofunctional inducer 1,2-dithiole-3-thione elevates quinone reductase enzyme levels equally in parental and mutant cell lines. Since Ah-dependent Phase I enzymes can activate procarcinogens to their ultimate reactive forms, we anticipate that monofunctional inducers would be more desirable candidates for chemoprevention in man.

Early studies with alkylethers of hydroquinones in vivo and diphenols in vitro suggested that diphenols, such as BHA, mediate their inductive effects via a chemical signal (23,24). It was also suggested that bifunctional inducers induce Phase II enzymes in part via a "metabolic cascade" wherein bifunctional inducers were metabolized by induced Phase I enzymes to species resembling monofunctional inducers. Talalay et al. (31) have identified the chemical signal present in all monofunctional inducers: the presence or acquisition of an electrophilic center. Many compounds are Michael reaction acceptors (e.g., an olefin conjugated to an electron withdrawing group) and potency is generally paralleled by their efficiency as Michael reaction acceptors and their ability to act as substrates for glutathione *S*-transferases (30). These generalizations can account for the inducer activity of many types of chemoprotectors and have led to the identification of other novel classes of inducers including acrylates, fumarates, maleates, vinyl ketones, and vinyl sulfones.

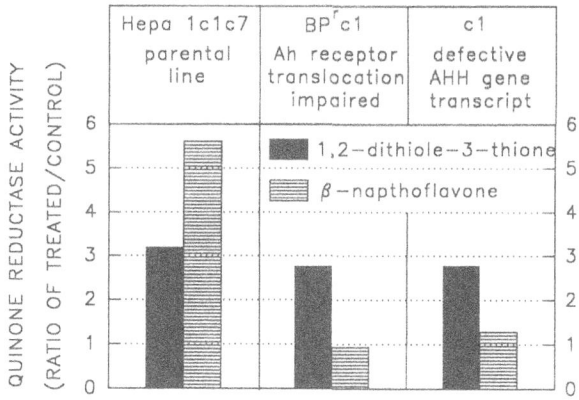

Figure 1. Effects of β-naphthoflavone (5 μM) or 1,2-dithiole-3-thione (30 μM) on the quinone reductase activity of Hepa 1c1c7 (wild-type), BP^{r}c1 (defective translocation of Ah receptor-ligand complex into nucleus), and c1 (defective cytochrome P-450 IA1 gene) cell lines. Adapted from Prochaska and Talalay (25).

The laboratories of Pickett and Daniel have recently identified several regulatory elements controlling the expression and inducibility of the Ya subunit of murine and rat glutathione *S*-transferases by bifunctional and monofunctional inducers (14,28). A 41 bp element in the 5'-flanking region of the rat glutathione S-transferase Ya gene, termed the "Antioxidant Response Element" (ARE), and a homologous element in the 5'-region of the murine Ya gene, designated the "Electrophile Response Element" (EpRE), have been identified using a series of S' deletion mutants fused to the chloramphenicol acetyltransferase gene and then transfected into hepatoma cells. Both elements are responsive to monofunctional inducers such as *tert*-butyl hydroquinone, and induction does not require the presence of functional Ah receptors or aryl hydrocarbon hydroxylase activity. By contrast, bifunctional inducers (e.g., β-naphthoflavone) were able to act through this sequence only if the cells contained functional Ah receptors and aryl hydrocarbon hydroxylase activity. Thus, the differences and similarities (e.g., conversion of bifunctional inducers to metabolites resembling monofunctional inducers) for mechanisms of induction by mono- and bi-functional inducers derived by classical pharmacological approaches are being borne out by molecular analyses. DNA footprinting and gel shift assays have recently established specific interactions between nuclear proteins and the ARE and EpRE regulatory elements. However, the identity and exact role of these proteins in the induction pathway remain to be elucidated. Undoubtedly, a full understanding of the expression and regulation of Phase II enzymes will permit the de novo synthesis or isolation from plants of highly selective and potent chemicals that may serve as potential chemoprotective agents.

INHIBITION OF AFLATOXIN HEPATOCARCINOGENESIS BY DITHIOLETHIONES

Experimental hepatocarcinogenesis in rodents can be inhibited by a number of antioxidants and is particularly suited for mechanistic studies. A brief discussion of the impact of dithiolethiones on aflatoxin-induced liver cancers will serve to illustrate some of the enzyme-inducing and anticarcinogenic properties of this newest class of chemoprotective antioxidants. Dithiolethiones are five-membered cyclic sulfur-containing compounds with radioprotective, chemoprotective, and chemotherapeutic activities. For example, the drug oltipraz [4-methyl-5-(2-pyrazinyl)-1,2-dithiole-3-thione] shows significant antischistosomal activity in experimental animals and in humans. During the course of studies on the mechanisms of schistosomicidal activity of oltipraz, Bueding et al. (5) noted that administration of this drug as well as several analogues to mice resulted in marked elevations of the activities of Phase II enzymes in hepatic and extrahepatic tissues. These findings led Bueding to predict that dithiolethiones such as oltipraz might be excellent candidate compounds for cancer chemoprotection studies. As recently summarized elsewhere (4,17), oltipraz has subsequently proven to be an effective anticarcinogen in breast, colon, pancreas, lung, forestomach, skin, bladder, and liver tumor models. As a consequence, oltipraz is presently undergoing Phase I clinical trial in the United States (18). It is important to note that dithiolethiones represent the first class of compounds for which anticarcinogenic activity was predicted from biochemical properties, namely, their ability to induce Phase II enzymes.

Aflatoxin is a potent hepatotoxin and carcinogen in a wide variety of animals and is linked epidemiologically with a high incidence of primary hepatocellular carcinoma in humans (6). Elimination of aflatoxins from the human food supply throughout the world will be extremely difficult, and chemoprotection offers an attractive alternative for populations at high risk for aflatoxin-induced diseases. A number of classic chemoprotective agents, notably BHA, BHT, and ethoxyquin, inhibit aflatoxin carcinogenesis in rats when fed simultaneously with the carcinogen (7,33). A search for protective agents more amenable for use in man led us to evaluate oltipraz in this rat model. After feeding male F344 rats either purified diet or diet supplemented with 0.075% oltipraz for 1 week, the animals received 10 intragastric doses of aflatoxin B_1 (25 µg/rat/day; days 8-12 and 15-19). One week after cessation of dosing, all animals were restored to the control diet and maintained until they became moribund or the study terminated at 23 months. This 10-dose exposure to aflatoxin B_1 produced an 11% incidence of hepatocellular carcinomas in the control animals, while an additional 9% of these rats had hyperplastic nodules in their livers. This incidence of hepatic disease mirrors the lifetime incidence of hepatocellular carcinoma in humans in high-risk areas of China, Southeast Asia, and Africa (6,34). In the rat intervention study, dietary oltipraz afforded complete protection against aflatoxin-induced hepatocellular carcinomas and hyperplastic nodules. Further, no tumors were seen in either group at extrahepatic sites, indicating that oltipraz did not serve to merely shift target organ specificity from the liver to other tissues.

These protective actions of oltipraz (as well as the food antioxidants) are thought to result primarily from an altered balance between the activation and detoxication of aflatoxin in the hepatocyte. Aflatoxin undergoes metabolic activation to a highly reactive 8,9-epoxide that can bind to macromolecules, particularly DNA. This epoxide can also be conjugated to glutathione through the actions of glutathione S-transferases yielding a detoxication product. In the rat, the induction of glutathione S-transferases through transcriptional activation is a prominent biochemical effect of dietary oltipraz treatment (9). Correspondingly, aflatoxin treatment of rats maintained on oltipraz-supplemented diets results in large increases in the rate of biliary elimination of aflatoxin-glutathione conjugates, demonstrating that altered metabolic processing of aflatoxin has occurred in vivo. Of more direct consequence to anticarcinogenesis, treatment with oltipraz leads to greatly diminished levels of aflatoxin modification of hepatic (and extrahepatic) DNA following either single or multiple exposures to this carcinogen (16). Presumably, the reduction in DNA adduct formation leads to comparable reductions in aflatoxin-induced mutations in potentially critical genes such as K-*ras* and p53, although this has not as yet been directly examined in vivo. The overall role for induction of Phase II enzymes in protection against aflatoxin hepatocarcinogenesis is highlighted by the finding that there exists a striking correlation between the degree of induction of hepatic glutathione S-transferases by structurally distinct antioxidants and the degree of chemoprotection as judged by reduced levels of the major aflatoxin-DNA adduct, aflatoxin-N^7-guanine, in rat liver DNA (15).

Structure-activity studies indicate that the enzyme inductive and chemoprotective properties of oltipraz are embodied in the dithiolethione nucleus. Thus, the unsubstituted 1,2-dithiole-3-thione is a potent, monofunctional enzyme inducer in vivo and is an effective inhibitor of aflatoxin B_1-induced tumorigenesis (12). Dietary concentrations of 0.03% engender > 99% reductions in the hepatic burden of preneoplastic lesions, notably, glutathione S-transferase P and γ-glutamyltranspeptidase-positive foci. Comparable effects are seen with feeding of equimolar concentrations of oltipraz (16). As shown in Fig. 2, feeding of 0.03% 1,2-dithiole-3-thione reduces the levels of hepatic aflatoxin-N^7-guanine adducts by 80% at both 2 and 24 hours after dosing with aflatoxin. The 2-hr post-dosing time represents the time of maximal levels of hepatic aflatoxin-DNA adducts. As is evident from the lower levels of DNA adducts at 24 hours, substantial amounts of the aflatoxin-N^7-guanine adduct are excised and are eliminated in the urine without further metabolic processing. The reduction in the amount of aflatoxin adducts excreted into the urines of these animals fed 1,2-dithiole-3-thione is remarkably similar to the degree of diminution in hepatic DNA adduct levels observed in these rats, and presumably directly reflects the decreased formation of aflatoxin-N^7-guanine in the liver, and possibly other tissues. We are presently evaluating whether this urinary biomarker may serve as a short-term, noninvasive prognosticator of chemoprotective efficacy in reducing risk of aflatoxin-induced disease.

PROSPECTS

It is well established that inducers of Phase II enzymes can serve as effective anticarcinogens in a multitude of animal models. As studies with dithiolethiones described above highlight, substantial protection against acute toxicity and cancer can be achieved in these models with the concurrent administration of monofunctional enzyme inducers that appear to exhibit a high degree of specificity in their mechanisms of action. An additional attribute is that this feature minimizes the potential complications of enhanced procarcinogen activation through the ancillary induction of certain Phase I enzymes such as cytochrome P-450s, as can occur with bifunctional inducers. Additional issues of efficacy, potency, low toxicity, and human safety suggest a potential role for drugs such as oltipraz in initial human chemoprotection interventions. Toward this end, oltipraz or other dithiolethiones may be particularly attractive compounds for interventions in individuals at high risk for aflatoxin exposure. A key feature of this exposure group has been the concurrent develop-

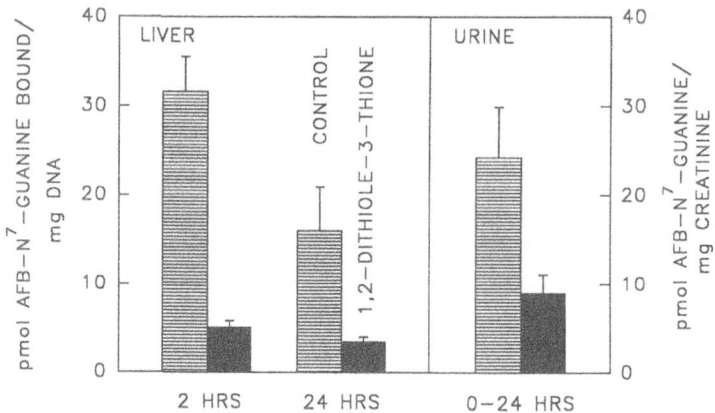

Figure 2. Levels of aflatoxin-N[7]-guanine in liver and urine of rats fed either control or 1,2-dithiole-3-thione-supplemented (0.03%) diets for 7 days prior to gavage with 250 µg aflatoxin B_1/kg/rat. Values represent means ± S.E. of determinations on 4 rats. Aflatoxin-N[7]-guanine levels were determined by antibody affinity chromatography followed by HPLC as described elsewhere (27).

ment of noninvasive biomarkers to determine levels of internal dose, metabolites, and genotoxicity of the aflatoxins. With such analytical tools it will be possible to design short-term clinical trials to directly test the hypothesis that induction of Phase II enzymes in man leads to a reduced risk for neoplasia. Over the long run, the development of Phase II enzyme inducers as chemoprotective agents in humans will be greatly facilitated by an increased understanding of the roles of these enzymes in carcinogenesis and the mechanisms that regulate their expression. Although such agents have pronounced efficacy as single agents, it also seems likely that their ultimate utility in humans will come in combination chemoprotection regimens employing agents with diverse mechanisms of action.

ACKNOWLEDGEMENTS

The authors gratefully acknowledge financial support for our studies on cancer chemoprotection from the National Institutes of Health (CA 39416 and CA 44530) and the American Cancer Society (SIG-3). T.W.K. and J.D.G. are recipients of NIH Research Career Development Awards CA 01230 and CA 01517, respectively.

REFERENCES

1.   Benson, A.M., R.P. Batzinger, S.-Y.L. Ou, E. Bueding, Y.-N. Cha, and P. Talalay (1978) Elevation of hepatic glutathione *S*-transferase activities and protection against mutagenic metabolites of benzo(a)pyrene by dietary antioxidants. *Cancer Res.* 38:4486-4495.

2.   Benson, A.M., Y.-N. Cha, E. Bueding, H.S. Heine, and P. Talalay (1979) Elevation of extrahepatic glutathione *S*-transferases and epoxide hydratase activities by 2(3)-*tert*-butyl-4-hydroxyanisole. *Cancer Res.* 39:2971-2977.

3.   Benson, A.M., M.J. Hunkeler, and J.F. Morrow (1984) Kinetics of glutathione transferase messenger RNA, and reduced nicotinamide adenine dinucleotide (phosphate): Quinone reductase induction by 2(3)-*tert*-butyl-4-hydroxyanisole in mice. *Cancer Res.* 44:5256-5261.

4.   Boone, C.W., G.J. Kelloff, and W.E. Malone (1990) Identification of candidate cancer chemopreventive agents and their evaluation in animal models and human clinical trials. *Cancer Res.* 50:2-9.

5.   Bueding, E., P. Dolan, and J.P. Leroy (1982) The antischistosomal activity of oltipraz. *Res. Comm. Chem. Pathol. Pharmacol.* 37:293-303.

6.   Busby, W.F., Jr., and G.N. Wogan (1984) Aflatoxins. In *Chemical Carcinogens, Second Edition,* C.E. Searle, ed. Amer. Chem. Soc., Washington, D.C., pp. 945-1136.

7.   Cabral, J.R.P., and G.E. Neal (1983) The inhibitory effects of ethoxyquin on the carcinogenic action of aflatoxin B_1 in rats. *Cancer Lett.* 19:125-132.

8.   Conney, A.H. (1982) Induction of microsomal enzymes by foreign chemicals and carcinogenesis by polycyclic aromatic hydrocarbons: G.H.A. Clowes Memorial Lecture. *Cancer Res.* 42:4875-4917.

9.   Davidson, N.E., P.A. Egner, and T.W. Kensler (1990) Transcriptional control of glutathione S-transferase gene expression by the chemoprotective agent 5-(2-pyrazinyl)-4-methyl-1,2-dithiole-3-thione (oltipraz) in rat liver. *Cancer Res.* 50:2251-2255.

10. De Flora, S., and C. Ramel (1988) Mechanisms of inhibitors of mutagenesis and carcinogenesis. Classification and overview. *Mutat. Res.* 202:285-306.

11. De Long, M.J., A.B. Santamaria, and P. Talalay (1987) Role of cytochrome P_1-450 in the induction of NAD(P)H:quinone reductase in a murine hepatoma cell line and its mutants. *Carcinogenesis* 8:1549-1553.

12. Egner, P.A., P. DeMatos, J.D. Groopman, and T.W. Kensler (1990) Effect of 1,2-dithiole-3-thione, a monofunctional enzyme inducer, on aflatoxin-DNA adduct formation in rat liver. *Proc. Amer. Assoc. Cancer Res.* 31:119.

13. Fisher, J.M., L. Wu, M.S. Denison, and J.P. Whitlock, Jr. (1990) Organization and function of a dioxin-responsive enhancer. *J. Biol. Chem.* 265:9676-9681.

14. Friling, R.S., A. Bensimon, Y. Tichauer, and V. Daniel (1990) Xenobiotic-inducible expression of murine glutathione S-transferase Ya subunit gene is controlled by an electrophile-responsive element. *Proc. Natl. Acad. Sci., USA* 87:6258-6262.

15. Kensler, T.W., P.A. Egner, M.A. Trush, E. Bueding, and J.D. Groopman (1985) Modification of aflatoxin B_1 binding to DNA *in vivo* in rats fed phenolic antioxidants, ethoxyquin and a dithiolthione. *Carcinogenesis* 6:759-763.

16. Kensler, T.W., P.A. Egner, P.M. Dolan, J.D. Groopman, and B.D. Roebuck (1987) Mechanism of protection against aflatoxin tumorigenicity in rats fed 5-(2-pyrazinyl)-4-methyl-1,2-dithiol-3-thione (oltipraz) and related 1,2-dithiol-3-thiones and 1,2-dithiol-3-ones. *Cancer Res.* 47:4271-4277.

17. Kensler, T.W., J.D. Groopman, and B.D. Roebuck (1991) Chemoprotection by oltipraz and other dithiolethiones. In *Cancer Chemopreventive Agents,* M. Lipkin and L. Wattenberg, eds. CRC Press, Boca Raton, Florida (in press).

18. Malone, W.F., G.J. Kelloff, C.W. Boone, and V.E. Steele (1991) Chemoprevention clinical trials. *Proc. Amer. Assoc. Cancer Res.* 32:478-479.

19. Monoharan, T.H., R.B. Puchalski, J.A. Burgess, C.B. Pickett, and W.E. Fahl (1987) Promoter-glutathione S-transferase Ya cDNA hybrid genes: Expression

and conferred resistance to an alkylating molecule in mammalian cells. *J. Biol. Chem.* 262:3739-3745.

20.   Miller, E.C., and J.A. Miller (1985) Some historical perspectives on the metabolism of xenobiotic chemicals to reactive electrophiles. In *Bioactivation of Foreign Compounds,* M.W. Anders, ed. Academic Press, New York, New York, pp. 1-28.

21.   Pearson, W.R., J.J. Windle, J.F. Morrow, A.M. Benson, and P. Talalay (1983) Increased synthesis of glutathione *S*-transferase in response to anticarcinogenic antioxidants. Cloning and measurement of messenger RNA. *J. Biol. Chem.* 258:2052-2062.

22.   Pearson, W.R., J. Reinhart, S.C. Sisk, K.S. Anderson, and P.N. Adler (1988) Tissue-specific induction of murine glutathione transferase mRNAs by butylated hydroxyanisole. *J. Biol Chem.* 263:13324-13332.

23.   Prochaska, H.J., H.S. Bregman, M.J. DeLong, and P. Talalay (1985) Specificity of induction of cancer protective enzymes by analogues of *tert*-butyl-4-hydroxyanisole (BHA). *Biochem. Pharmacol.* 34:3909-3914.

24.   Prochaska, H.J., M.J. De Long, and P. Talalay (1985) On the mechanisms of induction of cancer-protective enzymes: A unifying proposal. *Proc. Natl. Acad. Sci., USA* 82:8232-8236.

25.   Prochaska, H.J., and P. Talalay (1988) Regulatory mechanisms of monofunctional and bifunctional anticarcinogenic enzyme inducers in murine liver. *Cancer Res.* 48:4776-4782.

26.   Puchalski, R.B., and W.E. Fahl (1990) Expression of recombinant glutathione *S*-transferase π, Ya, or Yb_1 confers resistance to alkylating agents. *Proc. Natl. Acad. Sci., USA* 87:2443-2447.

27.   Roebuck, B.D., Y.-L. Liu, A.E. Rogers, J.D. Groopman, and T.W. Kensler (1992) Protection against aflatoxin B_1-induced hepatocarcinogenesis in rats by 5-(2-pyrazinyl)-4-methyl-1,2-dithiole-3-thione (oltipraz): Predictive role for short-term molecular dosimetry. *Cancer Res.* (in press).

28.   Rushmore, T.H., R.G. King, K.E. Paulson, and C.B. Pickett (1990) Regulation of glutathione *S*-transferase Ya subunit gene expression: Identification of a unique xenobiotic-responsive element controlling inducible expression by planar aromatic compounds. *Proc. Natl. Acad. Sci., USA* 87:3826-3830.

29.   Seidegård, J., R.W. Pero, D.G. Miller, and E.J. Beattie (1986) A glutathione transferase in human leukocytes as a marker for the susceptibility to lung cancer. *Carcinogenesis* 7:751-753.

30.   Spencer, S.R., L. Xue, E.M. Klenz, and P. Talalay (1991) The potency of inducers of NAD(P)H:(quinone-acceptor) oxidoreductase parallels their efficiency as substrates for glutathione transferases. Structural and electronic considerations. *Biochem. J.* 273:711-717.

31.   Talalay, P., M.J. De Long, and H.J. Prochaska (1988) Identification of a common chemical signal regulating the induction of enzymes that protect against chemical carcinogenesis. *Proc. Natl. Acad. Sci. USA* 85:8261-8265.

32.   Wattenberg, L.W. (1985) Chemoprevention of cancer. *Cancer Res.* 45:1-8.

33.   Williams, G.M., T. Tanaka, and Y. Maeura (1986) Dose-related inhibition of aflatoxin$_1$ induced hepatocarcinogenesis by the phenolic antioxidants, butylated hydroxytoluene and butylated hydroxyanisole. *Carcinogenesis* 7:1043-1050.

34.   Zhu, Y.-R., J.-G. Chen, and X.-Y. Huang (1988) Hepatocellular carcinoma in Qidong County. In *Primary Liver Cancer,* Z.-Y. Tang, N.-C. Wu, and S.-S. Xia, eds. China Academic Publishers, Beijing, pp. 204-222.

PROTECTIVE ROLE OF GLUTATHIONE, THIOLS, AND ANALOGUES

IN MUTAGENESIS AND CARCINOGENESIS

Bengt Jernström, Ralf Morgenstern, and Peter Moldéus

Department of Toxicology
Karolinska Institute
Box 60400
S-104 01 Stockholm, SWEDEN

INTRODUCTION

Glutathione, γ-L-glutamyl-L-cysteinyl-glycine (GSH), is the most important nonprotein thiol present in animal cells as well as in most plants and bacteria (3, 27). GSH is characterized by its reactive thiol group and its γ-glutamyl bond which makes it resistant to normal peptidase activity. In mammalian cells, the concentration of GSH is often very high (in the millimolar range); most of the intracellular GSH exists in the thiol form, although mixed disulfides (mainly G-SS-protein), thioethers, and, to a lesser extent, GSH disulfide (GSSG) contribute to the total cellular pool of GSH. The bulk of intracellular GSH is found in the cytosol, but the existence of minor mitochondrial and nuclear pools of GSH has been demonstrated. Moreover, both GSH and GSSG are present in various body fluids, including plasma, bile, and the glomerular filtrate, although at much lower concentrations.

The interest in research on GSH has increased markedly during recent years. This is mainly due to the discovery of the involvement of this tripeptide in a number of important cell functions. Thus, known physiological functions of GSH now include the maintenance of membrane integrity and cytoskeletal organization, involvement in protein and DNA synthesis, and the modulation of protein conformation and enzyme activity, as well as promotion of neurotransmitter release (25).

One of the most important functions of GSH is related to its role in drug biotransformation. GSH, or rather its thiolate anion, is a strong nucleophile, and electrophilic metabolites of drugs and carcinogens are often inactivated by the formation of GSH conjugates. GSH also serves as a reductant in the metabolism of various peroxides and free radicals. These reactions may occur spontaneously, but are usually catalyzed by glutathione transferases (GST) and glutathione peroxidases, respectively (Tab. 1).

GLUTATHIONE TRANSFERASE-CATALYZED REACTIONS

It is established that a crucial step in chemical carcinogenesis involves covalent binding of electrophilic metabolites to critical nucleophilic targets in DNA (51, 53). Furthermore, it is established that GSH, or rather the corresponding thiolate anion, can act as a competing nucleophile (4,8) and thus offers protection against potentially genotoxic products.

GSH may react spontaneously with electrophiles or, which is more common with respect to the chemical nature of DNA-binding intermediates, in conjunction with GSTs. The GSTs are products of a supergene family and exist as homo- or heterodimeric combinations of a number of subunits. For instance, in the rat, which is the most studied species with regard to GSTs, 12 different subunits with molecular weights of about 24-25 kD have been characterized and more forms are likely to be discovered. The different subunits can be grouped into different classes according to the homology of their amino acid sequence. At present, four classes of GSTs (alpha, mu, pi, and theta) have been established. Combinations of the subunits within each class form the catalytically active enzymes in which each subunit functions independently of the others (24,29,31). The subunits characterized in rat and human are compiled in Tab. 2.

In addition, a membrane-bound GST, which shares the functional characteristics of but is structurally distinct from the cytosolic enzymes, has been described (35). GST activity has been found in all animal species and tissues studied, and the activity as well as the composition of the GST isoenzymes vary considerably. In general, the liver shows high activity in accordance with the detoxifying function of this organ. Particular subunits may be more abundant in certain tissues than in others. For instance, GST-containing subunit 7 in the rat or the corresponding pi-class form in man, GST π, is lacking in normal liver but present in high concentrations in extrahepatic tissues, such as the lung, skin, and intestine, and in certain blood cells like leukocytes (24,29). The activity of GST and the isoenzymes may also vary within a tissue as well as within a cell. For instance, the area around the central vein in the liver exhibits a different distribution of isoenzyme than the periportal region (41). Similarly, the cell nucleus contains a different spectrum of GST isoenzymes than the cytosol (22).

A number of substances, naturally occurring as well as synthetic, have been shown to induce GSTs in the liver as well as in extrahepatic tissues. The synthetic inducers include polycyclic aromatic hydrocarbons (PAH) and antioxidants, such as butylated hydroxytoluene, butylated hydroxyanisole, and ethoxyquin; and these compounds preferentially increase the concentration of GSTs containing subunits 1 and 3. Naturally occurring inducers include derivatives of indoles and coumarins. In many cases substances that induce GSTs have been shown to inhibit carcinogenesis (24).

The GST system catalyzes the conjugation of GSH with a great number of electrophilic substrates. These include alkene- and arene oxides, alkyl- and aryl halides or other compounds with good leaving groups, substances with carbonyl functions, and compounds that either spontaneously or after metabolism form carbocations (4,8). The broad substrate specificity of the GST system and the

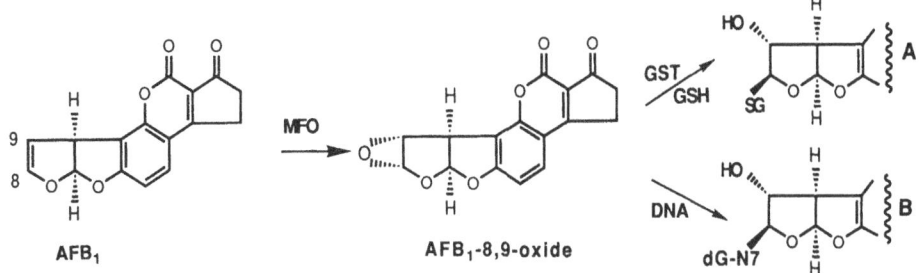

Figure 1. Metabolic activation of aflatoxin B_1 (AFB_1) to AFB_1-8,9-oxide. MFO: mixed function oxidase. GSH: glutathione. GST: glutathione transferase. A: GSH-conjugate. B: the principal DNA-adduct.

adaptive feature of the system strongly implicate an important protective role for it against potentially toxic and carcinogenic electrophiles.

Surprisingly little information is available on the protective role of GST/GSH against biologically relevant substances in normal cells and tissues. A few examples are, however, known. The fungal metabolite aflatoxin B_1 (AFB_1), a well-documented liver carcinogen in various animal species and man, is tumorigenic after metabolic conversion to the corresponding AFB_1-8, 9-epoxide and the subsequent covalent binding to N-7 in deoxyguanosine (dG) in critical DNA-sequences (Fig. 1) (10). In rodents, the extent of DNA-binding and incidence of liver tumors can be substantially reduced by treating the animals with the antioxidant ethoxyquin. This compound preferentially induces GST subunits 1 and 3. The inhibitory effect is, in part, ascribed to an increased rate of removal of AFB_1-8, 9-epoxide, due to the induction of GST 1-1: the isoenzyme demonstrating the highest catalytic efficiency with AFB_1-8, 9-epoxide (9,21).

Another class of carcinogens against which the GST/GSH system seems to provide protection is PAH. In general, the ultimate carcinogens of these compounds have been identified as bay- or fjord-region diolepoxides (DE) and, in particular, those with R, S epoxide absolute structure (14,53). The mutagenic and tumorigenic effects of PAH are closely associated with covalent binding of the DE to the exocyclic amino groups, N^2 and N^6 in dG or deoxyadenosine (dA), respectively. The preference for dG or dA depends on the actual DE-derivative. Previous studies have shown that the extent of DNA-binding derived from benzo(a)pyrene (BP) (Fig. 2) in isolated hepatocytes can be significantly increased by pretreating the cells with diethylmaleate, a compound that effectively depletes the cells of GSH (5,19). These results are consistent with a protective role of GST.

The ultimate carcinogen of BP has been identified as (+)-*anti*-BP-7, 8-dihydrodiol 9, 10-epoxide [(+)-*anti*-BPDE] (Fig. 2), and studies employing purified GSTs from rat liver, have demonstrated that all GST isoenzymes are active towards this electrophile although with a great difference in catalytic efficiency. The Mu-class GST, isoenzymes containing subunit 4, are most efficient in catalyzing the reaction with *anti*-BPDE and in particular the carcinogenic (+)-enantiomer (>90%

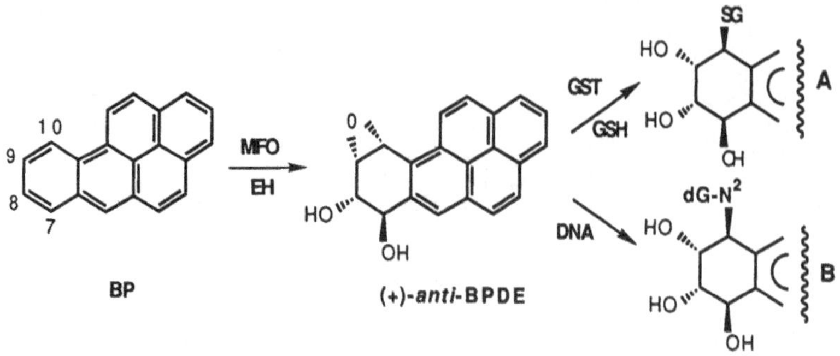

Figure 2. Metabolic activation of benzo(a)pyrene (BP) to (+)-*anti*-BP-7,8-dihydrodiol-9,10 epoxide [(+)-*anti*-BPDE]. MFO: mixed function oxidase. EH: epoxide hydrolase. GSH: glutathione. GST: glutathione transferase. A: GSH-conjugate. B: the principal DNA-adduct.

preference). Overall, the most active GST towards *anti*-BPDE in the rat is GST 7-7 (20,42).

More direct proof of the involvement of GST in protection against electrophilic products has been provided by transfection of cultured mammalian cells with cloned GST c-DNA or employing mammalian cell lines that differ with regard to overall GST activity and isoenzyme composition. With the first approach, transfection with cloned GST π, GST 1-1 or 3-3 confers partial resistance to cytotoxicity of *anti*-BPDE and the alkylating anti-cancer drugs chlorambucil, melphalan, and cisplatin (36,40). GST π demonstrated the highest protective effect against *anti*-BPDE, as expected considering the high catalytic efficiency of this human GST isoenzyme with *anti*-BPDE compared to the rat isoenzymes GST 1-1 and 3-3 (20,42,43). With the second approach, the role of GST in protection against mutations was studied by incubating the (-)-*trans*-BP-7, 8 dihydrodiol (the precursor of (+)-*anti*-BPDE) with a low GST-containing cell line (MCF-7) or a high GST-containing cell line (H4IIE) in the presence of V79 cells (devoid of significant amounts of GST). Both MCF-7 and H4IIE have the capacity to metabolize the dihydrodiol to *anti*-BPDE and contain similar concentrations of GSH. The results clearly demonstrate that H4IIE cells generate much less mutagenic product, as revealed by the number of mutations induced in V79 cells, than the MCF-7 cells (44). The results are easily explained by the much higher content of GST, and in particular GST 7-7, in the H4IIE-cells and, thus, efficient removal of *anti*-BPDE by conjugation with GSH.

To our knowledge, no direct proof of the protective role of GST against carcinogenesis in humans is available. However, based on epidemiological data, it has been suggested that cigarette-smoking individuals lacking mu-class GST isoenzymes are at a higher risk of developing lung cancer (48). More recently, it was demonstrated that a significantly greater proportion of individuals with adenocarcinoma of the stomach and colon were lacking mu-class GST (51). Taken together, these results implicate an important protective role of GST in certain human cancers.

140

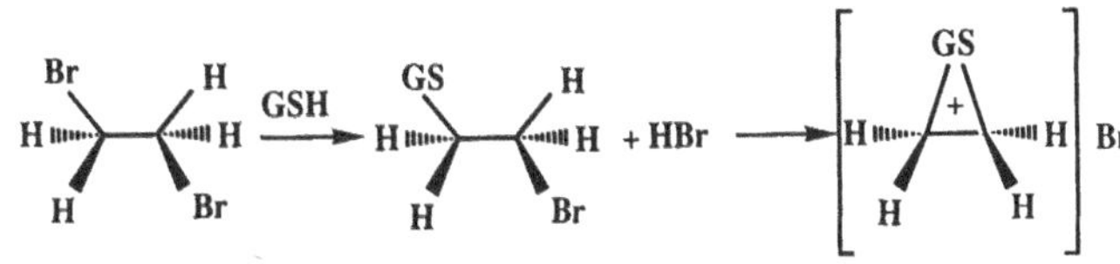

Figure 3. Glutathione (GSH)-dependent activation of dibromoethane to episulfonium anion.

Although GST-catalyzed conjugation with GSH is normally regarded as a detoxifying metabolic pathway, there are now several examples of the reverse, for example, formation of GSH conjugates that are more reactive and toxic than the compounds from which they are formed. One group of substrates that may undergo GSH-dependent metabolic activation is the dihaloalkanes (1). Dibromoethane has, for instance, been shown to be metabolized by a GSH-mediated reaction to yield the 1-glutathionyl-2-bromoethane intermediate. Through internal displacement of the second bromine atom, a reactive episulfonium ion is formed which then serves as the alkylating intermediate (Fig. 3).

It appears that it is primarily the kidney that is susceptible to the toxic effect of GSH and cysteine conjugates. Thus, GSH conjugates of the halogenated hydrocarbons hexachlorobutadiene (HCBD), chlorotrifluoro ethylene (CTFE), tetrafluoro ethylene, and 2-bromohydroquinone are nephrotoxic. A necessary step in the nephrotoxicity of these compounds is the renal processing to the corresponding S-substituted cysteine conjugate. The S-substituted cysteine conjugates may directly form reactive episulfonium ions that probably are the ultimate damaging agents (1,14).

Another mechanism that has received increased attention during the last few years is the one involving the renal cysteine conjugate β-lyase. This enzyme, which is also located in the gut microflora, catalyzes the cleavage of the C-S bond of certain cysteine conjugates, a reaction that may result in the formation of reactive thiols. A mechanism involving the β-lyase has been confirmed for the nephrotoxic cysteine conjugates of HCBD, CTFE, and trichloroethylene (14).

GSH AS A REDUCTANT

Oxygen radicals and hydroperoxides have been implicated in all stages of tumorigenesis (7,26,47). Furthermore, a constant oxidative DNA damage inherent to aerobic life does occur in vivo (47). It is, therefore, highly probable that the GSH-dependent reduction of hydrogen peroxide as well as lipid and DNA hydroperoxides serves an important protective role. For instance, GSH has been reported to inhibit the promotion of skin tumors by phorbol ester (38).

The enzymes that utilize GSH to reduce hydroperoxides can be divided into Selenium (Se)-dependent and non-Se-dependent GSH peroxidases. Two forms of cytosolic Se-dependent GSH peroxidase have been described (6,52). These can reduce hydrogen peroxide and lipid hydroperoxides. One of the forms, the Se-dependent phospholipid hydroperoxide GSH peroxidase, can also reduce phospholipid hydroperoxides and cholesterol hydroperoxides directly in the membrane and lipoprotein particles and prevent lipid peroxidation (52,57).

The non-Se-dependent GSH peroxidase activity is carried to varying degrees by GSTs that do not use hydrogen peroxide as a substrate (55). The cytosolic enzymes are characterized by their inability to reduce phospholipid hydroperoxides but can act on fatty acid hydroperoxides released from membranes by the action of phospholipase A_2 (49). In addition, cytosolic (and nuclear) GSTs can reduce DNA hydroperoxides (23), which can be mutagenic in the presence of metal ions (54). The membrane-bound "microsomal" GST reduces lipid and phospholipid hydroperoxides and has been suggested to play a role in the GSH-dependent inhibition of lipid peroxidation observed in liver microsomes (37). In addition, GSTs can conjugate the highly cytotoxic and mutagenic hydroxyalkenols that are formed during the peroxidative breakdown of lipids (29).

In conclusion, GSH is used in a whole array of enzymatic functions to detoxify mediators of oxidative stress that can potentially contribute to the carcinogenic process. GSH may also nonenzymatically serve as a reductant toward free radical intermediates. For instance, GSH has been shown to reduce the oxygen radicals O_2 . and ·OH in in vitro systems (45). This reaction leads to the formation of thiol radicals and eventually to GSSG formation. Whether or not this type of nonenzymatic free radical reduction occurs in vivo has not yet been determined.

PROTECTIVE EFFECTS OF EXOGENOUS THIOLS

Attempts have been made to design chemical protective agents that augment the function of GSH. These agents may be roughly divided into two groups: agents that compete with GSH for chemical reaction with electrophiles, and agents that augment the activity of GSH by supporting its intracellular biosynthesis. An example of the latter case is the cyclic cysteine precursor oxathiozolidine (OTZ), which has been shown to elevate intracellular GSH levels in many cell types. For instance, OTZ increases the resistance of bovine pulmonary endothelial cells to oxidative insult (56).

An example of an agent that may potentially fill both of the above protective roles is the thiol drug N-acetylcysteine (NAC). This drug is used both orally and intravenously in a variety of clinical therapies, including the treatment of chronic lung diseases such as bronchitis (3), and as an antidote to paracetamol poisoning (28). This drug has also proved to be efficient in relatively high intravenous doses in the treatment of a variety of experimental pathologies resembling human disease states that exhibit endothelial toxicity as a result of oxidative stress, including acute pulmonary oxygen toxicity (39) and septicemia and endotoxin shock (2).

In vitro NAC has been shown to protect cells from the toxicity of a variety of toxic agents including both electrophiles and free radical intermediates (34). In addition, NAC has been demonstrated to inhibit the mutagenicity of a wide variety of both direct-acting carcinogens and procarcinogens (11,13). NAC has also been shown to markedly inhibit the induction of lung tumors in mice by urethane (12).

The mechanism of NAC protection in vitro is most likely due to direct scavenging on reduction or reactive intermediates.

Direct action of NAC in vivo, however, is not likely, since the bioavailability of NAC is extremely low when given orally. A more relevant mechanism for the protective effect in vivo may be that NAC acts as a precursor of GSH and, thus, promotes its biosynthesis. GSH will then serve as the protective agent and detoxify reactive species both enzymatically and nonenzymatically.

MUTAGENICITY OF THIOLS

GSH and cysteine have been shown to be mutagenic in several strains of *Salmonella typhimurium* but only in the presence of kidney post-mitochondrial supernatant (S9) fraction (16,46). The identity of the mutagenic species has not been determined, but it appears to be the same for both GSH and cysteine. Thus, if γ-glutamyl transpeptidase catalyzed hydrolysis is inhibited, GSH is not mutagenic (46). The involvement of oxygen radicals formed through oxidation of cysteine has been implicated due to a protective effect of metal chelators. However, recent studies in which the mutagenicity spectra of cystein and GSH in various strains of *typhimurium* were shown to be different from those of reactive oxygen species refute this hypothesis (17). NAC in high concentration has also been shown to be mutagenic in *typhimurium* in the presence of kidney S9 fraction (18). No gene mutation induced by either thiol was, however, found in V79 Chinese hamster cells, even at very high concentration (16). Thus, the relevance of GSH-, cysteine-, or NAC-induced mutagenicity in vivo is highly questionable.

ACKNOWLEDGEMENTS

The studies from the authors' laboratories were supported by the Swedish Cancer Society, the Swedish Medical Research Council, and the Swedish Work Environment Protection Agency.

REFERENCES

1. M.W. Anders, and L.R. Pohl (1985) Halogenated alkanes. In *Bioactivation of Foreign Compounds*, (M.W. Anders, ed. New York, Academic Press, pp. 284-306.
2. G.R. Bernard, W.D. Lucht, M.L. Niedermeyer, J.R. Snapper, and K.L. Brigham (1983) Effect of *N*-acetylcysteine on the pulmonary response to endotoxin in the awake sheep and upon in vitro granulocyte function. *J. Clin. Invest.* 73:1772-1784.
3. G. Boman, U. Bäcker, S.Larsson, B. Melander, and L. Wåhlander (1983) Oral *N*-acetylcysteine reduces exacerbation rate in chronic bronchitis; report of a trial organised by the Swedish Society for Pulmonary Diseases. *Eur. J. Respir. Dis.* 64:405-415.
4. E. Boyland, and L.F. Chasseaud (1969) The role of glutathione and glutathione S transferases in mercapturic acid biosynthesis. *Adv. Enzymol.* 32:173-219.
5. M.D. Burke, H. Vadi, B. Jernström, and S. Orrenius (1977) Metabolism of benzo(a)pyrene with isolated hepatocytes and the formation and degradation of DNA-binding derivatives. *J. Biol. Chem.* 252:6424-6431.

6. R.F. Burk, M.J. Trumble, and R.A. Lawrence (1980) Rat hepatic cytosolic glutathione-dependent enzyme protection against lipid peroxidation in the NADPH-microsomal lipid peroxidation system. *Biochim. Biophys. Acta* 618:35-41.

7. P.A. Cerutti (1985) Pro-oxidant states and tumor promotion. *Science* 227:375-381.

8. L.F. Chasseaud (1979) The role of glutathione and glutathione transferases in the metabolism of carcinogens and other electrophilic reagents. *Adv. Cancer Res.* 29:175-274.

9. B. Coles, D.J. Meyer, B. Ketterer, C.A. Stanton, and R.C. Garner (1985) Studies on the detoxication of microsomally activated aflatoxin B_1 by glutathione and glutathione transferases in vitro. *Carcinogenesis* 6:693-697.

10. R.G. Croy, J.M. Essigman, V.N. Reinhold, and G.N. Wogan (1978) Identification of the principal aflatoxin B_1-DNA adduct formed in vivo in rat liver. *Proc. Natl. Acad. Sci., USA* 74:1745-1749.

11. S. De Flora, C. Bennicelli, P. Zanacchi, A. Camoirano, A. Morelli, and A. De Flora (1984) In vitro effects of N-acetylcysteine on the mutagenicity of direct-acting compounds and procarcinogens. *Carcinogenesis* 5:505-510.

12. S. De Flora, M. Astengo, D. Serra, and C. Bennicelli (1986) Inhibition of urethane-induced lung tumors in mice by dietary N-acetylcysteine. *Cancer Lett.* 32:235-241.

13. S. De Flora, G.A. Rossi, and A. De Flora (1986) Metabolic, desmutagenic and anticarcinogenic effects of N-acetylcysteine. *Respiration* 50(1):43-49.

14. A. Dipple (1985) Polycyclic aromatic hydrocarbon carcinogenesis: An introduction. In *Polycyclic Hydrocarbons and Carcinogenesis, ACS Symposium Series 283*, R.G. Harvey, ed. American Chemical Society, Washington, D.C. pp. 1-17.

15. A.A. Elfarra, and M.W. Anders (1984) Renal processing of glutathione conjugates. Role in nephrotoxicity. *Biochem. Pharmacol.* 33:3729-3732.

16. H.R. Glatt, M. Protic-Sabljie, and F. Oesch (1983) Mutagenicity of glutathione and cysteine in the Ames test. *Science* 220:961-963.

17. H. R. Glatt (1989) Mutagenicity spectra in *Salmonella typhimurium* strains of glutathione, L-cysteine and active oxygen species. *Mutagenesis* 4:221-227.

18. H. Glatt (1990) Endogenous mutagens derived from amino acids. *Mutat. Res.* 238: 235-243.

19. W.B. Jakoby, B. Ketterer, and B. Mannervik (1984) Glutathione transferases: Nomenclature. *Biochem. Pharmacol.* 33:2539-2540.

20. B. Jernström, J.R. Babson, P. Moldéus, A. Holmgren, and D.J. Reed (1982) Glutathione conjugation and DNA-binding of ($\pm$)-*trans*-7,8-dihydroxy-9a, 10a-epoxy 7,8,9,10-tetrahydrobenzo[a]pyrene in isolated rat hepatocytes. *Carcinogenesis* 3:861-866.

21. B. Jernström, M. Martinez, D.J. Meyer, and B. Ketterer (1985) Glutathione conjugation of the carcinogenic and mutagenic electrophile (+)-7b,8a-dihydroxy-9a, 10a-oxy-7,8,9,10-tetrahydrobenzo[a]pyrene catalyzed by purified rat liver glutathione transferases. *Carcinogenesis* 1:37-47.

22. T.W. Kensler, P.A. Egner, N.E. Davidson, B.D. Roebuch, A. Pikul, and J.D. Groopman (1986) Modulation of aflatoxin metabolism, aflatoxin-N^7-guanyl formation and hepatic tumorigenesis in rats fed ethoxyquin: Role of induction of glutathione S-transferase. *Cancer Res.* 46:3924-3931.

23. B. Ketterer, D.J. Meyer, J.B. Taylor, S. Pemble, B. Coles, and B. Fraser (1990) GSTs and protection against oxidative stress. In *Glutathione S-Transferases and*

Drug Resistance, J.D. Hayes, C.B. Picket, and T.J. Mantle, eds. Taylor & Francis, London, pp. 97-109.

24. B. Ketterer, K.H. Tan, D.J. Meyer, and B. Coles (1990) Glutathione transferases: A possible role in the detoxication of DNA and lipid hydroperoxides. In *Glutathione S-Transferases and Drug Resistance*, J.D. Hayes, C.B. Pickett, and T.J. Mantle, eds. Taylor & Francis, London. pp. 149-163.

25. B. Ketterer, and G.J. Mulder (1990) Glutathione conjugation. In *Conjugation Reactions in Drug Metabolism*, G.J. Mulder, ed. Taylor & Francis, London, pp. 308-364.

26. N.S. Kosower, and A.M. Kosower (1978) The glutathione status of cells. *Int. Rev. Cytol.* 54:109-160.

27. W.J. Kozumbo, and P.A. Cerutti (1986) Antioxidants as antitumor promoters. *Basic Life Sci.* 39:491-506.

28. A. Larsson, S. Orrenius, A. Holmgren, and B. Mannervik, eds. (1983) In *Functions of Glutathione, Biochemical, Physiological, Toxicological and Clinical Aspects*, Raven Press, New York.

29. B.H. Lautenberg, G.B. Corcoran, and J.R. Mitchell (1983) Mechanisms of N-acetylcysteine in the protection against hepatotoxicity of acetaminophen in rats in vivo. *J. Clin. Invest.* 71:980-991.

30. B. Mannervik, and U.H. Danielson (1988) Glutathione transferases: Structure and catalytic activity. *CRC Crit. Rev. Biochem.* 23:283-337.

31. B. Mannervik, Y.C. Awasthi, P.G. Board, J.D. Hayes, C. De Ilio, B. Ketterer, I. Listowsky, R. Morgenstern, M. Muramatsu, W. Pearson, C.B. Pickett, K. Sato, M. Widersten, and C.R. Wolf (1992) Nomenclature for human glutathione transferases. *Biochem. J.* (in press).

32. A. Meister, and M.E. Anderson (1983) Glutathione. *Ann. Rev. Biochem.* 52:711-760.

33. D.J. Meyer, B. Coles, S.E. Pemble, K.S. Gilmore, G.M. Fraser, and B. Ketterer (1991) Theta, a new class of glutathione transferases purified from rat and man. *Biochem. J.* 274:409-414.

34. J.P. Moldéus, I.A. Cotgreave, and M. Berggren (1986) Lung protection by a thiol-containing antioxidant: *N*-acetylcysteine. *Respiration* 50:31-42.

35. R. Morgenstern, and J.W. DePierre (1988) In *Glutathione conjugation: Its mechanism and biological significance*, B. Ketterer and H.V. Sies eds. Academic Press Ltd., London, pp. 157-174.

36. J.A. Moscow, A.J. Townsend, and K.J. Cowan (1989) Elevation of π class glutathione S-transferase activity in human breast cancer cells by transfection of the GST π gene and its effect on sensitivity to toxins. *Molec. Pharmacol.* 36:22-28.

37. E., Mosialou, and R. Morgenstern (1989) Activity of rat liver microsomal glutathione transferase towards products of lipid peroxidation and studies on the effect of inhibitors on glutathione-dependent protection against lipid peroxidation. *Arch. Biochem. Biophys.* 275:289-294.

38. J.-P. Perchellet, M.D. Owen, T.D. Posey, D.K. Orten, and B.A. Schneider (1985) Inhibitory effects of glutathione level-raising agents and D-alpha-tocopherol on ornithine decarboxylase induction and mouse skin tumor promotion by 12-O-tetradecanoylphorbol-13-acetate. *Carcinogenesis* 6:567-573.

39. C.E. Petterson, J.A. Butler, F.D. Bryne, and M.L. Rhodes (1985) Oxidant lung injury: Intervention with sulfhydryl reagents. *Lung* 163:23-32.

40. R.B. Puchalski, and W.E. Fahl (1990) Expression of recombinant glutathione S transferase p, Ya, or Yb_1 confers resistance to alkylating agents. *Proc. Natl. Acad. Sci., USA* 87:2443-2447.

41. J.A. Redick, W.B. Jakoby, and J. Baron (1982) Immunohistochemical localization of glutathione S-transferases in livers of untreated rats. *J. Biol. Chem.* 257:15200-15203.

42. I.G. Robertson, H. Jenssen, B. Mannervik, and B. Jernström (1986) Glutathione transferases in rat lung: the presence of transferase 7-7, highly efficient in the conjugation of glutathione with the carcinogenic (+)-7b,8a-dihydroxy-9a,10a-oxy-7,8,9,10-tetrahydrobenzo[a]pyrene. *Carcinogenesis* 7:295-299.

43. I.G. Robertson, C. Guthenberg, B. Mannervik, and B. Jernström (1986) Differences in stereoselectivity and catalytic efficiency of three human glutathione transferases in the conjugation of glutathione with 7b,81-dihydroxy-9a,10a-oxy-7,8,8,10-tetrahydrobenzo[a]pyrene. *Cancer Res.* 46:2220-2224.

44. L. Romert, L. Dock, D. Jenssen, and B. Jernström (1989) Effects of glutathione transferase activity on benzo[a]pyrene 7,8-dihydrodiol metabolism and mutagenesis studied in a mammalian cell co-cultivation assay. *Carcinogenesis* 10:1701-1707.

45. D. Ross, I. Cotgreave, and P. Moldéus (1985) The interaction of reduced glutathione with active oxygen species generated by PMA stimulated guinea pig peritoneal macrophages. *Biochem. Biophys. Acta* 841:278-282.

46. D. Ross, P. Moldéus, H. Sies, and M.T. Smith (1986) Mechanism and relevance of glutathione mutagenicity. *Mutat. Res.* 175:127-131.

47. R. Saul, and B. Ames (1986) Background levels of DNA damage in the population. *Basic Life Sci.* 38:529-535.

48. J. Seidegård, R.W. Pero, M.M. Markowitz, G. Rousch, D.G. Miller, and E. Beattie (1990) Isoenzyme(s) of glutathione transferase (class Mu) as a marker for the susceptibility to lung cancer: A follow-up study. *Carcinogenesis* 11:33-36.

49. A. Sevanian, S. Muakkassah-Kelly, and S. Montestruque (1983) The influence of phospholipase A_2 and glutathione peroxidase on the elimination of membrane lipid peroxides. *Arch. Biochem. Biophys.* 223:441-452.

50. P. Sims, and P.L. Grover (1974) Epoxides in polycyclic aromatic hydrocarbon metabolism and carcinogenesis. *Adv. Cancer Res.* 20:165-275.

51. R.C. Strange, B. Matharoo, G.C. Faulder, P. Jones, W. Cotton, J.B. Elder, and M. Deakin (1992) The human glutathione S-transferases: A case-control study of the incidence of the GST 10 phenotype in patients with adenocarcinoma. *Carcinogenesis* 12:25-28.

52. K.H. Tan, J. Meyer, J. Belin, and B. Ketterer (1984) Inhibition of lipid peroxidation by glutathione and glutathione transferases B and AA. *Biochem. J.* 220:243-252.

53. D.R. Thakker, H. Yagi, W. Levin, A.W. Wood, A.H. Conney, and D.M. Jerina (1985) Polycyclic aromatic hydrocarbons: Metabolic activation to ultimate carcinogens. In *Bioactivation of Foreign Compounds*, M.W. Anders, ed. Academic Press, pp. 177-242.

54. H.F. Thomas, R.M. Herriott, B.S. Hahn, and S.Y. Wang (1976) Thymine hydroperoxide as a mediator in ionising radiation mutagenesis. *Nature* 259:341.

55. J.P. Thomas, M. Maiorino, F. Ursini, and A.W. Girotti (1990) Protective action of phospholipid hydroperoxide glutathione peroxidase against membrane-damaging lipid peroxidation: In situ reduction of phospholipid and cholesterol hydroperoxides. *J. Biol. Chem.* 265:454-461.

56. M.F. Tsan, E.H. Danis, P.J. del Vecchio, and C.L. Rosano (1985) Enhancement of intracellular glutathione protects endothelial cells from oxidant damage. *Biochem. Biophys. Res. Comm.* 127:270-276.

57. F. Ursini, M. Maiorino, M. Valente, L. Ferri, and C. Gregolin (1982) Purification from pig liver of a protein which protects liposomes and biomembranes from peroxidative degradation and exhibits glutathione peroxidase activity on phosphatidylcholine hydroperoxides. *Biochim. Biophys. Acta* 710:197-211.

N-ACETYLCYSTEINE INHIBITS DIESEL EXTRACT MUTAGENICITY

IN THE AMES TEST AND SCE INDUCTION IN HUMAN LYMPHOCYTES

R. Barale, R. Micheletti, C. Sbrana, I. Glussich,
C. Scapoli,* and I. Barrai*

Dipt. Scienze dell'Ambiente e del Territorio
Università di Pisa
Pisa, ITALY
* Dipt. Biologia Evolutiva
Università di Ferrara
Ferrara, ITALY

SUMMARY

N-Acetylcysteine (NAC) has been reported to decrease genotoxicity induced by several mutagens. In this paper, the desmutagenic effect of NAC on a complex mixture, such as diesel extract, has been analyzed. Studies have been carried out in vitro with the Ames test (reverse mutations on TA98, TA100, and TA104 strains) and sister chromatid exchanges assay (SCE) in human lymphocytes. NAC inhibits diesel genotoxicity in both assays. NAC also inhibits the mutagenicity of 1,8-dinitropyrene (1,8-DNP) and 1-nitropyrene (1-NP) known to be present in diesel exhaust and to be activated by cellular O-transacetylases and nitropyrene reductases. NAC inhibits also the induction of SCE in human lymphocytes by diesel extract. These results, and those obtained by the preincubation of NAC with cells, suggest that the inhibition also takes place inside the cell.

INTRODUCTION

N-Acetylcysteine (NAC) has been in therapeutic use for several years for its mucolytic action (20). Moreover, NAC is usefully employed for the treatment of acute poisoning with heavy metals and chemicals such as halothane, acroleine, paracetamol, and bromobenzene (3). Several studies have shown that NAC is able to inhibit or modulate the mutagenic activity of some direct and indirect mutagens (7,19). The carcinogenic activity of urethane in mouse is also reduced by NAC administration (4). The antitoxic, desmutagenic, and anticarcinogenic activities of NAC have been attributed to the direct inactivation of reactive electrophiles and free radicals through reduction and conjugation (14). Moreover, NAC has been shown to act as a precursor of glutathione (GSH) or to facilitate its formation (4,9,14). This latter thiol has been found to reduce mutagenicity of complex mixtures such as those present in diesel exhaust in the Ames test (18). Therefore, these features prompted us to evaluate whether NAC was able to exert desmutagenic activity on diesel extracts and, possibly, to elucidate its mode of action.

MATERIALS AND METHODS

Diesel exhaust was obtained from a diesel-powered car and collected on fiberglass filters, following dilution and air cooling to 40°C, by a Hi-Vol Staplex sampler at a flow rate of approximately 1 m3/min. The filter was sonicated for 20 min in dichloro-methane (DCM) and overnight Soxhlet extracted, always in DCM. The solvent was then evaporated with a rotavapor under a nitrogen stream and finally the dried organic material was resuspended in dimethylsulphoxide (DMSO) for mutagenicity assay. Part (¾) of the total diesel exhaust organic extract (DEOE) was also separated into acid, basic, and neutral fractions according to the procedure of Hopke et al.

The standard Ames test (11) was applied using *Salmonella typhimurium* strains TA98, TA100, and TA104 without metabolic activation. Hycanthone (Winthrop, Spain) 20 µg/plate and formaldehyde (Carlo Erba, Italy) 40 µg/plate were used as positive controls for checking strain sensitivity. The nitroreductase-deficient strain TA98NR and the O-transacetylase-defective strain TA98/1, 8DNP6 were also used, and their resistance to 250 ng/plate of 1-NP (12) and to 4 ng/plate of 1,8-DNP (13,16) was confirmed.

The lymphocyte assay was applied as recommended by Carrano and Natarajan (2). Peripheral blood samples were donated by a healthy, non-smoking, 45-year old male. Blood (0.3 ml) was added to 4.11 ml HAM'S F10 (Flow Laboratories), 0.5 ml fetal calf serum (Flow Laboratories), 0.1 ml antibiotic (Flow Laboratories), and 0.075 ml phytohaemagglutinin (Wellcome). Cultures with BrdUdr (Sigma) in a final concentration of 9 µg/ml were incubated for 72 hrs at 37°C; BrdUdr was always present during the culture time. Lymphocytes were treated for 4 hrs from the 44th to the 48th hour of culture in the following medium: 4.7 ml PBS (Phosphate Buffer Solution), 0.1 ml antibiotics, 0.1 ml of NAC diluted in PBS (when NAC was not required, only PBS was added), and 50 µl of DEOE diluted in DMSO. Colchicine (0.8 µg/ml) was added 2 hrs before cell harvesting. SCEs were evidenced by means of a differential staining method that was carried out as follows: slides were kept for 15' in Hoechst 33258 (Sigma) buffered pH 7.4 Sorensen solution, then dropped with SSC 0.5X solution and lighted by a BLB 18W lamp for 30'. Slides were heated (about and not over 58°C) in Na_2HPO_4 0.06M solution for 20' and, finally, stained in a Giemsa solution (3%) in Sorensen buffer pH 6.8 for 10'. Air-dried slides were mounted with Eukitt.

Statistics

a) Ames Test. The effect of NAC on DEOE mutagenicity was assessed according to the methods of Shaeffer et al. (17) and Barale et al. (1). Both methods are based on the difference between the response obtained from the mixtures of two substances and the sum of the responses at different doses of single substances. The tests are such that a negative value of the Student's t test, if significant, indicates antagonism between the two substances. In this case, since antagonism was determined by the above test, a finer resolution of the inhibitory effect of NAC was obtained, regressing the response at constant values of DEOE on increasing doses of NAC. DEOE was also tested at different doses (25-400 µg/plate). Polinomial regressions above the first degree were fitted to the response, but degrees above the first had a weak although significant effect. In general, the linear component removed the largest part of the variability. We also studied the variation of the linear regression coefficients measuring the inhibitory effect of NAC, as a function of DEOE dose.

b) SCE Test. The effect of NAC on DEOE genotoxicity was assessed with and without preincubation of lymphocytes with NAC. Twenty-five metaphases/dose were scored at four doses of NAC and four doses of DEOE. In the first case, we studied the SCE number/cell as a function of two variables: a) presence or absence of NAC and b) presence or absence of DEOE. In the second treatment schedule, we also examined the effect of length of NAC preincubation with lymphocytes. As a methodology, we used multiple regression, where the dependent variable was the arcsine of the SCE number/cell, and the independent variables were the other variables described above. Some of the possible pairwise comparisons were studied with Student's t test.

RESULTS AND DISCUSSION

Mutagenic Characterization of Diesel Exhaust Organic Extract

a) Ames Test. The mutagenicity of DEOE was assessed on TA98, TA98NR, TA98/1,8DNP6, TA100, and TA104 Salmonella strains. In Fig. 1, we report the results obtained with the strain TA98 and its derivatives TA98NR and TA98/1,8DNP6. The dose-response is quite linear over the range of tested doses with all strains used. We observed that about 25% and 35% of the total mutagenicity on the TA98 strain can be attributed to mutagens not effective on the deficient strains TA98NR and TA98/1,8DNP6, respectively. As a whole, these classes of mutagens appear to account for a least 60% of the total direct mutagenicity on the TA98 strain. These findings are quite in agreement with the literature (15). Among the three chemical fractions obtained from DEOE, acid and neutral ones were almost equally active, whereas the basic fraction, which also contains polycyclic aromatic hydrocarbons, gave a very weak response without S9.

b) SCE Test. DEOE determined SCE increases in human lymphocytes (Fig. 2) with a biphasic dose-response trend: after a sharp and linear increase at the lower doses (linear regression coefficient (b = 0.16, P<0.001), the curve continues to grow linearly, but less steeply (b = 0.022, P<0.001). This trend might be due to the presence of lymphocyte sub populations having different responsiveness.

We can conclude that the DEOE examined contains a mixture of direct mutagens able to induce reverse mutations in all Salmonella strains tested and SCE in human lymphocytes.

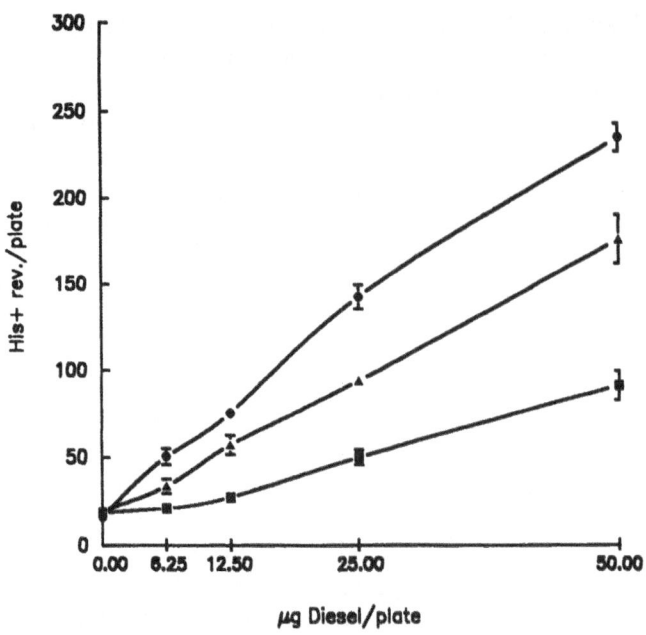

Figure 1. Mutagenicity of organic extract from diesel exhaust on TA98 (●), TA98NR (▲), and TA98/1,8-DNP6 (■). Vertical bars: standard deviations.

Desmutagenic Effect of NAC on DEOE

a) Ames Test. Initially we have evaluated the possible NAC toxicity on Salmonella strains over the range of 3.06-61.3 µmol/plate. In standard treatment conditions, NAC resulted in a toxic effect because of its acidity, with the production of microcolonies and filamentous cells (x400 magnification) above the doses of 22-36.8 µmol/plate. Therefore, neutralization of NAC (pH 6.5 with NaOH, 5N) was applied to avoid misleading results. After neutralization, no toxicity was observed by bacterial background lawn evaluation at low and high magnification (x30 and x400). No reduction, below the control number of colonies, was observed by plating diluted cell suspensions onto complete agar plates containing up to 61.3 µmoles of NAC, nor were variations in revertant numbers around the spontaneous level typical for all Salmonella strains used in this study detected, except for TA104 strain that showed a statistically significant reduction of his+ revertants NAC dose-related (F = 11.70, P<0.01; t = -3.42, P<0.01). Similar findings have been repeatedly reported (5,6).

In preliminary experiments on the possible desmutagenic activity of NAC on DEOE we compared two possible assay protocols: a) pouring onto plates simultaneously NAC, DEOE, and bacteria, or b) preincubating NAC with DEOE for 30, 60, and 120 min before adding bacteria and plating. With both procedures we obtained the same inhibitory effects. These findings suggest that the possible scavenger activity of NAC on DEOE mutagens outside the cells may be only one of the ways in which NAC inhibits DEOE. With other direct-acting mutagens, such as 4-nitroquinoline-N-oxide and sodium dichromate, preincubation was preferred for studying the inhibitory activity of NAC (3). This difference can be explained by the fact that in the present case we are dealing with a mix of hundreds or thousands of

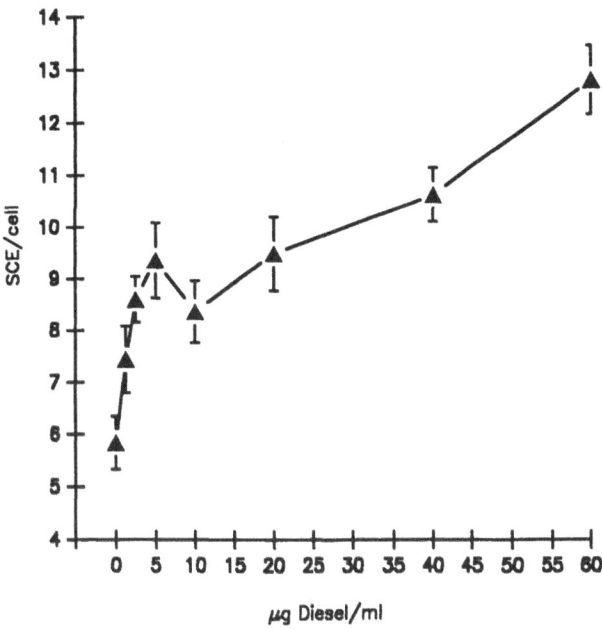

Figure 2. SCE induction in human lymphocytes by organic extract from diesel exhaust. Vertical bars: standard deviations.

chemicals whose mechanism of action and interaction with NAC might be very complex.

In Fig. 3 we report the inhibitory effects of NAC on DEOE mutagenicity tested on TA98, TA100, and TA104 strains. The inhibition, evaluated by the methods proposed by Shaeffer et al. (17) and Barale et al. (1), was highly significant (P<0.001) with both methodologies for all the strains used and shows common inhibitory trends. Inhibition is strong at low doses of NAC, then it flattens from 0.46-1.86, depending on the strains, up to 3.68 μmol/plate.

These findings indicate at least two possibilities. First, a variety of mutagens may be present in DEOE and only some of them are sensitive to the inhibitory effect of NAC. Second, in the case of NAC acting intracellularly, only a limited amount enters the cell. In order to explore which hypothesis was more plausible, we preincubated DEOE with NAC before adding Salmonella cells. It appears that NAC preincubation length has no effect on the inhibition of DEOE mutagenicity. This latter effect remains practically constant at all times tested (data not reported).

As shown, a large fraction of DEOE mutagenicity was due to nitro- and dinitropyrenes. Therefore, we studied the inhibitory effect of NAC on pure nitropyrenes such as 1-NP and 1,8-DNP. The inhibitory effect of NAC on these standards is reported in Fig 4. By applying the methods of Shaeffer or Barale, the statistical significance of inhibition appears high for both mutagens (P<0. 001). It has been shown that bacterial nitroreductases can convert 1-NP to a mutagenic hydroxylamine (16) that can be effectively detoxified by GSH (10). A possible similar action of NAC can occur inside the cell as well. To further explore this possibility

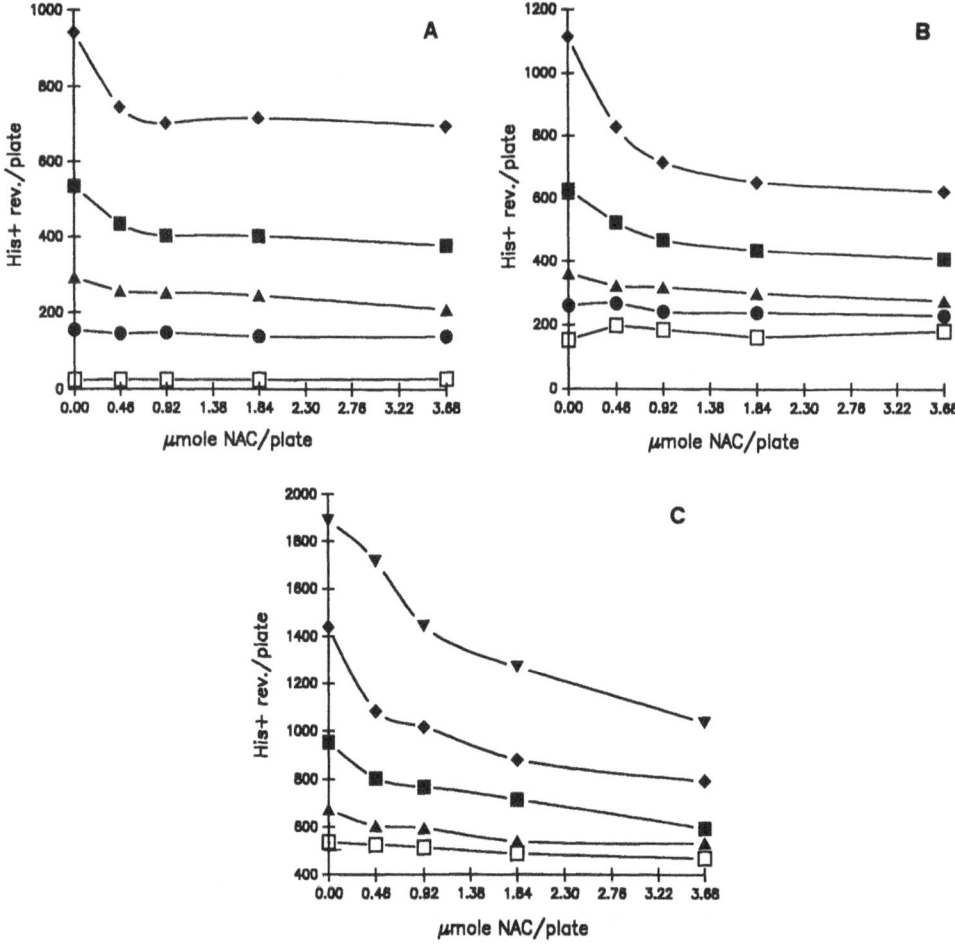

Figure 3. Inhibitor effect of NAC different amounts (0.46–3.68) μmoles/plate on the mutagenicity of four doses of diesel extract on a) TA98; b) TA100; and c) TA104. DEOE doses: 25 μg/plate = ●; 50 μg/plate = ▲; 100 μg/plate = ■; 200 μg/plate = ♦; 400 μg/plate = ▼; NAC alone = □.

we performed a preincubation assay in which 1,8-DNP was incubated with NAC for 15, 30, 60, and 120 min before adding bacteria and plating. The results (Fig. 5) show that there are no differences of inhibition between standard plate tests (0 min) and preincubation tests. The multiple regression analysis shows a strong mutagenic effect due to 1,8-DNP treatment (t = 34.72, P<0.001), a significant inhibitory effect of NAC (t = -7.17, P<0. 001), and no effect due to preincubation length of 1,8-DNP with NAC (t = -0.76; n.s.). These findings are in agreement with results obtained with DEOE and further support our previous conclusions.

We have also tested NAC on acid and neutral DEOE fractions in order to investigate whether NAC was active on particular classes of mutagens. Both fractions were almost equally inhibited by NAC (Fig. 6). To represent and compare the intensity of NAC inhibitory effects on the mutagenicity of crude extract, acid, and

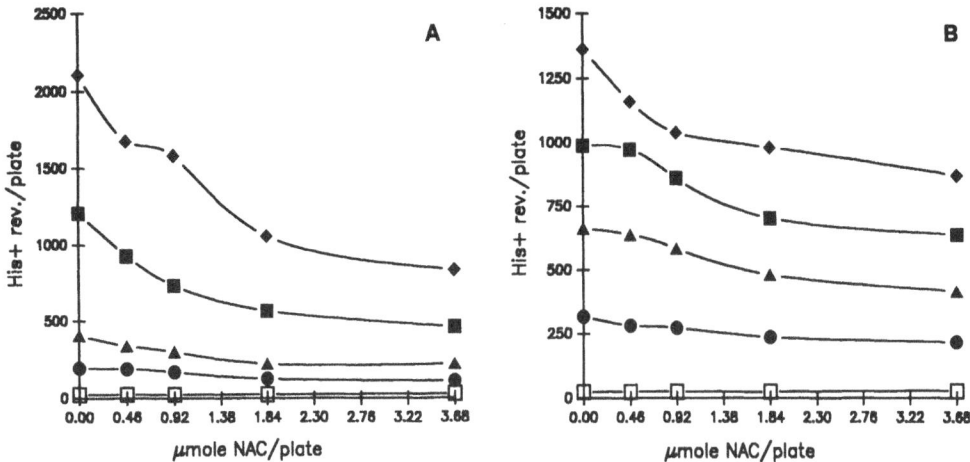

Figure 4. NAC inhibitor effect on the mutagenicity on TA98 of a) 1NP (16 nmole/plate = ◆; 8 nmole/plate = ■; 4 nmole/plate = ▲; 2 nmole/plate = ●; NAC alone = □) and b) 1,8-DNP (0.027 nmole/plate = ◆; 0.020 nmole/plate = ■; 0.013 nmole/plate = ▲; 0.007 nmole/plate = ●; NAC alone = □).

neutral fractions, we report in the ordinate the linear inhibition, and in the abscissa the dose of extracts (Fig. 7). The largest inhibition is observed with the total extract, whereas fractions appear to be less inhibited, obviously containing fewer mutagens. In any case, the higher the doses, the stronger the NAC inhibition. At present, we have no reasonable explanation of this result.

A final experiment in which the inhibition of NAC and GSH was compared showed that the responses were overlapping (Fig. 8). The linear coefficients of the polinomial regressions were not significantly different, suggesting considerable similarity of action of the two molecules.

b) SCE Test. First experiments with lymphocytes essentially showed similar kinetics of NAC inhibition on DEOE mutagenicity to those obtained with the Ames test. The maximal inhibition was reached at the lowest dose of NAC tested (0.61 μmol/ml), staying constant at higher doses (data not reported). Multiple regression analysis of the arcsine transformation of SCE/cell on NAC and DEOE doses confirms that the increase of NAC inhibitory effect was not significant (t = -1.35, P<0.2), whereas the mutagenic effect of DEOE was significantly dose-dependent (t = 7.82, P<0. 001).

The range between 0 and lowest dose of NAC (0.61 μmol/ml) was then explored, and the results are given in Fig. 9. It appears that inhibition begins at very low doses such as 0.08 μmol/ml (P<0.025) with a linear effect up to 0.15 μmol/ml (P<0.005) and stays constant afterward.

Finally, we explored the possible effect of preincubating NAC with lymphocytes for 30, 60, and 20 min (Fig. 10). Multiple regression analysis indicates that the effect of preincubation length is not significant (t = -1.63, P~ 0.10). The

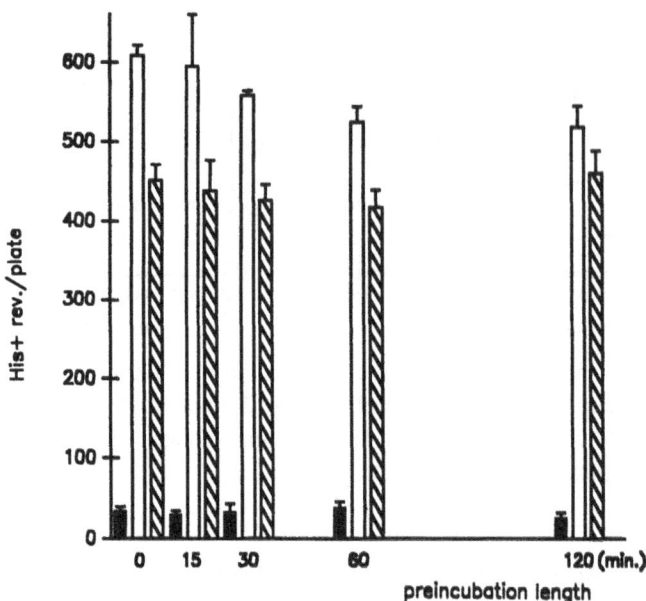

Figure 5. Effects of NAC preincubation (15–120 min) with 1,8-DNP: inhibition of mutagenicity (TA98). Blank bar = 1,8-DNP (4 ng/plate); solid bar = NAC alone (1.84 μmole/plate); hatched bar = NAC with 1,8-DNP.

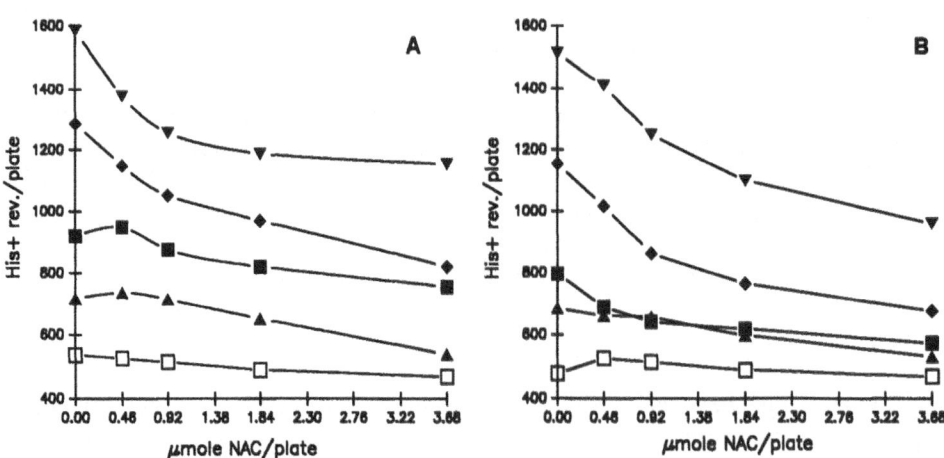

Figure 6. NAC inhibitor effect on the mutagenicity on TA104 of a) neutral fraction of diesel exhaust and b) acid fraction (50 μg/plate = ▲; 100 μg/plate = ■; 200 μg/plate = ◆; 400 μg/plate = ▼; NAC alone = ☐).

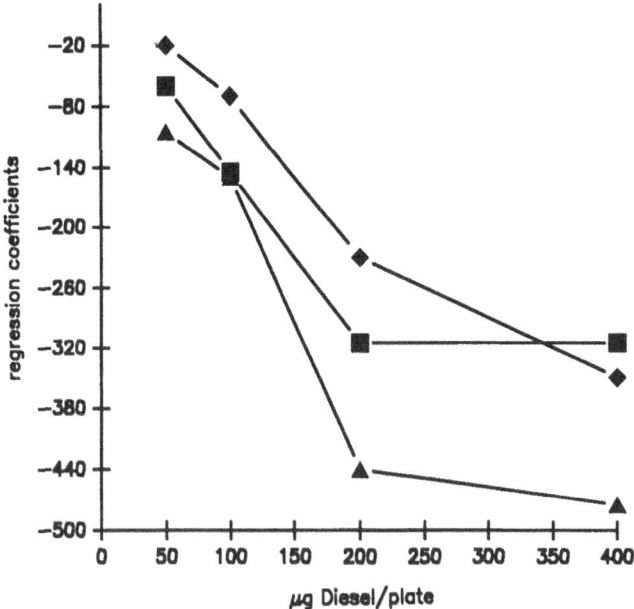

Figure 7. Comparison of NAC inhibition efficiency on DEOE mutagenicity (TA104) expressed as negative linear regression coefficients of polynomial regression (see text). Total DEOE = ▲; acid fraction = ■; neutral fraction = ♦.

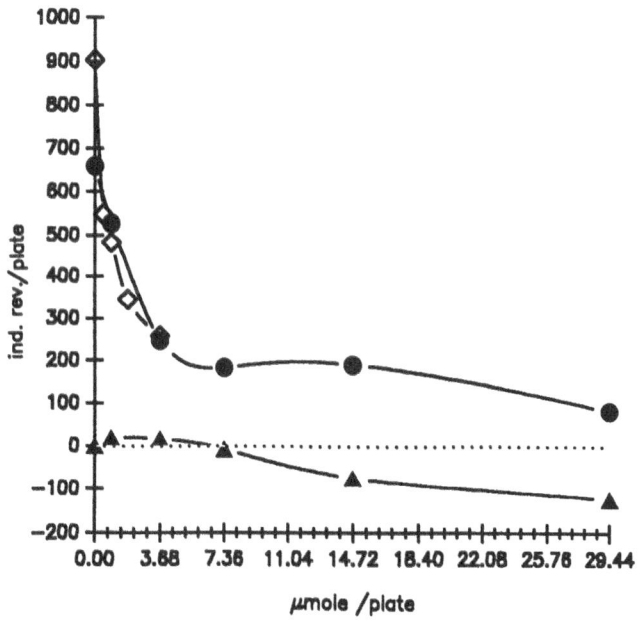

Figure 8. GSH and NAC inhibitor effect on the mutagenicity of diesel exhaust (200 μg/plate) on TA104. NAC = ◊; GSH = ●; GSH alone = ▲.

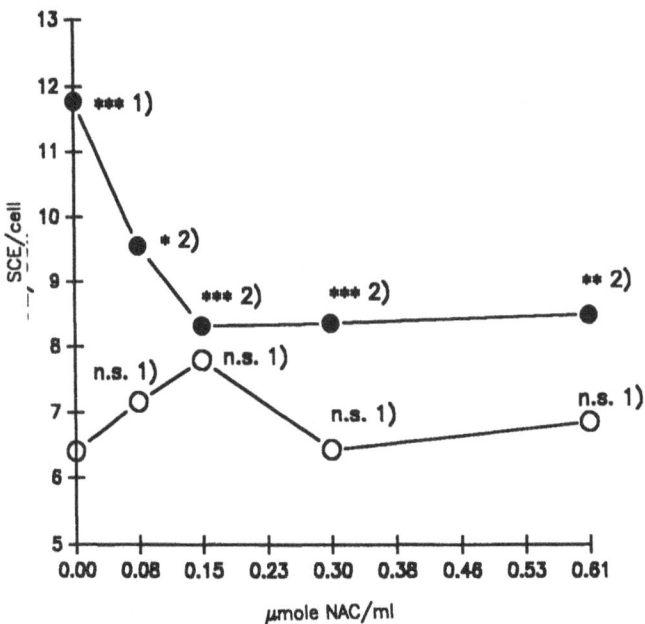

Figure 9. NAC reduction of SCE induced by DEOE in human lymphocytes. (O) NAC alone, negative control; (●) NAC with 5 µg/ml DEOE. Significance levels: * = P < 0.025; ** = P < 0.005; *** = p < 0.001; n.s. = not significant. 1) = differences between treated and control cultures (without NAC and DEOE); 2) = differences between DEOE treatments in the absence or presence of NAC.

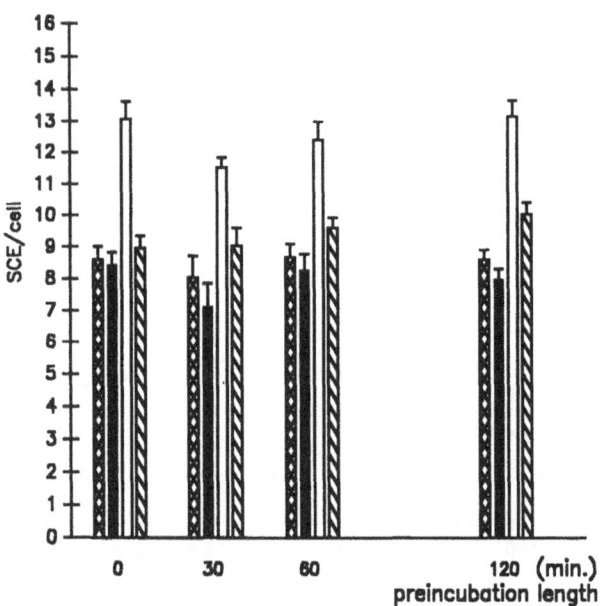

Figure 10. Different times of lymphocyte NAC preincubation (0.3 µmole/ml) before treatments with 40 µg/ml DEOE: effects on SCE induction. Crosshatched bar = control cultures; solid bar = NAC alone, negative control; blank bar = DEOE, positive control; hatched bar = DEOE and NAC.

158

strong mutagenic effect of DEOE and its inhibition by NAC are very evident (t = 10.5 and -7.2, respectively). The effectiveness of lymphocyte pretreatment with NAC on DEOE inactivation can be indicative of a cellular uptake of NAC and, consequently, of its intracellular mode of action as a scavenger of active mutagenic forms or as a GSH precursor. However, we cannot exclude more indirect mechanisms, since it has been shown that NAC promotes also cysteine uptake and its rapid utilization for cellular GSH biosynthesis (9).

CONCLUSION

In this study we have shown that NAC is able to inhibit in vitro DEOE genotoxicity evaluated in two different organisms, Salmonella and human lymphocytes, by the analysis of two different genetic endpoints, reverse mutation, and SCE. Results speak in favor of a preferential intracellular mode of action of NAC, and the observed inhibitory effect is very close to that determined by GSH at the same stoichiometric doses. Because DEOE contains many mutagens, possibly hundreds, their inhibition, even if not complete, suggests that NAC is able to modulate the mutagenicity of a variety of mutagens besides specific ones known to be present in DEOE, such as 1-nitro and 1,8-dinitropyrene.

In lymphocyte studies NAC was effective at doses as low as 0.08 μmol/ml. Therefore, it seems of great theoretical and practical concern to confirm this inhibitory activity in vivo as well.

ACKNOWLEDGEMENTS

This study was supported by Zambon Group, Bresso, Milano, Italy.

REFERENCES

1. Barale, R., A. Marrazzini, C. Betti, V. Vangelisti, N. Loprieno, and I. Barrai (1990) Genotoxicity of two metabolites of benzene: Phenol and hydroquinone show strong synergistic effects in vivo. *Mutat. Res.* 244:15-20.
2. Carrano, A.V., and A.T. Natarajan (1988) Considerations for population monitoring using cytogenetic techniques. *Mutat. Res.* 204:379-406.
3. De Flora, S., C. Bennicelli, P. Zanacchi, A. Camoirano, A. Morelli, and A. De Flora (1984) In vitro effects of N-acetylcysteine on the mutagenicity of direct-acting compounds and procarcinogens. *Carcinogenesis* 5:505-510.
4. De Flora, S., G.A. Rossi, and A. De Flora (1986) Metabolic, desmutagenic and anticarcinogenic effects of N-acetylcysteine. *Respiration* 50 (Suppl.) 1:43-49.
5. De Flora, S., C. Bennicelli, A. Camoirano, D. Serra, C. Basso, P. Zanacchi, and C.F. Cesarone (1987) Inhibition of mutagenesis and carcinogenesis by N-acetylcysteine. In *Anticarcinogenesls and Radiation Protection*, P.A. Cerutti, O. Nygaard, and M.G. Simic, eds. Plenum Press, New York and London, pp. 373-386.

6. De Flora, S., C. Bennicelli, P. Zanacchi, F. D'Agostini, and A. Camoirano (1989) Mutagenicity of active oxygen species in bacteria and its enzymatic or chemical inhibition. *Mutat. Res.* 214:153-158.

7. De Flora, S., A. Izzotti, F. D'Agostini, and C.F. Cesarone (1991) Antioxidant activity and other mechanisms of thiols in chemoprevention of mutation and cancer. *Am. J. Med.* 91 (Suppl.) 3C:122-130.

8. Hopke, P.K., M.J. Plewa, P.L. Stapleton, and D.L. Weaver (1984) Comparison of the mutagenicity of Sewage Sludges. *Env. Sci. Technol.* 18:909-916.

9. Issels, R.D., A. Nagele, K. Eckert, and W. Wilmanns (1988) Promotion of cystine uptake and its utilization for glutathione biosynthesis induced by cysteamine and N-acetylcysteine. *Biochem. Pharm.* 37:881-888.

10. Ketterer, B. (1988) Protective role of glutathione and gluthatione transferases in mutagenesis and carcinogenesis. *Mutat. Res.* 202:343-361.

11. Maron, D., and B.N. Ames (1983) Revised method for the Salmonella mutagenicity test. *Mutat. Res.* 113:173-215.

12. McCoy, E.C., M. Anders, and H.S. Rosenkranz (1983) The basis of insensitivity of *Salmonella typhimurium* strain TA98/1,8-DNP6 to the mutagenic action of nitroarenes. *Mutat. Res.* 121:17-23.

13. Mermelstein, R., D.K. Kiiazides, M. Butler, E.C. McCoy, and H.S. Rosenkranz (1981) The extraordinary mutagenicity of nitropyrenes in bacteria. *Mutat. Res.* 89:187-196.

14. Moldéus, P., I.A. Cotgreave, and M. Berggren (1986) Lung protection by a thiol-containing antioxidant: N-acetylcysteine. *Respiration* 50 (Suppl.) 1:31-42.

15. Rosenkranz, H.S. (1982) Direct-acting mutagens in diesel exhausts: Magnitude of the problem. *Mutat. Res.* 101:1-10.

16. Rosenkranz, H.S., and R. Mermelstein (1983) Mutagenicity and genotoxicity of nitroarenes: All nitro-containing chemicals were not created equal. *Mutat. Res.* 114:217-267.

17. Shaeffer, D.J., W.R. Glave, and K.G. Janardan (1982) Multivariate statistical methods in toxicology. III. Specifying joint toxic interaction using multiple regression analysis. *J. Toxicol. Env. Health* 9:705-718.

18. Wang, Y., E.R. Talcot, D.A. Seid, and T.E. Wei (1981) Antimutagenic properties of liver homogenates, proteins and glutathione on diesel exhaust particulates. *Canc. Lett.* 11:265-275.

19. Wilpart, M., P. Mainguet, D. Geeroms, and M. Roberfroid (1985) Desmutagenic effects of N-acetylcysteine on direct and indirect mutagens. *Mutat. Res.* 142:169-172.

20. Ziment, I. (1986) Acetylcysteine: A drug with an interesting past and a fascinating future. *Respiration* 50 (Suppl.) 1:26-30.

SULFOTRANSFERASE- AND ACETYLTRANSFERASE-MEDIATED

ACTIVATION OF CARCINOGENIC N-HYDROXYARYLAMINES

IN MAMMALS AND BACTERIA, AND THEIR MODULATION BY

THIOLS

Yasushi Yamazoe, Medhat Abu-Zeid, Norma Staiano,*
and Ryuichi Kato

Department of Pharmacology
School of Medicine
Keio University
35 Shinanomachi, Shinjuku-ku, Tokyo 160 JAPAN

*Dipartimento di Biochimica e Biotecnologie Medicine
II Facolta di Medicina
Universit di Napoli
Via Sergio Pansini, 5 1-80131 Napoli, ITALY

INTRODUCTION

Although carcinogenic arylamines can be activated by a single step of activation, such as prostaglandin H synthase-mediated oxidation to react with biological macromolecules, most arylamines are activated by two successive reactions, N-oxidation and O-esterification, to form their reactive intermediates. The former reaction is catalyzed mainly by microsomal cytochrome P-450, while the latter is mediated by cytosolic transferases in the liver. Among several transferases, acetyltransferase, and sulfotransferase are believed to be the main contributors to this activating reaction, O-esterification of N-hydroxylarylamides and N-hydroxyarylamines. Several heterocyclic arylamines isolated from protein and food pyrolysates are now known to be highly mutagenic in *Salmonella* and carcinogenic in rodents (10). For most of the heterocyclic arylamines, the results of the mutagenic effects obtained with bacteria and mammals are largely consistent but in some cases discordant quantitative results are seen between the two systems: For example, IQ is highly mutagenic in *Salmonella*, but is less potent in mammalian cells. Recent studies on chemopreventive aspects indicate that thiol compounds such as N-acetylcysteine and oltipraz effectively reduced the incidence of tumors induced by chemical carcinogens in experimental animals (6). However, several thiols have been shown to enhance

Table 1. Requirement for cytosolic sulfotransferase-mediated binding of N-hydroxyaryl-amines and -amide in rat liver.

System	Amounts bound to DNA	
	(pmol/mg DNA/mg protein/min)	(%)
N-Hydroxy-Glu-P-1		
Complete	19.8 ± 1.9	(100)
minus PAPS	1.0 ± 0.3	(5.1)
minus cytosol	1.0 ± 0.4	(5.1)
N-Hydroxy-IQ		
Complete	2.1 ± 0.2	
minus PAPS	<0.5	
N-Hydroxy-2-acetyl-aminofluorene		
Complete	53.60 ± 1.9	(100)
minus PAPS	17.97 ± 1.4	(33.5)
minus cytosol	0.91 ± 0.2	(1.7)

The binding to calf thymus DNA was measured at 20 μM for each one of the N-hydroxyaryl compounds as described (17) using liver cytosol obtained from adult male rats. Data are presented as the mean ± SD. Relative percents to the respective complete system are shown in parentheses.

the number of revertants induced by heterocyclic arylamines in *Salmonella* mutagenesis tests (2,9). These results suggest possible differences in *Salmonella* and mammalian systems in the activating or inactivating mechanism of N-hydroxyarylamines, and the possible existence of modulating factors. Therefore, we have studied the capacity of sulfotransferase- and acetyltransferase-mediated binding of N-hydroxyarylamines to calf thymus DNA is cytosols of *Salmonella* and mammalian species (including humans) to assess the species differences in the metabolic activating pathways, and to detect differences in the susceptibility of the two pathways to thiols.

SULFATION OF N-HYDROXYARYLAMINES

Requirements for cytosolic sulfation of N-hydroxaryl compounds are shown in Tab. 1. In the presence of phosphoadenosine-5'-phosphosulfate (PAPS), and N-hydroxy derivative of a glutamic acid-pyrolysate, Glu-P-1, was converted in rat liver cytosol to the presumed N-sulfate, which rapidly bound to calf thymus DNA. The binding was dependent on both cytosol and PAPS; it was lowered to 5.1% of the complete system by the omission of either one of them. Another pyrolysate-derived N-hydroxyarylamine, N-hydroxy-IQ, was also activated by sulfotransferase in rat livers, although the amount bound was nearly ten times less than that of N-hydroxy-Glu-P-1. As reported previously (3,7), a representative hepatocarcinogenic N-hydroxyarylamide, N-hydroxy-2-acetylaminofluorene, was converted to the DNA-

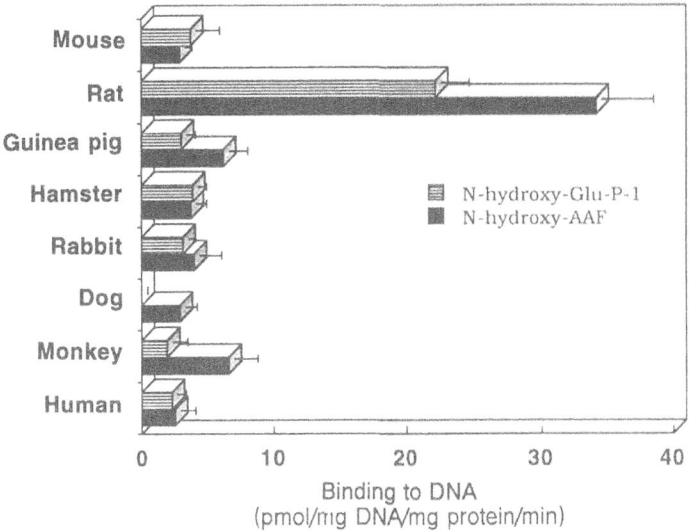

Figure 1. Species difference in PAPS-dependent DNA binding of N-hydroxy-Glu-P-1 and N-hydroxy-AAF. Covalent binding of N-hydroxy-Glu-P-1 and N-hydroxy-AAF was determined using at least four different individual livers for each species. Data are presented as the mean (column) $\pm$ SD (bar). Cytosol was prepared from adult male animals, except that human cytosols were prepared from livers of both sexes.

binding product in rat liver cytosol. The maximal binding was observed in the presence of PAPS, and the extent of the binding was roughly 2.5-fold higher than that of N-hydroxy-Glu-P-1 at 20 μM concentration. Although a considerable amount of binding was detected even in the absence of PAPS, this is probably due to cytosolic N,O-acetyltransfer of N-hydroxyarylacetamide (1). No appreciable binding was observed in the absence of hepatic cytosol.

Several sulfotransferases catalyzing sulfation of diverse types of chemicals are present in liver cytosols and classified into several groups depending on the substrate specificities (5,16). Multiple or plural forms of phenol sulfotransferase (arylsulfotransferase) have been purified from rat and human livers (4,5). Some phenol sulfotransferases are also shown to catalyze the O-sulfation of N-hydroxyarylamides (5). We have recently shown that hepatic N-hydroxarylamine sulfation and levels of two sulfotransferases are under endocrine influence in rats (18,19). Susceptibility of hepatic xenobiotic metabolizing activities to endocrine factors varies markedly among animal species. Thus, the capacity of sulfotransferase-mediated activation of N-hydroxyarylamines was examined with six other experimental animal species and humans. Covalent binding of calf thymus DNA of N-hydroxy-Glu-P-1 was detected in liver cytosols of all the species examined, except dogs, as shown in Fig. 1. The extent of the binding was highest in rats, at least 5-fold higher than those in other species. The activating ability of human livers was roughly the same as those of monkey, mouse, hamster, guinea pig, and rabbit. The binding of N-hydroxy-IQ was also examined using similar systems (data not shown), but no

Table 2. Activation of N-hydroxy-Glu-P-1 by rat liver and *Salmonella* cytosols. The activities were determined as the reductive formation of Glu-P-1 fromN-acetoxy or N-sulfonyloxy-Glu-P-1 (12,19). Cytosols were prepared from *Salmonella typhimurium* TA98 and adult male rats. The datum represents the mean of a duplicate determination using 20 M N-hydroxy-Glu-P-1 and 0.25 mM each of the cofactors.

Cytosol source	Sulfotransferase	Acetyltransferase	
	PAPS	AcetylCoA	CoA plus acetylphosphate
	(pmol/mg protein/min)		
Salmonella TA98	<5.0	153.1	135.6
Rat liver	147.0	197.3	12.1
Salmonella plus **rat liver**	133.0	-----not determined------	

The activities were determined as the reductive formation of Glu-P-1 from N-acetoxy or N-sulfonyloxy-Glu-P-1 (12, 19). Cytosols were prepared from *Salmonella typhimurium* TA98 and adult male rats. The data present the mean of the duplicate determination using 20 μM N-hydroxy-Glu-P-1 and 0.25 mM each cofactor.

significant binding was detected in livers of species other than rats. For N-hydroxy-AAF, all the species examined supported the PAPS-dependent activation for DNA binding. The extent of the binding of N-hydroxy-AAF to DNA was highest in the rat, and lowest in the human, but the level in human livers was not significantly different from those of mouse and dog. *p*-Nitrophenol has been used as a diagnostic substrate to determine the level of biological sulfating activity. Thus, we also measured that activity in these preparations. No clear relationship, however, was detected between sulfation of *p*-nitrophenol (2 mM substrate concentration) and N-hydroxy derivatives of Glu-P-1 and AAF. Instead, we observed a good correlation between the sulfation of these N-hydroxy compounds, suggesting that the PAPS-dependent activation of both N-hydroxy-Glu-P-1 and N-hydroxy-AFF is mediated by the same (or a closely related) enzyme throughout the species.

ACETYLATION OF N-HYDROXYARYLAMINES

As shown in Tab. 2, O-sulfation of N-hydroxy-Glu-P-1 was supported by liver cytosol of rats, but no appreciable binding was detected with cytosol obtained from *S. typhimurium* TA98. The addition of the bacterial cytosol to liver cytosol did not reduce the extent of the esterification, indicating that poor catalysis by the bacterial cytosol is not due to the presence of suppressive mechanism or inhibitors. These results indicate a clear difference between *Salmonella* and mammalian livers in their sulfating capacity. In *Salmonella*, acetyltransferase catalyzes the DNA binding of N-hydroxy-Glu-P-1 as reported previously (14,15). The acetyltransferase-mediated O-acetylation occurs in the presence of acetyl CoA as the acetyl donor. O-Acetylation of N-hydroxy-Glu-P-1 was also supported by acetyl phosphate plus CoA in *Salmonella* cytosol, but was only slightly supported by the liver cytosolic system. To further

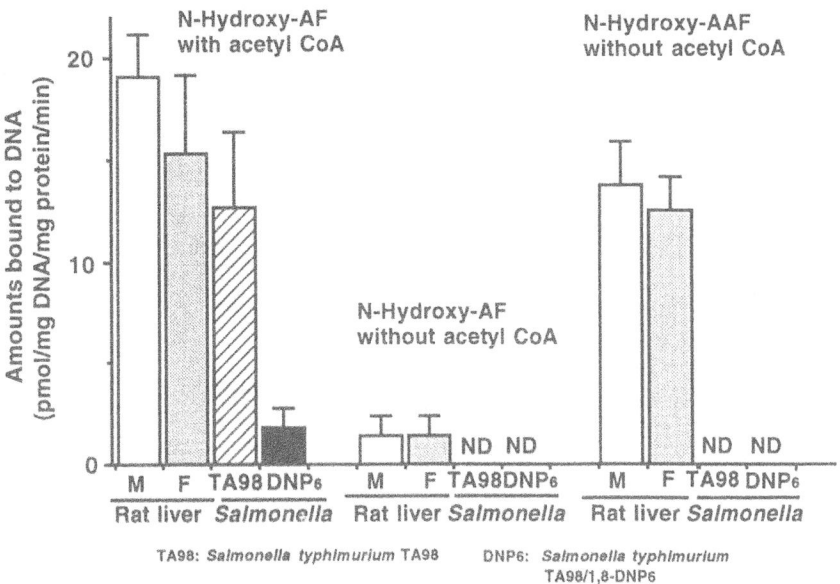

Figure 2. Covalent binding to DNA of N-hydroxy-AF and N-hydroxy-AAF in rat hepatic and bacterial cytosols. Covalent binding of N-hydroxy-AF and N-hydroxy-AAF to DNA was determined using liver cytosols of male (M) and female (F) rats, cytosols of *S. typhimurium* TA98 (TA98) and TA91 1,8-DNP_6 (DNP_6). Data are presented as the mean (column) $\pm$ SD (bar). ND, not detectable (less than 1.0 pmol/mg DNA/mg protein/min).

assess the differences in the characteristics of acetyltransferase in *Salmonella* and mammalian tissues, the activating capacities for N-hydroxy-AF and N-hydroxy-AAF were examined with cytosols of *Salmonella* and rat livers (Fig. 2). The acetyltransferase-mediated binding of N-hydroxy-AF to calf thymus DNA occurred in the system containing liver cytosols of either sex or of *S. typhimurium* TA98. Consistent with previous reports (11,14), strain TA98 1,8-DNP_6 contained no detectable amounts of cytosolic acetyltransferase. Thus, low levels of binding of N-hydroxy-AF were detected with cytosols from strain TA98 1,8-DNP_6. Covalent binding of N-hydroxy-AAF was detectable with liver cytosols from rats without the addition of acetyl CoA in amounts similar to those with N-hydroxy-AF in the presence of acetyl CoA. No binding, however, was detected in *Salmonella* cytosol systems. These results confirmed the idea suggested by the results of the *Salmonella* mutagenesis test (8) that *Salmonella* has no N,O-acetyltransferase activity of N-hydroxyarylacetamides. As shown in Fig. 3, acetyl CoA-dependent covalent bindings of N-hydroxy-Glu-P-1 and N-hydroxy-IQ were detected in both rat and *Salmonella* cytosol systems. The relative ratio of the binding of N-hydroxy-Glu-P-1 and N-hydroxy-IQ were detected in both rat and *Salmonella* cytosol systems. The relative ratio of the binding of N-hydroxy-Glu-P-1 between rat and *Salmonella* systems was not much different, but the ratio of the binding of N-hydroxy-IQ was more than 5-fold higher in the cytosol of *Salmonella* than in rats. These results indicate that, although N-hydroxyarylamines are activated by a common enzymatic system (acetylation) in *Salmonella* and mammalian tissues, the substrate specificity of acetyltransferases differs clearly between *Salmonella* and rat livers.

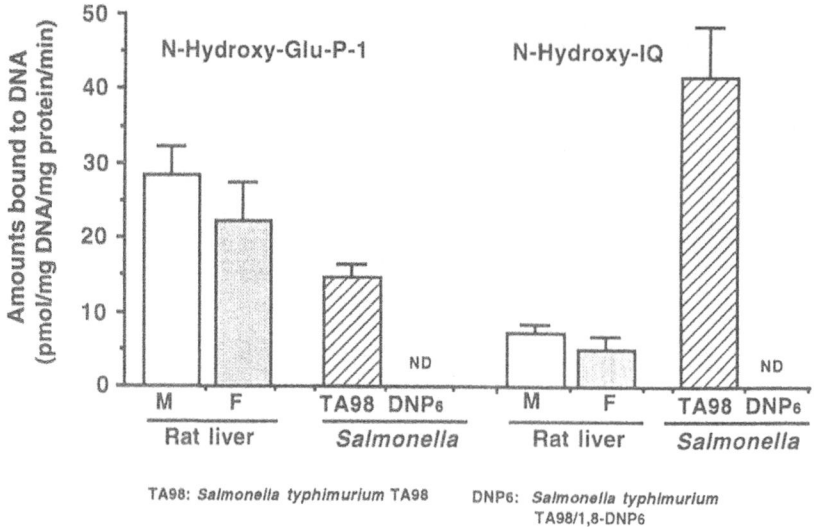

Figure 3. Acetyl CoA-dependent DNA binding of N-hydroxy derivatives of Glu-P-1 and IQ. Experimental details are the same as those in Fig. 2.

MODULATION OF ENZYMATIC ACTIVATION

Several chemicals have been shown to modulate the mutagenic effect of mutagens in various detecting systems. Hayatsu (9) and De Flora (2) showed that chemicals containing a thiol moiety enhanced the number of revertants induced by heterocyclic arylamines in the *Salmonella* mutagenesis test. We also reported that a glutathione conjugate formed by the reaction with N-hydroxy-Trp-P-2 in rat liver cytosol induced higher numbers of revertants than did N-hydroxy-Trp-P-2 in *Salmonella* (13). Increased formation of the reactive intermediate for DNA interaction and modulation of post-DNA damage is suggested as the cause of the enhancements, but few studies have been reported on the mechanism of the modulation of the enzymatic activation. Therefore, we have examined the modulating effects of chemicals on the covalent binding of N-hydroxy-Glu-P-1 to investigate the interaction with activating enzymes. As described in Tab. 3, chemicals containing a free thiol group reduced the amounts of DNA binding when added at 10 mM. In addition, 1 mM *tert*-butylhydroxytoluene (BHT) showed a strong inhibitory effect, although the addition of 10 mg/ml bovine serum albumin had no appreciable effect on the binding. Addition of BHT, dithiothreitol (DTT), or 2-mercaptoethanol to an acetyl CoA-dependent system also resulted in decreased binding, whereas bindings were somewhat increased by the addition of N-acetyl-L-cysteine, L-cysteine, or glutathione in the reaction mixture. The extents of the increases were highest with N-acetyl-L-cysteine, followed by L-cysteine and glutathione, but no stimulatory effect was observed with cystine, the oxidized form of cysteine. Analysis of the N-hydroxy-

Table 3. Effects of chemicals on the cytosolic acetyl CoA- and PAPS-dependent binding of N-hydroxy-Glu-P-1 to DNA.

Chemicals added	(mM)	Amounts bound to DNA (pmol/mg DNA/mg protein/30 min)	
		PAPS	Acetyl CoA
None		140±7 (100)	472±11 (100)
GSH	(10)	36±5 (25.7)	581±8 (123.1)
N-Acetyl-L-cysteine	(10)	96±9 (68.6)	1,400±123 (296.6)
L-Cysteine	(10)	113±14 (80.7)	969±54 (205.3)
L-Cystine	(10)	161±8 (115.0)	451±20 (95.6)
2-Mercaptoethanol	(10)	65±3 (46.4)	2±1 (0.4)
DTT	(10)	84±2 (60.0)	68±3 (14.4)
Methionine	(10)	174±7 (124.2)	282±7 (59.7)
BHT	(1)	19±2 (13.5)	52±4 (11.0)
BSA		130±6 (92.8)	315±9 (66.7)

GSH; glutathione, DTT; dithiothreitol, BHT; *tert*-butylhydroxytoluene, and BSA; bovine serum albumin. BSA was added at the concentration of 10 mg/ml. Data are shown as the mean ± SD of triplicate determinations. Numbers in parentheses indicate the percentages relative to non-treated controls. Other experimental details are the same as described in Table. 2.

Glu-P-1-modified nucleosides did not show any difference in chemical and physical properties after hydrolysis of the modified DNA. In addition, no appreciable difference was observed on the rate of the esterification (12,19). These results suggest that cysteinyl thiols interact in some ways with the putative N-acetoxy-arylamine intermediate to enhance the efficiency of the binding of DNA by stabilizing the intermediate or by another unknown mechanism. The results obtained in the present study, of course, did not exclude the possibility of cysteinyl thiols for protection against suicide inactivation of acetyltransferase by cysteinyl thiols.

CONCLUSION

During the past decade, our understanding of carcinogenesis and mutagenesis has improved, and the results are reflected in the ideas of anticarcinogenesis and antimutagenesis. However, the mechanism of the activation of environmental carcinogens differs markedly between human and experimental animals or bacteria. These differences are caused partly by the differences in the substrate specificity of the activating enzyme and susceptibility to modulating chemicals as described in this paper. Our understanding of even the first step of carcinogenesis, metabolic activation, is still limited. Therefore, a more precise understanding of the difference of the mechanisms of the activation/inactivation and modulation by chemicals between humans and an experimental system is necessary in order to predict the usefulness of a chemopreventive agent in the human situation.

REFERENCES

1. Bartsch, H., M. Dworkin, J.A. Miller, and E.C. Miller (1972) Electrophilic N-acetoxyaminoarenes derived from carcinogenic N-hydroxy-N-acetoxyaminoarenes by enzymatic deacetylation and transacetylation in liver. *Biochem. Biophys. Acta* 286:272-298.

2. De Flora, S., C. Bennicelli, P. Zanacchi, A. Camoirano, A. Morelli, and A. De Flora (1984) In vitro effects of N-acetylcysteine on the mutagenicity of direct-acting compounds and procarcinogens. *Carcinogenesis* 5:505-510.

3. DeBaun, J.R., J.Y. Rowley, E.C. Miller, and J.A. Miller (1968) Sulfotransferase activation of N-hydroxy-2-acetylaminofluorene in rodent livers susceptible and resistant to this carcinogen. *Proc. Soc. Exp. Biol. Med.* 129:268-273.

4. Gong, D., S. Ozawa, Y. Yamazoe, and R. Kato (1991) Purification of hepatic N-hydroxyarylamine sulfotransferases and their regulation by growth hormone and thyroid hormone in rats. *J. Biochem.* 110:226-231.

5. Jakoby, W.B., M.W. Duffel, E.S. Lyon, and S. Ramaswamy (1984) Sulfotransferases active with xenobiotics-Comments on mechanism. In *Progress in Drug Metabolism*, J.W. Bridges and L.F. Chasseaud, eds. Taylor & Francis, Ltd., London, pp. 11-33.

6. Kensler, T., N. Davidson, J. Groopman, H. Prochaska, and P. Talay (1992) Chemoprotection by inducers of electrophile detoxication enzymes. Abstract from these proceedings, p. 8.

7. King, C.M., and B. Philips (1968) Enzyme-catalyzed reactions of the carcinogen N-hydroxy-2-fluorenylacetamide with nucleic acid. *Science* 159:1351-1353.

8. McCoy, E.C., G.D. McCoy, and H.S. Rosenkranz (1982) Esterification of arylhydroxylamines: Evidence for a specific gene product in mutagenesis. *Biochem. Biophys. Res. Comm.* 108:1362-1367.

9. Negishi, T., and H. Hayatsu (1979) The enhancing effect of cysteine and its derivatives on the mutagenic activities of the tryptophan-pyrolysis products. *Biochem. Biophys. Res. Comm.* 88:97-102.

10. Ohgaki, H., S. Takayama, and T. Sugimura (1990) Carcinogenicity of food mutagens and risk assessment. In *Mutagens in Food: Detection and Prevention*, H. Hayatsu, eds. CRC Press, Boca Raton, Florida, pp. 219-228.

11. Saito, K., A. Shinohara, T. Kamataki, and R. Kato (1985) Metabolic activation of mutagenic N-hydroxyarylamines by O-acetyltransferase in *Salmonella typhimurium* TA98. *Arch. Biochem. Biophys.* 239:286-295.

12. Saito, K., A. Shinohara, T. Kamataki, and R. Kato (1986) A new assay for N-hydroxyarylamine O-acetoxyarylamines. *Analyt. Biochem.* 152:226-231.

13. Saito, K., Y. Yamazoe, T. Kamataki, and R. Kato (1983) Activation and detoxication of N-hydroxy-Trp-P-2 by glutathione and glutathione transferases. *Carcinogenesis* 4:1551-1557.

14. Saito, K., Y. Yamazoe, T. Kamataki, and R. Kato (1983) Mechanism of activation of proximate mutagens in Ames' tester strains: The acetyl CoA-dependent enzyme in *Salmonella typhimurium* TA98 deficient in TA98/1,8-DNP$_6$ catalyzes DNA binding as the cause of mutagenicity. *Biochem. Biophys. Res. Comm.* 116:141-147.

15. Shinohara, A., K. Saito, Y. Yamazoe, T. Kamataki, and R. Kato (1986) Acetyl coenzyme A-dependent activation of N-hydroxy derivative of carcinogenic

arylamines; Mechanism of activation, species difference, tissue distribution and acetyl donor specificity. *Cancer Res.* 46:4362-4367.

16. Singer, S.S. (1984) Glucocorticoid sulphotransferases in rats and other animal species. *Biochem. Soc. Trans.* 12:35-39.

17. Yamazoe, Y., M. Abu-Zeid, G. Dawei, N. Staiano, and R. Kato (1989) Enzymatic acetylation and sulfation of N-hydroxyarylamines in bacteria and rat livers. *Carcinogenesis* 10:1675-1679.

18. Yamazoe, Y., D. Gong, N. Murayama, and R. Kato (1989) Regulation of hepatic cortisol sulfotransferase by pituitary hormone. *Molec. Pharmacol.* 35:707-712.

19. Yamazoe, Y., S. Manabe, N. Murayama, and R. Kato (1987) Regulation of hepatic sulfotransferase catalyzing the activation of N-hydroxyarylamide and N-hydroxyarylamine by growth hormone. *Molec. Pharmacol.* 32:536-541.

INDUCTION OF RAT LIVER GSH TRANSFERASES BY 1,2-DITHIOLE-3-THIONE ILLUSTRATES BOTH ANTICARCINOGENIC AND TUMOR-PROMOTING PROPERTIES

David J. Meyer,[1] Brian Coles,[1] Jonathan Harris,[1] Kim S. Gilmore,[1] Kevin Raney,[2] Thomas M. Harris,[2] F. Peter Guengerich,[2] Thomas W. Kensler,[3] and Brian Ketterer[1]

[1] Cancer Research Campaign Molecular Toxicology Research Group
Department of Biochemistry and Molecular Biology
University College and Middlesex School of Medicine
London W1P 6DB, ENGLAND

[2] Center for Molecular Toxicology
Biochemistry Department
Vanderbilt University
Nashville, Tennessee 37232-0146 USA

[3] The Johns Hopkins University
School of Hygiene and Public Health
Department of Environmental Health Sciences
Baltimore, Maryland 21205-2179 USA

INTRODUCTION

1,2-Dithiole-3-thione is an antioxidant showing protective effects in rodents against carcinogens such as aflatoxin B_1 (1,2,13,37). It increases hepatic GSH and several GSH-dependent enzymes including GSH transferases (GSTs) (2,14) that catalyze the detoxification of aflatoxin B_1-8,9-oxide, the ultimate carcinogen. We had previously concluded (5) that in phenobarbitol-induced rat liver, GSTs 1-1 and 1-2 are most important among soluble GSTs in this reaction. We have since shown that these fractions contain not only two forms of GST subunit 1, namely 1a(Ya1) and 1b(Ya2) (25), which are distinct gene products (9) corresponding, respectively, to the cDNAs pGTR 261 (16) and pGTB 38 (28) but also present is a small amount of GST subunit 10 (23). It has been shown by Hayes et al. (9) that feeding aflatoxin-induced subunit 1b more than 1a suggsting that 1b may be more active in detoxication of the oxide. In the present report, the induction of GST subunits in rat liver by 1,2-dithiole-3-thione is quantitated and the relative activity of GST subunits 1a, 1b, 2, and 10 toward aflatoxin B_1-8,9-oxide is determined.

EXPERIMENTAL PROCEDURES

Sprague-Dawley rats (160-200 g) were maintained for 5 days on control diet, diet containing 0.075% (w/w) 1,2-dithiole-3-thione, or control diet with 0.1% (w/v) sodium phenobarbitone in their drinking water. They were killed by cervical dislocation. Livers were perfused with 0.25 M sucrose (0-4 C) and homogenized in KCl (1.15%, w/w), 2 mM dithiothreitol, 25 μM phenylmethanesulphonyl fluoride, 10 mM Na phosphate, pH 7.0. The soluble supernatant fraction was obtained by centrifugation at 105,000 g for 1 hr. Two experiments were carried out. First, 3 groups of 4 rats were treated as above and the soluble supernatants combined prior to analysis or for purification of GST 1-1. To assess the significance of differences observed, a second experiment was carried out with control and 1,2-dithiole-3-thione treatment with 3 animals per group and the analyses were carried out separately on soluble supernatant from each liver. Results reported for control and 1,2-dithiole-3-thione treatments are from the second experiment and for phenobarbitone are from the first experiment.

GSH Transferase Analysis

Portions of soluble supernatant were applied to GSH-agarose (36) and the bulk of GSH transferases purified by elution with 25 mM GSH, 1 mM dithiothreitol, 10% (v/v) glycerol, 0.1 M tris-NaOH, pH 9.6 at 4°C. GSH transferase subunits were then quantitated by reverse phase hplc (21). GST subunit 10 (23) was not adequately separated from subunit 2 by reverse phase hplc, so the combined 2/10 peak was subjected to SDS PAGE, and the relative amounts of subunits 2 and 10 so separated were determined from staining with coomassie blue.

The material not retained by GSH-agarose that contained GSH transferases 5-5, 12-12, and traces of other GSH transferases was dialyzed for 16 h (30 vol, 1 change) against 2 mM EDTA, 5 mM 2-mercaptoethanol, 10% (v/v) glycerol, 10 mM Na phosphate, pH 7.0. The content of GSH transferase 5-5 was assayed from the activity with p-nitrophenethyl bromide (8,22), and GST 12-12 was determined from activity with menaphthyl sulphate.

Purification of GSH Transferase

In order to determine the relative contribution of GST subunits 1a, 1b, 2, and 10 GSH conjugation of aflatoxin B_1-8,9-oxide, portions of liver soluble supernatant from control and 1,2-dithiole-3-thione-fed rats were dialyzed against 20 mM tris-HC, pH 8.4, and applied to DEAE-cellulose equilibrated with this buffer. The fraction neither retained nor retarded ("DE flo-thru") was concentrated and transferred into 2 mM dithiothreitol, 5% (v/v) glycerol, 40 mM Na phosphate, and fractionated by phosphate gradient elution from hydroxyapatite using an HPHT column (Bio-Rad, Richmond, CA, USA) operated at 0.35 ml/min by an fplc system (Pharmacia-LKB, Uppsala, Sweden). GSTs 1-1 and 2-2 were finally separated from subunit 10 by

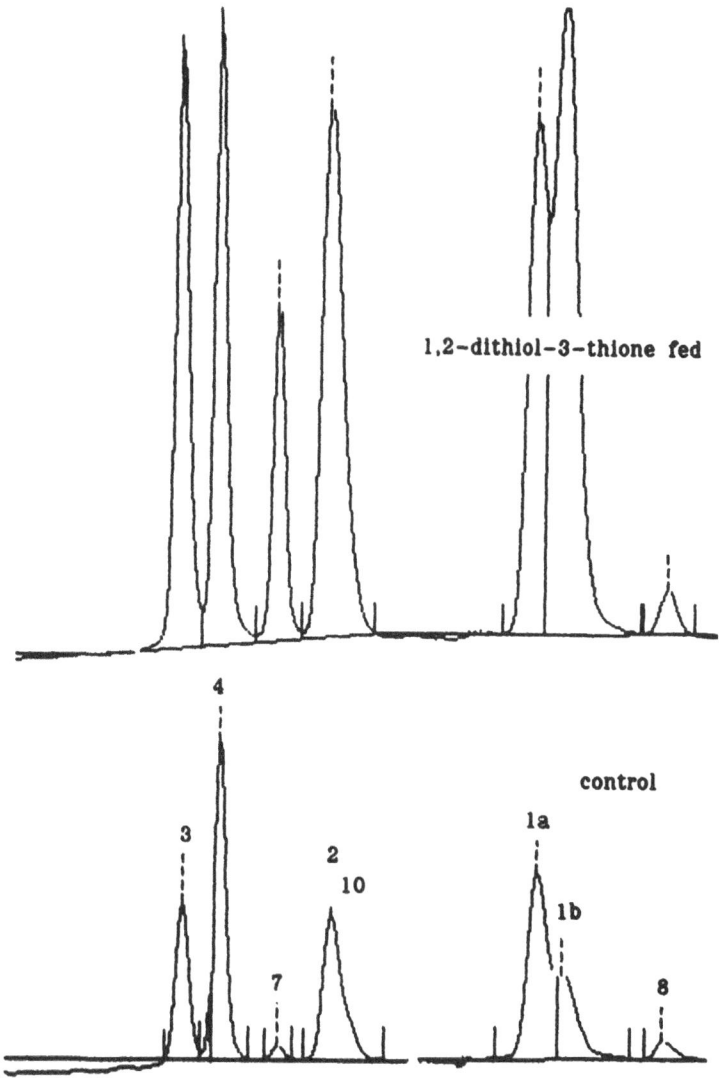

Figure 1. GST subunit analysis by reverse phase hplc.

cation-exchange fplc on a Mono S column equilibrated with 10% (v/v) glycerol, 0.2 mM dithiothreitol, 40 mM (2-[N-morpholino]ethane-sulphonic acid) hemisodium salt, pH 6.1, eluting with a gradient of NaCl. The GST subunit content of the DE flo-thru and the purified enzymes was determined by reverse phase hplc as described above.

Enzyme Activity

Purified GSTs and the DE flo-thru fractions were dialysed against 50 mM Na phasphate, pH 7.0 and their activity toward aflatoxin B_I-8,9-oxide was assayed at 37°C in this buffer with 5 mM GSH and 50 μM of the oxide. After 30s samples were acidified and the glutathione conjugate was quantitated by reverse phase hplc.

Table 1. GST subunit composition of control, 1,2-dithiol-3-thione- and phenobarbitone-treated rat liver soluble fraction.

GST subunit	Class	Control	1,2-Dithiol-3-thione		Phenobarbitone	
		GST subunit content in µg/g liver			(fold increase)	
1a	α	125 ± 66	567 ± 129	(4.5)	286	(2.3)
1b	α	69 ± 16	771 ± 26	(11.2)	581	(8.4)
2	α	178 ± 7	591 ± 29	(3.3)	553	(2.6)
3	µ	191 ± 5	671 ± 41	(3.5)	1313	(6.9)
4	µ	292 ± 8	559 ± 23	(1.9)	791	(2.7)
5	θ	9 ± 1	29 ± 6	(3.1)	13	(1.4)
7	π	14 ± 7	286 ± 22	(20.4)	8	(0.6)
8	α	23 ± 2	55 ± 7	(2.4)	88	(3.8)
10	α	13 ± 4	87 ± 9	(6.7)	nd	
12	θ	481	631	(1.3)	nd	

1,2-dithiol-3-thione, 0.075% (w/w); Na phenobarbitone 0.1% (w/v) in water; nd, not determined.

Immunohistochemistry

Portions of control and 1,2-dithiole-3-thione-induced liver were frozen in liquid nitrogen. Frozen sections were fixed with acetone and probed with rabbit antiserum to GST 7-7 (Medlabs, Dublin, Eire). Following treatment with horseradish peroxidase-coupled antirabbit IgG antiserum, the antigen was visualized using diaminobenzidine with hematoxylin counterstaining.

RESULTS

Induction of GSH Transferases by 1,2-Dithiole-3-thione

Examples of reverse phase hplc analyses of GSTs purified by affinity chromatography from a control and a 1,2-dithiole-3-thione-fed liver are shown in Fig. 1. There is a clear increase in the content of GST subunits in the experimental group, the most notable being subunits 7, 1b and 10. The quantitation of these data together with the determination of GSTs 5-5 and 12-12 that do not bind to GSH-agarose are given in Tab. 1, where they are compared with the inducing effects of phenobarbitone. Based on many GST analyses of rat liver samples in this laboratory, a 20% difference should be considered a significant change. Apart from subunit 7, the effect of 1,2-dithiole-3-thione and phenobarbitone on the hepatic GSTs is similar. The dramatic induction of GST 7-7 was examined by immunohistochem-istry. In the control liver (Fig. 2a), GST subunit 7 is localized primarily to the biliary epithelia and the lining of the hepatic portal vein. A faint staining was present in the hepatocytes.

In contrast, the 1,2-dithiole-3-thione induced liver (Fig. 2b) shows a strong even staining in all hepatocytes such that the bile ducts are much less distinct.

Relative Activity of GST Subunits la, 1b, 2, and 10 Toward Aflatoxin B_1 8,9-Oxide

In order to determine whether the induction of GST subunit 1b contributes significantly to the detoxification of aflatoxin B_1, GST 1-1 containing different proportions of subunits 1a and 1b was purified from control and 1,2-dithiole-3-thione-treated samples. The ratios of 1b to 1a were 0.38 and 2.0, respectively. The activity of the preparation enriched in subunit 1b toward aflatoxin B_1-8,9-oxide was about twice that of the control at both 25 μM and 50 μM substrate (Tab. 2). Assuming that each monomer in GST 1-1 dimers acts independently in the catalysis of aflatoxin B_1-8,9-oxide, as is the case with model substrates (6), subunit 1b is deduced to be 11.5 and 8.9 times as active as subunit 1a at 25 μM and 50 μM respectively of initial substrate concentration.

DISCUSSION

Induction of GSH Transferases

The results show that the previously observed hepatic induction of GSH transferase activity by 1,2-dithiole-3-thione (14) is associated with increased levels of most of the GST isoenzymes, the largest changes being in subunits 7, 1b, and 10. The induction of subunit 1b also occurs with phenobarbitone and is likely to be due to the "antioxidant-responsive element" present in the upstream, regulatory region of this gene (27,32). It has been proposed that such induction occurs with a diverse group of compounds, including some antioxidants, which are all Michael addition acceptors (33,34). A similar regulator has been found in a murine alpha class GST gene and

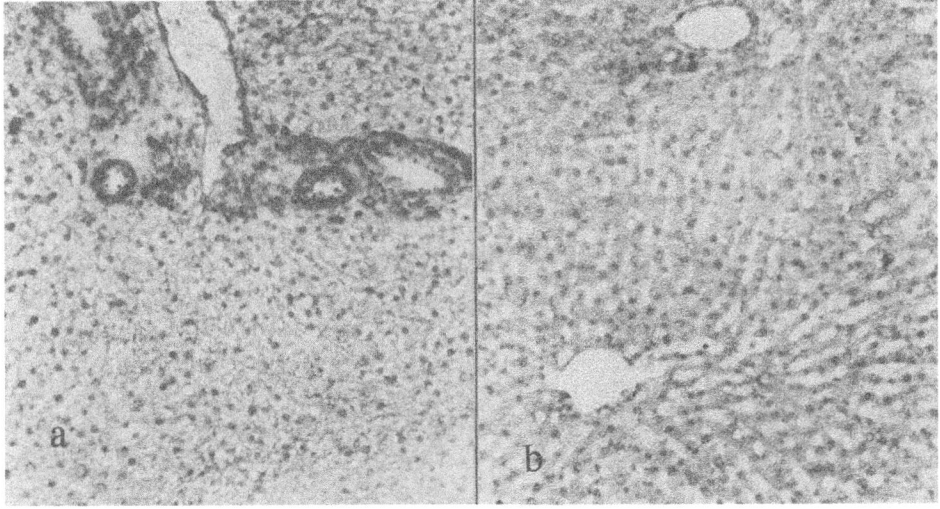

Figure 2. Distribution of GST subunit 7 by immunohistochemistry in control (a) and 1,2-dithiole-3-thione-induced (b) liver.

Table 2. Contribution of GST subunits 1a, 1b, 2, and 10 to GSH conjugation of aflatoxin B_1-8,9-oxide.

Fraction	GST Subunit Composition (%)	AFB-SG (pmol/mg)[a]
GST 1-1 control	1a,72.5; 1b,37.5	1710
GST 1-1 DTT[b]	1a,33.2; 1b,66.8	1060
GST 2-2	2,100	470
DE flo-thru control	1a,25.1; 1b, 13.3; 2,61.0; 10,0.3	460
DE flo-thru DTT	1a,22.1; 1b,30.4; 2,44.3; 10,3.2	2498

[a]Note that activity is in pmol AFB-SG recovered after 30s reaction per mg total protein. The DE flo-thru fractions contain proteins other than GSTs.
[b]DTT, isolated from liver of 1,2-dithiole-3-thione-fed rat.

termed an "electrophile-responsive element" (EpRE) (7). Presumably, induction by such Michael acceptors involves reaction with critical nucleophilic thiols in *trans*-acting proteins associated with the EpRE. Alternatively, glutathione conjugates of the electrophiles might act in a regulatory role as in the case of leukotrienes.

The large, rapid induction of GST subunit 7 in hepatocytes by 1,2-dithiole-3-thione is novel. Subunit 7 is present at very low levels in mature normal rat hepatocytes and is well known to be greatly elevated in preneoplastic foci (15) and hepatoma (19). Induction in normal hepatocytes has only been seen in the short term with lead nitrate treatment (31). Expression of subunit 7 in neoplasia is thought to be controlled by an upstream phorbol ester-responsive element (6,26) that acts through protein kinase C (4). It has been shown that this kinase, isolated from the brain, is activated by low levels of lead ions (18) and is, therefore, probably also involved in the action of lead nitrate.

The induction of GST subunit 10 is also novel, but it may easily have been missed in previous studies of GST induction. Higher levels of subunit 10 are present in neonatal and juvenile rat liver (20). Furthermore, both subunits 10 and 7 predominate in fetal hepatocytes (35), but the significance of this is unclear.

Effect of 1,2-Dithiole-3-thione on GSH Conjugation of Aflatoxin B_1-8,9-Oxide

Resistance to aflatoxin B_1 acute toxicity and carcinogenic effects appears to be related to capability to detoxify aflatoxin B-8,9-oxide by GSH conjugation catalyzed by GSTs. Thus, protection is increased by inducers of GSTs such as 1,2-dithiole-3-(12) thione, its derivative Oltipraz (14), ethoxygen (13), phenobarbitone (10,17), and butylated hydroxytoluene (11). The results presented here show that the induction of GST subunits 1a, 1b, 2, and particularly 10 will significantly increase the capacity of rat liver to detoxify aflatoxin B_1-8,9-oxide. The high activity of GST subunit 10 is in accord with observations of Neal (24) who reported that the GSH conjugation of aflatoxin B_1-8,9-oxide was much higher in 3-week than in 8-week rat liver that corresponds developmentally only with subunit 10 (20). Second, the murine alpha class GST that shows high activity toward aflatoxin B_1-8,9-oxide (29) shows structural homology with subunit 10 (3).

Although it is clear that alpha class GSTs are associated with high activity toward aflatoxin B_1-8,9-oxide in the rat and mouse, in man, GST μ is more active than the known human alpha class enzymes (30). It is of interest to determine whether or not a human orthologue of rat GST subunit 10 exists.

REFERENCES

1. Ansher, S.S., P. Dolan, and E. Bueding (1983) Chemoprotective effects of two dithiolthiones and of butyl-hydroxy anisole against carbon tetrachloride and acetaminophen toxicity. *Hepatology* 3:932-935.

2. Ansher, S.S., P. Dolan, and E. Bueding (1986) Biochemical effects of dithiolthiones. *Food Chem. Toxicol.* 25:405-415.

3. Beutler, T.M., and D.L. Eaton (1992) Complementary DNA cloning, messenger RNA expression and induction of α-class glutathione S-transferase in mouse tissues. *Cancer Res.* 52:314-318.

4. Castagna, M., Y. Takai, K. Kaibuchi, K. Sano, U. Kikkawa, and Y. Nishizuka (1982) Direct activation of calcium-activated, phospholipid-dependent protein kinase by tumour-promoting phorbol esters. *J. Biol. Chem.* 257:7847-7851.

5. Coles, B., D.J. Meyer, B. Ketterer, C.A. Stanton, and R.C. Garner (1985) Studies on the detoxication of microsomally-activated aflatoxin B_1 by glutathione and glutathione transferases in vitro. *Carcinogenesis* 6:693-697.

6. Cowell, I.G., K.H. Dixon. S.E. Pemble, B. Ketterer, and J.B. Taylor (1985) The structure of the glutathione S-transferase π gene. *Biochem. J.* 255:79-83.

7. Friling, R.S., A. Bensimon, Y. Tichauer, and V. Daniel (1990) Xenobiotic-inducible expression of murine glutathione S-transferase Ya subunit gene is controlled by an electrophile-responsive element. *Proc. Natl. Acad. Sci., USA* 87:6258-6262.

8. Habig, W.B., M.J. Pabst, and W.B. Jakoby (1974) Glutathione S-transferases. The first step in mercapturic acid formation. *J. Biol. Chem.* 249:7130-7139.

9. Hayes, J.D., L.A. Kerr, D.J. Harrison, A.D. Cronshaw, A.G. Ross, and G.E. Neal (1990) Preferential overexpression of the class alpha rat Ya2 glutathione S-transferase subunit in livers bearing aflatoxin-induced pre-neoplastic nodules. *Biochem. J.* 268:295-302.

10. Holeski, C.J., D.L. Eaton, D.H. Monroe, and G.M. Bellamy (1987) Effects of phenobarbitol on the biliary excretion of aflatoxin P1-glucuronide and aflatoxin B_1-S-glutathione in the rat. *Xenobiotica* 17:139-153.

11. Jhee, E.-C., L.L. Ho, K. Tsuji, P. Gopalan, and P.D. Lotlikar (1989) Effect of butylated hydroxyanisole pretreatment on in vitro hepatic aflatoxin B_1-DNA binding and aflatoxin B_1-glutathione conjugation in rats. *Cancer Res.* 49:1357-1360.

12. Kensler, T.W., P.A. Egner, M.A. Trush, and E. Bueding (1985) Modification of aflatoxin B_1 binding to DNA in vivo in rats fed phenolic antioxidants and a dithiolthione. *Carcinogenesis* 6:759-763.

13. Kensler, T.W., P.A. Egner, N.E. Davidson, B.D. Roebuck, A. Pikul, and J.D. Groopman (1986) Modulation of aflatoxin metabolism, aflatoxin-N^7-guanine

formation and hepatic tumorigenesis in rats fed ethoxyquin: Role of induction of glutathione S-transferases. *Cancer Res.* 46:3924-3931.

14. Kensler, T.W., P.A. Egner, P.M. Dolan, J.D. Groopman, and B.D. Roebuck (1987) Mechanism of protection against aflatoxin tumorigenicity in rats fed 5, 2-pyrinazyl-4-methyl-1, 2-dithiol-3-thione (oltipraz) and related 1,2-dithiol-3-thiones and 1,2-dithiol-3-ones. *Cancer Res.* 47:4271-4277.

15. Kitahara, A., K. Satoh, K. Nishimura, T. Ishikawa, K. Ruike, K. Sato, H. Tsuda, and N. Ito (1984) Changes in molecular forms of rat hepatic GSH S-transferase during hepatocarcinogenesis. *Cancer Res.* 44:2698-2703.

16. Lai, H.C.J., N.Q. Li, M.J. Weiss, C.C. Reddy, and C.-P.D. Tu (1984) The nucleotide sequence of a rat liver glutathione S-transferase subunit cDNA clone. *J. Biol. Chem.* 259:5536-5542.

17. Lotlikar, P.D., H.G. Raj, L.S. Bohm, L.L. Ho, E.-C. Jhee, K. Tsuji, and P. Gopalan (1989) A mechanism of inhibition of aflatoxin B_1-DNA binding in the liver by phenobarbital treatment. *Cancer Res.* 49:951-957.

18. Markovac, J., and G.W. Goldstein (1988) Picomolar concentrations of lead stimulate brain protein kinase C. *Nature* 347:71-73.

19. Meyer, D.J., D. Beale, K.H. Tan, B. Coles, and B. Ketterer (1985) Glutathione transferase in primary rat hepatomas. *FEBS. Lett.* 184:139-143.

20. Meyer, D.J., K.H. Tan, and B. Ketterer (1985) Recent studies of selenium-independent GSH peroxidases (GSH transferases) from the rat. In *Free Radicals in Liver Injury,* G. Poli, K.H. Cheeseman, M.U. Dianzani, and T.F. Slater, eds. IRL Press, Oxford, pp. 221-224.

21. Meyer, D.J., E. Lalor, B. Coles, A. Kispert, P. Alin, B. Mannervik, and B. Ketterer (1989) Single step purification and hplc analysis of glutathione transferase 8-8 in rat tissues. *Biochem. J.* 260:785-788.

22. Meyer, D.J., B. Coles, S.E. Pemble, K.S. Gilmore, G.M. Fraser, and B. Ketterer (1991) Theta, a new class of glutathione transferases purified from rat and man. *Biochem. J.* 274:414-419.

23. Meyer, D.J., K.S. Gilmore, B. Coles, K. Dalton, P.B. Hulbert, and B. Ketterer (1991) Structural distinction of rat GSH transferase subunit 10. *Biochem. J.* 274:619.

24. Neal, G.E. (1989) GST and resistance to aflatoxin B_1. In *Glutathione S-Transferases and Drug Resistance,* J.D. Hayes, C.B. Pickett, and T.J. Mantle, eds. Taylor and Francis, London, pp. 341-346.

25. Ostlund Farrants, A.-K., D.J. Meyer, B. Coles, C. Southan, A. Aitken, P.J. Johnson, and B. Ketterer (1987) The separation of glutathione transferase subunits by using reverse phase hplc. *Biochem. J.* 245:423-428

26. Okuda, A., M. Imagawa, Y. Maeda, M. Sakai, and M. Muramatsu (1989) Structural and functional analysis of an enhancer GPE1 having a phorbol-12-0-tetradecanoate 13-acetate responsive element-like sequence found in the rat glutathione transferase P gene. *J. Biol. Chem.* 264:16919-16926.

27. Paulsen, K.E., J.E. Darnell, T. Rushmore, and C.B. Pickett (1990) Analysis of the upstream elements of the xenobiotic compound-inducible and positionally regulated glutathione S-transferase Ya gene. *Molec. and Cell Biol.* 10:1841-1852.

28. Pickett, C.B., C.A. Telakowski-Hopkins, G.J.-F. Ding, L. Argenbright, and A.Y.H. Lu (1984) Complete nucleotide sequence of a glutathione S-transferase mRNA and the regulation of the Ya, Yb and Yc mRNAs by 3-methylcholanthrene and phenobarbital. *J. Biol. Chem.* 259:5182-5188.

29. Ramsdell, H.S., and D.L. Eaton (1990) Mouse liver glutathione transferase isoenzyme activity toward aflatoxin B_1-8,9-epoxide and benzo[a]pyrene-7,8-dihydrodiol-9,19-epoxide. *Toxicol. Appl. Pharmacol.* 105:216-225.

30. Raney, K.D., D.J. Meyer, B. Ketterer, T.M. Harris, and F.P. Guengerich (1992) Glutathione conjugation of aflatoxin B_1 *exo* and *endo* epoxides by rat and human glutathione S-transferases. *Chem. Res. Toxicol.* (in press).

31. Roomi, M.W., A. Columbano, G.M. Ledda-Columbano, and D.S.R. Sarma (1987) Induction of the placental form of glutathione S-transferase by lead nitrate administration in rat liver. *Toxicol. Pathol.* 15:202-205.

32. Rushmore, T.H., R.G. King, K.E. Paulsen, and C.B. Pickett (1990) Regulation of glutathione S-transferase Ya subunit gene expression: Identification of a unique xenobiotic-responsive element controlling inducible expression by planar aromatic compounds. *Proc. Natl. Acad. Sci., USA* 87:3826-3830.

33. Spencer, S.R., C.A. Wilczak, and P. Talalay (1990) Induction of glutathione transferases and NAD(P)H:quinone reductase by fumaric acid derivatives in rodent cells and tissues. *Cancer Res.* 50:7871-7875.

34. Talalay, P., M.J. De Long and H.J. Prochaska (1988) Identification of a common chemical signal regulating the induction of enzymes that protect against chemical carcinogenesis. *Proc. Natl. Acad. Sci., USA* 85:8261-8265.

35. Tee, L.B.G., K.S. Gilmore, D.J. Meyer, B. Ketterer, Y. Vandenberghe, G.C.T. Yeoh (1992) Expression of glutathione S-transferase during rat liver development. *Biochem. J.* 282:209-218.

36. Vander Jagt, D.L., L.A. Hunsaker, K.B. Garcia, and R.E. Royer (1985) Isolation and characterization of the multiple glutathione S-transferases from human liver. *J. Biol. Chem.* 260:11603-11610.

37. Wattenberg, L.W., and E. Bueding (1986) Inhibitory effects of 5-(2-pyrinazyl)-4-methyl-1,2-dithiol-3-thione (Oltipraz) on carcinogenesis induced by benzo[a]pyrene, diethylnitrosamine and uracil mustard. *Carcinogenesis* 7:1379-1381.

ORGAN-SPECIFIC MODIFICATION OF CARCINOGENESIS

BY ANTIOXIDANTS IN RATS

Masao Hirose, Tomoyuki Shirai, Satoru Takahashi,
Kumiko Ogawa, and Nobuyuki Ito

First Department of Pathology
Nagoya City University Medical School
1 Kawasumi, Mizuho-cho
Mizuho-ku, Nagoya 467 JAPAN

INTRODUCTION

The synthetic antioxidant butylated hydroxyanisole (BHA) has been shown to induce forestomach carcinomas in rats and hamsters. It strongly enhances second-stage forestomach and urinary bladder carcinogenesis in rats pretreated with carcinogens, but inhibits liver, lung, and mammary carcinogenesis (13). BHA at high doses rapidly induces cell proliferation and an increase in DNA synthesis in the forestomach epithelium of both rats and hamsters, these presumably playing an important role in the observed carcinogenic and promoting effects. Recently, some phenolic antioxidants, i.e., caffeic acid, sesamol, and catechol, have also been shown to induce strong cell proliferation in either the forestomach or glandular stomach epithelium of rats (9) and hamsters (6). These antioxidants similarly appear to exert carcinogenic potential in their target organs or enhance stomach carcinogenesis. Furthermore, simultaneous treatment with diethylmaleate (DEM), a known glutathione depleter, significantly reduced BHA-induced forestomach cell proliferation (7), indicating a role for glutathione in the BHA-induced cell proliferation.

In the present paper, carcinogenicities of BHA, caffeic acid, sesamol, catechol, and hydroquinone, as well as modification of carcinogenesis by antioxidants and the effect of DEM on carcinogenesis induced by BHA are reviewed.

MATERIALS AND METHODS

Carcinogenicities of Phenolic Antioxidants in Rats

Groups of 30 or 55 male six-week-old F344 rats (Charles River Japan Inc., Kanagawa) were treated with 2% BHA (Wako Pure Chemical Industries, Osaka, purity ⟩ 98%), 2% caffeic acid (Tokyo Kasei Kogyo Co, Tokyo, purity ⟩ 98%), 2% sesamol (Fluka Chemie, AG, Switzerland, purity ⟩ 98%), 0.8% catechol or 0.8% hydroquinone (Wako Pure Chemical Industries, Osaka, purity ⟩ 99%) in Oriental MF powdered basal diet, or basal diet alone, for 104 weeks. All surviving animals were killed under ether anesthesia and subjected to complete autopsy at the end of week 104. Six and three sections were cut from forestomach and glandular stomach, respectively. Tissues were processed in the usual way for histopathological examination.

Enhancing Effects of Phenolic Antioxidants on Stomach Carcinogenesis in Rats Pretreated with MNNG

Groups of 15 to 20 male, six-week-old F344 rats were given a single ig administration of 150 mg/kg bw of N-methyl-N'-nitro-N-nitrosoguanidine (MNNG). Starting 1 week later, they were given a diet containing 1% BHA, 1% caffeic acid, 0.8% catechol, 0.8% hydroquinone, or basal diet alone for 51 weeks. Other groups of 15-20 animals each were treated with 1% BHA, 1% caffeic acid, 0.8% catechol, 0.8% hydroquinone, or basal diet alone, for 51 weeks without MNNG pretreatment. Animals were killed at the end of week 52.

Modifying Effects of Antioxidants on Second-Stage Carcinogenesis in Various Organs

Separate groups of six- to seven-week-old F344 male or SD female rats were treated with carcinogens as initiators as follows: N,N-dibutylnitrosamine (DBN, 0.05% in drinking water, 4 weeks), N-methyl-N-amylnitrosamine (MNAN, 25 mg/kg bw sc, 1/week X 3), N-methylnitrosourea (MNU, 20-50 mg/kg bw ip, 2/week X 2-4), 1,2-dimethylhydrazine (DMH, 20 mg/kg bw sc, 1/week X 4), N-ethyl-N-hydroxyethylnitrosamine (EHEN, 0.1% in drinking water, 2 weeks), 2,2'-dihydroxy-di-n-propylnitrosamine (DHPN, 0.1% in drinking water, 2 weeks), N-butyl-N-(4-hydroxybutyl)nitrosamine (BBN, 0.05% in drinking water, 4 weeks) or 7,12-dimethylbenz(a)anthracene (DMBA, 50 mg/kg bw ig, single). Then they were treated with a diet containing 1-2% BHA, 0.8% catechol or 0.8% hydroquinone for 32-54 weeks. Further groups of 10-15 animals were treated with a carcinogen, a test chemical, or the basal diet alone. The animals were sacrificed under ether anesthesia at weeks 32-54 and the target organs of each carcinogen were examined histopathologically.

Antagonistic Effect of Diethylmaleate on the Promotion of Forestomach Carcinogenesis by BHA in Rats Pretreated with MNNG

At 6 weeks of age, 60 male F344 rats were given a single ig administration of 150 mg/kg body weight of MNNG. Starting 1 week later, groups of 15 rats were administered a diet containing 1% BHA plus 0.2% DEM, 1% BHA, 0.2% DEM, or basal diet alone for 51 weeks. Further groups of 15 animals were treated similarly

Table 1. Neoplastic and preneoplastic lesions of the stomach and kidney

Chemicals	No. of rats	Forestomach		Glandular Stomach		Kidney	
		Papilloma	Carcinoma	Adenoma	Carcinoma	Hyperplasia	Adenoma
BHA	52	52(100)***	18(34.6)***	0	0	0	0
Caffeic acid	30	23(77)**	17(57)***	1(3)	0	21(70)***	4 (13)
Sesamol	29	10(34)***	9(31)***	1(3)	0	0	0
Catechol	28	2(7)	0	28(100)***	15(54)***	0	0
Hydroquinone	30	0	0	0	0	30(100)***	14(47)***
Basal diet	30	0	0	0	0	1(3)	0

Significantly different from control values at **P<0.01, ***P<0.001

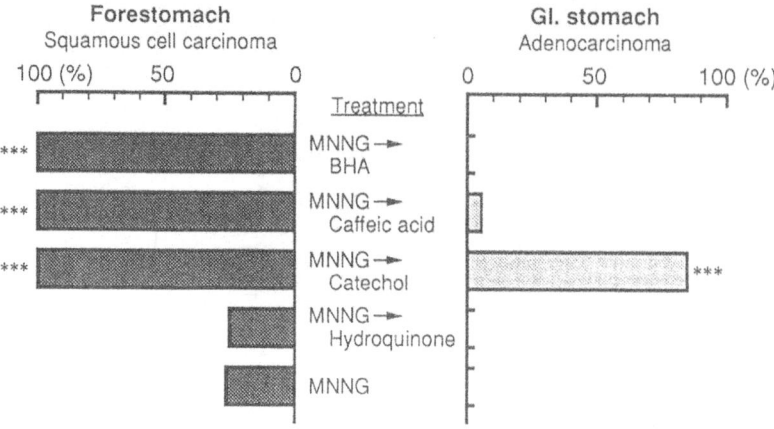

Figure 1. Incidence of forestomach squamous cell carcinomas and glandular stomach adenocarcinomas in rats treated with a single intragastric administration of 150 mg/kg bw MNNG followed by phenolic antioxidants. ■, forestomach squamous cell carcinomas; , glandular stomach carcinomas. ***, $P < 0.001$ vs MNNG → Basal diet group value

without MNNG pretreatment. Once every 2-4 weeks, the animals were weighed and the food intake was measured. All animals were killed under ether anesthesia at the end of week 52.

**Inhibition by Diethylmaleate of
BHA-induced Forestomach Carcinogenesis**

Groups of 20-22 male, six-week-old F344 rats were treated with powdered diet containing 2% BHA plus 0.2% DEM, 2% BHA, 0.2% DEM, or basal diet alone for 104 weeks and then killed for autopsy. During the experiment, the animals were weighed and food intake was measured once every 2-4 weeks.

Student's t-test and the Fisher extract test were used for statistical evaluation of the data of each experiment.

RESULTS

Carcinogenicities of Phenolic Antioxidants in Rats

At the end of the experiment, body weights of animals treated with antioxidants were generally lower than those of the controls. The results are summarized in Tab. 1. Significant increases in the incidences of forestomach papillomas and squamous cell carcinomas were apparent in rats treated with BHA (100% and 34.6%), caffeic acid (77% and 57%), and sesamol (34% and 31%), respectively. Although catechol also induced hyperplasia, no significant increase in tumor incidence was found. In the pyloric region of glandular stomach, catechol induced ademonas in all rats, and adenocarcinomas in 54% of the rats. In the

Table 2. Modifying effects of antioxidants on carcinogenesis in rats pretreated with different carcinogens

Target organ (carcinogen)	Antioxidants		
	BHA	Catechol	Hydroquinone
Esophagus (DBN, MNAN)	→	↑	↑
Colon (DMH)	→	↓	NE
Liver (DEN, EHEN)	↓	→	→
Lung (DHPN)	↓	↓	→
Kidney (EHEN)	↓	↓	→
Urinary bladder (BBN, MNU, DBN)	↑	→	→
Mammary gland (DMBA)	↓	↓	NE
Thyroid gland (MNU, DHPN)	→	→	NE

NE, not examined

↑, enhancement; →, no effect; ↓, inhibition;

kidneys, significant increases in the incidences of tubular hyperplasia were found in rats treated with hydroquinone or caffeic acid. In addition, ademonas were induced at incidences of 47% and 13%, respectively, in rats fed these compounds (5).

Enhancing Effects of Phenolic Antioxidants on Stomach Carcinogenesis in Rats Pretreated with MNNG

As shown in Fig. 1, all animals treated with 1% BHA, 1% caffeic acid, or 0.8% catechol after MNNG exposure had forestomach squamous cell carcinomas (SCC) whereas only 26% of animals treated with MNNG alone had SCC. The incidence of SCC in animals treated with hydroquinone (25%) was not different from this control value. Development of glandular stomach adenocarcinomas was significantly enhanced only in rats treated with MNNG followed by catechol (94.7% vs 0% in control) (4,8,10).

Modifying Effects of Antioxidants on Second-stage Carcinogenesis in Various Organs

Generally, in organs other than those targeted for carcinogenicity, different modifying effects (promotion and inhibition) were observed depending on the individual organ, as shown in Tab. 2. For example, BHA promoted urinary bladder carcinogenesis whereas it inhibited carcinogenesis in the liver, lung, and mammary

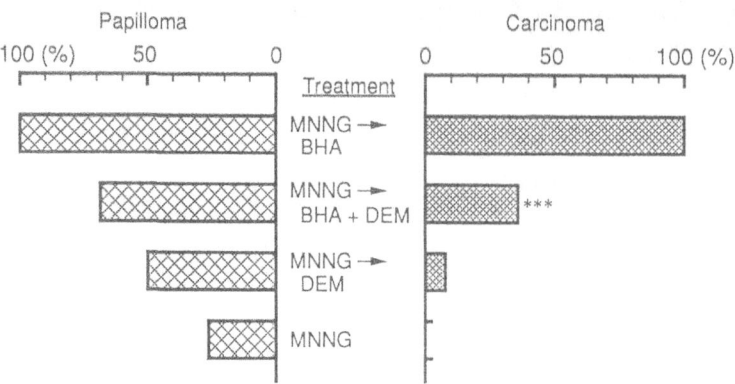

Figure 2. Incidences of forestomach tumors in rats treated with a single intragastric administration of 150 mg/kg bw MNNG followed by 1% BHA, 1% BHA plus 0.2% DEM, 0.2% DEM, or basal diet alone for 51 weeks. ▨, papillomas; ▨, squamous cell carcinomas; *** P ⟨ 0.001 vs MNNG → BHA group value.

gland; catechol promoted esophageal carcinogenesis but inhibited colon, lung, and mammary gland carcinogenesis; hydroquinone enhanced kidney carcinogenesis and, in addition, weakly promoted esophageal carcinogenesis (12-14,18).

Antagonistic Effect of Diethylmaleate on the Promotion of Forestomach Carcinogenesis by BHA in Rats Pretreated with MNNG

At the end of the experiment, a slight reduction in body weight gain was observed in rats treated with MNNG followed by BHA plus DEM as compared with MNNG followed by BHA alone. Tumors observed in the forestomach are summarized in Fig. 2. In the groups receiving MNNG, the incidence of papillomas was significantly increased by subsequent treatment with BHA, with no effect of additional DEM. while BHA markedly enhanced the development of squamous cell carcinomas (100% vs 0%), simultaneous treatment with DEM significantly reduced the incidence of these lesions to 35.7% (P ⟨ 0.001) (7).

Inhibition of Diethylmaleate on
BHA-induced Forestomach Carcinogenesis

Final average body weights of animals treated with BHA plus DEM were slightly lower than those treated with BHA alone, but food consumption was not significantly different between these two groups. In animals treated with BHA, the incidence of squamous cell carcinomas was 20%, additional treatment with DEM slightly reducing this to 4.5% (difference not statistically significant). On the other hand, development of papillomas was significantly reduced by the additional treatment with DEM (85% vs 27.3%, P ⟨ 0.001).

DISCUSSION

In the experiments reviewed above, antioxidants were demonstrated to have a wide variety of actions with regard to chemical carcinogenesis. BHA, caffeic acid, and sesamol proved to be carcinogenic to rat forestomach epithelium, whereas catechol caused neoplasia in the rat glandular stomach epithelium, and caffeic acid and hydroquinone were tumorigenic for the rat kidney. These chemicals are distributed widely in our environment, exposure to man being likely through foods, cigarette smoke, and cosmetic application (11,17). On balance, catechol would appear to be the most important, since the human stomach resembles the rat glandular stomach. It is of direct relevance that catechol or its conjugates are excreted in human urine at levels up to 30 mg/day (1). In the two-stage carcinogenesis models, antioxidants also showed inhibitory or enhancing effects on different organs. Generally, they enhanced carcinogenesis in their carcinogenic target organs, BHA and caffeic acid promoting forestomach carcinogenesis initiated with MNNG, and hydroquinone promoting renal carcinogenesis after EHEN treatment. These enhancing effects on the second stage of carcinogenesis are presumably closely related to cell proliferation and require a far weaker stimulus than that necessary for carcinogenicity (7,17).

It is of interest that DEM potently inhibited BHA-induced forestomach cell proliferation (17) as well as BHA-induced forestomach carcinogenesis and promotion of carcinogenesis. DEM has been shown to deplete reduced glutathione (GSH) level in several organs after intraperitoneal administration (16). Previously, we measured tissue GSH levels in rat forestomach epithelium treated with 2% BHA with or without simultaneous administration of DEM for two weeks. The result was that BHA increased forestomach GSH, and additional treatment with DEM caused an additional elevation. This might reflect an "overshoot" since following a gavage dose of ethyl acrylate, continuous treatment of which induced toxicity followed by cell proliferation and squamous cell carcinomas in rat forestomach epithelium, GSH content of forestomach epithelium was rapidly depleted 6 hours after dosing, but it increased thereafter (2,3). Recently, t-butylquinone has been demonstrated as a metabolite of BHA in forestomach epithelium treated with BHA (15). Quinones readily bind to tissue SH groups. Previously, we showed that BHA covalently binds to protein but not DNA or RNA in the rat forestomach epithelium treated with BHA. Therefore, it is possible that quinone metabolite(s) of BHA bind to tissue GSH and this conjugate(s) may play a role in BHA-induced toxicity, cell proliferation, and carcinogenesis.

REFERENCES

1. Carmella, S.G., E.J. La Voie, and S.S. Hecht (1988) Quantitative analysis of catechol and 4-methylcatechol in human urine. *Food Chem. Toxicol.* 20:587-590.
2. D'Souze, R.W., W.R. Francis, and M.E. Anderson (1988) Physiological model for tissue glutathione depletion and increased resynthesis after ethylene dichloride exposure. *J. Pharm. Exp. Ther.* 245:563-568.
3. Frederick, C.B., G.A. Hazelton, and J.D. Frantz (1990) The histopathological and biochemical response to the stomach of male F344/N rats following two weeks of oral dosing with ethyl acrylate. *Tox. Path.* 18:247-256.
4. Hirose, M., S. Fukushima, Y. Kurata, H. Tsuda, M. Tatematsu, and N. Ito (1988) Modification of N-methyl-N'-nitro-N-bnitroguanidine-induced forestomach and glandular stomach carcinogenesis by phenolic antioxidants in rats. *Cancer Res.* 48:5310-5315.

5.   Hirose, M., S. Fukushima, T. Shirai, R. Hasegawa, T. Kato, H. Tanaka, E. Asakawa, and N. Ito (1990) Stomach carcinogenicity of caffeic acid, sesamol and catechol in rats and mice. *Japan. J. Cancer Res.* 91:207-212.

6.   Hirose, M., T. Inoue, M. Asamoto, Y. Tagawa, and N. Ito (1986) Comparison of the effects of 13 phenolic compounds in induction of proliferative lesions of the forestomach and increase in the labeling indices of the glandular stomach and urinary bladder of Syrian golden hamsters. *Carcinogenesis* 7:1285-1289.

7.   Hirose, M., T. Inoue, A. Masuda, H. Tsuda, and N. Ito (1987) Effects of simultaneous treatment with various chemicals on BHA-induced development of rat forestomach hyperplasia: Complete inhibition of diethylmaleate in a 5-week feeding study. *Carcinogenesis* 8:1555-1558.

8.   Hirose, M., M. Kagawa, K. Ogawa, A. Yamamoto, and N. Ito (1989) Antagonistic effect of diethylmaleate on the promotion of forestomach carcinogenesis by butylated hydroxyanisole (BHA) in rats pretreated with N-methyl-N'-nitro-N-nitrosoguanidine. *Carcinogenesis* 10:2223-2226.

9.   Hirose, M., A. Masuda, K. Imaida, M. Kagawa, H. Tsuda, and N. Ito (1987) Induction of forestomach lesions in rats by oral administration of naturally occurring antioxidants for 4 weeks. *Japan. J. Cancer Res.* 78:317-321.

10. Hirose, M., S. Yamaguchi, S. Fukushima, R. Hasegawa, S. Takahashi, and N. Ito (1989) Promotion by dihydroxybenzene derivatives of N-methyl-N'-nitro-N-nitrosoguanidine-induced F344 rat forestomach and glandular stomach carcinogenesis. *Cancer Res.* 49:5143-5147.

11. IARC Monographs on the Evaluation of Carcinogenic Risk of Chemicals to Humans (1977) IARC, Lyon, Vol. 15, pp. 155-175.

12. Ito, N., and M. Hirose (1987) The role of antioxidants in chemical carcinogenesis. *Japan. J. Cancer Res.* 78:1011-1026.

13. Ito, N., and M. Hirose (1989) Antioxidants: Carcinogenic and chemopreventive properties. *Adv. Cancer Res.* 53:247-302.

14. Ito, N., M. Hirose, S. Fukushima, H. Tsuda, T. Shirai, and M. Tatematsu (1986) Studies on antioxidants: Their carcinogenic and modifying effects on chemical carcinogenesis. *Fd. Chem. Toxicol.* 24:1071-1082.

15. Morimoto, K., K. Tsuji, T. Iio, M. Miyata, A. Uchida, R. Osawa, H. Kitsutaka, and A. Takahashi (1991) DNA damage in forestomach epithelium from male F344 rats following oral administration of *tert*-butylquinone, one of the forestomach metabolites of 3-BHA. *Carcinogenesis* 12:703-708.

16. Plummer, J.L., B.R. Smith, H. Sies, and J.R. Bend (1981) Chemical depletion of glutathione in vivo. In *Methods in Enzymology*, W.B. Jakoby, ed. Academic Press, New York, Vol. 77, pp. 50-59.

17. Stich, F.F., and M.P. Rosin (1984) Naturally occurring phenolics as antimutagenic and anticarcinogenic agents. *Adv. Exp. Biol.* 177:1-19.

18. Yamaguchi, S., M. Hirose, S. Fukushima, R. Hasegawa, and N. Ito (1989) Modification by catechol and resorcinol of upper digestive tract carcinogenesis in rats treated with methyl-N-amylnitrosamine. *Cancer Res.* 49:6015-6018.

THE MODULATING EFFECTS OF ANTIOXIDANTS IN

RAT EMBRYOS AND SERTOLI CELLS IN CULTURE

Diana Anderson and Anne J. Francis

BIBRA Toxicology International
Woodmansterne Road
Carshalton
Surrey SM4 5DS, UNITED KINGDOM

ABSTRACT

The male and female reproductive systems are targets for the toxicity of a wide range of compounds. There is a paucity of information regarding the modulating effects of antioxidants in such systems.

Enzymically generated oxygen radicals have been shown to be toxic and/or mutagenic in a variety of in vitro test systems. It is known that vitamins C and E can modify responses in such systems.

Malformations and growth reductions have been observed in whole rat embryo cultures in this laboratory after treatment with the oxygen radical generating system of xanthine/xanthine oxidase. Groups of 9.5-day-old rat embryos were treated with this system with or without vitamin C or E. Vitamin C at the doses given totally abolished neural suture defects while vitamin E only partially did so. Vitamins C and E administered alone had no effect on the embryos.

Germ cell detachment has been shown to occur in mixed cultures of Sertoli and germ cells in response to some known in vivo testicular toxins. Such cultures were also treated with the oxygen radical generating system of xanthine/xanthine oxidase. There was an increase in germ cell detachment with this treatment which was reduced by vitamin C but not by vitamin E at the doses administered.

These findings would suggest that vitamin supplementation could protect somatic cells of reproductive systems against toxins that act through oxygen radical mechanisms.

INTRODUCTION

It is known that much of the toxicity, including genetic damage, caused by ionizing radiation and certain chemicals is mediated by active forms of oxygen, particularly oxygen-derived radicals. These are highly reactive species that have one or more unpaired electrons. Active oxygen species, including hydroxyl radicals, superoxide, and singlet oxygen, can cause tissue damage by reacting with unsaturated fatty acids in cellular membranes, critical sulfhydryl bases in proteins, and nucleotides in DNA. They can originate exogenously as components of tobacco smoke and air pollutants or indirectly through the metabolism of certain drugs, solvents, and pesticides or during radiation. They can also originate endogenously from normal metabolic processes or inflammatory reactions. Oxygen radical damage is thought to contribute to the etiology of many chronic health problems such as cancer, cardiovascular disease, and inflammatory responses (1,12,16,23,32).

Defenses against oxygen free radical damage include several metallo-enzymes such as glutathione peroxidase (selenium), catalase (iron), and superoxide dismutase (copper, zinc, manganese), proteins such as ceruloplasmin (copper), glutathione, uric acid, bilirubin, β-carotene, and the nutrient antioxidant vitamins such as tocopherols (vitamin E) and ascorbic acid (vitamin C).

Vitamin E is the major lipid soluble antioxidant that is present in all cellular membranes and protects against lipid peroxidation (22). It reacts with the lipid peroxy radical (LOO˙) and also with the hydroxyl radical (HO˙) (7, 24), the superoxide radical (O_2^{-}) (13,26), and singlet oxygen (10, 21). Vitamin C is water soluble and, with vitamin E, quenches singlet oxygen and free radicals. Vitamin C reacts directly with singlet oxygen (5), superoxide (18, 25), and hydroxyl radicals (4). In vivo, vitamin E functions as an antioxidant as shown by increased concentrations of aldehydes, peroxides, and lipfushin in the tissues of vitamin E-deficient animals (20).

The balance between the free radicals generated and the antioxidant protective defense system determines the extent of tissue damage. Due to the lack of effective therapies for many of the chronic diseases and resulting toxic damage, the usefulness of nutrient antioxidants in protecting against the adverse effect of oxidative injury warrants investigation.

We have previously shown that oxidative stress from oxygen radical generation can cause chromosome damage, cell mutation, and sister chromatid exchanges in mammalian cells in culture (3,27-29) and induces neural tube defects in developing embryos (2,19). Thus, toxic damage can be caused in somatic cells, including those associated with reproductive function. While the male and female reproductive systems are targets for the toxicity of a wide range of compounds, there is a paucity of information regarding the modulating effects of nutrient antioxidants in cells associated with these systems.

In the present investigations, we have examined the effects of vitamins C and E on cultured rat embryos and Sertoli cells in culture. The fetus in utero is one of the most widely recognized targets in reproductive toxicology, and the pivotal role played by the Sertoli cells within the testes in regulating the development of the germ cells makes them a potentially important target for toxicity. While in vitro systems might not exactly mimic in vivo events, they do give a good indication of how agents might interact with target tissues.

MATERIALS AND METHODS

Chemicals

Eagle's Minimal Essential Medium (EMEM) (Gibco, Paisley, UK) was supplemented with penicillin and streptomycin, 20 U/ml for embryos, 100 U/ml for Sertoli cells (Gibco, Paisley, UK); and L- glutamine, 2 mM for embryos, 4 mM for Sertoli cells (Gibco, Paisley, UK); and 0.1 mM nonessential amino acids for Sertoli cells (Gibco, Paisley, UK). Xanthine oxidase (grade III from buttermilk) (Sigma, Poole, UK) was used at 25 or 40 mU per ml EMEM. Xanthine (Sigma, Poole, UK) was 1 mg/ml in 0.5% sodium carbonate, and L-ascorbic acid (Sigma, Poole, UK) was 10 mM in Dulbecco's phosphate-buffered saline solution A, pH 7.3 [PBS(A)] or EMEM. DL-α-Tocopherol (Sigma, Poole, UK) was suspended at 0.95 mg/ml in PBS(A) or EMEM by sonication for 5-10 min and used within 2 hrs. Oxygen/carbon dioxide/nitrogen mixtures were obtained from the British Oxygen Company (London, UK). Hanks balanced salt solution without calcium and magnesium (HBSS) was obtained from Gibco (Paisley, UK) and trypsin and DNAase I from Sigma (Poole, UK).

Serum was prepared from Sprague-Dawley rats (Charles River, Margate, UK). Blood was immediately centrifuged (1,800 g, 5 min) and left to stand for 30 min before discarding red cells and fibrin clots. The serum was heat inactivated at 55°C for 30 min, sterilized by filtration, and stored at -20°C prior to use.

Embryo Culture

The embryo culture technique was essentially that of New (1978) and has been described previously by Jenkinson et al. (1986) (19). Briefly, female Sprague-Dawley rats were caged individually with sexually mature male rats. Matings were confirmed by the presence of vaginal plugs beneath the cages, and the day plugs were found was designated day 0 of gestation. Embryos were explanted on day 9 (9.5 days old), and the outermost membrane layer removed. The intact embryos, consisting of yolk-sac, amnion, and ectoplacental cone, were checked for damage. Undamaged embryos were assigned to the different treatment groups in equal proportions from different females. They were treated for 1 hr in 10 ml EMEM in 60 mm Petri dishes at 37°C, 5% CO_2 in air, before transfer in groups of 3-6 to 50 ml pyrex bottles containing 3 ml rat serum and 1 ml EMEM. Cultures were incubated for approximately 44 hr at 37°C, rotating at 40 rpm. Bottles were gassed for 5 min with $O_2/CO_2/N_2$ mixture at 0 hr (5:5:90), 18 hr (20:5:75), and 24 hr (40:5:55).

After 44 hrs, embryos with a healthy, well-formed yolk-sac and blood circulation were assessed for growth: yolk-sac diameter, crown-rump length, and head length were measured and the number of somites was counted. Embryos were also examined for abnormalities of the neural suture and graded according to severity: no abnormality = 0; mild abnormality (small deviation from the normal straight line) = 1; moderate abnormality (one or more sharp kinks) = 2; severe abnormality (incomplete change from dorsal to ventral concave flexure which should

Table 1. Effects of L-ascorbic acid on neural suture abnormalities in cultured rat embryos

Treatment	n	Neural Suture Abnormalities					CA trend test (chi-squared values)	
		None	Mild	Mod-erate	Severe	Turning failure	vs. X	vs. X/XO
		0	1	2	3	4		
X	13	9	4	0	0	0	-	14.1 ***
X/XO	22	3	5	5	7	2	14.1 ***	-
X/XO/AA 10 μM	13	3	3	1	2	4	9.4 **	0.0 ns
X/XO/AA 100 μM	10	0	4	3	1	2	12.3 ***	0.0 ns
X/XO/AA 1 mM	10	7	3	0	0	0	0.0 ns	11.8 ***
X/AA 10 μM	12	9	1	2	0	0	0.2 ns	11.3 ***
X/AA 100 μM	14	8	4	1	1	0	1.4 ns	9.6 **
X/AA 1 mM	10	7	3	0	0	0	0.0 ns	11.8 ***

occur at 10.5 days) = 4. Total protein content was determined by reaction of sonicated embryos with Coomassie Brilliant Blue G-250 (Sigma, Poole, UK) protein reagent and measurement of absorbance at 595 nm in a spectrophotometer (Cecil, Cambridge, UK) (6). Embryos with no circulation and showing poor growth and development of both yolk-sac and embryo were noted but growth parameters could not be measured as development had not continued for the full 44 hrs. Assessments were done without knowledge of treatment.

Sertoli/Germ Cell Culture

Mixed cultures of Sertoli cells and germ cells were prepared from the testes of 28-day-old Sprague-Dawley rats (Harlan Olac, Ltd., Bicester, Oxon) using the method described by Gray and Beamand (15). Briefly, testes from four 28-day-old rats were decapsulated and coarsely chopped with a scalpel. After washing once with HBSS, they were incubated with 27 ml 0.25% trypsin containing 10 μg DNAase I for 15 min at 37°C to separate the tissue into individual tubules. The trypsin solution was decanted through a 100 μm nylon mesh (Lockertex, Warrington, UK) and the retained tubules were washed with HBSS. The tubules were resuspended in 27 ml collagenase, 1 mg/ml, in HBSS and were incubated at 37°C until the tissue was reduced to 0.5-1 mm fragments of seminiferous tubules and aggregates of Sertoli cells and germ cells. This normally took 3.5 min. The suspension was filtered through a 75 μm nylon mesh and the retained tissue was washed with HBSS. Tissue was recovered from the filter by backwashing with 30 ml EMEM supplemented with 10% fetal calf serum and pipetted to produce a single-cell suspension. Then 30 mm Millicell filters (Millipore, Watford, UK) were coated with fibronectin (Sigma, Poole, UK) and placed in 6-well plates. The 1.5 ml of cell suspension, 5×10^6/ml, was placed in each Millicell and EMEM was placed in the outer wells. Cultures were incubated at 32°C in a humidified atmosphere of 5% CO_2 in air.

After incubation for 24 hrs, the culture medium was replaced with serum-free medium and cultures were maintained for a total of 5 days. The effects of treatments were measured as the ability to enhance detachment of germ cells from the Sertoli cell monolayer into the medium. Cultures were treated for 2 hrs and then refed with fresh serum-free medium. Cell counts were made at 2 hrs, 22 hrs later, and again after a further 24-hr period. The results presented are the total germ cell detachment over 48 hrs. Cells were fixed for 20 min in Bouin's fluid, washed in 70% ethanol, and stained with hematoxylin and eosin before being mounted on glass slides.

Statistics

Growth parameters in the embryo cultures and germ cell detachment in Sertoli cell cultures were tested for differences between treatments by analysis of variance and, if significant, by the Least Significant Difference (LSD) test, having first ensured that Bartlett's statistic for homogeneity of variance was not significant. The distribution of the severity of neural suture abnormalities was tested for differences between treatments by the Cochran Armitage trend test (30).

RESULTS

The Effect of L-Ascorbic Acid on Neural Suture Abnormalities

Treatment with xanthine/xanthine oxidase (X/XO) altered the distribution of neural suture abnormalities, causing a significant linear trend ($P \leq 0.001$) towards increasingly severe abnormalities when compared with control [xanthine (X) only] (Tab. 1). Low concentrations of AA (10 or 100 μM) added to cultures containing X/XO did not abolish this trend, which remained nonsignificantly different from cultures without AA. However, X/XO/1,000 μM AA caused a significant linear trend toward decreasingly severe abnormalities when compared with X/XO. The linear trend for distribution of abnormalities was not significantly different from X control.

When 10, 100, or 1,000 μM AA was added to X control cultures, it caused a linear trend in severity of abnormality that was not significantly different from control.

The Effect of DL-α-Tocopherol on Neural Suture Abnormalities

As in the AA series of experiments, X/XO caused a significant linear trend ($P \leq 0.001$) toward increasingly severe neural suture abnormalities when compared with X control (Tab. 2). In comparison with X control embryos, X/XO cultures with 95 μg/ml of AT still showed a significant linear trend toward increasingly severe

Table 2. Effects of DL-α-tocopherol on neural suture abnormalities in cultured rat embryos

Treatment	n	Neural Suture Abnormalities					CA trend test (chi-squared values)	
		None	Mild	Mod-erate	Severe	Turning failure	vs. X	vs. X/XO
		0	1	2	3	4		
X	18	13	4	0	0	1	-	19.7 ***
X/XO	16	1	2	1	2	10	19.7 ***	-
X/XO/AT 95 μg/ml	14	2	1	1	1	9	15.3 ***	0.1 ns
X/XO/AT 190 μg/ml	8	3	0	2	2	1	5.7 *	4.4 *
X/XO/AT 285 μg/ml	7	3	0	0	0	4	7.3 **	1.3 ns
X/AT 285 μg/ml	10	7	3	0	0	0	0.3 ns	16.2 ***

abnormalities. However, higher concentrations of AT reduced the significance of this trend, 190 μg/ml being slightly more effective than 285 μg/ml. This is a weaker effect than that seen with AA, where the highest level used completely abolished the effect of X/XO on abnormality distribution, returning it to the control pattern.

In comparison with X/XO-treated embryos, cultures with AT added at 190 μg/ml showed a significant linear trend ($P \leq 0.05$) toward decreasingly severe abnormalities, and, although a similar effect was seen in cultures with AT at 285 μg/ml, this was not significant. AT at 95 μg/ml had no effect.

With 285 μg/ml AT added to X control cultures, it caused a linear trend in severity of abnormality which was not significantly different from control.

Table 3 shows the distribution of neural suture abnormalities in embryos from the series of experiments in which XO was used at nearly double that of the previous studies (40 mU/ml). Embryos treated with X and XO at 40 mU/ml showed very poor growth and development, which resulted in a significant linear trend ($P \leq 0.001$) towards increasingly severe abnormalities when compared with control.

In comparison with X control embryos, the addition of AT to cultures containing X/XO did reduce this trend in a dose-dependent manner (i.e., X^2 decreased), but it was still significant in all groups at the 1% level.

In comparison with X/XO-treated embryos, the addition of 95 μg/ml AT to cultures containing X/XO did not cause any significant difference in linear trend in

Table 3. Effects of DL-α-tocopherol on neural suture abnormalities in cultured rat embryos

Treat-ment	n	Neural Suture Abnormalities					Poor develop-ment	CA trend test (chi-squared values)	
		None	Mild	Mod-er-ate	Sev-ere	TF		vs. X	vs. X/XO
		0	1	2	3	4	5		
X	18	16	1	0	0	1	0	-	30.3 ***
X/XO	18	0	1	1	0	0	16	30.3 ***	-
X/XO/AT 95 μg/ml	19	0	1	0	0	6	12	30.9 ***	0.2 ns
X/XO/AT 190 μg/ml	12	0	3	0	2	2	5	18.5 ***	4.3 *
X/XO/AT 285 μg/ml	8	2	0	0	2	2	2	12.6 ***	5.8 *

severity of abnormality, but higher levels (190 or 285 μg/ml) did cause a significant decrease ($P \leq 0.05$).

The Effect of L-Ascorbic Acid and DL-α-Tocopherol on Growth Parameters of Embryos

There were no biologically significant effects on growth parameters (yolk-sac diameter, crown-rump length, head length, somite number, and protein content), and data are not reported.

The Effect of L-Ascorbic Acid on Germ Cell Detachment in Sertoli/Germ Cell Co-Cultures

Treatment with X/XO in repeated studies significantly increased ($P \leq 0.05$; $P \leq 0.001$) germ cell detachment when compared with X (Tab. 4) without obvious morphological damage to the Sertoli cells. This detachment was still significantly increased in the presence of both doses of AA (1 mM, $P \leq 0.05$; 2 mM, $P \leq 0.01$). However, there was a significant decrease ($P \leq 0.001$) by comparison with X/XO at the highest dose of L-ascorbic acid. Treatment with AA alone had no significant effect on the system.

The Effect of DL-α-Tocopherol on Germ Cell Detachment in Sertoli/ Germ Cell Co-Cultures

As in the AA series of experiments, treatment with X/XO in repeated studies significantly increased ($P \leq 0.01$; $P \leq 0.001$) germ cell detachment when compared with

Table 4. Effects of L-ascorbic acid on germ cell detachment in Sertoli/germ cell co-cultures

Treatment				n	Germe cell detachment (48 h)		
X	XO	AT			$x \ 10^4$/ml	LSD test	
					mean	vs. X	vs. X/XO
μM	mU/ml	Pre-treat-ment, h	μg/ml				
0	0	0	0	8	185.9		
0	0	0	380	8	191.8	ns	
129	25	0	0	8	219.5	**	
129	25	0	380	8	248.4	***	*
0	0	0	0	8	73.2		
0	0	24	380	8	74.4	ns	
129	25	0	0	8	127.5	***	
129	25	24	380	8	120.8	***	ns

Table 5. Effects of DL-α-tocopherol on germ cell detachment in Sertoli/germ cell co-cultures

Treatment				n	Germe cell detachment (48 h)		
X	XO	AA			$x \ 10^4$/ml	LSD test	
					mean	vs. X	vs. X/XO
μM	mU/ml	mM					
0	0	0		4	96.0		
0	0	1		4	101.8	ns	
129	25	0		4	152.3	*	
129	25	1		4	153.1	*	ns
0	0	0		8	86.3		
0	0	2		8	77.8	ns	
129	25	0		8	125.1	***	
129	25	2		8	104.9	**	***

X (Tab. 5). This detachment was still significantly increased (P≤0.001) in the presence of AT (380 µg/ml) when given either as a pretreatment for 24 hrs or without pretreatment. AT did not cause a decrease by comparison with X/XO; in fact, there was an increase (P≤0.05) without pretreatment. Treatment with AT alone had no significant effect on the system.

DISCUSSION

Previously, Jenkinson et al. (18a or 19) showed that teratogenic effects were observed in cultured rat embryos treated with X and XO, and that these effects could be abolished by the addition of catalase or high concentrations of glutathione. The results of this study confirm that neural suture abnormalities can be induced in cultured rat embryos by X and XO and that the antioxidants AA and AT can abolish or reduce such effects, respectively.

The assessment of neural suture abnormalities provides a very sensitive index of the teratogenic potential of each of the different treatments in this system. In the AA series of experiments and in both series of AT experiments, a highly significant difference was observed between control embryos and those incubated with the oxygen radical generating system. AA at 1,000 µm was observed to have a strongly protective effect, completely abolishing the neural suture abnormalities caused by X/XO. This effect was not seen at lower concentrations (10 or 100 µM). AA did not damage control embryos at any of the three concentrations used. This correlates well with the findings of Varma et al. (34), who found that 1 mM Na-ascorbate protected cultured rat lenses from the action of hypoxanthine/XO, presumably by reaction with hydrogen peroxide, and that Na-ascorbate alone did not have a damaging effect. Faustman-Watts et al. (11) also found AA at 1 mm to be protective against radical-induced defects from reactive metabolites of 2-acetylamino-fluorine in cultured embryos.

In the present study, AT was observed to have only a weakly protective effect against X/XO-induced damage to cultured embryos, reducing but not abolishing the neural suture abnormalities caused by X/XO. The reason for the limited antioxidant effect of AT could be explained by its preparation as a suspension by sonication in PBS(A). This may have left a large proportion of AT as a micellar suspension and biologically unavailable (31). However, this was thought to be compensated for by the high concentrations used in the present study that compare favorably with in vivo levels of 5 µg/ml reported by Steele et al. (31). Weitberg et al. (33) failed to observe an acute protective effect of AT-succinate at concentrations up to 10 µM in a system that measured the induction of SCEs in CHO cells by hypoxanthine/XO. They did, however, find a protective effect if cultures were preincubated with AT for 72 hrs.

Detachment of germ cells from cultured Sertoli cells was increased by X/XO in the present study and only AA had a protective effect, as measured by a decrease in germ cell detachment. Again, AT was shown to be a less efficient antioxidant than AA in spite of the fact that cultures were pretreated for 24 hrs. Cave and Foster (7a or 8) showed that AA at 2 mM in dish cultures protected against germ cell detachment induced by 1,3-dinitrobenzene. Shedding of spermatocytes and spermatids from the germinal epithelium into the lumen of the seminiferous tubule and epididymis was evident within 24 hrs of administration of a single oral dose of di-n-pentyl phthalate (9). Similar results have been shown for other phthalate esters in vivo, and comparable results achieved in vitro (15). X/XO currently produces a similar detachment of germ cells in vitro.

CONCLUSION

In conclusion, the modulating and protective effect of AA and AT on oxygen radical-induced neural suture abnormalities in cultured rat embryos and of AA on oxygen radical-induced germ cell detachment in Sertoli cell cultures would suggest an important role for these nutrient antioxidants. Many toxins are known to be metabolized via free radical intermediates in somatic cells (17), and it has been reported that oxygen radicals may be generated during Leydig cell metabolism in the testis (14). Thus, in extrapolating to the *in vivo* human situation, the supplementation of diet with these vitamins would seem to be of obvious benefit.

ACKNOWLEDGEMENT

We wish to thank the Ministry of Agriculture, Fisheries and Food for their financial support.

REFERENCES

1. Ames, B.N. (1983) Dietary carcinogens and anticarcinogens (oxygen radicals and degenerative diseases). *Science* 221:1256-1264.
2. Anderson, D., P.C. Jenkinson, and S.D. Gangolli (1986) The effects of chemicals and radical species on rat embryos in culture. *Food and Chem. Toxicol.* 24:6-7, 637-638.
3. Anderson, D., and B. J. Phillips (1990) Effects of oxygen radicals and antioxidants on CHO cells and rat embryos in culture. In *Progress in Clinical & Biological Research*, M.L. Mendelsohn and R.J. Albertini, eds. Wiley, Liss, Vol. 340E, pp. 361-370.
4. Bielski, B.H. (1982) Chemistry of ascorbic acid radicals. Ascorbic acid: Chemistry, metabolism, and uses. *Adv. Chem. Ser.* 200:81-100.
5. Bodannes, R.S., and P.G. Chan (1979) Ascorbic acid as a scavenger of singlet oxygen. *FEBS Lett.* 105:195-196.
6. Bradford, M.M. (1976) A rapid and sensitive method for the quantitation of microgram quantities of protein utilizing the principle of protein-dye binding. *Analyt. Biochem.* 72:248-254.
7. Burton, G.W., D.O. Foster, B. Perly, T.F. Slater, I.C.P. Smith, and K.U. Ingold (1985) Biological antioxidants. *Philos. Trans. R. Soc. Lond. Biol. Sci.* 311:565-578.
8. Cave, D.A., and P.M.D. Foster (1990) Modulation of m-dinitro-benzene and m-nitrosonitrobenzene toxicity in rat Sertoli germ cell co-cultures. *Fund. Appl. Toxicol.* 14:199-207.
9. Creasy, D.M., J.R. Foster, and P.M.D. Foster (1983) The morphological development of di-n-pentyl phthalate induced testicular atrophy in the rat. *J. Path.* 139:309.

10. Fahrenholtz, S.R., F.H. Doleiden, A.M. Trozzolo, and A.A. Lamola (1974) On the quenching of singlet oxygen by alpha-tocopherol. *Photochem. Photobiol.* 20:505-509.

11. Faustman-Watts, E.M., M.J. Namkung, and M.R. Juchau (1986) Modulation of the embryotoxicity in vitro or reactive metabolites of 2-acetylaminofluorene by reduced glutathione and ascorbate and via sulfation. *Toxicol. Appl. Pharmacol.* 86:400-410.

12. Fridovich, I., and B. Freeman (1986) Antioxidant defenses in the lung. *An. Rev.Physiol.* 48:693-792.

13. Fukuzawa, K., and J.M. Gebicki (1983) Oxidation of alpha tocopherol in micelles and liposomes by the hydroxyl, perhydroxyl, and superoxide free radicals. *Arch. Biochem. Biophys.* 226:242-251.

14. Georgellis, A., M. Tsirigotis, and J. Rydstrom (1988) Generation of superoxide anion and lipid peroxidation in different cell types and subcellular fractions from rat testis. *Toxicol. Appl. Pharmacol.* 94:362-373.

15. Gray, T.J.B., and J.A. Beamand (1984) Effect of some phthalate esters and other testicular toxins on primary cultures of testicular cells. *Food Chem. Toxicol.* 22 (No. 2):123-131.

16. Grundy, S.M. (1986) Comparison of monounsaturated fatty acids and carbohydrates for plasma cholesterol lowering. *N. Engl. J.Med.* 4:745-748.

17. Halliwell, B., and J.M.C. Gutteridge (1985) *Free Radicals in Biology and Medicine.* Oxford, England: Clarendon.

18. Hemila, H., P. Roberts, and M. Wikstrom (1985) Activated polymorphonuclear leucocytes consume vitamin C. *FEBS Lett.* 178:25-30.

19. Jenkinson, P.C., D. Anderson, and S.D. Gangolli (1987) Malformations induced in cultured rat embryos by enzymatically generated active oxygen species. *Terat. Carcino. and Mutag.* (No. 6) 6:597-554.

20. Kunert, K.J., and A.L. Tappel (1983) The effects of vitamin C on in vivo lipid peroxidation in guinea pigs as measured by pentane and ethane production. *Lipids* 18:271-274.

21. Littarru, G.P., S. Lippa, P. De Sole, A. Oradei, F. Dalla Torre, and M. Macri (1984) Quenching of singlet oxygen by D-alpha-tocopherol in human granulocytes. *Biochem. Biophys. Res. Comm.* 119:1056-1061.

22. Machlin, L.J. (1980) *Vitamin E: A Comprehensive Treatise.* Marcel Dekker, New York .

23. Machlin, L.J., and A. Bendich (1987) Free radical tissue damage: Protective role of antioxidant nutrients. *FASEB*, Vol. 1, pp. 441-445.

24. McGay, P.B. (1985) Vitamin E: Interactions with free radicals and ascorbate. *Annu. Rev. Nutr.* 5:323-340.

25. Nishikimi, M. (1975) Oxidation of ascorbic acid with superoxide anion generated by the xanthine-xanthine oxidase system. *Biochem. Biophys. Res. Comm.* 63:463-468.

26. Ozawa, T., A. Hanaki, and M. Matsuo (1983) Reactions of superoxide ion with tocopherol and model compounds: Correlation between the physiological activities of tocopherols and the concentration of chromanoxyl-radicals. *Biochem. Int.* 6:685-692.

27. Phillips, B.J., T.E.B. James, and D. Anderson (1983) Chromosome damage caused by enzymatically generated active oxygen species. In *Development in the Science and Practice of Toxicology,* A.W. Hayes, R.C. Schnell, and T.S. Mija, eds. Elsevier Biomedical Press, North Holland, pp. 371-374.

28. Phillips, B.J., T.E.B. James, and D. Anderson (1984) Genetic damage in CHO cells exposed to enzymically generated active oxygen species. *Mutat. Res.* 126:265-271.

29. Phillips, B.J., D. Anderson, and S.D. Gangolli (1985) The respiratory burst of phagocyte cells as a potential source of mutagenic species. In *Free Radicals in Liver Injury,* G. Poli, K.H. Cheeseman, M.U. Diazuni, and T.F. Slater, eds. IRL Press, Oxford, pp. 21-25.

30. Snedecor, G.W., and W.G. Cochran (1968) *Statistical Methods.* Iowa State University Press, Iowa, pp. 246-248, 258-268, 271-272, 296-298.

31. Steele, C.E., E.H. Jeffery, and A.T. Diplock (1974) The effect of vitamin E and synthetic antioxidants on the growth in vitro of explanted rat embryos. *J. Reprod. Fertil.* 38:115-123 .

32. Tate, R.M., and J.E. Repine (1984) Phagocytes, oxygen radicals and lung injury. In *Free Radicals in Biology,* W.A. Pryor, ed. New York Academic, Vol. 6, pp. 199-212.

33. Weitberg, A.B., S.A. Weitzman, E.P. Clark, and T.P. Stossel (1985) Effects of antioxidants on oxidant-induced sister chromatid exchange formation. *J. Clin. Invest.* 75:1825-1841.

34. Varma, S.D., S.M. Morris, S.A. Bauer, and W.H. Koppenol (1986) In vitro damage to rat lens by xanthine-xanthine-oxidase: Protection by ascorbate. *Exp. Eye Res.* 43:1067-1076.

BLOCKING THE PLANT ACTIVATION OF PROMUTAGENIC

AROMATIC AMINES BY PEROXIDASE INHIBITORS

Michael J. Plewa

Institute for Environmental Studies
Department of Agronomy
and Department of Microbiology
University of Illinois at Urbana-Champaign
Urbana, Illinois 61801 USA

INTRODUCTION

Interest in the effects of toxic plant metabolites of environmental xenobiotics has grown since the demonstration of the plant activation of promutagens (20,30). Plant activation is the process by which a promutagen is activated into a mutagen by a plant system (18). A promutagen is a chemical that is not mutagenic in itself but can be biologically transformed into a mutagen. The capability of plants to bioconcentrate environmental agents and activate promutagens into toxic metabolites is significant when one realizes the immense diversity of xenobiotics to which plants are intentionally and unintentionally exposed.

Oxidative Metabolism in Higher Plant Systems

In mammalian systems the majority of enzymes participating in oxidative desulfuration, dealkylation, epoxidation, or ring hydroxylation are monooxygenases (equation 1). Currently it is unknown if microsomal cytochromes P-450 in plants have enzymatic characteristics similar to those of mammalian liver. The optical and magnetic properties of plant cytochrome P-450 are similar to those of hepatic microsomes (8). Although limited data exist about the inducibility of plant cyto-chrome P-450, it is unknown if there is an equivalent inducible system for hepatic monooxygenases.

$$R\text{-}NH_2 + O_2 + NADPH + H^+ \rightarrow R\text{-}NHOH + H_2O + NADP^+$$

Equation 1. The overall reaction in the monooxygenation of an aromatic amine by the cytochrome P-450 enzyme system.

Plant peroxidases catalyze the oxidation of a diverse class of xenobiotics (equation 2). Peroxidases are ubiquitous in plants; however, an understanding of their participation in the in vivo metabolism of foreign compounds is limited (22,27).

$$R\text{-}NH_2 + H_2O_2 \rightarrow R\text{-}NHOH + N_2O$$

Equation 2. The overall reaction in the oxidation of an aromatic amine by a plant peroxidase.

Review of Selected Plant Promutagens

We used two promutagenic aromatic amines as model substrates in our studies. 2-Aminofluorene (2-AF) is perhaps the most thoroughly studied plant-activated promutagen (27). The biochemistry of activation in mammalian systems, the activated products, and the DNA adducts of 2-AF are known. *m*-Phenylenediamine (*m*-PDA) has been well studied in our laboratory (11); it is chemically related to 2-AF as well as to a large class of pesticides. Both 2-AF and *m*-PDA are activated into potent frameshift mutagens by cultured plant cells (Figs. 1 and 2) (19).

MATERIALS AND METHODS

Chemicals

The reagents, buffers, and microbial and plant cell media have been described previously (21,22,31,32).

Cells

The microbial cells and the plant cells used in these studies have been described previously (21). *Salmonella typhimurium* tester strain YG1024 was provided by Dr. Takehiko Nohmi, National Institute of Hygienic Sciences, Tokyo, Japan.

Plant Cell/Microbe Coincubation Assay

The assay is based on employing living plant cells in suspension culture as the activating system and specific microbial strains as the genetic indicator organism (23). The plant and microbial cells are coincubated together in a suitable medium with a promutagen. The activation of the promutagen is detected by plating the microbe on selective media; the viability of the plant and microbial cells may be monitored as well as other components of the assay (21). Long-term plant cell suspension cultures of tobacco *(Nicotiana tabacum),* cell line TX1, were maintained in MX medium, a modified liquid culture medium of Murashige and Skoog (14). *S. typhimurium* strain TA98 was the genetic indicator organism used (13). A TX1 cell culture was grown at 28°C to early stationary phase, and the cells were washed and suspended in MX⁻ medium. MX- medium lacks plant growth hormone. The fresh weight of the plant cells was adjusted to 100 mg/ml, and the culture was stored on ice (≤30 min) until used. An overnight culture of *S. typhimurium* was grown from a single colony isolate in 100 ml of Luria broth (LB) at 37°C with shaking. The bacterial suspension was centrifuged and washed in 100 mM potassium phosphate

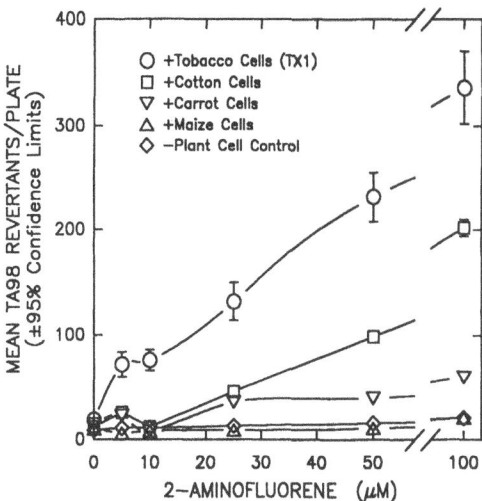

Figure 1. Concentration-response curve illustrating the activation of 2-AF by plant cells in suspension culture (O tobacco cells, □ cotton cells, ▽ carrot cells, △ maize cells) using the plant cell/microbe coincubation assay. The negative controls (◇) included reaction tubes without plant cells or with heat-killed plant cells.

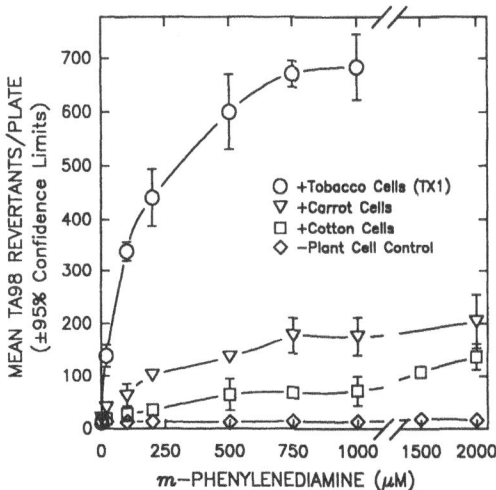

Figure 2. Concentration-response curve illustrating the activation of m-PDA by plant cells in suspension culture (O tobacco cells, ▽ carrot cells, △ cotton cells) using the plant cell/microbe coincubation assay. The negative controls (◇) included reaction tubes without plant cells or with heat-killed plant cells.

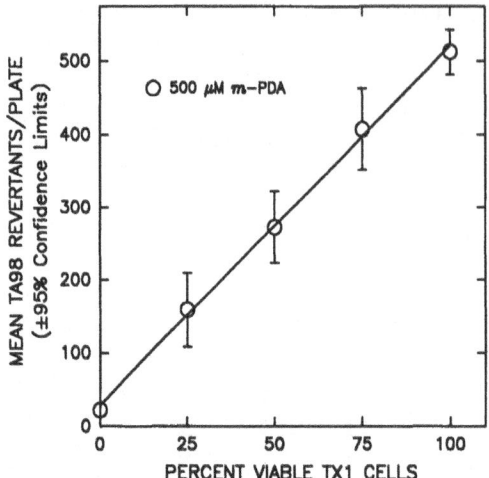

Figure 3. The effect of TX1 cell death on the induction of TA98 revertants induced by the TX1-cell activation of 500 μM m-PDA.

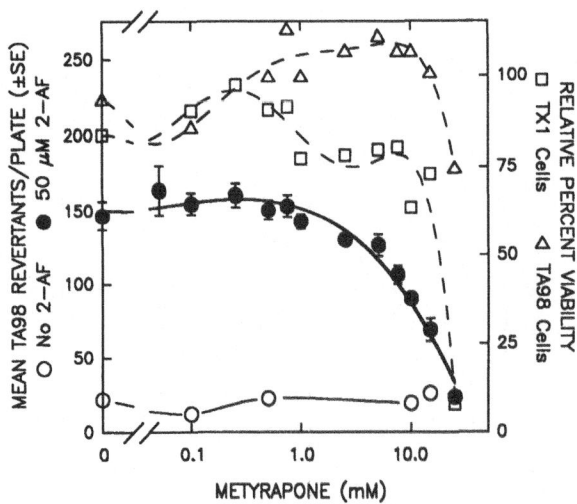

Figure 4. The inhibition of the TX1 cell-activation of 50 μM 2-AF by metyrapone. The concentration range of metyrapone for the inhibition curve is presented along the abscissa in a linear-log scale in a range of 0, and 50 μM - 50 mM (●). The concentration-response curve for the direct-acting mutagenicity of metyrapone in *S. typhimurium* was conducted in a range of 0, and 100 μM -15 mM (O). The toxicity of metyrapone was evaluated in TA98 cells (▵) or TX1 cells (□) in a concentration range of 0, and 100 μM - 50 mM.

buffer, pH 7.4. The titer of the suspension was determined spectrophotometrically at 660 nm and adjusted to 1×10^{10} cells/ml, and the culture was placed on ice. In the coincubation assay, each reaction mixture consisted of 4.5 ml of the plant cell suspension in MX⁻ medium, 0.5 ml of the bacterial suspension (5×10^9 cells), and a known amount of the promutagen in ≤25 μl dimethylsulfoxide. Concurrent negative controls consisted of plant and bacterial cells alone, heat-killed plant cells plus bacteria and the promutagen, and both buffer and solvent controls. The reaction tubes were incubated at 28°C for 1 hr with shaking at 150 rpm. After the treatment time, the reaction tubes were placed on ice. Triplicate 0.5 ml aliquots (~5×10^8 bacteria) were removed and added to molten Vogel Bonner (VB) top agar supplemented with 550 μM histidine and biotin. The top agar was poured onto VB minimal medium plates, incubated for 48-72 hr at 37°C, and revertant $his^{\pm}$ colonies were scored. The remainder of the reaction mixture was used to determine the viability of the plant and bacterial cells. One volume of cold 250 mM sodium citrate buffer, pH 7, was added to each reaction tube, which was then placed on ice. Of this suspension, 0.5 ml was removed and mixed with 2 ml of MX⁻ medium. The viability of the TX1 cells was immediately determined using the phenosafranin dye exclusion method (34). The viability of the bacterial cells was determined by adding 1 ml of the cold reaction mixture to 1 ml of cold 100 mM phosphate buffer, pH 7.4. A dilution series using phosphate buffer was conducted so that approximately 300-500 cells were added to each of 3 molten LB top agar tubes and poured upon LB plates. After incubation at 37°C for 24-36 hr the bacterial colonies were counted.

The Effect of Toxicity on the Plant Cell/Microbe Coincubation Assay

When analyzing antimutagens in the plant cell/microbe coincubation assay, it is essential to monitor the viability of the plant and microbial cells. Any agent that was toxic to the cells would cause a reduction of the induced mutant microbial cells by killing the activating system or the genetic indicator organism. This relationship is illustrated in Fig. 3 in which the TX1-cell populations were composed of a series of different ratios of live to heat-killed TX1 cells. These cells were exposed to 500 μM m-PDA under the conditions of the plant cell/microbe coincubation assay. The data indicated a direct linear relationship ($r = 0.99$) of the amount of TX1-cell activation of m-PDA and the percent viable TX1 cells in the reaction tube (Fig. 3). Thus, the coincubation assay is highly sensitive to toxicity in the cultured plant cells. Any test agent, enzyme inhibitor, or presumptive antimutagen that reduces the viability of the plant cells will cause a reduced yield in the number of microbial mutants per plate.

A similar experiment was conducted to evaluate the effect of varying the number of TA98 cells in reaction tubes that contained the normal titer of TX1 cells and 500 μM m-PDA. The number of TA98 cells varied from 5×10^9 to 1×10^6 per reaction tube. The data demonstrated that only after a reduction from 5×10^8 to 5×10^7 cells plated did a reduction in the resulting number of revertants per plate occur. This phenomenon is due to the number of induced and spontaneous plate mutants per round of cell division that arise after plating in supplemented top agar. Thus, the assay is much less sensitive to agents that exert toxic effects only on the bacteria (19).

Table 1.	Inhibitors Used to Study the Mechanisms of the Plant-Activation of Aromatic Amine Promutagens	
Inhibitor	Action/Effect on Plant Activation	Reference
Acetaminophen	• A co-substrate for horseradish peroxidase and prostaglandin H synthase Also can be metabolized by cytochrome P 450 isozymes • Concentrations above 2 5 mM or 10 mM inhibited the TX1 cell activation of m-PDA or 2-AF by TX1 cells, respectively	Nelson et al, 1981, Harvison et al, 1988, Raucy et al, 1989 Wagner et al, 1989, 1990
7,8 Benzoflavone	• Specific cytochrome P-448 inhibitor Inhibited 2-aminofluorene N hydroxylase • Did not inhibit the TX1 cell activation of m-PDA, however, low concentrations (10–250 μM) inhibited the plant activation of 2-AF	Ullrich et al, 1973, Razzouk et al, 1980 Wagner et al, 1989, 1990
(+) Catechin	• Antimutagen and antioxidant, and may function by binding to mutagenic metabolites and scavenging free radicals • Inhibited the TX1 cell activation of m-PDA at concentrations above 1 mM In a concentration range from 25 μM–2 5 mM it enhanced the plant activation of 2-AF, but caused inhibition at higher concentrations	Conn, 1981, Nagabhushan and Bhide, 1988, Nagabhushan et al, 1988 Wagner et al, 1989, 1990
Diethyldithio carbamate	• Metal chelator and reduced the concentration of cytochrome P-450 in mammals Specific inhibitor of plant peroxidases • At concentrations above 75 μM inhibited the TX1 cell activation of m-PDA and 2-AF due to the inhibition of TX1 cell peroxidase	Jensen et al, 1981, Hunter and Neal, 1975, Plewa et al, 1991 Wagner et al, 1989, 1990, Plewa et al, 1991
Methimazole	• A high-affinity flavin containing monooxygenase substrate • Inhibited the TX1 cell activation of m-PDA but enhanced the plant activation of 2-AF	Poulsen et al, 1974, Frederick et al, 1982 Wagner et al, 1989, 1990
Metyrapone	• Specific inhibitor of cytochrome P-450 in mammals and yeast Inhibited 2-AF hydroxylase • Did not inhibit the TX1 cell activation of m-PDA or 2 AF	Goujon et al, 1972, Carratore et al, 1986, Razzouk et al, 1980 Wagner et al, 1989, 1990
Potassium cyanide	• Inhibited horseradish peroxidase and peroxidase-type monooxygenase activity • Inhibited the TX1 cell activation of m-PDA but was refractory the activation of 2 AF	Wise et al, 1983, Dennis and Kennedy, 1986 Wagner et al, 1989, 1990

RESULTS AND DISCUSSION

Use of Inhibitors to Define Metabolic Pathways in Plant Activation

We studied seven inhibitors for their ability to affect the plant activation of 2-AF or m-PDA (Table 1). We also determined if each inhibitor was a direct-acting mutagen or a plant-activated promutagen (21,31,32). The endpoint of viability was included in the experimental design to investigate if the inhibitor, alone or in combination with TX1 cells and/or the promutagen, was toxic to the activating system or to the TA98 cells. Viability is crucial for data interpretation; a typical inhibition curve with decreasing numbers of TA98 revertants with increasing of the presumptive inhibitor or antimutagen concentrations could be due to toxicity in the plant cells, toxicity in the bacterial cells, toxicity in both cell types or to a true inhibition of plant cell activation. With viability as an endpoint, a significant alteration of the reversion frequency of TA98 could be interpreted as (*i*) a true amendment of TX1 cell activation, (*ii*) toxicity due to the inhibitor alone, or (*iii*) a toxic synergistic effect of the TX1 cells, inhibitor, and promutagen. The resolution of the plant cell/microbe coincubation assay is sufficiently high that the effect of μM to mM concentrations of specific inhibitors was easily detected.

The effectiveness of the use of inhibitors to identify biochemical pathways involved in the plant activation of aromatic amines may be illustrated by reviewing the results of four inhibitors, i.e., metyrapone, potassium cyanide, (+)-catechin, and diethyldithiocarbamate (31,32).

Metyrapone. Metyrapone is a specific cytochrome P-450 inhibitor in mammals and a weak inhibitor of 2-AF *N*-hydroxylase. Metyrapone at concentrations below 7.5 mM did not significantly inhibit the activation of 2-AF (Fig. 4). The activation of 2-AF was diminished by approximately 50% by 15 mM metyrapone ●. However, at this concentration, toxicity was beginning to be expressed in the TX1 cells (□). The reduction of the activation of 2-AF appears to be due to toxicity in the plant cells. Using *m*-PDA as the promutagen, metyrapone at concentrations below 7.5 mM did not significantly inhibit its TX1-cell activation (Fig. 5). The activation of *m*-PDA was diminished by approximately 50% by 10 mM metyrapone ●. At metyrapone concentrations above 1 mM, toxicity to TX1 cells was expressed (□). The reduction of the activation of m-PDA by TX1 cells appears to be a function of metyrapone toxicity in the plant cells. These data illustrate the necessity of monitoring viability when studying the inhibition of activation or mutagenesis. Metyrapone without the promutagen at concentrations above 10 mM was not toxic to TA98 (▲) or TX1 (□), and it was not mutagenic (O) (Figs. 4 and 5). These data indicate that metyrapone and 2-AF or *m*-PDA may interact synergistically to produce a toxin, or that metyrapone may inhibit a step in the TX1 metabolism of these agents that results in a phytotoxic, nonmutagenic intermediate.

Potassium Cyanide. Potassium cyanide (KCn) is an inhibitor of plant peroxidases. The effect of 5 μM - 5 mM KCn was studied in the activation of *m*-PDA (Fig. 6). A significant inhibition of mutation induction was observed at KCn concentrations above 750 μM ●. There was not a corresponding decrease in viability in either cell type (□, ▲). Thus, KCn inhibited *m*-PDA activation at nontoxic concentrations.

(+)-Catechin. Concentrations of (+)-catechin from 25 μM - 25 mM were titrated against TX1 and TA98 cells with 2-AF (Fig. 7). From 25 μM - 2.5 mM, (+)-catechin significantly enhanced the plant activation of 2-AF into a mutagen ●. The positive control (2-AF only) had 219.8 TA98 revertants per plate. The highest enhancement was induced with 750 μM (+)-catechin, resulting in 408.3 TA98 revertants per plate. (+)-Catechin was not a direct-acting or a plant-activated mutagen and it was not toxic to the bacterial cells (▲). The only toxicity to the plant

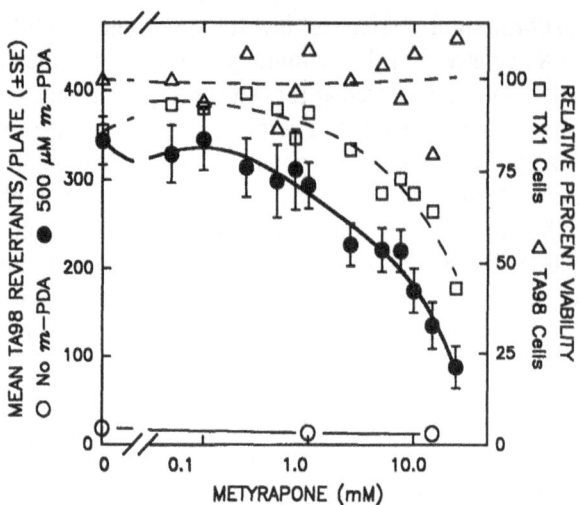

Figure 5. The inhibition of the TX1 cell-activation of 500 μM *m*-PDA by metyrapone. The concentration range of metyrapone for the inhibition curve is presented along the abscissa in a linear-log scale in a range of 0, and 50 μM - 25 mM ●. The concentration-response curve for the direct-acting mutagenicity of metyrapone in *S. typhimurium* was conducted in a range of 0, and 1-15 mM (O). The toxicity of metyrapone was evaluated in TA98 cells (▵) or TX1 cells (□) in a concentration range of 0, and 50 μM - 25 mM.

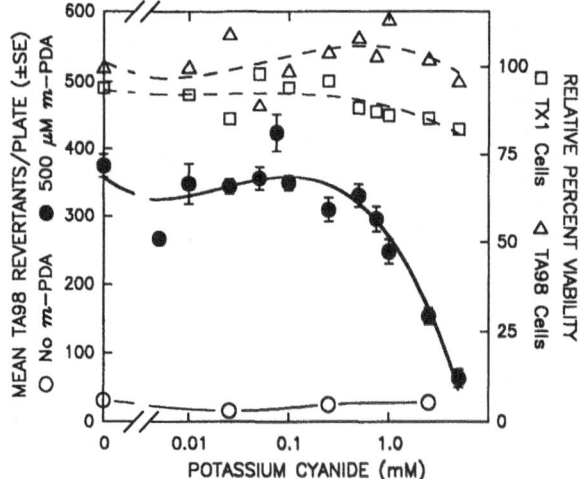

Figure 6. The inhibition of the TX1-cell activation of 500 μM *m*-PDA by potassium cyanide. The concentration range of potassium cyanide for the inhibition curve is presented along the abscissa in a linear-log scale in a range of 0, and 5 μM - 5 mM ●. The concentration-response curve for the direct-acting mutagenicity of potassium cyanide in *S. typhimurium* was conducted in a range of 0, and 25 μM - 2.5 mM (O). The toxicity of potassium cyanide was evaluated in *S. typhimurium* TA98 cells (▵) or *N. tabacum* TX1 cells (□) in a concentration range of 0, and 10 μM - 5 mM.

208

cells was observed at 25 mM ($\square$). (+)-Catechin concentrations above 5 mM significantly inhibited the activation of 2-AF. (+)-Catechin was titrated against TX1 and TA98 cells with 500 μM m-PDA (Fig. 8). There was no effect on the mean number of TA98 revertants at concentrations below 2.5 mM $\bullet$. From 2.5 mM - 25 mM there was a concentration-dependent reduction in the number of revertants. (+)-Catechin concentrations above 10 mM were toxic to TX1 ($\square$) and TA98 ($\triangle$) cells. From 1-10 mM, (+)-catechin inhibited the plant activation of m-PDA without any toxicity.

Diethyldithiocarbamate.   The   inhibition   of   2-AF   activation   by diethyldithiocarbamate (DEDTC) was analyzed in a concentration range from 50 μM- 50 mM (Fig. 9). The inhibition of TA98 revertants was a function of increased DEDTC concentration with a 50% inhibition observed at 1 mM $\bullet$. DEDTC at 25 mM completely inhibited the activation of 2-AF without any cellular toxicity ($\square$, $\triangle$). DEDTC in a concentration range of 25 μM - 50 mM inhibited the TX1 activation of 500 μM m-PDA (Fig. 10) $\bullet$. Under these conditions the viability of TX1 or TA98 cells was not affected ($\square$ and $\triangle$). DEDTC was not a direct-acting mutagen to TA98 ($\bullet$). A significant decline in activation occurred above 75 μM DEDTC with 50% inhibition between 1 and 1.5 mM.

We discovered that DEDTC suppressed the TX1 cell activation of aromatic amines by inhibiting intracellular peroxidases (22). Intact TX1 cells were exposed in vivo to DEDTC concentrations from 250 μM - 25 mM for 1 hr. After the cells were washed, TX1 cell homogenates were prepared and both peroxidase activity and protein content were measured. Peroxidase activity was measured by determining the oxidation of guaiacol by observing the change in absorbance at 470 nm (12, 22). One experiment is presented in Fig. 11. TX1 cells exposed to DEDTC express reduced peroxidase activities when normalized on a μg protein basis.   The DEDTC concentrations that caused a 50% reduction in TX1 cell peroxidase activity (750 μM - 2.5 mM) (Fig. 11) also caused a 50% inhibition of the TX1 cell activation of m-PDA and 2-AF (Figs. 9 and 10).

To understand the nature of the DEDTC inhibition of the intracellular TX1 peroxidase activity we incubated TX1 cells with 750 μM or 25 mM DEDTC. Concurrent controls were cells that were handled identically except they were not exposed to the inhibitor. Separate TX1 cell homogenates were prepared under identical conditions.   The peroxidase activities of triplicate samples of these homogenates were measured varying the concentration of the restrictive substrate (H_2O_2). Velocity was calculated from the change in absorbance in the linear portion of the curve and ϵ_{470} = 26.6 mM^{-1} cm^{-1} for tetraguaiacol (2).   From three independent experiments, the K_m value for the control cells was 2.79 $\pm$ 0.50 mM. The K_m values for the cells treated with 750 μM and 25 mM DEDTC were 2.31 $\pm$ 0.27 mM and 2.65 $\pm$ 0.62 mM, respectively. The mean K_m value for all the groups was 2.58 $\pm$ 0.23 mM. The K_m values among the control and treated groups did not differ significantly while the V_{max} values were different. The mean V_{max} value ($\pm$SE) for the control cells was 4.02 $\pm$ 0.26 nmoles tetraguaiacol/min/μg protein. The V_{max} values for the cells treated with 750 μM and 25 mM DEDTC were 3.41 $\pm$ 0.26 and 2.12 $\pm$ 0.15 nmoles tetraguaiacol /min/μg protein, respectively. These data indicate that DEDTC is a noncompetitive inhibitor of TX1-cell peroxidase (22).

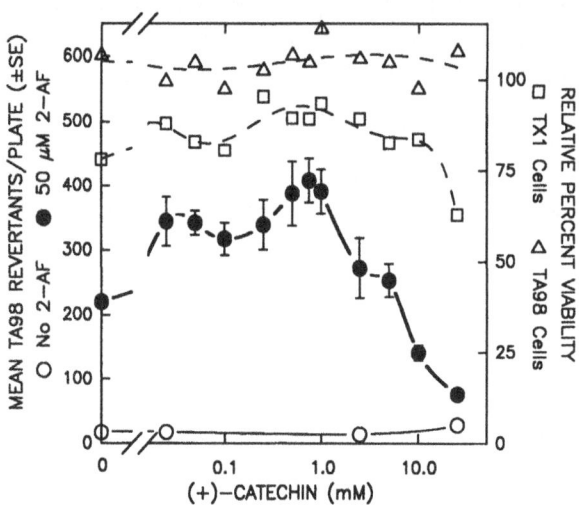

Figure 7. The inhibition of the TX1-cell activation of 50 μM 2-AF by (+)-catechin. The concentration range of (+)-catechin for the inhibition curve is presented along the abscissa in a linear-log scale in a range of 0, and 25 μM - 25 mM ●. The concentration-response curve for the direct-acting mutagenicity of (+)-catechin in *S. typhimurium* was conducted in a range of 0, and 250 μM - 25 mM (O). The toxicity of (+)-catechin was evaluated in *S. typhimurium* TA98 cells (▲) or *N. tabacum* TX1 cells (□) in a concentration range of 0, and 25 μM - 25 mM.

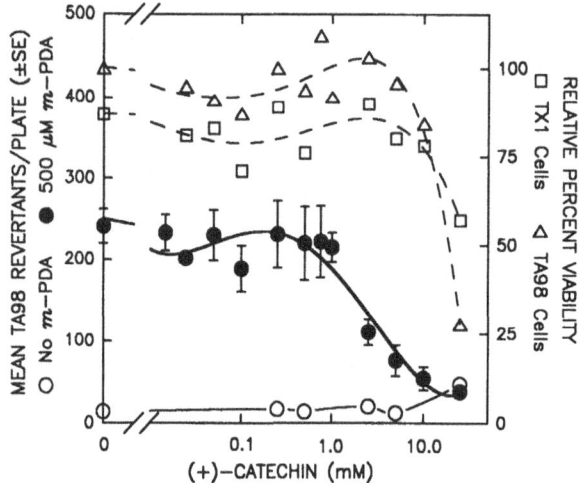

Figure 8. The inhibition of the TX1-cell activation of 500 μM *m*-PDA by (+)-catechin. The concentration range of (+)-catechin for the inhibition curve is presented along the abscissa in a linear-log scale in a range of 0, and 12.5 μM - 25 mM ●. The concentration-response curve for the direct-acting mutagenicity of (+)-catechin in *S. typhimurium* was conducted in a range of 0, and 250 μM - 25 mM (O). The toxicity of (+)-catechin was evaluated in *S. typhimurium* TA98 cells (▲) or *N. tabacum* TX1 cells (□) in a concentration range of 0, and 25 μM - 25 mM.

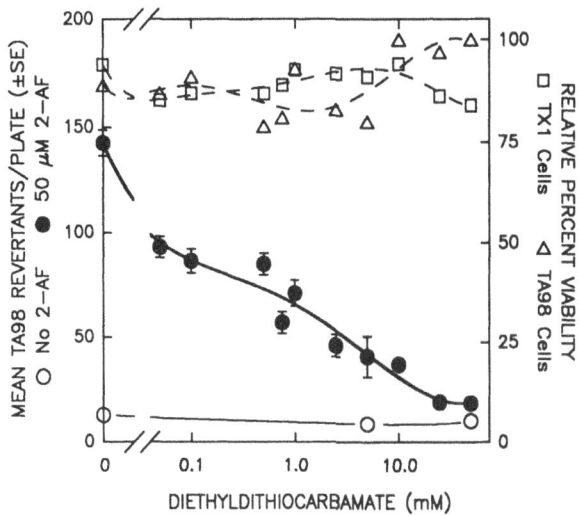

Figure 9. The inhibition of the TX1 cell-activation of 50 μM *2-AF* by DEDTC. The concentration range of DEDTC for the inhibition curve is presented along the abscissa in a linear-log scale in a range of 0, and 50 μM - 50 mM ●. The concentration-response curve for the direct-acting mutagenicity of DEDTC in *S. typhimurium* was conducted in a range of 0, and 5 mM - 50 mM (O). The toxicity of DEDTC was evaluated in TA98 cells (▲) or TX1 cells (□) in a concentration range of 0, and 50 μM - 50 mM.

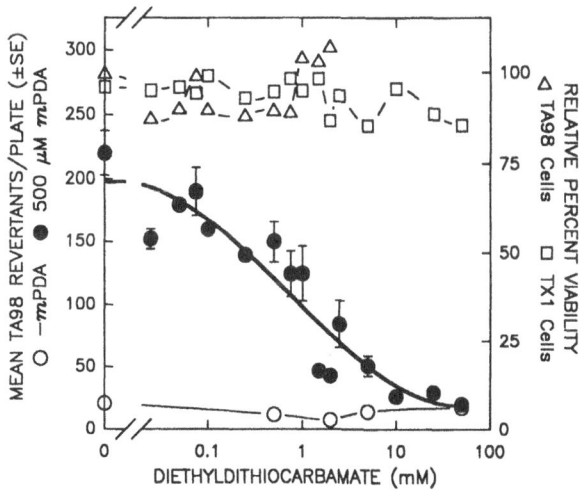

Figure 10. The inhibition of the TX1 cell-activation of 500 μM *m*-PDA by DEDTC. The concentration range of DEDTC for the inhibition curve is presented along the abscissa in a linear-log scale in a range of 0, and 25 μM - 50 mM (●). The concentration-response curve for the direct-acting mutagenicity of DEDTC in *S. typhimurium* was conducted in a range of 0, and 500 μM - 50 mM (O). The toxicity of DEDTC was evaluated in TA98 cells (▲) or TX1 cells (□) in a concentration range of 0, and 25 μM - 50 mM.

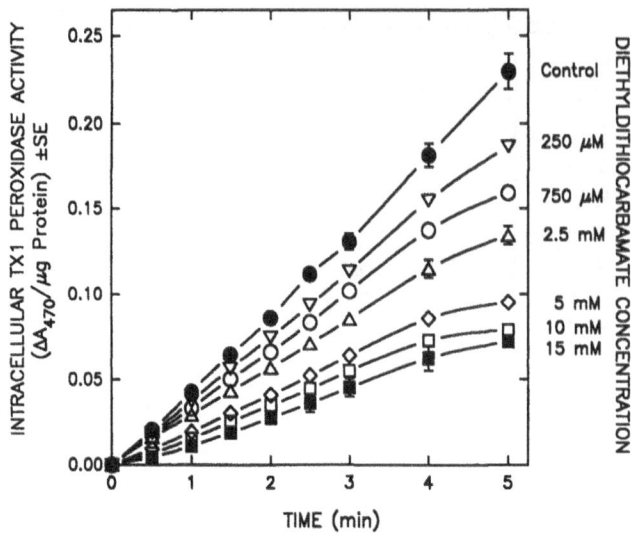

Figure 11. Peroxidase activity in 25 μl volumes of cell free homogenates from TX1 cells treated with DEDTC under in vivo conditions. The concentrations of DEDTC that were exposed to intact TX1 cells were 0 Control (●), 250 μM (▽), 750 μM (O), 2.5 mM (▲), 5 mM (◇), 10 mM (□), and 15 mM (■). Each point represents a mean $\Delta A_{470}/\mu$g protein value ±SE of three repeated measurements.

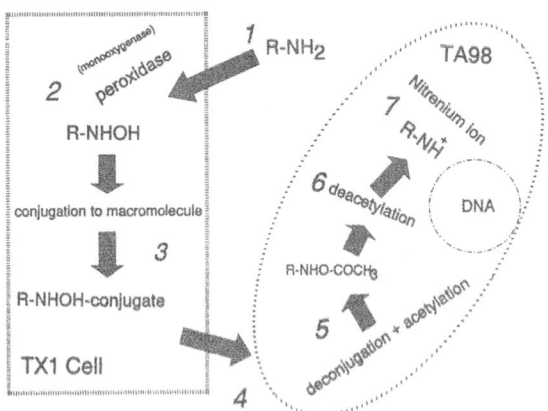

Figure 12. The current model for the plant-activation of promutagenic aromatic amines. The model has seven components: 1.) The aromatic amine (R-NH$_2$) is taken up by the plant (TX1) cell; 2.) TX1 intracellular peroxidase oxidizes the molecule (R-NHOH); 3.) the metabolite is conjugated to a macromolecule (R-NHOH-conjugate); 4.) the amine-conjugate is secreted into the extracellular medium; 5.) the conjugate is absorbed by the bacterial tester strain (TA98 or YG1024); 6. the molecule may be deconjugated and is acetylated (R-NHO-COCH$_3$) and deacetylated by the bacterial acetyl-Co A: N-hydroxyarylamine O-acetyltransferase; and 7. the deacetylation results in a highly reactive nitrenium ion (R-NH$^+$) that induces mutation in the bacterium.

Current Working Model

We developed a model of the TX1-cell activation of aromatic amines based on the data presented in this paper and from our previous studies. The model (Fig. 12), albeit simplistic and incomplete, integrates our data into a mechanistic framework and serves as a foundation for new experimental designs. The model has seven components. They are: 1.) The aromatic amine ($R-NH_2$) is transported into the plant (TX1) cell; 2.) TX1 intracellular peroxidase oxidizes the molecule (R-NHOH); 3.) the metabolite is conjugated to a macromolecule (R-NHOH-conjugate); 4.) the amine-conjugate is secreted into the extracellular medium; 5.) the conjugate is absorbed by the bacterial tester strain (TA98); 6.) the molecule may be deconjugated and is acetylated ($R-NHO-COCH_3$) and deacetylated by the bacterial acetyl-Co A: N-hydroxyarylamine O-acetyltransferase; and 7.) the deacetylation results in a highly reactive nitrenium ion ($R-NH^+$) (27).

Conjugation of the Aromatic Amine Metabolite(s). After the TX1-cell peroxidase-mediated metabolism of the aromatic amine, we propose that the product is conjugated to a macromolecule (such as a protein) and transported into the extracellular medium (Fig. 12, steps 3 and 4). The plant-activated product of 2-AF or m-PDA can be easily isolated from the extracellular medium. The activated promutagen is a very stable frameshift mutagen; media recovered from treated TX1 cells have retained their mutagenic characteristics after four months of storage at 4°C (28). Experiments using Centricon C-100 microconcentrators indicate that the plant-activated product is associated with a large molecule (Figs. 13 and 14). For our model, we propose that the activated promutagen is conjugated with a protein (Fig. 12, step 3) and then transported into the extracellular medium (Fig. 12, step 4). The molecular weight of the TX1-cell activated 2-AF or m-PDA mutagenic product is between 100 kd and 1,000 kd. This conclusion is based on the finding that the mutagenicity seen in the supernatant sample from treated TX1 cells is associated with the Centricon C-100 retentate fraction but not with the C-30 retentate fraction. Thus, the plant-activated product is associated with a large molecule.

Acetylation/Deacetylation of the Mutagen-Macromolecule Conjugate. We compared the mutagenic response in *Salmonella* tester strains TA98 and YG1024 of the TX1-cell activated products of 2-AF (Fig. 15) and m-PDA. The data demonstrate that the metabolite is a substrate for the bacterial acetyl-Co A: N-hydroxyarylamine O-acetyltransferase. This suggests that the plant-activated mutagen has an N-hydroxyamino functional group (Fig. 12, steps 5 and 6). Strain YG1024 is a derivative of TA98 and differs in that the former contains pYG219 with the cloned OAT gene. YG1024 has about 100 x higher O-acetyltransferase activity as compared to TA98 (33). We propose that acetylation followed by deacetylation of the plant-activated product causes the formation of a highly reactive aromatic nitrenium ion that can adduct to DNA and induce genetic damage (Fig. 12, step 7).

GENERAL CONCLUSIONS

Although the model presented here may not be specifically correct, it does in general, account for the data on the plant-activation of aromatic amines. From the experiments reviewed in this paper, we conclude the following: (*i*) Plant cell peroxidases play an integral role in the activation of these aromatic amines. (*ii*) An

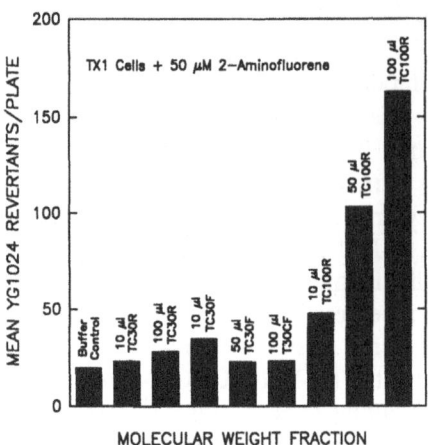

Figure 13. Isolation, partial purification by ultrafiltration, and mutagenic analysis of the TX1 cell-activated 2-AF product. The labels for each fraction indicate the following: Control, YG1024 buffer control. Fractions isolated from 2-AF-treated cell medium; TC100R--the retentate from passing medium from treated cells over a Centricon C-100 column (100-1,000 kd molecular weight fraction); TC30R--the retentate from passing the C-100 treated filtrate over a Centricon C-30 column (30-100 kd molecular weight fraction); TC30F--the treated filtrate from the Centricon C-30 column (<30 kd molecular weight fraction). The amounts, in μl, of each fraction assayed with *S. typhimurium*, YG1024 are listed at the top of each bar in the graph.

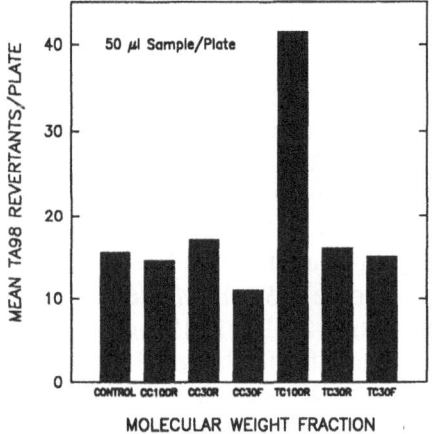

Figure 14. Isolation, partial purification by ultrafiltration, and mutagenic analysis of the TX1 cellactivated *m*-PDA product. The labels for each fraction indicate the following: Control, TA98 buffer control. Fractions isolated from untreated cell medium: CC100R--the retentate from passing medium from control cells over a Centricon C-100 column (100 kd - 1,000 kd molecular weight fraction); CC30R--the retentate from passing the C-100 control filtrate over a Centricon C-30 column (30 kd - 100 kd molecular weight fraction); CC30F--the control filtrate from the Centricon C-30 column (<30 kd molecular weight fraction). Fractions isolated from *m*-PDA-treated cell medium: TC100R--the retentate from passing medium from treated cells over a Centricon C-100 column (100 kd - 1,000 kd molecular weight fraction); TC30R--the retentate from passing the C-100 treated filtrate over a Centricon C-30 column (30 kd - 100 kd molecular weight fraction); TC30F--the treated filtrate from the Centricon C-30 column (<30 kd molecular weight fraction).

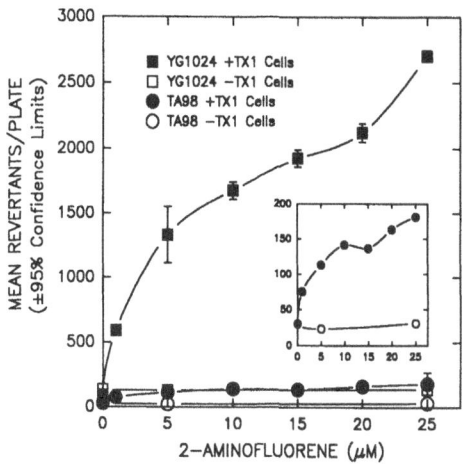

Figure 15. Comparison of the concentration-response curves of TX1 cell-activated 2-AF in *S. typhimurium* strains TA98 (●, O insert) and YG1024 (■,□).

AR-NHOH functional group is implicated in the plant-activated proximal mutagen by the differential response with *Salmonella* tester strains that express different levels of acetyl-Co A: *N*-hydroxyarylamine *O*-acetyltransferase. *(iii)* Finally, the isolation of a high molecular weight mutagenic product from treated cells indicates that plants may form stable mutagenic conjugants that are distinctly different from the metabolic products seen in animal systems.

ACKNOWLEDGEMENTS

This review was based in part on data generated by Ms. Elizabeth D. Wagner, Ms. Shannon Smith, Ms. Joan Riley, and Mr. Kwang-Young Seo. This research was supported by U.S. Air Force, Office of Scientific Research (grant No. AFOSR-88-0336) and by U.S. Environmental Protection Agency (grant No. R-815008).

REFERENCES

1. Carratore, R.D., E. Cundari, R. Vellosi, A. Galli, and G. Bronzetti (1986) Specific inhibitors of the monooxygenase system of *Saccharomyces cerevisiae* modified the mutagenic effect of 4-nitroquinoline 1-oxide and the deethylation activity of the yeast. *Carcinogenesis* 7:1127-1130.
2. Chance, B., and A C. Maehly (1955) Assays of catalases and peroxidases. In *Methods in Enzymology, Vol. 2*, S. Colowick and N. Kaplan, eds. Academic Press, New York, pp. 764-775.
3. Conn, H.O., ed. (1981) *International Workshop on (+)-Cyanidanol-3 in Diseases of the Liver*, The Royal Society of Medicine, International Congress and Symposium Series, No. 7, Academic Press, London.
4. Dennis, S., and I.R. Kennedy (1986) Monooxygenases from soybean root nodules: Aldrin epoxidase and cinnamic acid 4-hydroxylase. *Pes. Biochem. Physiol.* 26:29-35.

5. Frederick, C.B., J.B. Mays, D.M. Ziegler, F.P. Guengerich, and F. Kadlubar (1982) Cytochrome P-450- and flavin-containing monooxygenase-catalyzed formation of the carcinogen N-hydroxy-2-aminofluorene and its covalent binding to nuclear DNA. *Cancer Res.* 42:2671-2677.

6. Goujon, F.M., D.W. Nebert, and M. Gielen (1972) Genetic expression of aryl hydrocarbon hydroxylase induction, IV. Interaction of various compounds with different forms of cytochrome P-450 and the effect on benzo(α)pyrene metabolism in vitro. *Molec. Pharmacol.* 8:667-680.

7. Harvison, P.J., R.W. Egan, P.H. Gale, G.D. Christian, B.S. Hill, and S.D. Nelson (1988) Acetaminophen and analogs as cosubstrates and inhibitors of prostaglandin H synthase. *Chem. Biol. Interact.* 64:251-266.

8. Higashi, K (1988) Metabolic activation of environmental chemicals by microsomal enzymes of higher plants. *Mutat. Res.* 197:273-288.

9. Hunter, A.L., and R.A. Neal (1975) Inhibition of hepatic mixed-function oxidase activity in vitro and in vivo by various thiono-sulfur-containing compounds. *Biochem. Pharmacol.* 24:2199-2205.

10. Jensen, D.F., P.D. Lotlikar, and P.N. Magee (1981) The in vitro methylation of DNA by microsomally-activated dimethylnitrosamine and its correlation with formaldehyde production. *Carcinogenesis* 2:349-354.

11. Lhotka, M.A., M.J. Plewa, and J.M. Gentile (1987) Plant activation of m-phenylenediamine by tobacco, cotton, and carrot cells suspension cultures. *Env. and Molec. Mutagen.* 10:79-88.

12. Maehly, A.C., and B. Chance (1954) The assay of catalases and peroxidases. In *Methods of Biochemical Analysis, Vol. I*, D. Glick, ed. Interscience Publications, New York, pp. 357-424.

13. Maron, D.M., and B.N. Ames (1983) Revised methods for the *Salmonella* mutagenicity test. *Mutat. Res.* 113:173-215.

14. Murashige, T., and F. Skoog (1962) A revised medium for rapid growth and bioassays with tobacco cultures. *Physiol. Plant* 15:473.

15. Nagabhushan, M., A.J. Amonkar, U.J. Nair, U. Santhanam, N. Ammigan, A.V. D'Souza, and S.V. Bhide (1988) Catechin as an antimutagen: its mode of action. *J. Cancer Res. Clin. Oncol.* 114:177-182.

16. Nagabhushan, M., and S.V. Bhide (1988) Anti-mutagenicity of catechin against environmental mutagens. *Mutagenesis* 3:293-296.

17. Nelson, S.D., D.C. Dahlin, E.J. Rauckman, and G.M. Rosen (1981) Peroxidase-mediated formation of reactive metabolites of acetaminophen. *Molec. Pharmacol.* 20:195-199.

18. Plewa, M.J. (1978) Activation of chemicals into mutagens by green plants: A preliminary discussion. *Env. Health Persp.* 27:45-50.

19. Plewa, M.J. (1990) Activation of promutagens by plant cell systems. In *Mutation and the Environment: Environmental Genotoxicity, Risk and Modulation*, M.L. Mendelsohn and R.J. Albertini, eds. Wiley-Liss, New York, 385 pp.

20. Plewa, M.J., and J.M. Gentile (1976) Mutagenicity of atrazine: A maize microbe bioassay. *Mutat. Res.* 38:287-292.

21. Plewa, M.J., S.R. Smith, and E.D. Wagner (1991) Diethyldithiocarbamate suppresses the plant activation of aromatic amines by inhibiting tobacco cell peroxidase. *Mutat. Res.* 247:47-64.

22. Plewa, M.J., E.D. Wagner, and J.M. Gentile (1988) The plant cell/microbe coincubation assay for the analysis of plant-activated promutagens. *Mutat. Res.* 197:207-219.

23. Plewa, M.J., D.L. Weaver, L.C. Blair, and J.M. Gentile (1983) Activation of 2-aminofluorene by cultured plant cells. *Science* 219:1427-1429.

24. Poulsen, L.L., R.M. Hyslop, and D.M. Ziegler (1974) S-oxidation of thioureylenes catalyzed by a microsomal flavoprotein mixed-function oxidase. *Biochem. Pharmacol.* 23:3431-3440.

25. Raucy, J.L., J.M. Lasker, C.S. Lieber, and M. Black (1989) Acetaminophen activation by human liver cytochromes P450IIE1 and P450IA2. *Arch. Biochem. Biophys.* 271:270-283.

26. Razzouk, C., M. Mercier, and M. Roberfroid (1980) Induction, activation and inhibition of hamster and rat liver microsomal arylamide and arylamine N-hydroxylase. *Cancer Res.* 40:3540-3546.

27. Sandermann, H., Jr. (1988) Mutagenic activation of xenobiotics by plant enzymes. *Mutat. Res.* 197:183-194.

28. Seo, K.Y., and M.J. Plewa (1991) Partial isolation of the mutagenic product of m-phenylenediamine after activation by tobacco cells. *Env. Molec. Mutagen.* 17:67, Supp. 19.

29. Ullrich, V., U. Formmer, and P. Weber (1973) Differences in the O-dealkylation of 7-ethoxycoumarin after pretreatment with phenobarbital and 3-methylcholanthrene. *Hoppe-Seyler's Z. Physiol. Chem.* 354:514-520.

30. Veleminsky, J., and T. Gichner (1988) Mutagenic activity of promutagens in plants: Indirect evidence of their activation. *Mutat. Res.* 197:221-242.

31. Wagner, E.D., J.M. Gentile, and M.J. Plewa (1989) Effect of specific monooxygenase and oxidase inhibitors on the activation of 2-aminofluorene by plant cells. *Mutat. Res.* 216:163-178.

32. Wagner, E.D., M.M. Verdier, and M.J. Plewa (1990) The biochemical mechanisms of the plant activation of promutagenic aromatic amines. *Env. Molec. Mutagen.* 15:236-244.

33. Watenabe, M., M. Ishidate, and T. Nohmi (1990) Sensitive method for the detection of mutagenic nitroarenes and aromatic amines: New derivatives of *Salmonella typhimurium* tester strains possessing elevated O-acetyltransferase levels. *Mutat. Res.* 234:337-348.

34. Widholm, J.M. (1972) The use of fluorescein diacetate and phenosafranine for determining viability of cultured plant cells. *Stain Technol.* 354:514-520.

35. Wise, R.W., T.V. Zenser, and B.B. Davis (1983) Peroxidase metabolism of the urinary bladder carcinogen 2-amino-4-(5-nitro-2-furyl)thiazole. *Cancer Res.* 43:1518-1522.

CHEMOPREVENTION OF RAT LIVER CARCINOGENESIS BY

S-ADENOSYL-L-METHIONINE: IS DNA METHYLATION INVOLVED?

R.M. Pascale, M.M. Simile, M.A. Seddaiu, L. Daino,
M.A. Vinci, G. Pinna, S. Bennati, L. Gaspa, and F. Feo

Istituto di Patologia Generale dell'Università di Sassari
07100 Sassari, ITALY

INTRODUCTION

The best approach to the control of cancer incidence in a population at risk is the removal of the causative agent (primary prevention). However, when primary prevention cannot be applied, the occurrence of cancer can be prevented by administration, to populations at risk, of one or more chemical compounds (chemoprevention; Ref. 1-3). These compounds could interfere with cancer initiation by preventing the formation of ultimate carcinogens and carcinogen-DNA adducts, or by preventing carcinogens from reaching or reacting with critical targets. Once initiation occurs, chemopreventive agents may be used to kill preneoplastic cells or suppress their evolution to neoplastic cells. This latter goal could be reached by inducing preneoplastic cells to differentiate so that they remain viable but irreversibly lose their proliferative potential.

Cancer chemoprevention has been rendered possible by the discovery that carcinogenesis is a multistep process in humans as well as in laboratory animals (4-6). In experimental models, several steps of the carcinogenic process have been described (7,8). Unfortunately, all of these models are largely artificial and do not reproduce human conditions. The multistep approach is only a model situation which, however, permits the study of the mechanisms of tumor chemoprevention. Operatively, each step of the carcinogenic process may be envisioned as a "barrier" (9) that must be breached in order to induce the evolution of initiated cells to cancer. It may be assumed that chemoprevention should protect these barriers from breaching.

In many models of experimental liver carcinogenesis, at least four levels may be recognized, where chemoprevention can be carried out (Fig. 1). The first level corresponds to carcinogenesis initiation. Level 2 coincides with the clonal expansion of initiated cells, giving rise to a population of preneoplastic cells organized in centers of focal growth that rapidly increase in size becoming nodules visible to the naked

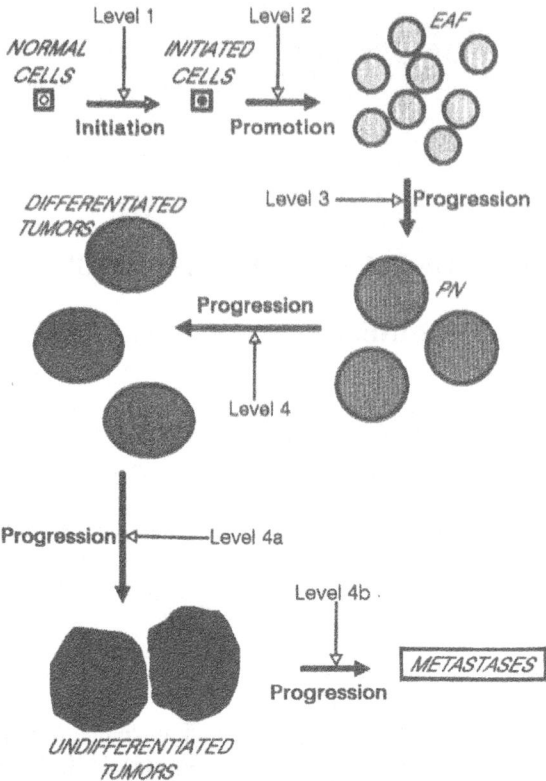

Figure 1. Schematic representation of the main steps of experimental hepato-carcinogenesis with four levels corresponding to steps at which chemopreventive treatments may be performed.

eye, larger than a liver lobule, and compressing surrounding parenchyma (7). The agents that suppress carcinogenesis at this level are called antipromoting agents.

Most preneoplastic foci and nodules do not undergo further evolution to cancer. In various models of multistage carcinogenesis these lesions remodel, under exhaustion or interruption of the promoting stimulus (10-13). During remodeling, focal cells tend to lose their biochemical markers and reverse to normal-appearing liver cells, many of which, however, presumably remain still initiated cells. Nonetheless, some lesions do not undergo spontaneous remodeling (neoplastic, hyperplastic, persistent nodules, adenomas). They acquire the capacity to persist and grow actively, even in the absence of a mitogenic stimulus (7), and progress to carcinomas (7). Progression of preneoplastic lesions to persistent nodules (PNs) and of PNs to carcinomas permits the identification of levels 3 and 4, respectively, where chemopreventive agents may interfere with the carcinogenic process (Fig. 1).

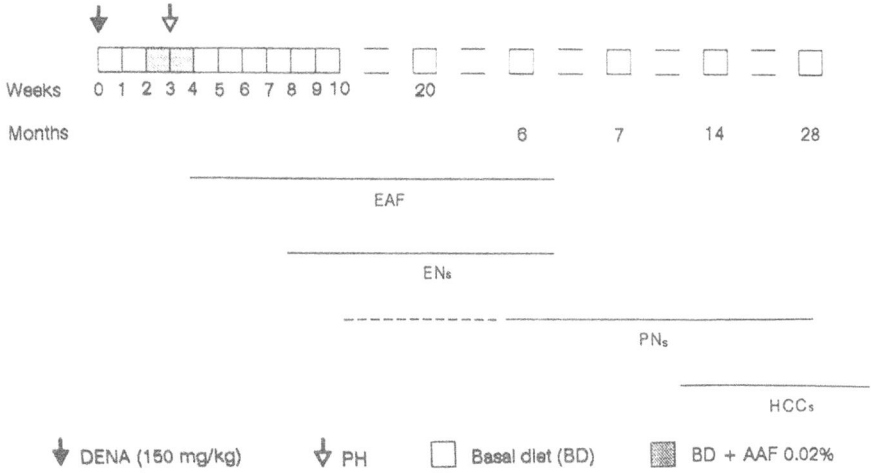

Figure 2. Schematic representation of the "resistant hepatocyte" model of experimental hepatocarcinogenesis. The horizontal lines roughly indicate the period of time when enzyme-altered foci (EAF), early nodules (ENs), persistent nodules (PNs), and hepatocellular carcinomas (HCCs) are present in the liver. The dotted line corresponds to a period of time in which relatively few persistent nodules may be found.

RELATIONSHIP BETWEEN LIPOTROPE CONTENT AND CARCINOGENESIS

A number of observations have proved that variations of the liver level of labile methyl groups is involved in hepatic carcinogenesis. Diets deficient in labile methyl groups are promoters or initiators of carcinogenesis (14-17). Methyl group insufficiency has been postulated in liver carcinogenesis by azodyes (18). A decrease in the major lipotropic compound, S-adenosyl-L-methionine (SAM), and in SAM/S-adenosylhomocysteine (SAH) ratio occurs in the liver during the development of preneoplastic foci as well as in nodules and carcinomas in rats subjected to initiation/selection treatments of experimental carcinogenesis (19), and fed a diet containing adequate lipotrope amounts (20,21). The administration of various lipotropic compounds, such as choline, betaine, methionine, and SAM, to rats subjected to various initiation/promotion treatments, inhibits the development of preneoplastic and neoplastic lesions (see below). Thus, according to available evidence a deficiency of labile methyl groups is associated with cancer, while a methyl donor's load is associated with a decrease in the development of preneoplastic and neoplastic tissue.

The observation that SAM liver content undergoes a great decrease during liver carcinogenesis (20,21) and the possibility of reconstituting SAM content and SAM/SAH ratios by exogenous SAM gave us the opportunity of developing an experimental system in which the effect of changing SAM level and SAM/SAH ratio in liver cells may be studied during hepatic carcinogenesis.

Chemopreventive Effect of Lipotropic Compounds

An experimental model profitably used for chemoprevention studies on rat liver carcinogenesis is represented by the so-called "resistant hepatocyte" or Solt/Farber model (Ref. 19, Fig. 2). In this model initiation is performed by a necrogenic dose of a carcinogen. After repair the animals are fed a diet containing mito-inhibitory amounts of N-acetylaminofluorene (AAF) for 15 days, with a partial hepatectomy (PH) at the midpoint of AAF feeding (selection). Putative preneoplastic foci, recognizable for their biochemical alterations (enzyme-altered foci, EAF) develop as early as 3.5-4 weeks after initiation. They increase in size and number, and then only in size, reaching their maximum development around the 8th-10th week (Fig. 2). At the 8th to 10th week, early nodules (ENs) develop. They exhibit high DNA and polyamine synthesis (20,22), but most of them undergo remodeling (10-13) while remaining nodules (PNs) maintain a high growth rate, do not lose their biochemical markers, and further progress to carcinomas.

We have observed that SAM injection at the dose of 384 μmol/kg/day, started at the end of selection (Fig. 2) and continued up to killing, greatly prevents the development of EAF, ENs, and PNs and hepatocellular carcinomas (HCCs, Ref. 13,21,23). These observations are in keeping with the findings from various laboratories indicating that other lipotropic compounds, such as methionine, choline, and betaine, prevent the development of mammary gland, skin, and liver cancers induced in rats and mice by various carcinogens, with/without successive administration of a promoting agent (24-28). Chemoprevention of spontaneous thymic lymphoma has been observed in AKR/J mice fed a lipotrope-enriched diet (29). Moreover, it has recently been demonstrated (23) that a relatively short treatment with SAM (6 mo), started at the end of selection, in rats subjected to initiation/selection followed by a diet containing 0.05% phenobarbital (PB) for 16 weeks, largely prevents the development of nodules and carcinomas 14 and 28 mo after initiation. Carcinomas, when present, were smaller in SAM-treated rats. In these animals, only trabecular carcinomas and adenocarcinomas developed, whereas undifferentiated carcinomas were present in some of the rats not treated with SAM. These observations indicate that, in SAM-treated rats, most initiated (preneoplastic) cells disappear or remain in a condition that does not permit further progression to cancer for a period of time (28 mo) roughly corresponding to 2/3 of the rat life-span. It was concluded that SAM has a strong antipromoting effect for liver carcinogenesis.

Weaker evidence of an antiprogression effect of lipotropic compounds has been obtained so far. This effect may be studied by subjecting animals to a long-term treatment with chemopreventive compounds, after the development of PNs and before the appearance of carcinomas. The incidence of carcinomas may then be evaluated. To our knowledge, only one short report describes the inhibition of PN progression to carcinoma in Fisher rats fed a diet containing extra-choline amounts (30). Similarly, SAM given to rats after the development of PNs inhibits the growth of these lesions (13,31,). However, in these experiments, SAM treatment was started 10 wk after initiation, when a mixture of early (remodeling) and persistent nodules are present in the liver (Fig. 2), and was continued for a maximum of 11 wk. No demonstration of a reduced carcinoma incidence was given.

How Does SAM Treatment Prevent Tumor Development?

According to Solt and Farber, in the rats subjected to initiation/selection treatments with/without PB, a phase of high DNA synthesis, between 4 and 9 wk after initiation, during which EAF and ENs develop, is followed by a progressive decrease. However, DNA synthesis is always at least 4-5 times higher in focal cells than in surrounding liver. Treatment with SAM, started after the selection treatment, causes a further fall in DNA synthesis of EAF (13,20). When SAM is given after development of nodules a decrease in DNA synthesis of nodule cells occurs (13,31).

These observations may explain the decrease in tumor yield in SAM-treated rats. However, most preneoplastic lesions disappear as a consequence of SAM treatment (13,23), suggesting a prevalence of cell loss in these lesions, during SAM treatment. This possibility has been evaluated by investigating the effect of SAM on remodeling and apoptosis.

In Wistar rats, treated according to the "resistant hepatocyte" model of experimental carcinogenesis, spontaneous remodeling of EAF and ENs is relatively low for at least 3-5 wk after the end of selection treatment (AAF feeding). Thereafter it progressively increases (13). PB administered after selection inhibits remodeling, which thus results in very low levels for at least 5 wk after selection (13). SAM administration (384 μmol/kg/day) to rats from the 5th to the 16th wk after initiation, markedly stimulates remodeling, either in the presence or absence of PB. This is associated with a progressive decrease in DNA synthesis and loss of biochemical markers resulting in a decrease in the complexity level of EAF and ENs (number of markers per lesion, Ref. 13).

When SAM treatment is started 10 (F344 rats) or 20 (Wistar rats) wk after initiation, nodule remodeling appears as early as one week after the onset of SAM administration. Longer treatments (3-8 wk) more severely affect nodule phenotype and cause a great inhibition of DNA synthesis in nodular cells (13,31).

Recent observations (Pascale et al., unpublished data) have shown a significant decrease in nodule size and number, in rats initiated with 1,2-dimethylhydrazine and promoted by orotic acid (OA) for 25-27 wks, when a 12- to 20-wk treatment with SAM was started after OA administration. Interestingly, different from the Solt/-Farber model, the OA model is characterized by the development of relatively few nodules, undergoing very slight or no spontaneous remodeling, in the absence of SAM.

Cell loss by apoptosis has been suggested to play a crucial role in the development of putative preneoplastic liver lesions (32-34). In the presence of a proliferative stimulus, either reparative (resistant hepatocyte model) or hyperplastic (PB or other promoters), during carcinogenesis, apoptosis is low in early preneoplastic lesions (13,32). However, after interruption of promoter administration, a sharp rise in the number of apoptotic bodies occurs in EAF and ENs (13,32). Present knowledge suggests a co-ordinate regulation of cell growth and cell death by apoptosis (35-38). Continuous stimulation of cell production and inhibition of cell death by mitogens could largely affect the homeostatic control of cell growth in putative preneoplastic foci and ENs. After interruption of PB treatment, re-establishment of homeostatic control presumably leads to a high cell loss by apoptosis, in the attempt to establish again (or with the result of establishing again) the original liver mass (see, for example, Ref. 39). In accordance with this interpretation, SAM stimulates apoptosis both in the presence or absence of PB, and accelerates the disappearance of EAF and ENs.

Different from the early lesions, high cell death is constitutively coupled with high DNA synthesis in late nodules (40). Homeostatic control mechanisms for the maintenance of liver mass operate in PNs (40), even though with a slight prevalence of cell production on cell loss, responsible for the slow growth and progression of nodules to carcinomas. Thus, high cell death by apoptosis could be interpreted as a response to rapid cell proliferation by these primary lesions, in an attempt to maintain the original liver mass (39). Relatively short (1 wk) SAM treatment was observed to sharply increase the number of apoptotic bodies in nodules (13). However, apoptosis is low in nodules after prolonged SAM treatment (13). This could depend on phenotypic reversion of nodules leading to the appearance of liver cells with an apparently normal phenotype, which are expected to exhibit low cell production and low cell loss.

In conclusion, our data support the view that cell loss by apoptosis and remodeling may contribute to the disappearance of preneoplastic lesions in SAM-treated rats.

MECHANISTIC STUDIES

The mechanisms underlying tumor chemoprevention by lipotropic compounds are not yet completely understood. A series of studies, aimed at elucidating the SAM chemopreventive effect, have recently been performed in our laboratory to analyze some possible biochemical and molecular mechanisms of the SAM effect.

Biochemical Approach

The first attempts to explain the chemopreventive effect of SAM were based on the hypothesis that an SAM load causes the accumulation of some growth inhibiting catabolites. A compound was identified, 5-methylthioadenosine (MTA), which is formed during the synthesis of spermidine and spermine by transfer of amynopryl residues from decarboxylated SAM to putrescine and spermidine, respectively. MTA is rapidly catabolized by MTA phosphorylase, into methylthioribose-1-phosphate and adenine. Methylthioribose-1-phosphate may be used for methionine resynthesis through a salvage pathway (41).

The observation that MTA inhibits spermine and spermidine synthetases in vitro (42,43) and ornithine decarboxylase activity in vivo in liver and preneoplastic tissue (44), suggests that MTA exerts a negative control on polyamine synthesis. This denies the possibility that MTA accumulation depends on active polyamine synthesis in growing tissues. An increase in MTA phosphorylase has been found in preneo-plastic and neoplastic liver tissue, in the rat (21). MTA accumulation during SAM treatment can occur, however, as a consequence of spontaneous splitting of SAM into

MTA and homoserine lactone, at physiologic temperature and pH (45). Thus, pharmacologic amounts of MTA could accumulate during SAM treatment and the reaction catalyzed by MTA phosphorylase could be insufficient to avoid this accumulation. However, recent data from our laboratory have definitively proved that MTA does not accumulate in the liver during SAM treatment (44). Exogenous MTA exhibits a chemopreventive effect on the development of liver preneoplastic lesions, but, different from SAM, it is not able to reconstitute SAM/SAH ratio and to induce DNA methylation (see below). It was concluded that exogenous SAM is not massively transformed into MTA.

SAM is a well-known cytoprotective agent against toxic liver injury (46). SAM and reduced glutathione (GSH) liver contents decrease during acute and chronic intoxication, and are reconstituted by exogenous SAM, which also prevents toxic liver injury (47,48). Our studies of the SAM chemopreventive effect do not seem to favor a role of SAM cytoprotective action. In fact, SAM was never given to rats during the administration of cytotoxic compounds, such as initiator or AAF, so that no interference between SAM and initiation/selection stages of carcinogenesis may be envisioned. SAM effect does not appear to be influenced by PB administration, indicating that no interaction between SAM and PB occurs during liver carcinogenesis (13). Moreover, although free radicals have been involved in the initiation and promotion stages of carcinogens in various experimental models (49), there is no evidence that free radicals are produced in the liver of rats during initiation and selection treatments or in the following steps. Lipid peroxidation is low in preneoplastic tissue (50). These observations, taken as a whole, tend to exclude the idea that cytoprotection by SAM plays a role in the chemopreventive effect of the lipotropic compound.

Molecular Approach

Chemical carcinogens inhibit, even though with some exceptions, DNA methylation (51). Hypomethylating agents are promoters (52) and, in some experimental systems, initiators (53) of carcinogenesis. Fast growth and resistance to xenobiotics are two hallmarks of preneoplastic tissue (7). Some growth-related genes (31,54) or genes of phase II enzymes (55) are hypomethylated and overexpressed in preneoplastic liver nodules. Finally, in vitro methylation of cloned Moloney sarcoma virus (56) or Ha-*ras* gene (57) limits their transforming activity after translation. These observations seem to indicate that DNA hypomethylation plays a role in initiation and promotion/progression stages of carcinogenesis. However the mechanisms through which DNA hypomethylation enhances carcinogenesis have not yet been elucidated. It has been suggested that hypomethylation inhibits chomosome condensation and causes chromosome mispairing and nondisjunction (58). Drug-induced hypomethylation can lead to decondensation and chromatin abnormalities (59). Alterations of methylation pattern may, thus, be important in promoting allelic deletions, which could play a role in the genesis of tumors (60).

If DNA hypomethylation is involved in the carcinogenic process, it may be inferred that factors regulating DNA methylation modulate this process. Regulation of DNA methylation is a complex and not yet completely known phenomenon involving chromatin arrangement, nucleotide sequence, protein binding, etc. (61). However, once the conditions for the methylation of a given site apply, the specificity and the activity of DNA methyltransferases play a crucial role. Manipulation of this ratio in vivo (62) and in vitro (63) modulates the activity of DNA methyltransferases: a fall in SAM/SAH ratio is inhibitory.

Great decreases in SAM/SAH ratio and overall DNA methylation may be observed in rat liver as early as 5 wk after initiation of hepatic carcinogenesis, and reach their minimum at the 7th wk (21,64,65). Both parameters recover in SAM-treated rats. A strong statistical correlation has been found between SAM/SAH ratio and DNA methylation, indicating that the changes in this ratio could be, at least in part, responsible for modifications of DNA methylation (Ref. 65 and Pascale et al., manuscript in preparation). However, other mechanisms, such as changes in DNA conformation and protein binding, and active demethylation cannot be excluded and need further evaluation to explain DNA hypomethylation in preneoplastic and neopalstic tissue. Low SAM/SAH ratio and DNA hypomethylation seem to be stable features of rat liver carcinogenesis. They may in fact be found in PNs and carcinomas (31,54,66,67). Decreases in SAM/SAH ratio (68) also occur in the liver of rats subjected to methyl-deficient diet, a treatment that leads to DNA hypomethylation (69-70). The mechanisms leading to a drop in SAM/SAH ratio in the early stages of hepatocarcinogenesis, in an experimental system not based on the administration of a methyl-deficient diet, are presently unknown. Perhaps, SAM decarboxylation for active polyamine synthesis in preneoplastic tissue is involved in this phenomenon. In regenerating liver, a decrease in SAM/SAH ratio occurs with a minimum in correspondence of the peak of SAM decarboxylase activity (21,71), SAM administration causes a recovery of SAM/SAH ratio and overall DNA methylation and inhibits growth. Interestingly during the restriction of the SAM pool, polyamine biosynthesis is metabolically conserved at the expense of methylation reactions (72). However, the possibility of changes in SAM synthetase activity and isozyme pattern in liver during the early stages of carcinogenesis should be taken into account.

One of the main features of the SAM chemopreventive effect is represented by inhibition of DNA synthesis in preneoplastic tissue. Accordingly, methionine, a SAM precursor added to a suspension of human bone marrow cells, impairs the utilization of labeled precursors for DNA synthesis (73). A role of DNA methylation in the inhibition of DNA synthesis in SAM-treated rats is strongly supported by the observation that the inhibitory effect of SAM is partially overcome by 5-azacytidine (74), a well-known inhibitor of DNA methyltransferases (75). This indicates that the recovery of DNA methylation in SAM-treated rats is not a mere consequence of growth inhibition. On the other hand, the inhibition of growth of liver preneoplastic tissue by MTA is not paralleled by the recovery of DNA methylation, even though treatment of rats with SAM plus MTA leads to DNA methylation, indicating the absence of an inhibitory effect of MTA on DNA methyltransferases (44).

We have further investigated the link between DNA methylation and growth by studying the effect of variations of SAM/SAH ratio and DNA methylation on the expression of some growth-related genes in the liver, during the development of preneoplastic tissue, and in nodules.

Gene Expression in Preneoplastic Liver. With some exceptions (76), increases in mRNA transcripts of c-*myc*, c-Ha-*ras*, and c-Ki-*ras* proto-oncogenes have been found in the early stages of liver carcinogenesis induced by a choline-deficient diet containing 0.1% ethionine (77), or by DENA in rats subjected to selection, according to the "resistant hepatocyte" model (Ref. 65 and Pascale et al., manuscript in preparation). An increase in mRNA transcripts of various genes, including, c-*myc*, c-Ha-*ras*, and c-Ki-*ras*, has also been described in regenerating liver with peaks at different times after partial hepatectomy. The return to the pre-PH situation was observed for all mRNA transcripts, within 30-72 hr after PH (31,54,78). When PH is performed in initiated rats during selection treatment, liver regeneration proceeds more slowly, being complete in 2-3 wk (5-6 wk after initiation, Ref. 79). This denies the possibility that increases in mRNA transcripts of c-*myc*, c-Ha-*ras*, c-Ki-*ras* proto-oncogenes in the liver, 7 wk after initiation (4 wk after PH), depend merely on reparative growth. Indeed, in uninitiated rats, subjected to PH plus AAF (selection), in which no EAF develop, mRNA transcripts of the three proto-oncogenes increase less than in initiated rats. Moreover, the increase is transitory and mRNA transcripts return to normal liver values in about one week (Ref. 65 and Pascale et al., manuscript in preparation). These results suggest that the presence of rapidly growing putative preneoplastic lesions in the liver of initiated rats could contribute to the increase in gene expression. Nagy et al. (80) have shown, by using Northern blot analysis and in situ hybridization, a marked increase in c-*myc* expression in preneoplastic foci and oval cells, with respect to surrounding hepatocytes, 3-5 wk after initiation, in the liver of rats treated according to the "resistant hepatocyte" protocol. However, the possibility that surrounding liver also contributes to the rise in gene expression, at least in the early stages of hepatocarcinogenesis, cannot be excluded at present.

Gene expression has also been studied during later phases of the hepato-carcinogenic process. An increase in the transcripts of c-*ras* family genes, c-*myc* and c-*fos* was shown in rat liver carcinoma nodules, induced by DENA (81,82). Enhanced expression of c-*raf* occurs in liver carcinomas induced by DENA and promoted by PB, as well as in neoplastic nodules showing moderate to severe atypism and focal areas suggestive of carcinomatous changes (83). Significant increases in mRNA transcripts for c-*myc*, c-Ha-*ras*, and c-Ki-*ras* proto-oncogenes have also been found in PN free of carcinomatous changes, induced in F344 rats treated according to the Solt/Farber model (31) or the OA model (54).

Thus, according to the available evidence, preneoplastic liver tissue exhibits, in different stages of its progression to cancer, an increased expression of some genes involved in growth regulation. An overexpression in cells surrounding preneoplastic lesions could also occur.

Effect of SAM on Gene Expression. The study of SAM administration on gene expression has clearly shown an inhibitory effect. SAM administration, started at the end of selection, results in a dose-dependent decrease in mRNA transcripts of c-*myc*, c-Ha-*ras*, and c-Ki-*ras* genes (65). A decrease in mRNA transcripts of these genes, correlated to the length of SAM treatment, was obtained when SAM administration was started after the development of PNs (10 wk after initiation in F344 rats, Ref. 31).

DNA hypomethylation has been shown to be a condition necessary, even if not sufficient, for gene expression (84). Variation in the methylating environment could modulate gene expression. Methylation of specific gene sequences has been analyzed in isolated liver PNs. Rao et al. (54) have found a hypomethylation of 5'-CCGG sequences of DNA isolated from rats initiated with 1,2-dimethylhydrazine and promoted by OA, hybridizing with c-*fos* and c-*myc* probes. The c-Ha-*ras* proto-oncogene showed hypomethylation only in the alternating 5'-GCGC sequence. This methylating pattern was not found in surrounding liver, regenerating liver, or liver of rats exposed to the initiator or promoter alone. Accordingly, we have demonstrated a marked hypomethylation of the 5'-CCGG sequences hybridizing with c-*myc* (exons 2 and 3), c-Ha-*ras*, and c-Ki-*ras* probes, in PNs induced in F344 rats according to the "resistant hepatocyte" model (31). No hypomethylation of these sequences was found in normal and regenerating liver between 0.5 and 30 hr after PH. Garcea et al. (31) were also able to modify the methylation pattern of 5'-CCGG sequences of the three proto-oncogenes studied by subjecting the rats to a 3-wk treatment with 384 μmol/kg/day of SAM. This treatment causes the appearance of a few methylated sites, of c-*myc*, c-Ha-*ras*, and c-Ki-*ras* genes, otherwise highly hypomethylated in nodules.

The possible role in hepatocarcinogenesis of co-ordinate changes in SAM/SAH ratio and gene methylation and expression is also substantiated by a number of observations on rats fed chronically methyl-deficient diets. This treatment leads to DNA and gene hypomethylation (66,67,85-88), enhances the expression of various protooncogenes (87), and causes cancer (4-7) in rat liver. High expression or amplification of c-Ha-*ras* proto-oncogenes, as well as of c-*myc* genes, has been found in HCCs arising from the livers of rats fed methyl-deficient diets (85,86). Hypomethylation of c-Ha-*ras* proto-oncogene has also been demonstrated in the liver of rats subjected to the methyl-deficient regimen, during the development of preneoplastic lesions (85,87). However, transfecting activity for NIH/T3T cells has only been observed with the DNA from neoplastic livers and was associated with the presence of an amplified c-Ha-*ras* gene (89). It was suggested that hypomethylation of c-Ha-*ras* is a condition predisposing to its activation and subsequent development of tumors (89).

Recent findings have proved that overall DNA hypomethylation occurs in the liver of rats fed a methyl-deficient diet as early as one week after starting this regimen (88). Furthermore, in rats fed this diet for 4 wk there occurred an increase in mRNA levels for c-*myc*, c-*fos*, and c-Ha-*ras*, while mRNAs for EGFR and EGF decreased, yielding a pattern of gene expression similar to that seen in liver tumors (90). Hypomethylation of specific 5'-CCGG sequences within c-*myc*, c-*fos*, and c-Ha-*ras* was also found. After returning to a diet containing adequate amounts of labile methyl groups, the overall extent of tRNA and DNA hypomethylation was rapidly reversed in about one week. Levels of mRNAs of the proto-oncogenes returned to normal by 3 wk, with the exception of c-*fos* which remained somewhat elevated at that time. Hypomethylation of 5'-CCGG sites also persisted for about 4 wk after returning to an adequate diet.

Thus, whereas a high SAM content is generally associated with DNA methylation and inhibition of gene expression, a low lipotrope content is associated with opposite effects. However, according to recent observations, an hypermethylated c-*myc* proto-oncogene is present in the carcinomas arising from choline-deficient rats, for example, in a hypomethylating environment (86). An increase in DNA methylation at the 5' end and a decrease at the 3' end of α-fetoprotein gene was also observed, a pattern resembling that of the active α-fetoprotein gene in fetal liver (91). Substantial hypermethylation has been also found within specific regions of human chromosomes in tumor cells (92). The above findings indicate that the relationships between DNA methylation and gene expression are rather complex. An attractive suggestion, in view of the increasing interest in suppressor genes in cancer pathogenesis (9), is that increased methylation of particular DNA spots could be involved in silencing suppressor genes (93).

Christy and Scangos (94) have provided evidence indicating that specific methylation sites, rather than the absolute level of 5mC residues, are important in determining the ability of bovine papilloma virus to transform mouse cells. Barr et al. (95) have shown that inactive but functionally intact thymidine kinase (TK) genes may be activated by methyl-N-nitrosoguanidine (MNNG) in Chinese hamster cells. It was proposed that activation depended on focal demethylation of TK gene resulting in low TK activity. In contrast, demethylation throughout the genome resulted in much higher activity. Thus, overall demethylation may also play a role in the regulation of gene expression. In addition, MNNG and 5-azacytidine were shown to induce a 50- to 100-fold rise in the reversion frequency of the mutant HeLa cells harboring a hypoxanthine phosphorybosyltransferase gene silenced by methylation (96). Thus, MNNG (and possibly other carcinogens) have epigenetic effects in addition to mutation activity. However, the role of changes in methylation pattern of regulatory DNA sequences in multistage hepatic carcinogenesis remains to be elucidated.

CONCLUSIONS

SAM is a strong chemopreventive agent for liver carcinogenesis. This naturally occurring, nontoxic and nonmutagenic (97) compound that enters liver cells in vivo and in vitro (98-103) is currently used in humans for the treatment of liver injury, such as acute and chronic ethanol (104) and contraceptive toxicity (105), cholestasis in pregnant women (106), liver cirrhosis (107), chronic hepatitis (108), etc. The cytoprotective effect of SAM may be at least in part linked to its ability to restore the GSH pool in GSH-depleted cells (47,48). Thus, the use of SAM for chemoprevention of hepatic carcinogenesis in populations at risk may be envisioned.

There is much evidence that DNA hypomethylation is involved in carcinogenesis, although the nature of this involvement is still unknown. On the other hand, a decrease in methylating environment (SAM/SAH ratio) may be associated with DNA hypomethylation. After reconstitution of the methylating environment by SAM, recovery of DNA methylation occurs. Thus, SAM administration during hepatic carcinogenesis abolishes one of the conditions linked to (or favoring) the carcinogenic process. There exists a strong negative correlation between SAM/SAH ratio and expression of some growth-related genes. However, although some recent results suggest that this phenomenon may be mediated by gene methylation, there is no definitive proof in favor of this conclusion. This possibility may be only hypothesized at present, on the basis of knowledge on the relationships between SAM/SAH ratio, DNA methylation, and gene expression.

Decrease in the expression of growth-related genes could be one of the mechanisms through which SAM administration inhibits DNA synthesis in preneoplastic tissue thus causing a shift of the equilibrium between cell production and cell loss in favor of cell loss. This mechanism could explain prevention by SAM of the progression of the majority of preneoplastic lesions to liver carcinoma, for a period of time roughly corresponding to more than 2/3 of the rat life-span. However, further work is necessary to fully explain the mechanisms of the SAM chemopreventive effect for hepatic carcinogenesis.

ACKNOWLEDGEMENT

This work was supported by funds from Associazione Italiana Ricerca sul Cancro, and MURST (programs 60% and 40%).

REFERENCES

1.   Wattenberg, L.W. (1985) Chemoprevention of cancer. *Cancer Res.* 45:1-8.
2.   Potzsch, J., and T. Schramm (1987) Chemoprevention of cancer. Present status, problems and trends. *Arch. Geschwulstforsch.* 57:249-260.
3.   Fiala, E.S., B.S. Reddy, and J.H. Weisburger (1985) Naturally occurring anticarcinogenic substances in foodstuffs. *Ann. Rev. Nutr.* 5:295-321.
4.   Foulds, L. (1958) The natural history of cancer. *J. Chron. Dis.* 8:2-57.
5.   Nowell, P. (1976) The clonal evolution of tumor cell population. *Science* 194:23-28.
6.   Sakamoto M., S. Hirohashi, and Y. Shimosato (1991) Early stages of multistep hepatocarcinogenesis: Adenomatous hyperplasia and early hepatocellular carcinoma. *Hum. Pathol.* 22:172-178.
7.   Farber, E., and D.S.R. Sarma (1987) Hepatocarcinogenesis: A dynamic cellular perspective. *Lab. Invest.* 56:4-22.
8.   Marks, F., G. Füstenberger, M. Gschwendt, M. Rogers, B. Schurich, B. Kaina, and G. Bauer (1988) The wound response as a key element for an understanding of multistage carcinogenesis in the skin. In *Chemical Carcinogenesis. Models and Mechanisms*, F. Feo, P. Pani, A. Columbano, and R. Garcea, eds. Plenum Publ. Corp., New York, pp. 217-234.
9.   Weinberg, R.A. (1989) Oncogenes, antioncogenes, and the molecular bases of carcinogenesis. *Cancer Res.* 49:3713-3721.
10.   Teebor, G.W., and F.F. Becker (1971) Regression and persistence of hyperplastic hepatic nodules induced by N-2-fluorenylacetamide and their relationship to hepatocarcinogenesis. *Cancer Res.* 31:1-3.
11.   Moore, M.A., J. Hacker, and P. Bannasch (1983) Phenotypic instability in focal and nodular lesions induced by a short term system in the rat liver. *Carcinogenesis* 4:595-603.
12.   Enomoto, K., and E. Farber (1982) Kinetics of phenotypic maturation of remodelling of hyperplastic nodules during liver carcinogenesis. *Cancer Res.* 42:2330-2335.
13.   Garcea, R., L. Daino, R. Pascale, M.M. Simile, M. Puddu, S. Frassetto, P. Cozzolino, M.A. Seddaiu, L. Gaspa, and F. Feo (1989) Inhibition of promotion and persistent nodule growth by S-adenosyl-L-methionine in rat liver carcinogenesis: Role of remodeling and apoptosis. *Cancer Res.* 49:1850-1856.
14.   Mikol, Y.B., K.L. Hoover, D. Creasia, and L.A. Poirier (1983) Hepatocarcinogenesis in rats fed methyl-deficient, amino acid-defined diets. *Carcinogenesis* 4:1619-1629.

15. Ghoshal, A.K., and E. Farber (1984) The induction of liver cancer by a dietary deficiency of choline and methionine without added carcinogens. *Carcinogenesis* 5:1367-1370.

16. Lombardi, B. (1988) The choline-devoid diet model of hepatocarcinogenesis in the rat. In *Chemical Carcinogenesis. Models and Mechanisms,* F. Feo, P. Pani, A. Columbano, and R. Garcea, eds. Plenum Publ. Corp., New York, pp. 563-581.

17. Newberne, P.M., and A.E. Rogers (1986) Labile methyl groups and the promotion of cancer. *Ann. Rev. Nutr.* 6:407-432.

18. Miller, J.A., and E.C. Miller (1953) The carcinogenic aminoazo dyes. *Adv. Cancer Res.* 1:339-396.

19. Solt, D.B., A. Medline, and E. Farber (1977) Rapid emergence of carcinogen-induced hyperplastic lesions in a new model for sequential analysis of liver carcinogenesis. *Am J. Pathol.* 88:595-618.

20. Feo, F., R. Garcea, L. Daino, R. Pascale, L. Pirisi, S. Frassetto, and M.E. Ruggiu (1985) Early stimulation of polyamine biosynthesis during promotion by phenobarbital of diethylnitrosamine-induced rat liver carcinogenesis. The effects of variations of the S-adenosyl-L-methionine cellular pool. *Carcinogenesis* 6:1713-1720.

21. Garcea, R., R. Pascale, L. Daino, S. Frassetto, P. Cozzolino, M.E. Ruggiu, M.G. Vannini, L. Gaspa, and F. Feo (1987) Variations of ornithine decarboxylase activity and S-adenosyl-L-methionine and 5'-methylthioadenosine contents during the development of diethylnitrosamine-induced liver hyperplastic nodules and hepatocellular carcinomas. *Carcinogenesis* 8:653-658.

22. Feo, F., R. Garcea, R. Pascale, L. Pirisi, L. Daino, and A. Donaera (1987) The variations of S-adenosyl-L-methionine content modulate hepatocyte growth during phenobarbital promotion of diethylnitrosamine-induced rat liver carcinogenesis. *Toxicol. Pathol.* 15:109-113.

23. Simile, M.M., V. Marras, R.M. Pascale, M.A. Seddaiu, L. Adaino, F. Feo, and G. Massarelli (1992) Chemoprevention of rat liver carcinogenesis by S-adenosyl-L-methionine: A long-term study. Submitted for publication.

24. Farber, E., and H. Ichinose (1958) The prevention of ethionine-induced carcinoma in the liver in rats by methionine. *Cancer Res.* 18:1209-1213.

25. Anisimov, V.N., G.I. Miretskii, E.V. Danetskaya, M.N. Troiskaya, and P.V. Ramzaev (1982) Inhibition by methionine of development of mammary gland carcinoma induced by 7,12-dimethylbenz(a)anthracene and N-nitrosomethylurea in rats. *Bull. Exp. Biol. Med.* 92:1424-1426.

26. Brada, Z., N.H. Altaman, M. Hill, and S. Bulba (1982) The effect of methionine on the progression of hepatocellular carcinoma induced by ethionine. *Res. Chem. Pathol. Pharmacol.* 38:157-160.

27. Brada, Z., J. Hillova, M. Hill, N.H. Altman, and S. Bulba (1986) Effect of methionine on development of benzopyrene (BP) induced sarcomas. *Proc. Amer. Assn. Cancer Res.*, 27:121.

28. Fullerton, F.R., K. Hoover, Y.B. Mikol, D.A. Creasia, and L.A. Poirier (1990) The inhibition by methionine and choline of liver carcinoma formation in male C3H mice dosed with diethylnitrosamine and fed phenobarbital. *Carcinogenesis* 11:1301-1305.

29. Akhtar, R., and A.E. Rogers (1989) Lipotrope deficiency and methotrexate in procarbazine-induced mammary carcinoma in male rats. *Proc. Amer. Assn. Cancer Res.* 30:175.

30. Farber, E. (1988) Approaches to cancer prevention: Introduction and perspective. *Proc. Amer. Assn. Cancer Res.* 29:527.

31. Garcea, R., L. Daino, R. Pascale, M.M. Simile, M. Puddu, M.E. Ruggiu, M.A. Seddaiu, G. Satta, M.J. Sequenza, and F. Feo (1989) Protooncogene methylation and expression in regenerating liver and preneoplastic liver nodules induced in the rat by diethylnitrosamine: Effect of variations of S-adenosylmethionine:S-adenosylhomocysteine ratio. *Carcinogenesis* 10:1183-1192.

32. Bursch, W., B. Lauer, I. Timmermann-Trosiener, G. Bartel, J. Schuppler, and R. Schulte-Hermann (1984) Controlled death (apoptosis) of normal and putative preneoplastic cell in rat liver following withdrawal of tumor promoters. *Carcinogenesis* 5:453-458.

33. Columbano, A., G.M. Ledda-Columbano, P.M. Rao, S. Rajalakshmi, and D.S.R. Sarma (1984) Occurrence of cell death (apoptosis) in preneoplastic and neoplastic liver cells. *Amer. J. Pathol.* 116:441-446.

34. Schulte-Hermann, R., W. Bursch, L. Fesus, and B. Kraupp (1988) Cell death by apoptosis in normal and neoplastic tissue. In *Chemical Carcinogenesis. Models and Mechanisms*, F. Feo, P. Pani, A. Columbano, and R. Garcea, eds. Plenum Publ. Corp., New York, pp. 263-276.

35. Columbano, A., G.M. Ledda-Columbano, P.P. Coni, M. Vargiu, G. Faa, and P. Pani (1984) Liver hyperplasia and regression after lead nitrate administration. *Toxicol. Pathol.* 12:89-95.

36. Bursch, W., H.S. Taper, B. Lauer, and R. Schulte-Hermann (1985) Quantitative histological and histochemical studies on the occurrence and stages of controlled cell death (apoptosis) during regression of rat liver hyperplasia. *Wirchovs Arch.* 50:153-166.

37. Nawaz, S., M.P. Lynch, P. Galand, and L.E. Gerschenson (1987) Hormonal regulation of cell death in rabbit uterine epithelium. *Am. J. Pathol.* 127:51-59.

38. Tessitore, L., G. Valente, G. Bonelli, P. Costelli, and F.M. Baccino (1989) Regulation of cell turnover in the livers of tumor-bearing rats: Occurrence of apoptosis. *Int. J. Cancer* 44:697-700.

39. Farber, E., J. Rotstein, L. Harris, G. Lee, and Z.-Y. Chen (1988) Cell proliferation and cell loss in progression in liver carcinogenesis: A new hypothesis. In *Chemical Carcinogenesis. Models and Mechanisms*, F. Feo, P. Pani, A. Columbano, and R. Garcea, eds. Plenum Publ. Corp., New York, pp. 167-172.

40. Rotstein J., D.S.R. Sarma, and E. Farber (1986) Sequential alterations in growth control and cell dynamics of rat hepatocytes in early precancerous steps in carcinogenesis. *Cancer Res.* 46:2377-2385.

41. Tisdale, M.J. (1983) Methionine synthesis from 5'-methylthioadenosine by tumour cells. *Biochem. Pharmacol.* 32:2915-2920.

42. Pegg, A.E., R.T. Borchardt, and J.K. Coward (1981) Effect of inhibitors of spermidine and spermine synthesis on polyamine concentrations and growth of transformed mouse fibroblasts. *Biochem. J.* 194:79-89.

43. Raina, A., K. Tuomi, and R.L. Pajula (1982) Inhibition of the synthesis of polyamines and macromolecules by 5'-methylthioadenosine and 5'-alkylthiotubercidins in BHK21 cells. *Biochem. J.* 4:697-703.

44. Pascale, R.M., M.M. Simile, G. Satta, M.A. Seddaiu, L. Daino, G. Pinna, M.A. Vinci, L. Gaspa, and F. Feo (1991) Comparative effects of L-methionine, S-adenosyl-L-methionine and 5'-methylthioadenosine on the growth of

preneoplastic lesions and DNA methylation in rat liver during the early stages of hepatocarcinogenesis. *Anticancer Res.* (in press).

45. Wu, S.E., W.P. Huskey, R.T. Borchardt, and R.L. Schowen (1983) Chiral instability at sulfur of S-adenosylmethionine. *Biochemistry* 22:2828-2832.

46. Chawla, R.K., H.L. Bonkovsky, and J.T. Galambos (1990) Biochemistry and pharmacology of S-adenosyl-L-methionine and a rationale for its use in liver disease. *Drug* 40 (Supplement 3):98-110.

47. Feo, F., R. Pascale, R. Garcea, L. Daino, L. Pirisi, S. Frassetto, M.E. Ruggiu, C. DiPadova, and G. Stramentinoli (1986) Effect of the variations of S-adenosyl-L-methionine liver content on fat accumulation and ethanol metabolism in ethanol-intoxicated rats. *Toxicol. Appl. Pharacol.* 83:331-341.

48. Pascale, R., L. Daino, R. Garcea, S. Frassetto, M.E. Ruggiu, M.G. Vannini, P. Cozzolino, and F. Feo (1989) Inhibition by ethanol of rat liver plasma membrane (NA$^+$,K$^+$)ATPase: Protective effect of S-adenosyl-L-methionine, L-methionine and N-acetylcysteine. *Toxicol. Appl. Pharacol.* 97:216-229.

49. Perera, M.I.R., A.J. Demetris, S.L. Katyal, and H. Shinozuka (1985) Lipid peroxidation of liver microsome membranes induced by choline-deficient diets and its relationships to diet-induced promotion of the induction of γ-glutamyl-transpeptidase-positive foci. *Cancer Res.* 45:2533-2538.

50. Gravela, E., F. Feo, R.A. Canuto, R. Garcea, and L. Gabriel (1975) Functional and structural alterations of liver ergastoplasmic membranes during DL-ethionine hepatocarcinogenesis. *Cancer Res.* 35:3041-3047.

51. Ruchirawat, M., F.F. Becker, and J.N. Lepeyre (1984) Interaction of DNA methyltransferase with aminofluorene and N-acetylaminofluorene modified poly(dC-dG). *Nucleic Acids Res.* 12:3357-3372

52. Carr, B.I., J.G. Reilly, S.S. Smith, C. Winberg, and A. Riggs (1984) The tumorigenicity of 5-azacytidine in male Fisher rats. *Carcinogenesis* 5:1897-1902.

53. Cavaliere, A., A. Bufalari, and R. Vitali (1987) 5-Azacytidine carcinogenesis in Balb/c mice. *Cancer Lett.* 37:51-58.

54. Rao, P.M., A. Antony, S. Rajalakshmy, and D.S.R. Sarma (1989) Studies on hypomethylation of liver DNA during early stages of chemical carcinogenesis in rat liver. *Carcinogenesis* 10:933-937.

55. Koo, P., and E. Farber (1991) Hypomethylation as a possible mechanism in gene overexpression unique in precancerous hepatocyte nodules. *Proc. Amer. Assn. Cancer Res.* 32:135.

56. McGeady, M.L., C. Jhappan, R. Ascinoe, and G.F. Wande Woude (1983) In vitro methylation of specific regions of the cloned Moleny sarcoma virus genome inhibits its transforming activity. *Molec. Cell Biol.* 3:305-314.

57. Borrello, M.G., M.A. Pierotti, I. Bongarzone, R. Donghi, P. Mondellini, and G. Della Porta (1987) DNA methylation affecting the transforming activity of the human Ha-*ras* oncogene. *Cancer Res.* 47:75-79.

58. Goelz, S.E., B. Vodelstein, S.R. Hamilton, and A.P. Feinberg (1985) Hypomethylaton of DNA from benign and malignant human colon neoplasms. *Science* 228:187-190.

59. Bianchi, N.O., M. Larramendy, and M.S. Bianchi ù (1988) Mitomycin c-induced damage and repair in human and pig lymphocytes. *Mutat. Res.* 197:151-156.

60. Vogelstein, B., E.R. Fearon, S.R. Hamilton, S.E. Kern, A.C. Preisinger, M. Leppert, Y. Nakamura, R. White, A.M.M. Smits, and J.L. Bos (1988) Genetic

alterations during colorectal tumor development. *N. Engl. J. Med.* 319:525-532.

61. Selker, E.U. (1990) DNA methylation and chromatin structure: A view from below. *Nature* 15:103-107.

62. Woodcock, D.M., J.K. Adams, R.G. Allan, and I.A. Cooper (1983) Effect of several inhibitors of enzymatic DNA methylation on the in vivo methylation of different classes of DNA sequences in a cultured human cell line. *Nucl. Acids Res.* 11:489-499.

63. Zappia, V., R. Zydek-Cwick, and F. Schlenk (1969) The specificity of S-adenosyl-methionine derivatives in methyl transfer reactions. *J. Biol. Chem.* 244:4499-4509.

64. Feo, F., R. Garcea, L. Daino, R. Pascale, S. Frassetto, P. Cozzolino, M.G. Vannini, M.E. Ruggiu, M.M. Simile, and M. Puddu (1988) S-adenosylmethionine antipromotion and antiprogression effect in hepatocarcinogenesis. Its association with DNA methylation and inhibition of gene expression. In *Chemical Carcinogenesis. Models and Mechanisms*, F. Feo, P. Pani, A. Columbano, and R. Garcea, eds. Plenum Publ. Corp., New York, pp. 407-423.

65. Pascale, R.M., M.M. Simile, G. Satta, L. Daino, M.A. Seddaiu, M.J. Sequenza, and F. Feo (1991) Effects of L-methionine, S-adenosyl-L-methionine (SAM) and 5'-methylthioadenosine (MTA) on the growth of preneoplastic lesions and DNA methylation in rat liver carcinogenesis. *Proc. Amer. Assn. Cancer Res.* 32:127.

66. Lepeyre, J.N., and F.F. Becker (1979) 5-Methylcytosine content of nuclear DNA during chemical hepatocarcinogenesis and in carcinomas which result. *Biochem. Biophys. Res. Comm.* 87:698-705.

67. Kanduc, D., A. Ghoshal, E. Quagliariello, and E. Farber (1988) DNA hypomethylation in ethionine-induced rat preneoplastic hepatocyte nodules. *Biochem. Biophys. Res. Comm.* 150:739-744.

68. Shivapurkar, N., and L.A. Poirier (1983) Tissue levels of S-adenosylmethionine and S-adenosylhomocysteine in rats fed methyl-deficient, amino acid-defined diets for one to five weeks. *Carcinogenesis* 4:1051-1057.

69. Wilson, M.J., N. Shivapurkar, and L. Poirier (1984) Hypomethylation of hepatic nuclear DNA in rats fed with a carcinogenic methyl-deficient diet. *Biochem. J.* 218:987-990.

70. Locker, J.D., V.R. Tirimuru, and B. Lombardi (1986) DNA methylation and hepatocarcinogenesis in rat fed a choline-devoid diet. *Carcinogenesis* 7:1309-1312.

71. Sturman, J.A., and G.E. Gaull (1976) Changes in subcellular distribution of S-adenosylmethionine decarboxylase in regenerating and in developing liver. *Biochim. Biophys. Acta* 428:70-77.

72. Kramer, D.L., J.R. Sufrin, and C.W. Porter (1982) Relative effects of S-adenosylmethionine depletion on nucleic acid methylation and polyamine biosynthesis. *Biochem. J.* 247:259-265.

73. Taguchi, H., and I. Chanarin (1978) The effect of homocysteine, methionine, serine and glycine on DNA synthesis by human normoblastic and megaloblastic bone marrow cells. *J. Nutr. Sci. Vitaminol.* 24:83-89.

74. Pascale, R.M., M.M. Simile, M.E. Ruggiu, M.A. Seddaiu, G. Satta, M.J. Sequenza, L. Daino, M.G. Vannini, P. Lai, and F. Feo (1991) Reversal by

5-azacytidine of the S-adenosyl-L-methionine-induced inhibition of the development of putative preneopalstic foci in rat liver carcinogenesis. *Cancer Lett.* 56:259-265.

75. Chiak, A., J. Veseley, and J. Skoda (1985) Azapyrimidine nucleosides: Metabolism and mechanisms. *Adv. Enz. Regula.* 24:335-354.

76. Beer, G., M. Schwartz, N. Sawada, and H.C. Pitot (1986) Expression of H-*ras* and c-*myc* protooncogenes in isolated γ-glutamyltranspeptidase rat hepatocytes and in hepatocellular carcinomas induced by diethylnitrosamine. *Cancer Res.* 46:2435-2441.

77. Yaswen, P., M. Goyette, P.R. Shank, and N. Fausto (1985) Expression of c-Ki-*ras*, c-Ha-*ras*, and c-*myc* in specific cell types during hepatocarcinogenesis. *Molec. Cell Biol.* 5:780-786.

78. Thompson, N.L., J.E. Mead, L. Braun, M. Goyette, P.R. Shank, and N. Fausto (1986) Sequential protooncogene expression during rat liver regeneration. *Cancer Res.* 46:3111-3117.

79. Feo, F., R. Garcea, L. Daino, and R. Pascale (1988) Mechanism of the inhibition of liver hepatocarcinogenesis promotion by S-adenosyl-L-methionine. In *Experimental Hepatocarcinogenesis*, M.B. Roberfroid and V. Préat, eds. Plenum Publ. Corp., New York, pp. 197-207.

80. Nagy, P., R.P. Evarts, E. Marsedn, J. Roach, and S.S. Thorgeirsson (1988) Cellular distribution of c-*myc* transcripts during chemical hepatocarcinogenesis in rats. *Cancer Res.* 48:5522.

81. Corcos, D., N. Defer, M. Raymondjean, B. Paris, M. Corral, L. Tichonicky, J. Khru, D. Glaise, A. Saulnier, and C. Guguen-Guillouzo (1984) Correlated increase of the expression of the c-*ras* genes in chemically induced carcinogenesis. *Biochem. Biophys. Res. Comm.* 122:259-264.

82. Corral, M., L. Tichonicky, C.P.L. Guguen-Guillouzo, D. Corcos, M. Raymondjean, B. Paris, J. Khru, and N. Defer (1985) Expression of c-*fos* oncogene during hepatocarcinogenesis, liver regeneration, and in synchronized HCT cells. *Exptl. Cell Res.* 160:427-434.

83. Beer, D.G., M.J. Neveu, D.L. Paul, U.R. Rapp, and H.C. Pitot (1988) Expression of c-*raf* protooncogene, γ-glutamyltranspeptidase, and gap junction protein in rat liver neoplasms. *Cancer Res.* 48:1610-1617.

84. Adams, L.P., and R.H. Burdon (1985) *Molecular Biology of DNA Methylation*. Springer Verlag, New York.

85. Bhave, M.R., M.J. Wilson, and L.A. Poirier (1988) c-Ha-*ras* gene hypomethylation in the livers and hepatomas of rats fed a choline-devoid and a choline supplemented diet. *Carcinogenesis* 9:259-263.

86. Chandar, P.K., B. Lombardi, and J. Locker (1989) C-*myc* gene amplification during hepatocarcinogenesis by a choline-devoid diet. *Proc. Natl. Acad. Sci., USA* 86:2703-2707.

87. Hsieh, L.L., E. Wainfan, S. Hishina, and I.B. Weinstein (1989) Altered expression of retrovirus-like sequences and cellular oncogenes in mice fed methyl-deficient diet. *Cancer Res.* 49:3795-3799.

88. Wainfan, E., M. Dizik, M. Stender, and J.K. Christman (1989) Rapid appearance of hypomethylated DNA in livers of rats fed cancer-promoting, methyl-deficient diets. *Cancer Res.* 49:4094-4097.

89. Poirier, L.A. (1988) Methyl deficiency in carcinogenesis. In *Chemical*

Carcinogenesis. Models and Mechanisms, F. Feo, P. Pani, A. Columbano, and R. Garcea, eds. Plenum Publ. Corp., New York, pp. 583-589.

90.   Wainfan, E., M. Dizik, G. Sheikhnejad, and J.K. Christman (1991) Rapid and reversible changes in gene expression and methylation of DNA in rats fed a cancer-promoting, methyl-deficient diet. *Proceedings of the Sixth Sardinian International Meeting: Genetic and Determinants of Premalignant and Malignant Phenotype*. Abstracts. Tampacolor, s.r.l., Sassari, p. 33.

91.   Locker, J., S. Hutt, and B. Lombardi (1987) Alpha-fetoprotein gene methylation and hepatocarcinogenesis in rats fed a choline-devoid diet. *Carcinogenesis* 8:241-246.

92.   De Brustos, A., B.D. Nelkin, A. Silvernam, G. Ehrlich, B. Poiesz, and S.B. Baylin (1988) *Proc. Natl. Acad. Sci., USA* 85:5693-5697.

93.   Jones, P.A., and J.D. Buckley (1990) The role of DNA methylation in cancer. *Adv. Cancer Res.* 54:1-23.

94.   Christy, B.A., and G.A. Scangos (1986) In vitro methylation of bovine papillomavirus alters its ability to transform mouse cells. *Molec. Cell Biol.* 6:2910-2915.

95.   Barr, F.G., S. Rajagopalan, C.A. MacArthur, and M.W. Lieberman (1986) Genomic hypomethylation and far-5' sequence alterations are associated with carcinogen-induced activation of the hamster thymidine kinase gene. *Molec. Cell Biol.* 6:3023-3033.

96.   Ivarie, R., and J.A. Morris (1986) Activation of a nonexpressed hypoxantine phosphoribosyltransferase allele in mutant H23 HeLa cells by agents that inhibit DNA methylation. *Molec. Cell Biol.* 6:97-104.

97.   Pezzoli, C., M. Galli-Kienle, and G. Stramentinoli (1987) Lack of mutagenic activity of adomethionine in vivo and in vitro. *Arzneim.-Forsch.* 37:826-829.

98.   Stekol, J.A., E.L. Anderson, and S. Weiss (1958) S-adenosyl-L-methionine in the synthesis of choline, creatinine, and cysteine in vivo and in vitro. *J. Biol. Chem.* 233:425-429.

99.   Zappia, V., P. Galletti, M. Porcelli, G. Ruggiero, and A. Andreana (1978) Uptake of adenosylmethionine and related sulfur compounds by isolated rat liver. *FEBS Lett.* 90:331-335.

100.   Pezzoli, C., G. Stramentinoli, M. Galli-Kienle, and E. Pfaff (1978) Uptake and metabolism of S-adenosyl-L-methionine by isolated rat hepatocytes. *Biochem. Biophys. Res. Comm.* 85:1031-1038.

101.   Farooqui, J.Z., H.W. Lee, S. Kim, and W.K. Paik (1983) Studies on compartmentation of S-adenosyl-L-methionine in *Saccharomyces cerevisiae* and isolated hepatocytes. *Biochim. Biophys. Acta* 757:342-351.

102.   Engstrom, M.A., and N.J. Benevenga (1987) Rate of oxidation of methionine and S-adenosylmethionine methyl carbons in isolated hepatocytes. *J. Nutr.* 117:1820-1826.

103.   Giulidori, P., M. Galli-Kienle, E. Cato, and G. Stramentinoli (1984) Transmethylation, transsulfuration, and aminopropylation reactions of S-adenosylmethionine. *J. Biol. Chem.* 259:4205-4211.

104.   DiPadova, C., R. Tritapiede, P. Rovagnati, M. Pezzoli, and G. Stramentinoli (1984) Decreased blood levels of ethanol and acetaldehyde by S-adenosyl-L-methionine in humans. *Arch. Toxicol.* 7:240-242.

105.   Frezza, M., G. Pozzato, G. Pison, C. Zalateo, L. Chiesa, and C. DiPadova (1987) S-adenosylmethionine counteracts oral contraceptive hepatotoxicity in women. *Amer. J. Med. Sci.* 30:234-238.

106. Frezza, M., G. Pozzato, L. Chiesa, G. Stramentinoli, and C. DiPadova (1984) Reversal of intrahepatic cholestasis of pregnancy in women after high dose of S-adenosyl-L-methionine administration. *Hepatology* 4:274-278.
107. Vendemiale, G., E. Altomare, R. Altavilla, C. Le Grazie, and C. Di Padova (1989) S-adenosyl-L-methionine (SAMe) improves acetaminophen metabolism in cirrhotic patients. *J. Hepatol.* 9:S240.
108. Micali, M., D. Chiti, and V. Balestra (1983) Double-blind controlled clinical trial of SAM administered orally in chronic liver diseases. *Curr. Ther. Res.* 33:1004-1013.

MOLECULAR CONTROL OF HUMAN PAPILLOMAVIRUS

RNA EXPRESSION IN NEOPLASIA

J.A. DiPaolo and C.D. Woodworth

Laboratory of Biology
National Cancer Institute
Bethesda, Maryland 20892 USA

INTRODUCTION

Animal and animal-derived (cell culture) models have been used to study diverse classes of chemical carcinogens relevant to humans. Furthermore, the interaction of various agents has resulted in cocarcinogenesis as well as anticarcinogenesis. Unfortunately, parallell studies utilizing human cells have been of limited value. Single carcinogenic agents, such as benzo[a]pyene (16), SV-40 (15) and radiation (8) have immortalized normal human cells, but, progression to malignancy did not occur. As a result, human cancer has been investigated primarily with epidemiologic studies.

Thus, until recently, tumorigenesis in human usually was studied indirectly. In multistep carcinogenesis, each step represents a physiological barrier that must be unblocked for a cell to progress towards malignancy. Primary questions must be addressed: What are the cellular and molecular mechanisms that govern multistep tumorigenesis? How can the physiological controls be overridden to produce the deregulation of malignancy? How can normal cell growth be restored? The fundamental question is: what controls normal and aberrant differentiation?

Cervical cancer is an ideal model for studying human carcinogenesis because the normal, benign, and malignant stages can be distinguished both in vivo and in vitro. Squamous cell carcinomas are derived from a specific region of the cervix referred to as the transformation zone. Squamous cell carcinoma of the cervix is a major public health problem worldwide; as a cause of cancer death in women, it is number two after breast cancer (1). The first documented data concerning marital or sexual events was published by Rigoni-Stern in 1842 (13) who concluded, from a statistical analysis of death records, that cancer of the uterus occurred rarely in virgins and nuns. Since then, evidence has been accumulating consistent with the premise that cervical cancer behaves as a venerally transmitted disease. Two important factors in the development of cancer appear to be the age of onset of sexual activity and the number of partners that a woman has had. Today, it is accepted that human papillomaviruses (HPV) are linked with neoplastic events in the epithelia of the female genitalia. In fact, over 95% of invasive cervical cancers are associated with a specific subset of HPV DNAs (12). One concern is the control of differentiation of papilloma cells by biological modifiers that can also be considered to be anticarcinogens or tumor suppressors.

RESULTS AND DISCUSSION

Primary epithelial cells derived from neonatal foreskins and the squamo-columnar zone between the endo- and exo-cervix were obtained from healthy individuals. All tissue was HPV negative. Epithelial cells were isolated by a collagenase float technique in serum-free medium (MCDB153-LB) (4). The medium is a modification of that originally formulated by Ham.

At the present time, papillomaviruses cannot be grown in vitro. Therefore, recombinant DNA technology was used to transfect a variety of HPV DNAs into human genital cells (cervical- or foreskin-derived) to determine their role in carcinogenesis. Cells grown in a serum free medium were transfected by either a calcium phosphate or lipofection method. Cells expressing the transfected DNA were selected by adding G418 (100 µg per ml for 48 hours) or by repeatedly subculturing the cultures until the normal nontransfected cells had senesced, leaving only colonies of HPV-containing cells. Only HPV DNAs such as 16, 18, 31, and 33 that have a high or moderate association with cervical malignancy were capable of inducing immortality (20). All derived cell lines contained integrated and rearranged DNA and transcriptionally active viral genes. Under similar circumstances, those HPV DNAs with low or no oncogenic potential for cervical cancer, such as HPV1a, 5, 6b, and 11, induced only small colonies after G418 selection, and were subsequently senesced. It was demonstrated that the colonies prior to senescence contained only episomal virus DNA. Thus, integration of HPV sequences can be concluded to be important for immortalization of genital cells. However, of the cell lines with HPV DNAs associated with cervical cancer, none produced tumors when injected into immune deficient mice. Therefore, because integration is important for immortal-ization it was concluded that the virus-encoded immortalization function must contribute to the pathogenesis of cervical carcinomas. Although all cell lines express a number of virus mRNAs, one cell line expressed only transcripts from the intact E6 and E7 genes. This was the first indication that these genes are responsible for immortalization.

A significant development regarding DNA tumor virus oncogenes has been the observation that the adenovirus E1A oncogene (19) product and the SV40 large T antigen can complex with the retinoblastoma antioncogene product referred to as RB-1 (1). The working hypothesis is that the binding of the viral oncoprotein to the RB protein effectively results in the loss of RB protein function and, thus, is indirectly responsible for stimulation of cell proliferation. This type of RB protein complex has now been extended to include the HPV type 16, E7 protein (22). Immortal lines as well as tumor lines examined have normal amounts of RB expression and protein, suggesting that the immortalization process can be obtained without interaction with or suppression of the RB 1 gene.

It is possible to demonstrate that epithelial cells immortalized by HPV DNAs exhibit dysplastic differentiation in vivo. Normal human cervical and foreskin epithelial cells and cells immortalized by the common HPV DNAs were transplanted beneath a skin-muscle flap in athymic nude mice (7). With this technique, normal human epidermal cells formed a well-differentiated, stratified squamous epithelium. Cells of confluent cultures were changed to a medium containing 2.0 mM calcium to induce intracellular attachment and stratification. After further incubation, intact monolayers could be removed by digestion with dispase. The dislodged epithelium was covered with a rectangular piece of Silastic medical sheeting and inverted to support the epithelium basal side on the plastic. The epithelium with Silastic support was deposited, epithelial side up, on the dorsal musculature of a nude mouse. The degree of differentiation was assessed by killing the mice two to three weeks post-transplantation. The histological preparations classified as normal demonstrated epithelium stratification, basal cell polarity and were devoid of nuclear abnormalities. Grafts classified as dysplastic had abnormal mitosis, multinucleated cells, increased nuclear cytoplasmic ratio, or absence of basal cell polarity in the upper stratum. Involucrin was localized in individual sections by the indirect peroxidase method using an involucrin immuno-kit. The immortalized cell lines exhibited defective terminal differentiation in culture compared to the normal that formed keratinizing, stratified squamous epithelium two to three weeks after transplantation. Foreskin-derived cells could be distinguished from cervical-derived cells in that the latter had thickened squamous epithelia and did not keratinize. Therefore, the cells from either source were morphologically similar to those from which they had been derived.

Cell lines immortalized by HPV18 DNA and grafted at relatively early passage (under 65 population doublings) formed dysplastic epithelia resembling cervical intraepithelial neoplasia (CIN) grades 1-3. Such grafts were characterized by altered mitosis, increased nuclear cytoplasmic ratio, and frequently by total lack of squamous differentiation. They cannot be classified as malignancies because invasion of the basement membrane was not observed, and they did not form tumors. Grafts immortalized by the other HPVs such as 16, 31, or 33 DNAs formed thin squamous epithelia in which dysplastic changes were not evident. With late passage cultures (greater than 200 population doublings), lines immortalized by HPV16 31 or 33 DNAs formed thick dysplastic epithelia in nude mice. The severity of aberrant differentiation varied greatly with different HPV immortalized cell lines and often involved total absence of normal squamous maturation and, therefore, resembled CIN 3. When the histopathology sections were examined using involucrin, a precursor of the cornified envelope, as a marker for the early stages of squamous differentiation, involucrin expression was found to be confined through the superbasal layers and grafts of normal cervical or foreskin epithelial cells but it altered in grafts containing HPV-immortalized cell lines.

The more severely dysplastic grafts often stained variably with a mosaic pattern. This mosaic pattern of involucrin expression resembled that of an invasive cervical carcinoma cell line. A similar alteration in the pattern of involucrin expression has been reported to occur during the development of dysplastic cervical differentiation in situ.

241

Concurrently, replicate cultures of the immortalized cell lines used in vivo for transplantation studies and involucrin localization were examined for alteration and expression of genes associated with growth as well as squamous cell differentiation (7). The steady state levels for involucrin and keratin I RNAs were increased in normal cells induced to differentiate in culture. Late passage of immortal cells, however, often had altered and reduced expression of these genes.

Epidermal growth factor receptor, myc and HPV 16 E6/E7 genes, were assessed in both early and late passage cells by Northern analysis. Normal foreskin keratinocytes induced to differentiate in culture expressed increased involucrin and keratin 1 RNAs (markers for squamous differentiation). RNA for laminin, which is down regulated during epidermal differentiation, decreased dramatically. Cultures (early passages) immortalized by HPV also showed increased steady-state levels of involucrin and keratin I RNAs that increased when cells were induced to differentiate in culture. Levels of keratin I and involucrin RNAs were reduced in several late passage HPV immortalized foreskin lines, although laminin RNAs remained at relatively high levels. The findings that immortalized cell lines have decreased expression of involucrin and that keratin I RNAs in vitro also formed severely dysplastic epithelial xenografts suggest a correlation between in vitro and in vivo gene expression.

Additional studies were focused on TGFβ (21), γinterferon and leukoregulin (23). TGFβs are 25 kd disulfide cross-linked homodimers that occur in multiple forms (14). TGFβ1 stimulates the growth of some cells, primarily fibroblasts and osteoblasts. Other cells, such as hemopoietic and epithelial cells, are inhibited by TGFβ that is produced by cultured cervical cells. Both TGFβ 1 and 2 dramatically reduced the transcription of HPV 16 RNA in cells immortalized by HPV (21). In contrast, laminin B1 chain and β actin RNA transcripts were not decreased by either TGFβ 1 or 2. Specifically, loss of the E6 and E7 protein expression was both time- and dose-dependent. One experiment demonstrated that TGFβ inhibition of HPV 16 RNA expression was at the transcriptional level. Addition of TGFβ induced a 6- to 7-fold increase in TGFβ 1 RNA. Examination of spontaneous cervical carcinomas and immortalized cells maintained for extended intervals in culture revealed partial resistance to the inhibitory effects of TGFβ 1 on cell growth and on HPV gene expression. Therefore, the TGFβ may function in an autoregulatory manner to regulate HPV gene expression in infected genital epithelial cells.

Leukoregulin as well as gamma interferon (IFNγ) are natural products secreted by lymphocytes. Leukoregulin is a 50 kd lymphokine that increases plasma membrane permeability and susceptibility of many tumor cells to destruction by natural killer cells as well as by lymphokine activated killer cells (5,6). The interferons represent a different class of lymphokines in that they have direct antiproliferative (10) and antiviral activities and can stimulate immunological effector cells (9,18). The contrast between the two lymphokines made it possible to compare their ability to regulate gene expression in an immortalized human cervical epithelial cell line. The HPV 16 human cervical epithelial cell line, CX16-2, chosen for analysis is nontumorigenic in nude mice.

The lymphokines, leukoregulin, and r-IFNγ inhibit cell proliferation (23). Both can completely inhibit clonal growth of secondary cultures of normal epithelial cells derived from either foreskin or from cervix. Immortalized cells obtained by transfection with recombinant HPV16 DNA at early passage exhibited resistance to the two lymphokines as shown by their ability to form colonies at low frequencies in the presence of the lymphokines. This resistance increased with continuous subculturing of the cells. Malignant cell lines obtained by further transfection with v-Ha-*ras* oncogene (2) or herpes simplex virus type 2 Bgl II N fragment (3) were more resistant to the lymphokines than were the immortal cell lines from which they had been derived. Furthermore, spontaneous cervical carcinoma tumor lines such as SiHa and QGH also developed partial resistance. Therefore, the progression to malignancy of cervical-derived cells is accompanied by increased resistance to growth inhibition. The effect on growth inhibition was also reflected in the reduced expression of the E6 and E7 proteins required to maintain proliferation of cultured cervical cells in vitro. Both r-IFNγ and leukoregulin treatment for 48 hours decreased the levels of the E6, E7 proteins relative to untreated cells. Therefore, inhibition of cell growth of immortalized cells was reflected in the downregulated expression of both HPV E6 and E7 proteins. Examination of cultures inhibited by confluence showed that they had no changes in E6 and E7 expression. Therefore, the change in HPV RNA expression are attributed to direct action of the lymphokines. The same two lymphokines increased the expression of class 1 histocompatibility surface antigens (HLA) involved in immune recognition and antigen presentation to cyotoxic T lymphocytes (17) relative to untreated cells that express only very low levels of these antigens. The r-IFNγ also induced the expression of HLA class 2 antigens that were not apparent in untreated cells.

To determine the molecular mechanisms by which these lymphokines regulated virus and cellular gene expression, RNA was analyzed. The cervical line CX16-2 ordinarily expresses three different HPV16 RNAs, all of which include the E6 and E7 viral oncogenes. Lymphokines induced time-dependent reductions in steady state levels of E6/E7 RNAs. Forty eight hour treatment with leukoregulin or r-IFNγ caused reductions of 21% to 29%, respectively, relative to the values of untreated controls; exposure for an additional two days caused further reduction of E6/E7 RNAs. This inhibition of E6/E7 was reversed by removing the lymphokines and allowing the cells to recover for 48 hours in fresh medium.

The immortalized CX16-2 cells expressed low levels of HLA class 1 RNA, and class 2 RNA was not detectable. A 24- or 48-hour exposure to either lymphokine enhanced the steady state of HLA class 1 RNA. However, HLA class 2 RNA was upregulated significantly by the r-IFNγ and only slightly induced by leukoregulin. Additional experiments were done with late passage immortal cells as well as with tumorigenic CX16-2 cells obtained from transfection of the v-Ha-*ras* gene. Minimal effects were caused by the lymphokines of the steady state levels of the HPV16 E6/E7 RNAs in the relatively late passage of the CX16-2. The cell line malignancy transformed by subsequent transfection with ras gave r-IFNγ upregulated expression of cellular HLA class 1 RNA in both the premalignant and the malignant cells. However, most cell lines were resistant to upregulation of HLA RNA by leukoregulin. On this basis, the two lymphokines can be discriminated. The ability to demonstrate upregulation of class 1 gene expression by r-IFNγ indicated that these cells retained functional receptors for interferon. Thus, resistance to rIFNγ is not due solely to the absence of receptors at the cell surface.

CONCLUSION

The burden of evidence indicates that HPV is an etiologic agent of cervical cancer. Laboratory data confirm this proposition. Genital epithelial cells transfected with recombinant HPVs associated with cervical cancer can induce dysplastic lesions similar to preinvasive cervical neoplasia; malignancy, however, does not result. The HPV DNA sequences are stably integrated into the immortalized cell lines and, although the integration sites may vary, it generally occurs within an E1 or E2 open reading frame thereby deregulating the expression of E6 and E7 open reading frames for cervical cancers that are positive for HPV DNA. Aberrant differentiation that occurs after HPV transfection, part of the multistage process leading to invasive squamous cell carcinomas can be modulated by using natural biological modifiers. The TGFβ 1, γIFN, and LR downregulated viral gene expression of the E7 protein. The γIFN and the LR also induced expression of RNAs encoding for class 1 and 2 major histocompatability antigens and can be considered to be anticarcinogens. Thus, cytokines and TGFβ 1 should be considered a suppressor gene since it controls proliferation of normal epithelial cells. Therefore, HPV gene expression is differently regulated by specific cytokines produced by cervical epithelia or by leukocytes in filtrating cervical dysplasia and carcinoma.

REFERENCES

1. DeCaprio, J.A., J.W. Ludlow, J. Figge, J. Shew, C. Huang, W. Lee, E. Marsilio, E. Paucha, and D.M. Livingston (1988) SV40 large T antigen forms a specific complex with the product of the retinoblastoma susceptibility gene. Cell 54:275-283.
2. DiPaolo, J.A., C.D. Woodworth, N.C. Popescu, V. Notario, and J. Doniger (1989) Induction of human cervical squamous cell carcinoma by sequential transfection with human papillomavirus 16 DNA and viral Harvey ras. Oncogene 4:395-399.
3. DiPaolo, J.A., C.D. Woodworth, N.C. Popescu, D.L. Koval, J.V. Lopez, and J. Doniger (1990) HSV2-induced tumorigenicity in HPV16-immortalized human genital keratinocytes. Virology 177:777-779.
4. Durst, M., R.T. Dzarlieva-Petrusevska, P. Boukamp, N.E. Fusenig, and L. Gissmann (1987) Molecular and cytogenetic analysis of immortalized primary human keratinocytes obtained after transfection with human papillomavirus type 16 DNA. Oncogene 1:251-256.
5. Evans, C.H., S.C. Barnett, B.A. Gelleri, P.M. Furbert-Harris, P.A. Sheehy, J.A. Barker, P.D. Baker, A.C. Wilson, E.K. Farley, and F. D'Alessandro. Biological and molecular characteristics of leukoregulin action. In Mechanisms of Action and Therapeutic Applications of Biologicals in Cancer and Immune Deficiency Disorders, pp. 315-329. Alan R. Liss, Inc., New York.
6. Furbert-Harris, P.M., C.H. Evans, C.D. Woodworth, and J.A. DiPaolo (1989) Loss of leukoregulin up-regulation of natural killer but not lymphokine-

activated killer lymphocytotoxity in human papillomavirus 16 DNA-immortalized cervical epithelial cells. J. Natl. Cancer Inst. 81:1080-1085.

7. Munger, K., B.A. Werness, N. Dyson, W.C. Phelps, E. Harlow, and P.M. Howley (1989) Complex formation of human papillomavirus E7 proteins with the retinoblastoma tumor suppressor gene product. EMBO 8:4099-4105.

8. Namba, M., K. Nishitani, F. Fukushima, T. Kimoto, and K. Nose (1986) Multistep process of neoplastic transformation of normal human fibroblasts by Co gamma rays and Harvey sarcoma viruses. Int. J. Cancer. 37:419-423.

9. Nawa, A., Y. Nishiyama, N. Yamamoto, K. Maeno, S. Goto, and Y. Tomoda (1990) Selective supression of human papillomavirus 18 mRNA level in HeLa cells by interferon. Biochem. Biophys. Res. Comm. 170:793-799.

10. Nickoloff, B.J., T.Y. Basham, T.C. Merigan, and V.B. Morhenn (1984) Antiproliferative effects of recombinant and γ interferons on cultured human keratinocytes. Lab. Invest. 51:697-701.

11. Peto, R. (1986) Geographic patterns and trends. Viral etiology of cervical cancer. Banbury Rep., Cold Spring Harbor, New York, 21:3-15.

12. Pfister, H. (1987) Human papillomaviruses and genital cancer. Cancer Res. 48:113-147.

13. Rigoni-Stern, D.G. (1842) Fatti statistici relativi alle malattie cancerose che servirono di base alle poche cose dette dal dott. Serv. Progr. Pathol. Therap. 2:507-517.

14. Roberts, A.B., and M.B. Sporn (1990) The transforming growth factor-β's. Handb. Exp. Pharmacol. 45:419-472.

15. Shein, H., and J.F. Enders (1962) Transformation induced by Simian virus 40 in human renal cell cultures. I. Morphology and growth characteristics. Proc. Natl. Acad. Sci., USA 48:1164-1172.

16. Stampfer, M.R., and J.C. Bartley (1985) Induction of transformation and continuous cell lines from normal human mammary epithelial cells after exposure to benzo[a]pyrene. Proc. Natl. Acad. Sci., USA 82:2394-2398.

17. Tanaka, K., T. Yoshioka, C. Bieberich, and G. Jay (1988) Role of the major histocompatibility complex class 1 antigens in tumor growth and metastasis. Ann. Rev. Immunol. 6:359-380.

18. Turek, L.P., J.C. Byrne, D.R. Lowy, I. Dvoretzky, R.M. Friedman, and P.M. Howley (1982) Interferon induces morphologic reversion with elimination of extrachromosomal viral genomes in bovine papillomavirus-transformed mouse cells. Proc. Natl. Acad. Sci., USA 79:7914-7918.

19. Whyte, P., K.J. Buchkovich, J.M. Horowitz, S.H. Friend, M. Raybuck, R.A. Weinberg, and E. Harlow (1988) Association between an oncogene and an antioncogene: the adenovirus E1A proteins bind to the retinoblastoma gene product. Nature 334:124-129.

20. Woodworth, C.D., J. Doniger, and J.A. DiPaolo (1989) Immortalization of human foreskin keratinocytes by various human papillomavirus DNAs corresponds to their association with cervical carcinoma. Journal of Virology 63:159-164.

21. Woodworth, C.D., V. Notario, and J.A. DiPaolo (1990) Transforming growth factors beta 1 and 2 transcriptionally regulate human papillomavirus (HPV) type 16 early gene expression in HPV-immortalized human genital epithelial cells. J.Virol. 64:4767-4775.

22. Woodworth, C.D., S. Waggoner, W. Barnes, M.H. Stoler, and J.A. DiPaolo (1990) Human cervical and foreskin epithelial cells immortalized by human papillomavirus DNAs exhibit dysplastic differentiation in vivo. Cancer Res. 50:3709-3715.

23. Woodworth, C.D., U. Lichti, S. Simpson, C.H. Evans, and J.A. DiPaolo (1992) Leukoregulin and interferon gamma inhibit human papillomavirus (HPV) type 16 gene transcription in HPV-immortalized human cervical Cells. Cancer Res. 52:456-463.

THE TWO *umuDC*-LIKE OPERONS, *samAB* AND *umuDC$_{ST}$*, IN *SALMONELLA TYPHIMURIUM*: THE *umuDC$_{ST}$* OPERON MAY REDUCE UV-MUTAGENESIS-PROMOTING ABILITY OF THE *samAB* OPERON

Takehiko Nohmi,[1] Atsushi Hakura,[1] Yasuharu Nakai,[2] Masahiko Watanabe,[1] Masami Yamada,[1] Somay Y. Murayama,[3] and Toshio Sofuni[1]

[1]Division of Genetics and Mutagenesis
Biological Safety Research Center
National Institute of Hygienic Sciences
Tokyo 158, JAPAN

[2]Safety Research Department
Teijin Limited
Tokyo 191, JAPAN

[3]Department of Bacteriology
Teikyo University School of Medicine
Tokyo 173, JAPAN

INTRODUCTION

Salmonella typhimurium, especially its derivatives containing pKM101 plasmid, has been widely used in the Ames test for the detection of environmental mutagens and carcinogens (1,12). It is known, however, that if the pKM101 plasmid is eliminated, *S. typhimurium* itself shows a much weaker mutagenic response to UV and some chemical mutagens than does *Escherichia coli* (12,22,25,32). In fact, certain potent base-change type mutagens, such as furylfuramide and aflatoxin B$_1$, are nonmutagenic to *S. typhimurium* in the absence of pKM101, whereas they are strongly mutagenic to *S. typhimurium* in the presence of pKM101 plasmid as well as to *E. coli* (12). The low mutability can be restored to levels comparable to *E. coli* by introducing the plasmid carrying the *E. coli umuDC* operon or the pKM101 plasmid carrying *mucAB* operon (11-13,32). *Salmonella typhimurium* has an SOS regulatory system which resembles that of *E. coli* (10,12,17,18,20,22). Thus, it was suggested that *S. typhimurium* is deficient in the function of *umuDC* operon, which plays an essential role in UV and most chemical mutagenesis in *E. coli* (7,28,33).

Antimutagenesis and Anticarcinogenesis Mechanisms III,
Edited by G. Bronzetti *et al.*, Plenum Press, New York, 1993

Recent genetic experiments suggested, however, that *S. typhimurium* has a gene functionally homologous to the *E. coli umuC* gene, since UV mutability of *S. typhimurium* TA2659 was increased when the plasmid carrying the *E. coli umuD$^+$C$^-$* but not that carrying the *E. coli umuD$^-$C$^+$* was introduced into this strain (5). Furthermore, the DNA sequence of *S. typhimurium* LT2 that can hybridize to the *E. coli umuDC* sequence has been cloned (31) and its nucleotide sequence has been determined (27,30). The cloned *umuDC* of *S. typhimurium* LT2 and the *E. coli umuDC* are 71% homologous at the nucleotide level. The plasmid carrying the *umuDC* operon of LT2 restored UV mutability to both *umuD* and *umuC* mutants of *E. coli* (27). These findings raised the new question of why *S. typhimurium* shows poor mutability despite its functional *umuDC* operon.

It is known that expression of some mutant polypeptides can reduce the activity of the wild-type protein (6). In fact, we have reported that the introduction of the multi-copy-number plasmids carrying the mutant *umuD* genes reduce UV mutability of a wild-type strain of *E. coli* (15). This antimutagenic effect is probably due to dominant negative actions of the mutant UmuD proteins encoded by the plasmids. The mutant UmuD proteins could titrate out another component essential to mutagenesis such as wild-type UmuD, UmuC, or RecA, thereby reducing the UV mutability of a wild-type strain. By analogy, we suggest that the investigation of the relationship between a *umuDC* gene and the poor mutability of *S. typhimurium* might lead to finding a new mechanism of antimutagenesis, for example, *S. typhimurium* might have another *umuDC*-like operon in the same cell, and the products of two *umuDC*-like operons might titrate out each other, leading to the less mutable phenotype of *S. typhimurium*.

In order to clarify the implications of *umuDC* genes in mutagenesis and antimutagenesis in typhimurium, we have independently screened the *umuDC*-like genes of *S. typhimurium* TA1538. Consequently, we have cloned another *umuDC*-like operon which is 40% diverged from the aforementioned *umuDC* operon of *S. typhimurium* LT2 at the nucleotide level (16). We have termed the cloned DNA the *samAB* (Salmonella; mutagenesis) operon, and tentatively referred to the *umuDC* operon cloned from *S. typhimurium* LT2 (27,31) as the *umuDC$_{ST}$* operon. Based on the results of the Southern hybridization experiments, we concluded that the two sets of *umuDC*-like operons reside in the same cells of *S. typhimurium* LT2 and TA1538. Our results also suggested that the *umuDC$_{ST}$* operon reduces the UV-mutagenesis promoting ability of the *samAB* operon when the two operons are present on the same multi-copy-number plasmid.

CLONING OF THE GENES OF *S. TYPHIMURIUM* TA1538 THAT CAN RESTORE UV MUTABILITY TO A *umuC* MUTANT OF *E. COLI* (CLONING OF THE *samAB* OPERON)

Out of about 1,000 ampicillin-resistant transformants of an AB1157 *umuC122* :: Tn*5* strain transformed with the library DNA of TA1538 (34), we selected a candidate that apparently showed UV mutability in the patch mutagenesis assay. We have extracted a plasmid from the candidate and reintroduced it into a fresh *umuC122* :: Tn*5* and *umuD44* background. The plasmid, which we designated pYG8011, restored the UV mutability to both a *umuC122* :: Tn*5* strain and a *umuD44* strain to almost the same extent. The levels of UV mutagenesis were, however, about one-half or one-third of those observed for the same strains containing pSE117, which carries the *E. coli umuDC* operon. Introduction of pYG8011 also increased about five-fold the UV mutability of *S. typhimurium* TA2659.

A restriction map of pYG8011 was constructed by digesting the plasmid with several restriction enzymes. The plasmid (17.2 kb) was composed of pBR322 (4.4 kb) and genomic DNA of TA1538 (12.8 kb). To determine the region necessary for the suppression of the UV nonmutable phenotype of *umuDC* mutants of *E. coli*, we have partially digested pYG8011 with *Eco*RV and constructed a set of deletion derivatives. Upon checking the UV mutability of such deletion plasmids and their derivatives, we suggest that the 1.9-kb region is the minimum essential region for the suppression of the nonmutable phenotype of both *umuD44* and *umuC122* :: Tn*5* strains of *E. coli*.

DNA SEQUENCE OF THE *samAB* OPERON AND ITS SIMILARITY TO OTHER RELATED OPERONS

We have determined the nucleotide sequence of both strands of the 1.9-kb region. The nucleotide sequence of the 1.9-kb region contains two continuous open reading frames of 420 bp and 1,272 bp, which potentially encode proteins with calculated molecular weights of 15,523 and 47,726, respectively. We have identified the products of these genes by labeling experiments using the maxicell technique. A potential SOS box sequence for the binding of LexA has been found upstream of the genes. Thus, we have concluded that the cloned genes are analogs of the *E. coli umuDC* genes and termed the cloned genes *samAB*. The *samA* gene encodes the sequence of Ala-24 and Gly-25 at a putative cleavage site by an activated form of RecA, as do the *mucAB* and *impAB* operons (3,9,14,19,23,29). In addition, it encodes Ser-61 and Lys-98, which are highly conserved not only in UmuD, MucA, and ImpA (8,9,19) but also in LexA and the phage repressors (2,4). The nucleotide sequence of the *samAB* operon appeared in the DDBJ, EMBL, and GenBank Nucleotide Sequence databases under the accession number of D90202 (16).

On the other hand, Thomas et al. (30,31) and Smith et al. (26,27) have independently cloned and sequenced another analog of *umuDC* genes from *S. typhimurium* LT2, which we have tentatively referred to as the *umuDC_{ST}* operon. The *samAB* and the *umuDC_{ST}* operons were 60% homologous at the nucleotide level. The predicted amino acid sequences of the UmuD-like proteins encoded by the two operons were 49% homologous; those of the UmuC-like proteins encoded by them were 63% homologous. Interestingly, the *samAB* operon showed higher similarity to the *impAB* operon of TP110 plasmid (9,29) than to the other three related operons, whereas the *umuDC_{ST}* operon showed the highest similarity to the *E. coli umuDC* operon (8,19). Both the *samAB* and the *umuDC_{ST}* operons showed the lowest similarity to the *samAB* and the *umuDC_{ST}* operons showed the lowest similarity to the *mucAB* operon (19). These results indicated that the two *umuDC* operons from *S. typhimurium* are homologous but different.

SALMONELLA TYPHIMURIUM HAS TWO SETS OF umuDC-LIKE OPERONS

Since *S. typhimurium* TA1538 is a derivative of *S. typhimurium* LT2, there was a possibility that both *samAB* and *umuDC*$_{ST}$ operons were present in a single cell. To address the question of whether these two operons are present in an LT2 strain, multiple restriction enzyme digests of genomic DNA extracted from LT2 strain were subjected to Southern hybridization using a probe containing the *samAB* or *umuDC*$_{ST}$ sequence.

The probe DNA containing the *samAB* operon hybridized to the filters containing the digests of LT2. Relatively intense bands at 3.2-kb were observed in *Eco*RV digests of DNA of LT2 along with the control track of *Eco*RV digests of pYG8011 DNA, which carries the *samAB* operon. Characteristic bands at the molecular weights of 400 to 700 bp were observed in *Pst*I digests of DNA of LT2 along with the control track of *Pst*I digests of pYG8011. The probe DNA containing the *umuDC*$_{ST}$ operon also hydribized to the identical filter. Electrophoretic mobilities of the positive bands were consistent with those of the hybridization bands reported by Thomas and Sedgwick (31). Similar results were obtained using DNA of TA1538. Thus, we conclude that the two sets of *umuDC*-like operons, for example, the *samAB* and *umuDC* operons, reside in the same cells of *S. typhimurium* LT2 and TA1538.

THE samAB OPERON IS PRESENT IN THE 60-MDa CRYPTIC PLASMID OF SALMONELLA TYPHIMURIUM

The original line of *S. typhimurium* LT2 contained a specific plasmid that has been called the cryptic plasmid, the virulence plasmid, the 100-kb plasmid, the 60-MDa plasmid, pSLT, or pYQ100 (21). On the other hand, many bacterial plasmids encode analogs of the *E. coli umuDC* operon and thus have the effects of increasing the UV mutability of host strains (29). It is of interest, therefore, to see whether the *samAB* is located in the cryptic plasmid. To address this question, the cryptic plasmid, pYQ100, was digested with several restriction enzymes and subjected to Southern hybridization using the *samAB* sequence as a probe.

The probe DNA carrying the *samAB* strongly hybridized to the plasmid DNA bound on the filter. Electrophoretic mobilities of the positive bands all corresponded to those of the plasmid DNA bands visualized by ethidium bromide staining on agarose gel, suggesting that the positive bands were not due to the contaminated chromosome DNA but due to the plasmid DNA itself. One of us (S.M.) has already constructed a restriction map of pYQ100 using *Hind*III, *Sal*I, and *Eco*RI. By comparing the apparent molecular weights of the positive bands with those of the restriction fragments of pYQ100, we have assigned the *samAB* operon to a region around the junction between the H4 and H8 fragments of the cryptic 60-MDa plasmid (Fig. 1). In fact, the restriction map around H4 and H8 fragments of the cryptic plasmid was very similar to that of the *samAB* and its flanking region deduced from restriction enzyme analysis and Southern hybridization analysis. It is reported that the *umuDC*$_{ST}$ operon is located in a region between 35.9 min and 40 min on the *S. typhimurium* chromosome (27).

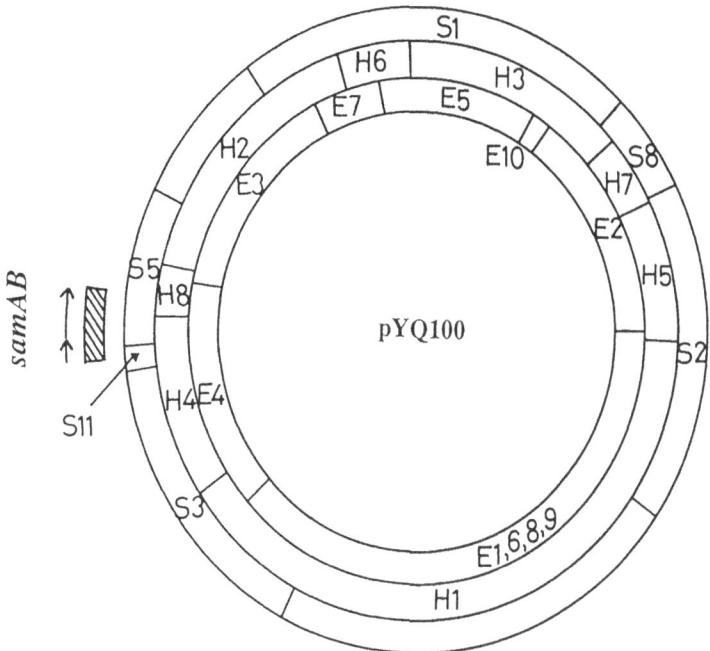

Figure 1. A restriction map of the 60-MDa cryptic plasmid, pYQ100, and a map position of the *samAB* operon. Restriction enzymes: H, *Hin*dIII; S, *Sal*I; E, *Eco*RI.

THE *umuDC$_{ST}$* OPERON MAY REDUCE THE ABILITY OF THE *samAB* OPERON TO PROMOTE UV MUTAGENESIS

Since the two *umuDC*-like operons reside in the same cell, there was a possibility that one operon might modify the ability of the other to promote mutagenesis in *S. typhimurium*. To examine this possibility, we have constructed the multi-copy-number plasmids carrying either the *umuDC$_{ST}$* operon or the *samAB* operon or both operons. We then introduced the plasmids into a *umuD44* strain and a *umuC122* Tn5 strain of *E. coli* and compared the UV mutabilities of the resulting strains (Fig. 2).

The strains harboring the plasmids carrying the *samAB* operon showed higher UV mutabilities than did the strains harboring the plasmids carrying the *umuDC$_{ST}$* operon in both *umuD⁻* and *umuC⁻* backgrounds. It seems reasonable that the *samAB* operon has a higher ability to promote UV mutagenesis than does the *umuDC$_{ST}$* operon since the *samAB* operon was cloned by the functional assay, for example, searching the genes that can restore UV mutability to a *umuC122* Tn5 strain while the *umuDC$_{ST}$* operon was isolated by hybridization using the *E. coli umuDC* operon as probe DNA. The control strains, a *umuD44* strain harboring pBR322 and a *umuC122* :: Tn5 strain harboring pBR322, were nonmutable even at a UV dose of 20 J per m^2 (data not shown). Interestingly the strains harboring the plasmid carrying both operons showed almost the same level of UV mutabilities as did the strains harboring the plasmids carrying the *umuDC$_{ST}$* alone in both *umuD⁻* and

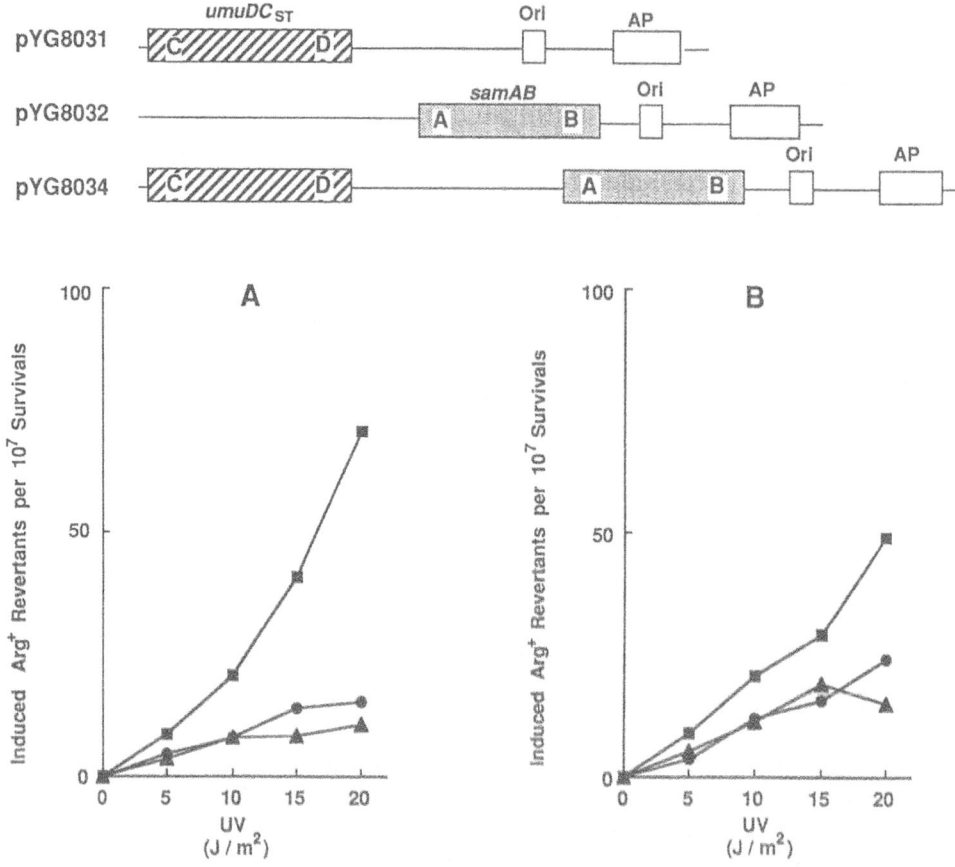

Figure 2. Pictorial representation of the plasmids carrying the $umuDC_{ST}$ operon alone (pYG8031), the *samAB* operon alone (pYG80323), or both operons (pYG8034) and their effects on UV mutagenesis in an AB1157 *umuD44* strain (A) and an AB1157 *umuC122* :: Tn5 strain (B) of *E. coli*. The pYG8031 plasmid was constructed by replacing the *Hin*dIII-*Sal*I region of pBR322 with the 2.2-kb DNA fragment carrying the $umuDC_{ST}$ operon (30). The pYG8032 plasmid was constructed by subcloning the 3.2-kb DNA carrying the *samAB* operon (16) at the *Pvu*II site of pBR322. The pYG8034 plasmid was constructed by subcloning the 3.2-kb DNA carrying the *samAB* operon at the *Pu*II site of pYG8031. ●--●, pYG8031; ■--■, pYG8032; ▲--▲, pYG80334; Ori, DNA replication origin; AP, ampicillin resistance gene.

$umuC^-$ background. These results suggest that $umuDC_{ST}$ operon is negatively dominant over the *samAB* operon when the two operons are present on the same plasmid. They also raise the possibility that the $umuDC_{ST}$ operon might reduce the UV mutagenesis promoting ability of the *samAB* operon in *S. typhimurium*. This possibility must be examined by carrying out more rigorous experiments before a conclusion can be reached. We are currently trying to create mutants of *S. typhimurium*, which lack either the $umuDC_{ST}$ or the *samAB* or both operons. These strains would be helpful in clarifying the implications of the two *umuDC*-like operons in mutagenesis and antimutagenesis in *S. typhimurium*.

REFERENCES

1. Ames, B.N., J. McCann, and E. Yamasaki (1975) Methods for detecting carcinogens and mutagens with the Salmonella/mammalian-microsome mutagenicity test. *Mutat. Res.* 31:347-364.
2. Battista, J.R., T. Ohta, T. Nohmi, W. Sun, and G.C. Walker (1990) Dominant negative *umuD* mutations decreasing RecA-mediated cleavage suggest roles for intact UmuD in modulation of SOS mutagenesis. *Proc. Natl. Acad. Sci., USA* 87:7190-7194.
3. Burckhardt, S.E., R. Woodgate, R.H. Scheuermann, and H. Echols (1988) UmuD mutagenesis protein of *Escherichia coli*: Overproduction, purification, and cleavage by RecA. *Proc. Natl. Acad. Sci., USA* 85:1811-1815.
4. Eguchi, Y., T. Ogawa, and H. Ogawa (1988) Cleavage of bacteriophage phi80 CI repressor by RecA protein. *J. Molec. Biol.* 202:565-574.
5. Herrera, G., A. Urios, V. Aleixandre, and M. Blanco (1988) UV light-induced mutability in Salmonella strains containing the *umuDC* or the *mucAB* operon: Evidence for a *umuC* function. *Mutat. Res.* 198:9-13.
6. Herskowitz, I. (1988) Functional inactivation of genes by dominant negative mutations. *Nature* 329:219-222.
7. Kato, T., and Y. Shinoura (1977) Isolation and characterization of mutants of *Escherichia coli* deficient in induction of mutations by ultraviolet light. *Molec. Gen. Genet.* 156:121-131.
8. Kitagawa, Y., E. Akaboshi, H. Shinagawa, T. Horii, H. Ogawa, and T. Kato (1985) Structural analysis of the *umu* operon required for inducible mutagenesis in *Escherichia coli*. *Proc. Natl. Acad. Sci., USA* 82:4336-4340.
9. Lodwick, D., D. Owen, and P. Strike 91990) DNA sequence analysis of the *imp* UV protection and mutation operon of the plasmid TP110: Identification of a third gene. *Nucl. Acids Res.* 18:5046-5050.
10. MacPhee, D.G. (1973) Effect of *rec* mutations on the ultraviolet protecting and mutation-enhancing properties of the plasmid R-Utrecht in *Salmonella typhimurium*. *Mutat. Res.* 19:357-359.
11. MacPhee, D.G. (1973) Effect of an R factor and caffeine on ultraviolet mutability in *Salmonella typhimurium*. *Mutat. Res.* 18:367-370.
12. McCann, J., N.E. Springarn, J. Kobori, and B.N. Ames (1975) Detection of carcinogens as mutagens: Bacterial tester strains with R-factor plasmids. *Proc. Natl. Acad. Sci., USA* 72:979-983.
13. Mortelmans, K.E., and B.A.D. Stocker (1976) Ultraviolet light protection, enhancement of ultraviolet light mutagenesis, and mutator effect of plasmid R46 in *Salmonella typhimurium*. *J. Bact.* 128:271-282.
14. Nohmi, T., J.R. Battista, L.A. Dodson, and G.C. Walker (1988) RecA-

mediated cleavage activates UmuD for mutagenesis: Mechanistic relationship between transcriptional derepression and posttranslational activation. *Proc. Natl. Acad. Sci., USA* 85:1816-1820.

15. Nohmi, T., J.R. Battista, T. Ohta, V. Igras, W. Sun, and G.C. Walker (1990) Antimutagenic effect of *umuD* mutant plasmids: Isolation and characterization of *umuD* mutants reduced in their ability to promote UV mutagenesis in *Escherichia coli*. In *Antimutagenesis and Anticarcinogenesis Mechanisms II*, Y. Kuroda, D.M. Shankel, and M.D. Waters, eds. Plenum Press, New York and London, pp. 417-421.

16. Nohmi, T., A. Hakura, Y. Nakai, M. Watanabe, S.Y. Murayama, and T. Sofuni (1991) *Salmonella typhimurium* has two homologous but different *umuDC* operons: Cloning of a new *umuDC*-like operon (*samAB*) present in a 60-megadalton cryptic plasmid of *S. typhimurium*. *J. Bact.* 173:1051-1063.

17. Orrego, C., and Eisenstadt (1987) An inducible pathway is required for mutagenesis in *Salmonella typhimurium* LT2. *J. Bact.* 169:2885-2888.

18. Pang, P.P., and G.C. Walker (1983) The *Salmonella typhimurium* LT2 *uvrD* gene is regulated by the *lexA* gene product. *J. Bact.* 154:1502-1504.

19. Perry, K.L., S.J. Elledge, B.B. Mitchell, L. Marsh, and G.C. Walker (1985) *umuDC* and *mucAB* operons whose products are required for UV light- and chemical-induced mutagenesis: UmuD, MucA, and LexA proteins share homology. *Proc. Natl. Acad. Sci., USA* 82:4331-4335.

20. Pierre, A., and C. Paoletti (1983) Purification and characterization of *recA* protein from *Salmonella typhimurium*. *J. Biol. Chem.* 258:2870-2879.

21. Sanderson, K.E., and J.R. Roth (1988) Linkage map of *Salmonella typhimurium*, Edition VII. *Microb. Rev.* 52:485-532.

22. Sedgwick, S.G., and P.A. Goodwin (1985) Differences in mutagenic and recombinational DNA repair in enterobacteria. *Proc. Natl. Acad. Sci., USA* 82:4172-4176.

23. Shiba, T., H. Iwasaki, A. Nakata, and H. Shinagawa (1990) Proteolytic processing of MucA protein in SOS mutagenesis: Both processed and unprocessed MucA may be active in the mutagenesis. *Molec. Gen. Genet.* 224:169-176.

24. Shinagawa, H., T. Kato, T. ise, K. Makino, and A. Nakata (1983) Cloning and characterization of the *umu* operon responsible for inducible mutagenesis in *Escherichia coli*. *Gene* 23:167-174.

25. Skavronskaya, A.G., N.F. Stepanova, and I.V. Andreeva (1982) UV-mutable hybrids of *Salmonella* incorporating *Escherichia coli* region adjacent to tryptophan operon. *Molec. Gen. Genet.* 185:315-318.

26. Smith, C.M., and E. Eisenstadt (1989) Identification of a *umuDC* locus in *Salmonella typhimurium* LT2. *J. Bact.* 171:3860-3865.

27. Smith, M., W.M. Koch, S.B. Franklin, P.L. Foster, T.A. Cebula, and E. Eisenstadt (1990) Sequence analysis and mapping of the *Salmonella typhimurium* LT2 *umuDC* operon. *J. Bact.* 172:4964-4978.

28. Steinborn, G. (1978) Uvm mutants of *Escherichia coli* K12 deficient in UV mutagenesis. II. Further evidence for a novel function in error-prone repair. *Molec. Gen. Genet.* 175:203-208.

29. Strike, P., and D. Lodwick (1987) Plasmid genes affecting DNA repair and mutation. *J. Cell Sci.* 6(Suppl.):303-321.

30. Thomas, S.M., and S.G. Sedgwick (1989) Cloning of *Salmonella typhimurium* DNA encoding mutagenic DNA repair. *J. Bact.* 171:5776-5782.

31. Thomas, S.M., H.M. Crowne, S.C. Pidsley, and S.G. Sedgwick (1990)

Structural characterization of the *Salmonella typhimurium* LT2 *umu* operon. *J. Bact.* 172:4979-4987.

32. Walker, G.C. (1978) Inducible reactivation and mutagenesis of UV-irradiated bacteriophage P22 in *Salmonella typhimurium* LT2 containing the plasmid pKM101. *J. Bact.* 135:415-421.

33. Walker, G.C. (1984) Mutagenesis and inducible responses to deoxyribonucleic acid damage in *Escherichia coli. Microb. Rev.* 48:60-93.

34. Watanabe, M., M. Ishidate, Jr., and T. Nohmi (1989) A sensitive method for the detection of mutagenic nitroarenes: Construction of nitroreductase-overproducing derivatives of *Salmonella typhimurium* strains TA98 and TA100. *Mutat. Res.* 216:211-220.

FORMATION, INHIBITION OF FORMATION, AND REPAIR OF

OXIDATIVE 8-HYDROXYGUANINE DNA DAMAGE

H. Kasai,[1] M.-H. Chung,[2] F. Yamamoto,[1] E. Ohtsuka,[3] J. Laval,[4] A.P. Grollman,[5] and S. Nishimura[1]

[1]Biology Division
National Cancer Center Research Institute
Tokyo, JAPAN

[2]Department of Pharmacy, College of Medicine
Seoul National University
Seoul, KOREA

[3]Department of Bioorganic Chemistry
Faculty of Pharmaceutical Sciences
Hokkaido University
Sapporo, JAPAN

[4]Institut Gustave-Roussy
Villejuif, FRANCE

[5]Department of Pharmaceutical Sciences
State University of New York
Stony Brook, New York, USA

Oxygen radicals are produced by many carcinogenic agents and by ionizing radiation, and cause DNA damage in vitro and in vivo. Oxygen radicals are also produced in vivo during cellular metabolism of oxygen. Therefore, studies on DNA damage induced by oxygen radicals should be helpful in elucidating the mechanisms of induction of cancer by ionizing radiation and environmental chemicals as well as spontaneous carcinogenesis during aging.

Formation of 8-hydroxydeoxyguanosine (oh^{8}dG) by oxidative damage (Fig. 1) was first observed in 1983 (7,8) during a study on DNA modification in vitro by heated carbohydrates used as models of cooked foods. Various oxygen radical-forming agents (10), such as ionizing radiation, cigarette smoke condensate, asbestos, nickel compounds (14), and chewing tobacco components (18), have been found to induce formation of oh^{8}dG in DNA in vitro.

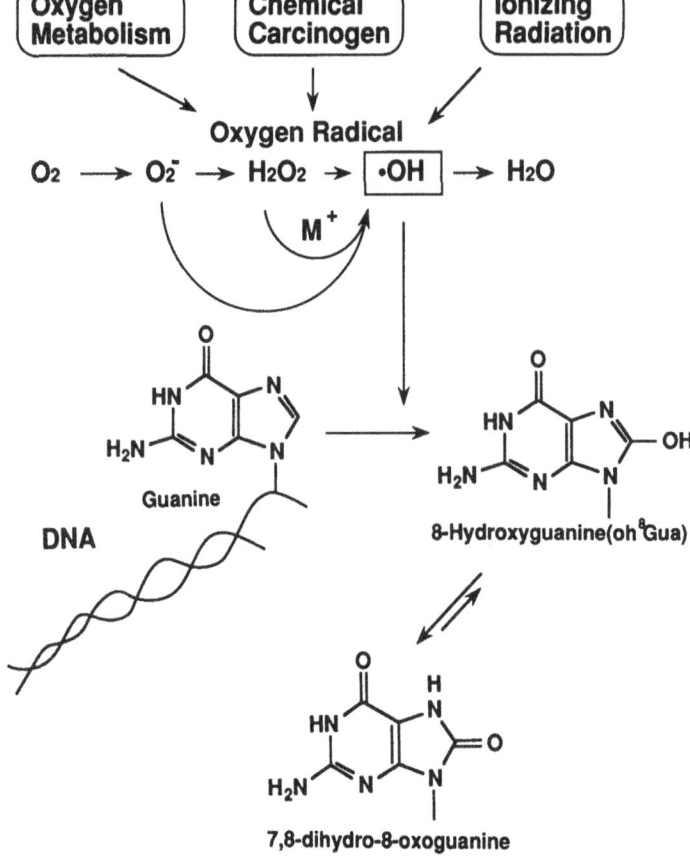

Figure 1. Formation of oh^{8}Gua in DNA by oxygen radical.

When deoxyguanosine or DNA was treated with an autooxidation system including unsaturated fatty acid, such as linolenic acid or linoleic acid, formation of oh^{8}dG was observed (9). Formation of oh^{8}dG in these model systems of lipid peroxidation was inhibited by addition of the metal chelators EDTA and DETAPAK, plant polyphenol, catechin, quercetin, or extracts of carrot, apricot, or prune. An inhibitor of oh^{8}dG formation was purified from this plant extract by thin-layer chromatography and high performance liquid chromatography (HPLC). By these chromatographic procedures, the inhibitory principle was separated from yellow β-carotin fraction. This inhibitor was identified as chlorogenic acid (Fig. 2), by UV- and mass-spectrum analyses.

Formation of oh^{8}dG was also observed in cellular DNA (11) by HPLC combined with an electrochemical detector (ECD) by the method of Floyd et al. (5). After oral administration to rats of the oxygen radical-forming renal carcinogen potassium bromate (KBrO$_3$) an increase of oh^{8}dG in the DNA of rat target organs was also observed after the administration of other carcinogens, such as peroxisome proliferators (liver) (13), 2-nitropropane (liver) (4,13), Fe-nitrilotriacetate (kidney) (22), choline-deficient diet (liver) (6,19), and 1-6-dinitropyrene (mammary gland) (3). As shown by these examples, oh^{8}dG can be used as a marker for monitoring cellular

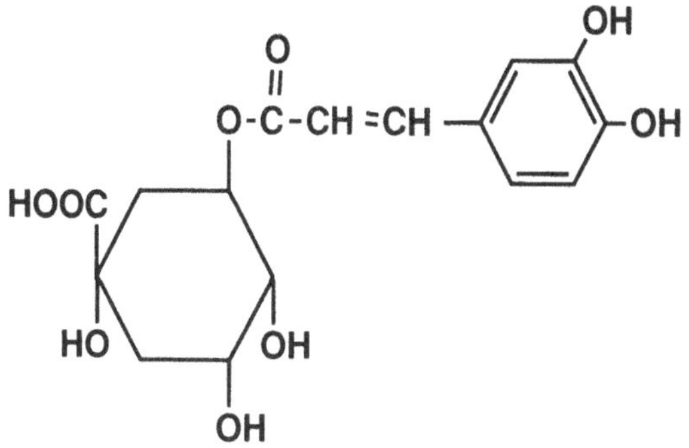

Figure 2. Structure of chlorogenic acid.

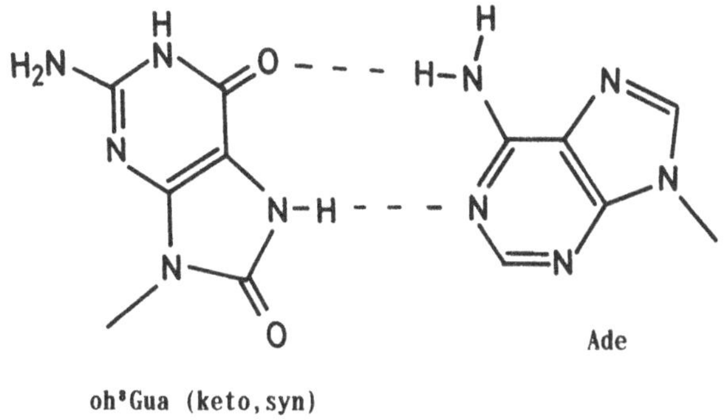

oh⁸Gua (keto,syn)

Figure 3. Mechanism of GC   TA transversion induced by oh^8Gua.

DNA damage by oxygen radicals and may be useful for predicting carcinogenic potency.

The oh^8dG residue in DNA induces misreading during DNA synthesis in vitro (16,20) and also mutations in an *Escherichia coli* system (1,17,23). Probably oh^8dG in the 8-*keto* form (Fig. 1) with the *syn* conformation forms a base pair with deoxadenosine during DNA replication and induces GC → TA transversions (Fig. 3). After treatment of mammalian cells with nickel compounds, most point mutations found in codon 12 of the K-*ras* oncogene were GC → TA transversions (15). The oh^8dG residue in DNA may induce these mutations.

Irradiation of mice with γ-rays resulted in formation of oh^8dG in liver DNA (11). However, the oh^8dG level decreased with time, suggesting the presence of a repair mechanism. A repair enzyme for oh^8dG in DNA, 8-hydroxyguanine (oh^8Gua) endonuclease has been purified from *E. coli* (2). This enzyme first removes the free

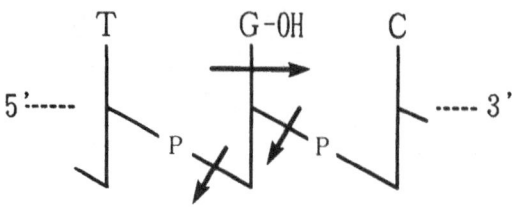

Figure 4. Cleavage sites of *E. coli* oh^{8}Gua endonuclease.

base oh^{8}Gua from DNA and then cleaves DNA 3' and 5' to the resulting AP site (Fig. 4). Release of oh^{8}Gua and cleavage of the DNA strand were also observed when DNA containing oh^{8}dG was incubated with *E. coli* Fapy glycosylase, which is known to remove the ring opened form of guanine (21). From these results, we conclude that oh^{8}Gua endonuclease is the same enzyme as Fapy glycosylase and that the true substrate of this enzyme may be DNA containing oh^{8}Gua. Similar oh^{8}Gua endonuclease activity was detected in mammalian cells such as human leucocytes, and cell of rat liver and brain.

Recently, J.H. Miller and his colleagues found that the DNA sequence of the *E. coli mut* M mutator locus, which generates GC → TA transversion, is the same as that of the Fapy glycosylase gene (personal communication). A *mut* M mutant that does not produce Fapy glycosylase was found to show GC → TA transversion at high frequency. From these results, we conclude that formation of oh^{8}dG in DNA is relevant to spontaneous mutation in *E. coli*.

REFERENCES

1. Cheng, K.C., D.S. Cahill, H. Kasai, S. Nishimura, and L.A. Loeb (1991) The oxygen free radical associated DNA adduct 8-oxoguanine causes G·C to T·A transversions during in vivo DNA replication. *Proceedings, Am. Assoc. Cancer Res.* 32:149.

2. Chung, M.H., H. Kasai, D.S. Jones, H. Inuoe, H. Ishikawa, E. Ohtsuka, and S. Nishimura (1990) An endonuclease activity of *Escherichia coli* that specifically removes 8-hydroxyguanine residues from DNA. *Mutat. Res.* 254:1-12.

3. Djuric, Z., and D.W. Potter (1990) Oxidative DNA damage by 1,6-dinitropyrene in vivo. *Proceedings, Am. Assoc. Cancer Res.* 31:146.

4. Fiala, E.S., C.C. Conaway, and J.E. Mathis (1989) Oxidative DNA and RNA damage in the liver of Sprague-Dawley rats treated with the hepatocarcinogen 2-nitropropane. *Cancer Res.* 49:5518-5522.

5. Floyd, R.A., J.J. Watson, P.K. Wong, D.H. Altmiller, and R.C. Rickard (1986) Hydroxyl free radical adduct of deoxyguanosine: Sensitive detection and mechanisms of formation. *Free Radicals Res. Comm.* 1:163-172.

6. Hinrichsen, L.I., R.A. Floyd, and O. Sudilovsky (1990) Is 8-hydroxydeoxyguanosine a mediator of carcinogenesis by a choline-devoid diet in the rat liver? *Carcinogenesis* 11:1879-1881.

7. Kasai, H., S. Nishimura (1983) Hydroxylation of the C-8 position of deoxyguanosine by reducing agents in the presence of oxygen. *Nucl. Acids Res. Symp. Ser.* 12:165-167.

8. Kasai, H., and S. Nishimura (1984) Hyroxylation of deoxyguanosine at the C-8 position by ascorbic acid and other reducing agents. *Nucl. Acids Res.* 12:2137-2145.

9. Kasai, H., and S. Nishimura (1988) Formation of 8-hydroxdeooxyguanosine in DNA by auto-oxidized unsaturated fatty acids. In *Medical, Biochemical and Chemical Aspects of Free Radicals*, O. Hayaishi, E. Niki, M. Kondo, and T. Yoshikawa, eds. Elsevier Science Publishers, pp. 1021-1023.

10. Kasai, H., and S. Nishimura (1991) Formation of 8-hydroxydeoxyguanosine in DNA by oxygen radicals and its biological significance. In *Oxidative Stress, Oxidants and Antioxidants*, H. Sies, ed. Academic Press Ltd., pp. 99-116.

11. Kasai, H., P.F. Crain, Y. Kuchino, S. Nishmura, A. Ootsuyama, and H. Tanooka (1986) Formation of 8-hydroxyguanine moiety in cellular DAN by agents producing oxygen radicals and evidence for its repair. *Carcinogenesis* 7:1849-1851.

12. Kasai, H., S. Nishimura, Y. Kurokawa, and Y. Hayashi (1987) Oral administration of the renal carcinogen, potassium bromate, specifically produces 8-hydroxydeoxyguanosine in rat target organ DNA. *Carcinogenesis* 8:1959-1961.

13. Kasai, H., Y. Okada, S. Nishimura, M.S. Rao, and J.K. Reddy (1989) Formation of 8-hydrooxydeoxyguanosine in liver DNA of rats following long-term exposure to a peroxisome proliferator. *Cancer Res.* 49:2603-2605.

14. Kasprzak, K.S., and L. Hernandez (1989) Enhancement of hydroxylation and deglycosylation of 2'-deoxyguanosine by carcinogenic nickel compounds. *Cancer Res.* 49:5964-5968.

15. Kasprzak, K.S., K. Higinbotham, B.A. Diwan, A.O. Perantoni, and J.M. Rice (1990) Correlation of DNA base oxidation with the activation of K-*ras* oncogene in nickel-induced renal tumors. *Free Rad. Res. Comm.* 9(Suppl. 1):172.

16. Kuchino, Y., F. Mori, H. Kasai, H. Inoue, S. Isai, K. Miura, E. Ohtsuka, and S. Nishimura (1987) Misreading of DNA templates containing 8-hydroxydeoxyguanosine at the modified base and at adjacent residues. *Nature* 327:77-79.

17. Moriya, M., C.O.V. Bodpudi, F. Johnson, M. Takeshita, and A.P. Grollman (1991) Site specific mutagenesis using a gapped duplex vector: A study of translesion synthesis past 8-oxodeoxyguanosine in *E. coli*. *Mutat. Res.* 254:281-288.

18. Nair, U.J., R.A. Floyd, J. Nair, V. Bussachini, M. Friesen, and H. Bartsch (1987) Formation of reactive oxygen species and of 8-hydroxyguanosine in DNA in vitro with betel quid ingredients. *Chem. Biol. Interact.* 63:157-169.

19. Nakae, D., H. Yoshiji, H. Maruyama, T. Kinugasa, A. Denda, and Y. Knoishi (1990) Production of both 8-hydroxydeoyguanosine in liver DAN and g-glutamyltransferase-positive hepatocellular lesion in rats given a choline-deficient, L-amino acid-defined diet. *Japan. J. Cancer Res.* 81:1081-1084.

20. Shibutani, S., M. Takeshita, and A.P. Grollman (1991) Insertion of specific bases during DNA synthesis past the oxidation-damaged base 8-oxodG. *Nature* 349:431-434.

21.   Tchou, J., H. Kasai, S. Shibutani, M.H. Chung, J. Laval, A.P. Grollman, and S. Nishimura (1991) 8-Oxoguanine (8-hydroxyguanine) DNA glycosylase and its substrate specificity. 88:4690-4694.

22.   Umemura, T., K. Sai, A. Takagi, R. Hasegawa and Y. Kurokawa (1990) Formation of 8-hydroxydeoxyguanosine (8-OH-dG) in rat kidney DNA after intraperitoneal administration of ferric nitrilotriacetate (Fe-NTA). *Carcinogenesis* 11:345-347.

23.   Wood, M.L., M. Dizdaroglu, E. Gajewski, and J.M. Essigmann (1990) Mechanistic studies of ionizing radiation and oxidative mutagenesis: Genetic effects of a single 8-hydroxyguanine (7-hydro-8-oxoguanine) residue inserted at a unique site in a viral genome. *Biochemistry* 29:7024-7032.

MODULATION BY RETINOIC ACID OF SPONTANEOUS AND

BENZO(A)PYRENE-INDUCED *C*-HA-*RAS* EXPRESSION

Devaki Nandan Sadhu and Kenneth Ramos

Department of Physiology and Pharmacology
College of Veterinary Medicine
Texas A & M University
College Station, Texas 77843-4466 USA

ABSTRACT

The effects of retinoic acid on the expression of the Ha-ras gene were studied in transformed rat hepatoma cells (H4IIE) and in rat aortic smooth muscle cells (ASMC) treated with benzo(a)pyrene (30 μM) in vitro. In H4IIE cells, a dose-dependent increase in steady state Ha-ras mRNA levels was observed upon exposure to retinoic acid for 24 hr. Exposure of ASMC to 10 μM retinoic acid under similar experimental conditions was also associated with increased Ha-ras expression. In contrast, retinoic acid (1 and l0 μM) inhibited benzo(a)pyrene-induced expression of Ha-ras in ASMC. These results suggest that retinoic acid modulates spontaneous and carcinogen-induced expression of Ha-ras in a differential manner.

INTRODUCTION

Uncontrolled cell proliferation is one of the chief underlying mechanisms of human diseases such as cancer and atherosclerosis. The ras p21 proteins, with their guanine nucleotide binding and GTPase activity, are believed to play an important role in signal transduction and cell proliferation (16). Aberrant expression of ras oncogenes is accompanied by pleiotropic changes involving morphological, biochemical, and growth-related abnormalities (8). Mutations in ras genes have been observed in a large percentage of spontaneous and chemically-induced tumors (1,2,6,9,12,14,15).

Recently, attention has been drawn to the role of retinoic acid as an anticarcinogen and a regulator of cellular processes such as gene expression (10) and cell proliferation (5,7). In view of the suggested involvement of ras in tumorigenicity and cancer (1,2,6,9,12,14,15), it would be of interest to determine if retinoic acid modulates ras expression. Therefore, the present studies were carried out to

Antimutagenesis and Anticarcinogenesis Mechanisms III,
Edited by G. Bronzetti *et al.*, Plenum Press, New York, 1993

263

investigate the effects of retinoic acid on ras MRNA levels in H4IIE and ASMC. Our results demonstrate for the first time that retinoic acid inhibits benzo(a)pyrene-induced ras expression in ASMC in vitro. The ability of retinoic acid to modulate ras expression was differentially expressed in H4IIE cells relative to ASMC.

MATERIALS AND METHODS

^{32}P-UTP was purchased from New England Nuclear; SP6 polymerase RNA labeling system was purchased from Pharmacia. Nitrocellulose and nylon membranes were purchased from Bio-Rad Laboratories. The X-ray film used for autoradiography was Kodak XAR-5. A 0.8 kb human c-Ha-ras cDNA containing exons I, II, III, and IV from Oncogene Science was used to generate radiolabeled complementary RNA using SP6 RNA polymerase. Rapid hybridization buffer was from Amersham. Retinoic acid was purchased from Sigma Chemical Company, and benzo(a)pyrene was obtained from Aldrich Chemicals. The H4IIE cell line was from American Type Culture Collection, Rockville, Maryland.

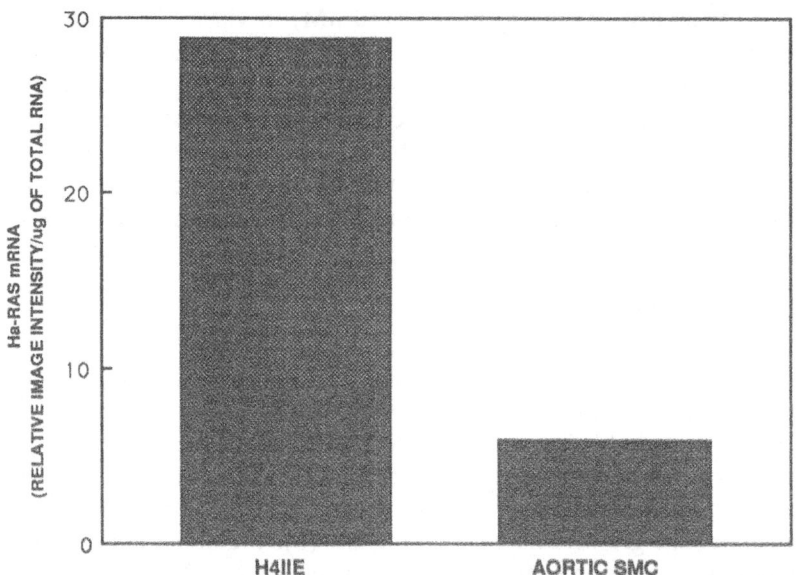

Figure 1. Steady state levels of c-Ha-ras mRNA in H4IIE and rat aortic smooth muscle cells. Values represent the mean of two samples containing three replicates each.

In Vitro Studies

H4IIE cells were seeded at a density of 4,000/cm^2 and exposed to retinoic acid (0.1 to 100 µM) for 24 hr. Rat ASMC cultures, established as described previously (13), were seeded at the same density as H4IIE cells. ASMC were exposed to retinoic acid (1 and 10 µM) for 24 hr and then challenged with benzo(a)pyrene (30 µM) for 6 hr. RNA was then extracted and analyzed as described below.

RNA Analysis

The method of extraction of total RNA from the cells was essentially as described by Chemczynski and Sacchi (4). Briefly, cells were rinsed twice with phosphate-buffered saline trypsinized, and rinsed again in phosphate-buffered saline before homogenization in solution D. Samples were extracted with phenol, chloroform, and isoamyl alcohol and precipitated twice with isopropanol and finally dissolved in 0.1% SDS. The concentrations of RNA in the samples were determined by UV spectrophotometric analysis.

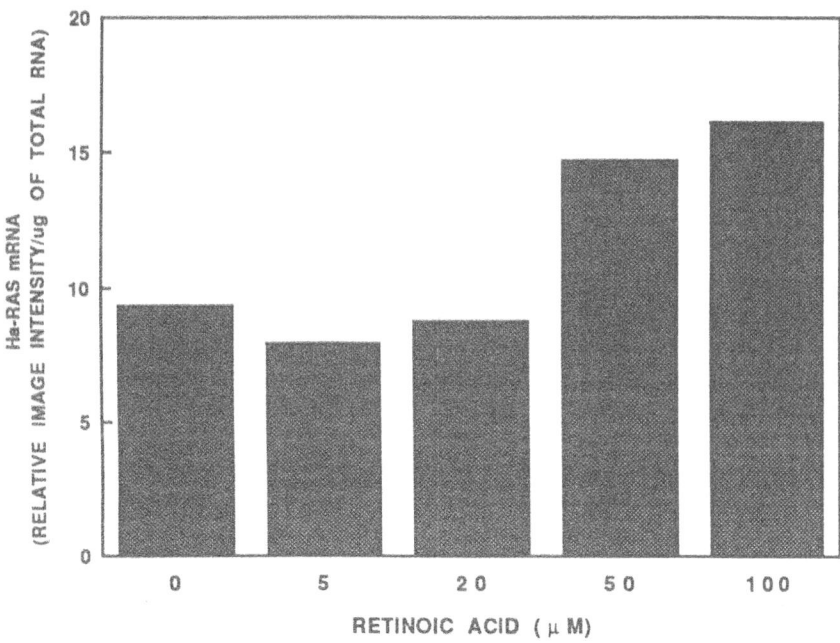

Figure 2. Steady state levels of ras mRNA in H4IIE cells treated with various concentrations of retinoic acid (RA) for 24 hrs. Values represent the mean of two samples containing three replicates each.

Transfer of RNA

RNA samples were prepared by heating RNA in 20X SSC and deionized formaldehyde at 60°C for 10 min, and various concentrations of RNA, namely 1, 0.5, 0.25, and 0.125 μg, were blotted onto a nylon or nitrocellulose membrane using a dot blot manifold from Bio-Rad. The blotted RNA was cross-linked to the membrane using a UV cross linker (Stratagene) according to manufacturer's specifications. Membranes were enclosed in seal-a-meal bags and hybridized to a c-Ha-ras cRNA probe in rapid hybridization buffer at 70°C for 2 hr following prehybridization for 30 min. Membranes were then washed sequentially in 2X SSC + 0.1% SDS two times, and in 0.2X SSC + 0.1% SDS twice at room temperature and at 65°C. Autoradiograms were developed by exposing X-ray film to the membranes for 2 to 6 hr at room temperature. The relative image intensity was quantified using a Visage 2000.

RESULTS AND DISCUSSION

Experiments in the present study were designed to evaluate the effect of retinoic acid on spontaneously high levels of ras expression in H4IIE cells and carcinogen-induced ras expression in ASMC in vitro. The levels of H-ras mRNA in H4IIE cells were five-fold higher than those in 1 ASMC (Fig. 1). Retinoic acid did not modulate ras MRNA abundance at concentrations up to 20 μM in H4IIE cells, but enhanced expression at 50 and 100 μM (Fig. 2). A dose- dependent increase in cell atrophy was observed in H4IIE cells at concentrations greater than 10 μM. A moderate increase in ras expression was observed in ASMC treated with benzo(a)pyrene in vitro for 6 hr (Fig. 3). Pre-treatment with retinoic acid for 24 hr inhibited benzo(a)pyrene-induced ras expression in ASMC (Fig 3). In fact, in the presence of retinoic acid and benzo(a) pyrene, ras expression was reduced below basal levels.

The high basal expression of ras MRNA in H4IIE cells relative to ASMC raises the possibility that enhanced ras expression is one of the characteristics of the transformed phenotype. The finding that retinoic acid treatment increased ras mRNA levels in H4IIE cells and ASMC under basal conditions, but reduced benzo-(a)pyrene-induced expression in ASMC, suggests that retinoic acid influences multiple cellular pathways.

The mechanism by which retinoic acid modulates ras expression is unknown. Niles (11) has reported that retinoic acid causes growth-arrest of melanoma cells in G_1 phase of the cell cycle. Hence, the induction of ras expression in H4IIE and ASMC may be due to interference with cell cycle-related events. This interpretation is consistent with other studies showing that increased ras expression occurs during the G_1 phase of the cell cycle (3). The differential response of ASMC treated with benzo(a)pyrene after retinoic acid exposure is intriguing. Because concomitant exposure to these agents reduced ras expression below basal levels, additional studies addressing the complexity of this interaction would be desirable.

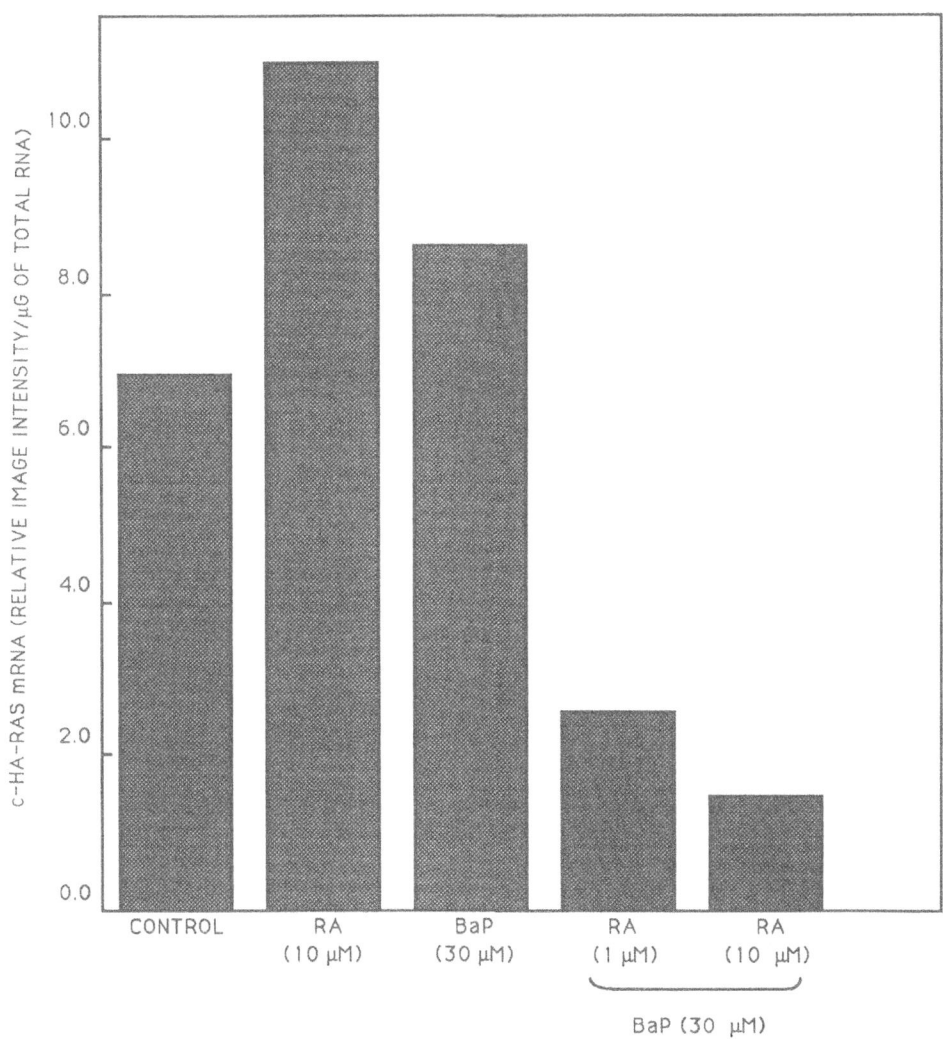

Figure 3. Steady state levels of ras mRNA in rat aortic smooth muscle cells exposed to 1 and 10 μM of retinoic acid for 24 hrs. and 30 μM of BaP for an additional 6 hrs. Values represent the mean of two samples containing three replicates each.

ACKNOWLEDGEMENTS

This work was supported in part by Research Development funds from Texas Agricultural Experiment Station and a grant to K. Ramos (ES 04849) from the National Institute of Environmental Health Sciences.

REFERENCES

1. Albino, A.P., D.M. Nanus, I.R. Mentle, Cordon-Cardo, U.C. McNutt, N.S. McNutt, J. Bressler, and M. Andreeff (1989) *Oncogene* 4:1363-1374.
2. Bos, J.L., E.R. Fearon, S.R. Hamilton, M. Verlaan-de Vries, J.H. van Boom, A.J. der Eb, and B. Vogelstein (1987) *Nature* 327:293-297.
3. Campisi, J., H.E. Gray, A.B. Pardee, M. Dean, and G.E. Sonnenshein (1984) *Cell* 36:241-243.
4. Chomczynski, P., and N. Sacchi (1987) *Analyt. Biochem.* 162:156-159.
5. Davis, B.H., R.T. Kramer, and N.O. Davidson, (1990) *J. Clin. Invest.* 86:2062-2070.
6. Farr, C.J., C.J. Marshall, D.J. Easty, N.A. Wright, S.C. Powel, and C. Paraskeva, (1988) *Oncogene* 3:673-678.
7. Frey, J.R., R. Peck, and W. Bollag (1991) *Cancer Lett.* 57:223-227.
8. Godwin, K.A., and M.W. Liberman (1990) *Oncogene* 5(8):1231-1241.
9. Janssen, J.W.G., A.C.M. Steenvoorden, J. Lyons, B. Anger, J.U. Bohlke, J.L. Bos, H. Seliger, and C.R. Bartram (1987) *Proc. Natl. Acad. Sci., USA* 84:9228-9232.
10. Munoz-Canoves, P., D.P. Vik, and B.F. Tack (1990) *J. Biol. Chem.* 265:20065-20068.
11. Niles, R.M. (1987) *In Vitro Cell Dev. Biol.* 23:803-804.
12. Polverino, A.J., B.P. Hughes, and G.J. Barritt (1990) *Biochem.* 271:309-315.
13. Ramos, K., and L.R. Cox (1987) *In Vitro Cell Dev. Biol.* 21:495-504.
14. Rodenhuis, S., M.L. van de Weterlng, W.J. Mooi, S.G. Evers, P.C. van Zandwijk, and J.L. Bos (1987) *New Eng.J. Med.* 317:929-935.
15. Rodenhuis, S., R.J.C. Slebos, A.J.M. Boot, S.G. Evers, W.J. Mooi, S.S. Wagenaar, P.C. van Bodegom, and J.L. Bos (1988) *Cancer Res.* 48:5738-5741.
16. Suzuki, Y., M. Orita, M. Shiraishi, K. Hayashi, and T. Sekiya (1990) *Oncogene* 5:1037-1043.

IDENTIFICATION OF NATURAL ANTIMUTAGENS

WITH MODULATING EFFECTS ON DNA REPAIR

Branka Vuković-Gačić and Draga Simić

Botanical Institute and Garden
University of Belgrade
Belgrade, YUGOSLAVIA

ABSTRACT

The results of a study of bioantimutagenesis, with emphasis on natural antimutagens from plant extracts with modulating effects on DNA repair in *Escherichia coli* bacteria are presented in this chapter. Comparative screening for spontaneous or induced mutagenesis, as well as expression of the SOS gene, *sfiA* was accomplished.

Antimutagenic capacity was obtained with nontoxic concentrations of the plant extracts; the same plant extract may decrease or increase the mutation rate, or even be ineffective, depending on the bacterial strain used and the concentration of the extract applied.

Since antimutagenic effects may be the consequence of either stimulation of error-free repair, inhibition of error-prone repair, or involvement of multiple mechanisms, the effects of several plant extracts on the level of UV-induced beta-galactosidase were screened (to monitor SOS induction in cells).

Reduction of the enzyme activity induced by UV was observed following addition of St. John's wort extract, while there was not reduction after thyme, aloe, camomile, or lime-tree and the level of UV-induced enzyme was even higher with sage extract.

Our results indicate that the antimutagenic effect of St. John's wort is probably due to suppression of error-prone repair. Moreover, we assume that an antimutagenic effect obtained with thyme, mint, and sage under certain conditions may be due to enhanced error-free repair.

INTRODUCTION

DNA repair systems are involved as causative, as well as ameliorative, factors in mutagenesis/carcinogenesis (7). Cellular repair is a dynamic process and can be modulated by many factors such as the relative rate of DNA replication and cell proliferation, DNA structure, the amplified or impaired transcription of selected genes, etc. Hypothetically, all factors that improve the fidelity of DNA repair and replication, might also be antimutagenic/anticarcinogenic.

According to the classification of inhibitors of mutagenesis based on their mechanisms of action given by De Flora and Ramel (5,6) such factors have been classified as *Modulators of DNA replication and repair*, for example, bioantimutagens, as suggested by Kada and coworkers (9,10). There is considerable evidence for antimutagenic effects of certain naturally-occurring and synthetic compounds affecting DNA repair in bacteria (11,12) and mammalian cells (3).

Moreover, as suggested by Clarke and Shankel (4), the same kinds of short-term tests that have been useful in detecting mutagens, might be effective in identifying antimutagens/anticarcinogens, and their "antimutagenic specificity" (19).

Results of a study of bioantimutagenesis, with emphasis on natural antimutagens from plant extracts with modulating effects on DNA repair in *E. coli* bacteria are presented in this chapter. Comparative screening for spontaneous or induced mutagenesis, as well as expression of SOS gene, *sfiA*, indicates the complex situation when the plant extracts are known or suspected to act through multiple mechanisms.

Table 1. *Escherichia coli* K12 (SY252) relevant markers.

Gene symbol	Mnemonic	Phenotypic trait affected	References
1. *mutH*	Mutator	Increased rates of frameshift and base substitution mutations	1
2. *mutL*	Mutator	High rate of AT ⇌ GC transitions	1
3. *mutS*	Mutator	High rate of AT ⇌ GC transitions	1
4. *uvrD*	Ultraviolet	Repair of UV damage to DNA; DNA-dependent ATPase; helicase II	1
5. *argE*	Arginine	Acetylornithine deacetylase	1
6. *argE*[λp (*sfiA::lacZ$^+$*) *c*Iind1]			This Volume

MATERIALS AND METHODS

Strains

The bacterial strains, all derivatives of *S. coli* K12, are listed in Tab. 1. For assaying SOS induction, the strains were lysogenized with λp (*sfiA::lacZ*$^+$)*c*I*ind*1 (8) obtained by courtesy of M. Radman.

Media and Growth Conditions

The bacteria were cultured in LB medium (5 g NaCl, 10 g bacto-tryptone, 5 g yeast-extract, 1,000 ml distilled water) at 37° with aeration, to the mid-log phase (5 x 10^8 cells/ml).

LA (LB containing 20 g of Difco agar per liter) supplemented with rifampicin purchased from Sigma (100 µg/ml) was used to selected Rif^r mutants.

The semi-enriched minimal agar medium (SEM) (25) for survival and mutation assays contained 1 g $(NH_4)_2SO_4$, 10 g KH_2PO_4, 0.1 g $MgSO_4$ x $7H_2O$, 0.5 g trisodium citrate x $2H_2O$ per liter, adjusted to pH 7.0 with 10 N NaOH, supplemented with 0.4% D-glucose, 1.5% Difco bacto agar, and 3% (v/v) NB solution (14). Top agar contained Difco agar (7 mg/ml) and NaCl (6 mg/ml).

Assay for Antimutagenic Effects on UV-induced Mutations

The bacteria were grown in LB solution at 37° C with gentle shaking to reach the mid-log phase (5 x 10^8 cells/ml), washed twice by centrifugation and resuspended in 0.01 M $MgSO_4$ giving a similar titer.

UV-irradiation conditions were the same as described in our previous report (20). Samples (0.1 ml) of UV-irradiated cells, appropriately diluted for determination of cellular viability and undiluted for determination of mutagenesis, were spread on SEM agar plates with or without plant extract. Numbers of revertants and viable cells were determined after incubation at 37° C for 72 hr.

Assay for Antimutagenic Effects on Spontaneous Mutants

Undiluted cell suspensions (0.1 ml) were added to 3 ml of molten top agar containing different concentrations of plant extracts, mixed and poured onto LA plates supple-mented with 100 µg/ml of rifampicin. For survival determinations appropriately diluted cell suspensions were used. Numbers of mutants and viable cells were determined after incubation at 37° C for 24 hr.

Assay of β-Galactosidase

UV-irradiated cells were centrifuged and resuspended in minimal medium supplemented with 1% Casamino acids Difco. β-galactosidase was assayed in duplicate as described previously (13).

Table 2. Antimutagenic effects of plant extracts on mut^- and mut^+ strains of *Escherichia coli* K12.

Plant extract	mg/ml[a]	% Inhibition or % Enhancement of Mutagenic Activity[b]				mut^+ mg/ml[a]	
		mutH	mutS	mutL	uvrD		
X-tea	0.6	(67)	(6)	(36)	(30)	0.8	34
	3.2	(52)	(32)	(16)	(18)	2.0	43
St. John's wort	0.6	27	53	62	32	0.4	62
	3.2	52	70	75	73	2.0	65
Thyme	0.6	(58)	(30)	0	(45)	0.8	(30)
	3.2	(13)	(11)	(10)	38	2.0	24
Aloe	0.6	7	16	(8)	18	0.2	(94)
	3.2	0	0	(33)	(18)	2.0	(32)
Camomile	0.6	16	15	(180)	(35)	0.4	(30)
	3.2	31	22	(280)	(25)	2.0	(52)
Nettle	0.6	16	10	(66)	(25)	0.8	(30)
	3.2	5	(8)	(66)	(43)	2.0	(122)
Mint	0.6	18	28	3	12	0.8	(53)
	3.2	(20)	0	10	0	2.0	39
Lime-tree	0.6	10	53	2	(12)	0.8	8
	3.2	50	63	15	(38)	2.0	(61)
Sage	0.6	16	30	9	33	0.8	(27)
	3.2	(35)	8	(9)	26	4.0	30

[a]See Materials and Methods. [b]A positive value represents the percentage inhibition of the effect induced by the mutagen in the short-term test system; a negative value () is the percentage enhancement of the effect; and a value of zero indicates no significant difference between the effect observed (24). Spontaneous ($Rif^s \rightarrow Rif^r$) or UV ($60 J/m^2$) induced ($Arg^- \rightarrow Arg^+$) mutants were screened. The mean values are from 2-3 experiments.

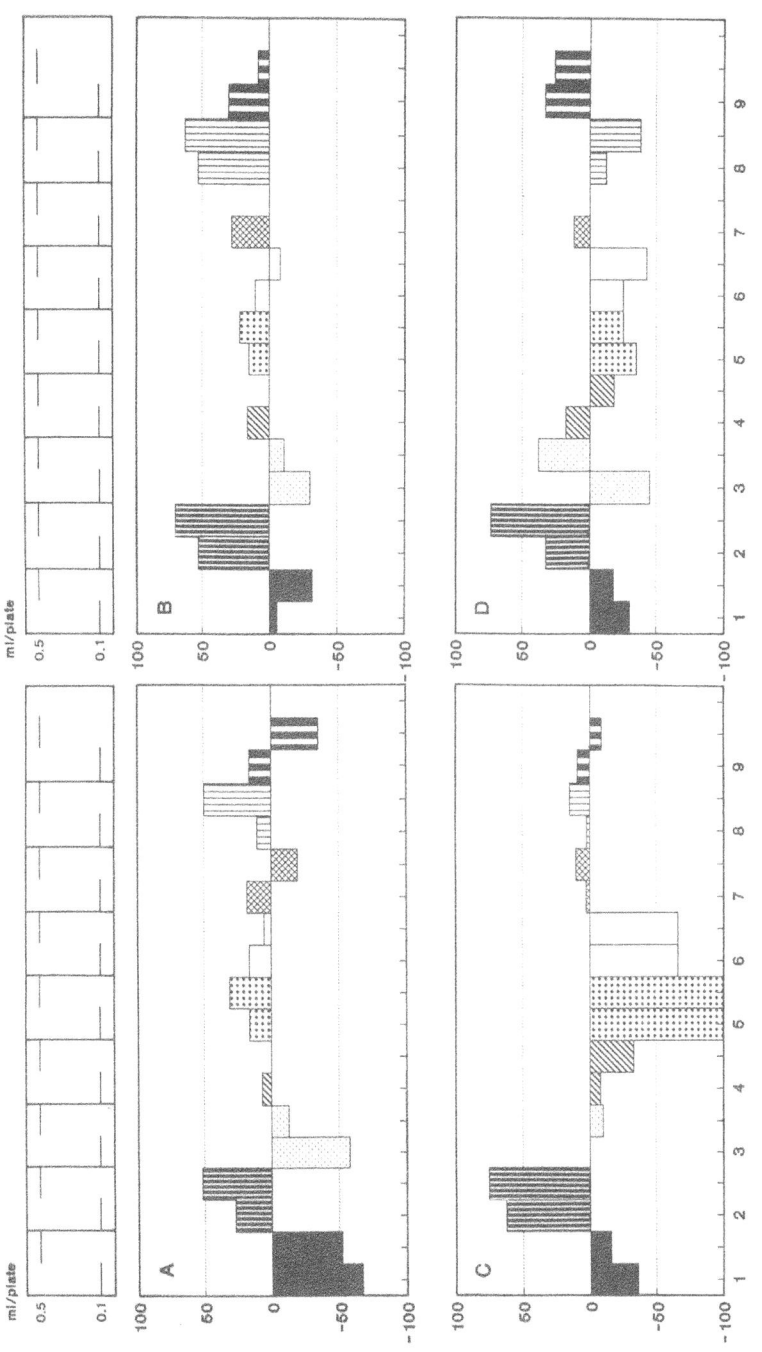

Figure 1. Genetic activity profiles of plant extracts. 1) X-tea; 2) St. John's wort; 3) thyme; 4) aloe; 5) camomile; 6) nettle; 7) mint; 8) lime-tree; and 9) sage. The upper bar graph represents the concentrations of investigated extracts (ml/plate). The lower graphs represent the percent of inhibition or enhancement of spontaneous mutagenic frequency as follows: A) in *mutH*;   B) in *mutS*; C) in *mutL*; and D) in *uvrD*.   (% inhibition or enhancement; x = [1 - (Nt/Nc)] x 100; Nt = number of mutants/plate with plant extract; and Nc = number of mutant/plate without plant extract.

Preparation of Plant Extracts

After mincing, dry plant material (15 g) was homogenized in 95% ethanol (150 ml). Homogenates were filtered and concentrated by a vacuum-rotary evaporator and resuspended in distilled water.

The concentrations of crude extracts used for the experiments were in the range of 0.4-8 mg/ml. These concentrations are nontoxic (the same number of colonies was found in treated and nontreated samples) (23).

RESULTS AND DISCUSSION

The assay-systems used in this work have been designed to exclude desmutagens, factors that act directly on mutagens or their precursors and inactivate them. In order to detect bioantimutagens, factors that reduce the apparent spontaneous and induced mutation frequency by interfering with cellular processes of mutation fixation (10,12) we measured the effects of different concentrations of plant extracts: a) on the frequency of spontaneous mutation of mutator mutants (Tab. 1 and Fig. 1), and b) on the frequency of induced mutation, by the use of UV-irradiation as a mutagen prior to the treatment with potential antimutagen (Tab. 2).

Antimutagenic capacity was obtained with nine plant extracts (nontoxic concentrations). Figure 1 shows the effects on spontaneous mutation frequencies of *E. coli* mismatch-repair deficient mutants. The same plant extract may decrease or increase the mutation frequency or even by ineffective, depending on the mutator strain used and the concentration of the extract applied. When these data were compared with UV-induced mutagenesis, screened by a reversion assay in an *E. coli* repair proficient strain (Tab. 2), the results illustrate the complex situation that is already postulated for antimutagens that can interfere by virtue of their multiple biological properties (2,5,6,17).

Analysis of the data from Tab. 2 shows a heterogeneous response of the plant extracts examined. With the exception of St. John's wort extract showing an antimutagenic effect on spontaneous and UV-induced mutagenesis, the eight other extracts may be classified according to their genetic activity as antimutagenic (AM), mutagenic (M), or ineffective (I) (Tab. 2): (i) M for spontaneous and AM for UV-induced mutagenesis (X-tea); (ii) M for spontaneous (*mutL, uvrD*) and UV-induced mutagenesis (aloe, camomile, nettle); (iii) M or I, but AM (*uvrD* and *mut*[+]) with higher concentrations applied (thyme, mint, sage); and (iv) AM for spontaneous mutagenesis (*uvrD* excepted), M for *uvrD* and UV-induced mutagenesis (lime-tree).

Comparative screening for potential bioantimutagenic factors from different plant extracts indicated that some extracts have a strong antimutagenic effect. Since it is known that antimutagenic effects may be the consequence of either stimulation of error-free repair, inhibition of error-prone repair, or involvement of multiple mechanisms, we screened the effects of several plant extracts ont he level of UV-induced β-galactosidase in the cells.

The β-galactosidase assay in the fusion of *lacZ* to the SOS-regulated gene *sfiA* (8) is a sensitive measure of SOS induction. It is well known that induction of the

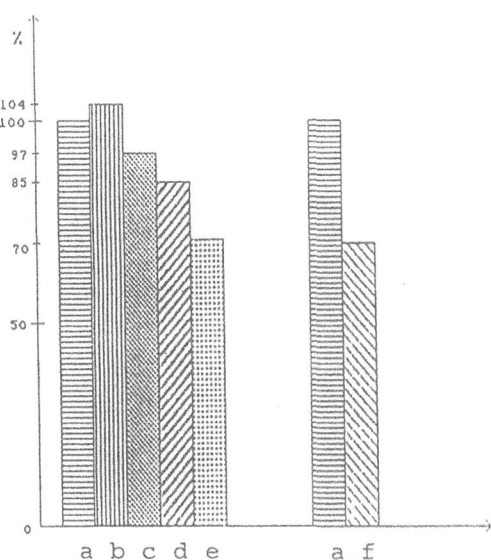

Figure 2. Effect of St. John's wort extract (*Hypericum perforatum* L) on the level of UV-induced beta-galactosidase (% of reduction): a) control; b-e) plant extract (ml/ml) as follows: b) 0.004; c) 0.02; d) 0.04; e) 0.08; and f) $CoCl_2$ (50 µg/ml). The enzyme was measured 20 min following UV irradiation (20 J/m^2).

SOS response in bacteria leads to an increased repair capacity of cells (16,21,22). As shown on Fig. 2, reduction of enzyme activity induced by UV was observed following the addition of the St. John's wort extract, indicating the inhibition of SOS induction. Taking together all results obtained with St. John's wort extract (Figs. 1 and 2, Tab. 2), we propose that its antimutagenic effect is due to the suppression of error-prone repair. Our results are in good agreement with those obtained for inhibitors of the SOS induction (15,18). Moreover, as compared to results obtained with cobalt chloride (Fig. 2), St. John's wort extract may correct the error-proneness of the DNA replicating enzyme(s) by improving its fidelity in DNA synthesis even after SOS induction (12).

It is difficult to distinguish whether the antimutagenic activity of crude extract tested is due to the acceleration of error-free (recombination, excision) or inhibition of error-prone (SOS) repair activity by monitoring the levels of β-galactosidase. There was no reduction of UV-induced enzyme after treatment with thyme, aloe, camomile, or lime-tree, respectively, and the level of UV-induced enzyme was even higher when the cells were treated with sage extract (data not shown).

We propose that antimutagenic effects obtained with thyme, mint, and sage shown on Tab. 2 may be due to enhanced error-free repair capacity.

ACKNOWLEDGEMENTS

We are grateful to Miroslav Radman for the strains and to Michael Simic for valuable help in preparing the manuscript. We also acknowledge the skillful assistance of Aida Ajanovic during the experimental work. This work was supported by YU Programme P22/BB3 and Project 8108.

REFERENCES

1. Bachmann, B.J. (1987) Linkage map of *Escherichia coli* K12, Edition 7. In *Escherichia coli and Salmonella typhimurium Cellular and Molecular Biology 2*, F.C. Neidhardt, J.L. Ingraham, K.B. Low, B. Magasanik, M. Schaechter, and H.E. Umbarger, eds. American Society for Microbiology, Washington, DC, pp. 807-876.

2. Bronzetti, G., A. Galli, and C. Ella Croce (1990) Antimutagenic effects of chlorophyllin. In *Antimutagenesis and Anticarcinogenesis Mechanism II*, Y. Kuroda, D.M. Shankel, and M.D. Waters, eds. Plenum Press, New York and London, pp. 463-468.

3. Boothman, D.A., R. Schlegel, and A.B. Pardee 91988) Anticarcinogenic potential of DNA-repair modulators. *Mutat. Res.* 202:393-411.

4. Clarke, C.H., and D.M. Shankel (1975) Antimutagenesis in microbial systems. *Bacteriol. Rev.* 39:33-53.

5. De Flora, S., and C. Ramel (1988) Mechanisms of inhibitors of mutagenesis and carcinogenesis. *Mutat. Res.* 202:285-306.

6. De Flora, S., and C. Ramel (1990) Classification of mechanisms of inhibitors of mutagenesis and carcinogenesis. In *Antimutagenesis and Anticarcinogenesis Mechanisms II*, Y. Kuroda, D.M. Shankel, and M.D. Waters, eds. Plenum Press, New York and London, pp. 461-462.

7. Haynes, R.H. (1988) Biological context of DNA repair. In *Mechanisms and Consequences of DNA Damage Processing*, Alan R. Liss, Inc., pp. 577-584.

8. Huisman, O., and R. D'Ari (1981) An inducible DNA replication-cell division coupling mechanism in *E. coli*. *Nature* (London) 290:797-799.

9. Kada, T., T. Inoue, and N. Namiki (1982) Environmental desmutagens and antimutagens. In *Environmental Mutagenesis and Plant Biology*, E.J. Klekowski, ed., Praeger, New York, pp. 137-151.

10. Kada, T., K. Kaneko, S. Matsuzaki, T. Matsuzaki, and Y. Hara (1985) Detection and chemical identification of natural bio-antimutagens. A case of the green tea factor. *Mutat. Res.* 150:127-132.

11. Kuroda, Y., and T. Inoue (1988) Antimutagenesis by factors affecting DNA repair in bacteria. *Mutat. Res.* 202:387-391.

12. Kuroda, Y. (1988) Antimutagenesis studies in Japan. In *Antimutagenesis and Anticarcinogenesis Mechanisms II*, Y. Kuroda, D.M. Shankel, and M.D. Waters, eds. Plenum Press, New York and London, pp. 1-22.

13. Miller, J. (1972) *Experiments in Molecular Genetics*, Cold Spring Harbor Laboratory, Cold Spring Harbor, New York.

14. Nunoshiba, T., and H. Nishioka (1987) Sodium arsenite inhibits spontaneous and induced mutations in *Escherichia coli*. *Mutat. Res.* 173:19-24.

15. Ohta, T., K. Watanabe, R. Tsukamoto, Y. Shirasu, and T. Kada (1986) Antimutagenic effect of 5-fluorouracil and 5-fluorodeoxyuridine on UV-induced mutagenesis in *Escherichia coli*. *Mutat. Res.* 173:19-24.

16. Peterson, K.R., N. Ossanna, A.T. Thliveris, D.G. Ennis, and D.W. Mount (1988) derepression of specific genes promotes DNA repair and mutagenesis in *Escherichia coli*. *J. Bact.* 170:1-4.

17. Ramel, C., U.K. Alekperov, B.N. Ames, T. Kada, and L.W. Wattenberg (1986) Inhibitors of mutagenesis and their relevance to carcinogenesis. Report by ICPEMC Expert Group on Antimutagens and Desmutagens. *Mutat. Res.* 168:47-65.

18. Sargentini, N.J., and K.C. Smith (1981) Much of spontaneous mutagenesis in *Escherichia coli* is due to error-prone DNA repair: Implications of spontaneous carcinogenesis. *Carcinogenesis* 9:863-872.

19. Shankel, D.M., and C.H. Clarke 1990) Specificity of antimutagens against chemical mutagens in microbial systems. In *Antimutagenesis and Anticarcinogenesis Mechanism II*, Y. Kuroda, D.M. Shankel, and M.D. Waters, eds. Plenum Press, New York and London, pp. 457-460.

20. Simić, D. J. Knežević, and B. Vuković (1985) Influence of the *recB21* mutation of *Escherichia coli* K12 on prophage lambda induction. *Mutat. Res.* 142:159-162.

21. Walker, G.C. (1984) Mutagenesis and inducible responses to deoxyribonucleic acid damage in *Escherichia coli*. *Microb. Rev.* 48:60-93.

22. Walker, G.C. (1985) Inducible DNA repair system. *Ann. Rev. Biochem.* 54:425-457.

23. Wall, M.E., M.C. Wani, T.J. Hughes, and H. Taylor (1990) Plant antimutagens. In *Antimutagenesis and Anticarcinogenesis Mechanisms II*, Y. Kuroda, D.M. Shankel, and M.D. Waters, eds. Plenum Press, New York and London, pp. 61-78.

24. Waters, M.D., A.L. Brady, H.F. Stack, and H.E. Brockman (199) The concept of activity profiles of antimutagens. In *Antimutagenesis and Anticarcinogenesis Mechanisms II*, Y. Kuroda, D.M. Shankel, and M.D. Waters, eds. Plenum Press, New York and London, pp. 87-104.

25. Witkin, E.M. (1956) Time, temperature and protein synthesis: A study of ultraviolet-induced mutation in bacteria. *Cold Spring Harbor Symposia on Quantitative Biology* 21:123-140.

MECHANISMS OF INHIBITION OF TUMOR PROGRESSION

Barbour S. Warren and Thomas J. Slaga

Research Division
M.D. Anderson Cancer Center
University of Texas
Smithville, Texas 78957 USA

INTRODUCTION

Experimental and epidemiologic studies have demonstrated that carcinogenesis is a multistage process that can be divided into three major stages: initiation, promotion, and progression (22). Progression encompasses the changes by which a tumor evolves from a contained, differentiated lesion of uncoordinated growth into a rapidly growing, invasive lesion with potential to spread to other areas of the body. This change is also associated with conversion from an easily removed growth to one requiring extensive therapy with variable rates of recovery and, thus, constitutes a phase of cancer development of critical clinical relevance. Nonetheless, this stage of carcinogenesis has remained the least understood.

TUMOR PROGRESSION IN MOUSE SKIN

The mouse skin model of carcinogenesis has inherent advantages for the study of tumor progression, and has served a primary role in the current understanding of the mechanisms of progression and antiprogression. Accordingly, this review will focus largely on this system; however, examples from other systems will also be given. These studies have focused on the initial invasive stages of tumor progression and do not deal with factors that affect metastasis specifically.

Mouse skin is a very flexible model system for the study of tumor progression, in part because tumors are produced on the exterior of the animal. This allows ready monitoring of changes in the occurrence, size, and type of tumors. A time line for the treatments and occurrence of tumors in a typical two-stage mouse skin carcinogenesis protocol is illustrated in Fig. 1. Animals are first treated with an initiating agent, 7,12-dimethylbenz(a)anthracene (DMBA) at time 0 and, after a two-week period, biweekly treatment with a promoting agent such as 12-O-tetradecanoylphorbol-13-acetate (TPA) is begun. Using a treatment protocol of this

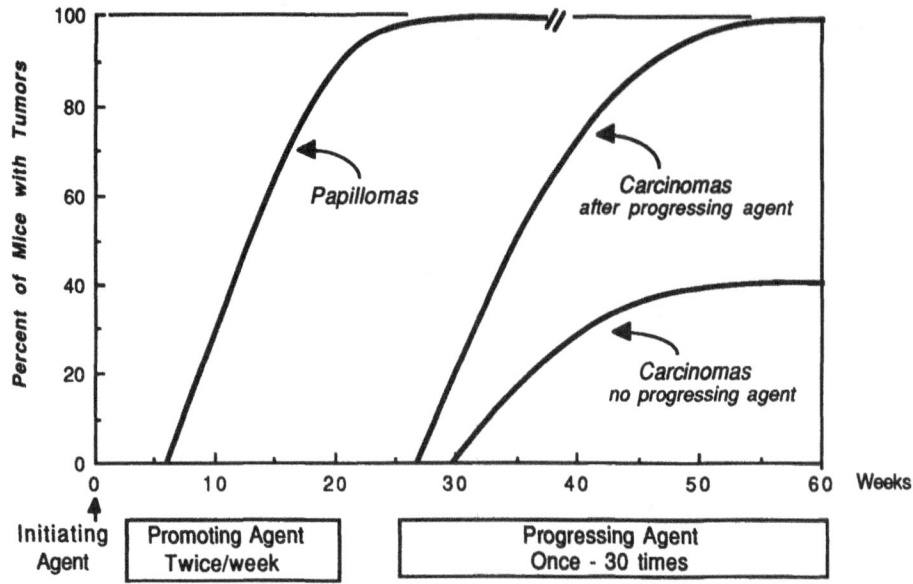

Figure 1. Time course for the generation of papillomas and carcinomas in two- and three-stage models of carcinogenesis in mouse skin.

type, papillomas, the benign lesion in this system, are first visible at 6 to 8 weeks, and reach a plateau at about 20 weeks. Carcinomas, the malignant lesion, are initially seen at approximately 30 weeks and reach a maximum number about 20 weeks later. Three stages are thus delineated: initiation, promotion, and progression, and the agents utilized in the stages are named accordingly: initiating, promoting, and progressing agents.

Initiating agents are carcinogens and will cause the formation of tumors in the absence of other agents, provided the dose or the frequency of administration is great enough. These agents are genotoxic and are considered to act through the activation or inactivation of cancer-related genes, oncogenes, and tumor suppressor genes, respectively (8). These changes represent an irreversible, heritable change that lasts for the lifetime of the animal. In the multistage protocols, the initiator treatment is given at a subeffective dose that does not produce any papillomas in the absence of further treatment. However, following repeated treatment with a promoting agent, such as TPA, all of the animals develop papillomas.

Promoting agents, in contrast to initiating agents, are not carcinogenic and produce few tumors in the absence of other treatments. In addition, the effects of promoting agents are reversible, and these compounds must be given at the proper dose and frequency of administration to be effective. Many activities have been described for this class of compounds; however, the ability to produce sustained hyperplasia correlates best with tumor promotion (5). For example, TPA treatment leads to a 5-fold increase in [^{3}H]-thymidine incorporation in the basal layer of epidermis and an increase in skin thickness approximately 4 to 5 times normal (2). Papilloma development is dependent on the doses of both the initiating and promoting agents, and a good dose-response relationship is observed in both

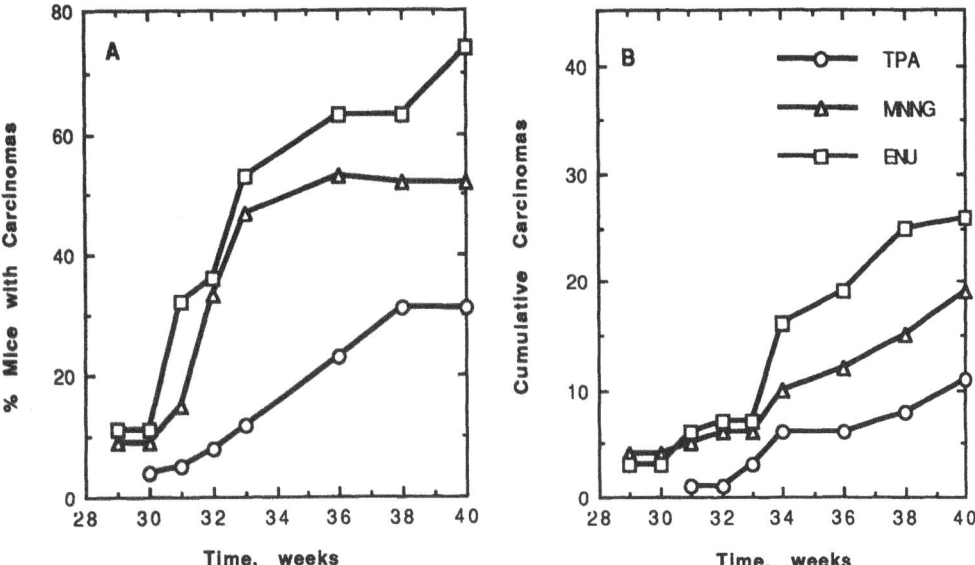

Figure 2. The percentage of mice with carcinomas (A) and the cumulative number of carcinomas per group (B) for mice treated during the progression stage with TPA, MNNG, or ENU as described in the text.

cases (10). With an optimal treatment protocol, the number of papillomas that develop on each mouse can be as high as 30.

The simplest model for the formation of papillomas in this system would be the following. The initiating agent leads to an oncogenic mutation in a limited number of cells within the skin. However, these cells remain in a repressed state due to interactions with adjoining normal cells. Promoting agents, while inducing cellular proliferation and hyperplasia, lead to a derepression and selective expansion of the initiated cell population, ultimately leading to the formation of a papilloma.

Carcinomas begin to arise at about 30 weeks into the time course of these experiments and develop in the vast majority of cases from existing papillomas (3). As would be expected, their incidence is dependent on the number of papillomas formed. However, there appears to be a certain tumor mass that an animal can support, and for carcinomas this sets a maximum of 3 to 4 carcinomas. The ratio of papillomas to carcinomas or percent conversion, thus will vary depending on the number of papillomas that are produced. The percent conversion can range from 5% in protocols that have a maximal papilloma response to 40% where fewer papillomas are produced (10).

AGENTS AND MECHANISMS THAT INCREASE TUMOR PROGRESSION

Complete carcinogenesis protocols in which animals are repeatedly treated with carcinogen produce fewer papillomas and a larger number of carcinomas (31).

In addition, the carcinomas arise at earlier time points and, in some cases, independent of papilloma formation. This finding suggested that carcinogens are able to generate lesions that lead to both papillomas and carcinomas, and led our group (22) and others (17,18) to ask if treatment of papilloma-bearing skin with carcinogens could lead to an increase in the conversion rate of papillomas to carcinomas. This proved to be the case, and, as shown in Fig. 2, treatment of papilloma-bearing skin with the direct-acting carcinogens ethylnitrosourea (ENU) or N-methyl-N'-nitro-N-nitrosoguanidine (MNNG) resulted in an earlier appearance of carcinomas, an increase in the percent of mice with carcinomas, and an increase in the cumulative number of carcinomas.

In these experiments, three groups of animals were initiated with DMBA (10 μmol) and treated with TPA (1 μg) biweekly for 20 weeks until the formation of papillomas had stabilized. At this point, TPA treatment was discontinued for two groups of the papilloma-bearing mice, and these animals were treated with 10 μmol of ENU or 1 μmol of MNNG twice weekly for two weeks; the third, control group received TPA continually. Following these treatments, the animals received biweekly TPA treatments until week 40. It should also be noted that there was no increase in the cumulative number of papillomas in these groups during the progression stage. Thus, the effect of the ENU or the MNNG was upon existing papillomas rather than the generation of new tumors. In summary, an initiation-promotion-initiation, such as this protocol, leads to formation of a larger number of carcinomas from existing papillomas.

Table 1 lists agents that have tumor progressing activity. As is shown in this table, some of the carcinogen dosage regimes rival complete carcinogenesis protocols, with repeated carcinogen treatments being carried out for periods as long as 30 weeks. Although the increases in carcinoma formation in these studies were clearly greater than those of acetone controls, mechanistic evaluation of such regimens is difficult. Notable exceptions are benzo(a)pyrene diol epoxide (BPDE) and cisplatin, both of which increased tumor progression following a single noncarcinogenic dose. Studies in our laboratory showed that a single noncarcinogenic dose (200 μmol) of BPDE led to a three-fold increase in the cumulative number of carcinomas formed. Hennings and coworkers also demonstrated that a single treatment of the cancer chemotherapeutic agent, cisplatin, at a noncarcinogenic dose resulted in carcinoma formation at a rate equivalent to studies using numerous carcinogen treatments (19). Although the dose-response relationship for tumor progression for any of the carcinogens has not been well studied, these results suggest that BPDE and cisplatin are able to efficiently induce specific genetic changes necessary for tumor progression and may provide means to identify specific progression-associated lesions.

If we expand the model presented earlier for papilloma formation, experimental results support the involvement of additional genetic lesions in the progression of papillomas to carcinomas. It would be predicted that this lesion occurs at a low frequency in papillomas but treatment with carcinogens leads directly or indirectly to an increase in the rate of its occurrence. This model predicts that carcinoma formation requires at least two oncogenic mutations, and, indeed, transfection studies have demonstrated that expression of two activated oncogenes can cause malignant progression (13).

Table 1. Agents that increase tumor progression in mouse skin

Agents	Treatment	Reference
Carcinogens		
Benzo(a)pyrene diol epoxide	200 nmol, topical, once	unpub. obs.
Cisplatin	335 nmol, IP, once	19
MNNG	1 μmol, topical, biweekly, 4X	22
MNNG	815 nmol, topical, weekly, 30X	17,18
Urethane	225 μmol, IP, weeky, 30X	17,18
Ethylnitrosourea	10 μmol, topical, weekly, 4X	22
4-Nitroquinoline-N-oxide	1.5 μmol, topical, weekly, 30X	17,18
Free Radical Generators		
Benzoyl Peroxide	85 μmol, topical, biweekly, 30X	23,30
Hydrogen Peroxide	15% Soln, topical, biweekly, 12X	23
Other		
Acetic Acid	670 μmol, topical, biweekly, 30X	16

IP = intraperitoneal; X = times (4X = 4 times); unpub. obs. = unpublished observations.

Surprisingly, agents without carcinogenic activity are also able to increase tumor progression. Benzoyl and hydrogen peroxides are such noncarcinogenic agents with tumor progressing activity. Studies in our laboratory examined benzoyl peroxide as a complete carcinogen, tumor initiator, and tumor promoter (30). It was found to have no activity as a complete carcinogen or tumor initiator but had good activity as a tumor promoter. An inordinate number of carcinomas were seen in these studies, and when benzoyl peroxide was examined as a progressing agent it was found to lead to a more than three-fold increase in the cumulative carcinoma formation (23). Further examination demonstrated that hydrogen peroxide displayed similar activity (25). The mechanism for enhanced progression by these peroxides is currently unclear, but two possibilities predominate. The peroxides are free radical generators and have been shown to produce single-strand DNA breaks (15,27) and DNA-protein cross-links (12) that may be genetic lesions sufficient to cause progression. As the peroxides do not have carcinogenic activity, such effects must not be sufficient for initiation or complete carcinogenesis. This difference may be due to a progressive increase in genetic instability that is observed during tumor progression (1) and may have an additive effect with that of the peroxides.

A second potential mechanism for the tumor progressing activity of the peroxides invokes selective toxicity of epidermal cells in response to these compounds. Keratinocyte cell lines developed from initiated skin or papillomas have

Table 2. Agents examined as inhibitors of tumor progression

Agent	Dose	Activity
Reduced glutathione	5 μmol, 25 μmol	moderate
Disulfiram	5 μmol	moderate
Butylated hydroxyanisole	50 μmol	+
Vitamin E	5 μmol	--
Copper(II)(3,5-diisopropyl-salicylate)2	5 μmol	--
Sodium benzoate	50 μmol	--
N-Acetyl cysteine	5 μmol, 25 μmol	--

-- = No activity observed; + = tumor progressing activity. All treatments were topical and biweekly.

been found to be resistant to DNA strand breakage and cytotoxicity by benzoyl peroxide. Within a group of 10 cell lines, there was an inverse correlation between malignant behavior and DNA strand breakage by benzoyl peroxide (16). The resulting selective toxicity could lead to elimination of cells without malignant potential, allowing malignant cells the space and nutrients for expansion.

Studies of other progressing agents in our laboratory also support selective toxicity as a potential mechanism involved in tumor progression. Acetic acid was found to have tumor progressing activity, as biweekly treatment of papilloma-bearing mice with acetic acid (667 μmol) led to about a two-fold increase in both the percentage of animals with carcinomas and the cumulative number of carcinomas formed (26). This dosage may well have had toxic effects. Earlier studies had demonstrated that acetic acid was a very weak tumor promoter (29). As is the case for all mouse skin tumor promoters, acetic acid produced an increase in epidermal DNA synthesis. However, the time for maximal DNA synthesis increased with the dose of acetic acid, suggesting the increase in DNA synthesis was the result of a progressive increase in toxicity. Overt toxicity was indeed observed at the highest dose (1,000 mmol) of acetic acid. This suggests that toxicity may well be occurring at the dose utilized and that it is playing a role in the increase in carcinoma formation. In this respect, a second, possibly more viable, interpretation also exists. Increased DNA synthesis has been associated with increases in mutation rates (4), and the increased DNA synthesis dependent or independent of the toxicity observed here could potentially lead to malignant lesions.

In addition to these studies in mouse skin, studies in other tumor systems indicate tumor progression occurs as the result of cumulative genetic lesions. Examples of such systems would include the colon (11), lung, and breast (8). Accordingly, antimutagenesis protocols will have good potential for the prevention of tumor progression.

AGENTS AND MECHANISMS THAT INHIBIT TUMOR PROGRESSION

As our studies with the peroxides demonstrated that oxidative stress has a potentially important role in tumor progression, we examined free radical scavengers and antioxidants as inhibitors of the tumor progression process. The use of such compounds is supported by the fact that these compounds also have activity as inhibitors of tumor initiation as well as tumor promotion (9). The agents examined, dosages utilized, and the resulting activities are listed in Tab. 2. For these studies, animals were initiated with DMBA and promoted with TPA for 16 weeks, at which point TPA treatment was terminated. Following four weeks with no treatment, the animals were assigned to groups adjusted for papilloma number, and papilloma-bearing skin was treated biweekly with the putative progression inhibitors for the remaining 30 weeks of the study. Only reduced glutathione and disulfuram were shown to have significant activities as suppressors of tumor progression. Pharmacokinetic effects could have contributed to these results, as only a single dose, which inhibited tumor promotion, was examined.

A 25 μmol dose of reduced glutathione led to a 50% reduction in the formation of carcinomas as is shown in Fig. 3A. We also found that administration of the glutathione depleting agent diethylmaleate (DEM) significantly increased the percentage of animals with carcinomas (77% vs 47%, Fig. 3B). GSH treatment of epidermal cells has been shown to increase intracellular GSH levels, and it thus appears that GSH, itself, has an inhibitory effect on tumor progression. GSH and the GSH transferases are considered to be important components in the prevention of mutagenesis (20) and may possibly be involved in this effect. In interesting contrast, a progressive increase in intracellular GSH levels is observed as different types of tumors, including papillomas, progress to malignancy (24). Further, the expression of γ-glutamyltranspeptidase, an enzyme thought to play a role in intracellular GSH transport and synthesis, is increased in late papillomas and carcinomas (21). Recent studies in our laboratory suggest that overexpresssion of this enzyme is sufficient to cause tumor progression. The mechanisms for such paradoxical actions associated with GSH are currently under investigation.

Disulfuram (DSF) was also shown to have moderate antiprogression activity as it reduced by approximately 50% the number of animals with carcinomas. DSF is a good inhibitor of tumor initiation in chemical carcinogenesis studies. Several pathways for this effect, including inhibition of oxidative metabolism (33), direct interaction with electrophiles (6), increases in cellular thiol levels, and induction of GSH transferase (7), have been described. These mechanisms are all potentially involved in the inhibition of tumor progression by DSF.

N-acetyl cysteine (NAC) would have been predicted to have activity similar to GSH, as cysteine pools are generally considered to be rate limiting for GSH synthesis (14). We observed no significant inhibition of tumor progression with NAC.

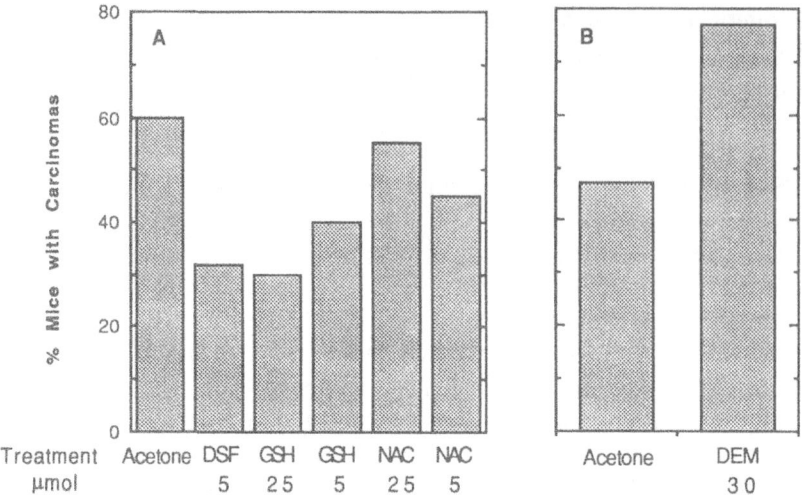

Figure 3. The percentage of mice with carcinomas for mice treated duing the progression stage with (A) DSF, GSH, or NAC or (B) DEM as described in the text.

Possibly the compound could not be absorbed through the skin or the dosages were insufficient to increase intracellular pools.

Unexpectedly, butylated hydroxyanisole (BHA) acted as a tumor progressing agent in these studies. BHA treatment led to a significant increase in the percentage of papillomas that became carcinomas, 11% vs. 7.8% for the carrier treated animals. This result was surprising, as BHA has inhibitory activity against both tumor initiation and promotion and has been shown to increase GSH transferase activity (11). However, in the forestomach model of carcinogenesis, BHA does act as a tumor promoter (28). These results underline the fact that our understanding of these processes is incomplete. It can be expected that chemical inhibition of tumor progression will follow the general principles found for the inhibition of the other stages of carcinogenesis where multiple variables including dosage, time, and tissue examined, determine the efficacy of inhibition.

PERSPECTIVES

Inhibition of the tumor progression stage of carcinogenesis has not been extensively studied. At this time few inhibitors of tumor progression have been described, and only modest activity of those is observed. As this is a clinically relevant stage of carcinogenesis, more extensive examination is warranted. Tumor progression has been well aligned with mutagenic changes, and initial studies should focus on compounds with antimutagenic activity.

Nonetheless, the inhibition of tumor progression and the other stages of carcinogenesis is foreseeable in the not too distant future. Numerous molecular lesions associated with carcinogenesis have been described, and as their biochemical functioning is understood, we can expect the parallel development of compounds that are able to inhibit or bypass their function. Accordingly, such compounds may be able to inhibit and possibly reverse this process.

REFERENCES

1. Aldaz, C.M., C.J. Conti, A.J.P. Klein-Szanto, and T.J. Slaga (1987) Progressive dysplasia and aneuploidy are hallmarks of mouse skin papillomas: Relevance to malignancy. Proc. Natl. Acad. Sci., USA 84:2029-2032.
2. Aldaz, C.M., C.J. Conti, I.B. Gimenez, T.J. Slaga, and A.J.P. Klein-Szanto (1985) Cutaneous changes during prolonged application of 12-O-tetradecanoylphorbol-13-acetate on mouse skin and residual effects after cessation of treatment. Cancer Res. 45:2753-2759.
3. Aldaz, C.M., C.J. Conti, A. Chen, A. Bianchi, S.B. Walker, and J. DiGiovanni (1991) Promoter independence as a feature of most skin papillomas in SENCAR mice. Cancer Res. 51:1045-1050.
4. Ames, B.N., and L.S. Gold (1990) Chemical carcinogenesis: Too many rodent carcinogens. Proc. Natl. Acad Sci., USA 87:772-776.
5. Argyris, T.S. (1985) Regeneration and the mechanism of epidermal tumor promotion. CRC Crit. Rev. Toxicol. 14:211-258.
6. Bertram, B., A.M. Tacchi, B.L. Pool, and M. Weisler (1982) *In vitro* formation of methyl-diethyldithiocarbamate after the reaction of nitrosoacetoxy-methylmethylamine or methylnitrosourea with disulfuram. Carcinogenesis 3:1361-1366.
7. Bertram, B., E. Frei, H.R. Scherf, J. Schumacher, A.M. Tacchi, and M.J. Wiessler (1985) Influence of a prolonged treatment with disulfuram and D(-)-penacillamine on nitrosodiethylamine induced biological and biochemical effects in rats. J. Cancer Res. Clin. Oncol. 109:9-15.
8. Bishop, J.M. (1991) Molecular themes in oncogenesis. Cell 64:235-248.
9. DiGiovanni, J. (1991) Inhibition of chemical carcinogenesis. In Chemical Carcinogenesis and Mutagenesis II, C.S. Cooper and P.L. Grover, eds. Springer-Verlag, Berlin, pp. 159-224.
10. Ewing, M.W., C.J. Conti, F.H. Kruszewski, T.J. Slaga, and J. DiGiovanni (1988) Tumor progression in SENCAR mouse skin as a function of initiator dose and promoter dose, duration and type. Cancer Res. 48:7048-7054.
11. Fearon, E.R., and B. Vogelstein (1990) A genetic model for colorectal carcinogenesis. Cell 61:759-767.
12. Gensler, H.L., and G.T. Bowden (1983) Evidence suggesting a dissociation of DNA strand scissions and late-stage promotion of tumor cell phenotype. Carcinogenesis 4:1507-1511.
13. Greenhalgh, D.A., D.J. Welty, A. Player, and S.H. Yuspa (1990) Two oncogenes, v-fos and v-ras, cooperate to convert normal keratinocytes to squamous cell carcinoma. Proc. Natl. Acad. Sci., USA 87:643-647.
14. Hanigan, M.H., and H.C. Pitot (1985) Gamma-glutamyl transpeptidase: Its role in hepatocarcinogenesis. Carcinogenesis 6:165-172.
15. Hartley, J.A., N.W. Gibson, L.A. Zwelling, and S.H. Yuspa (1985) The association of DNA strand breaks with accelerated terminal differentiation in mouse epidermal cells exposed to tumor promoters. Cancer Res. 45:4864-4870.

16. Hartley, J.A., N.W. Gibson, A. Kilkenny, and S.H. Yuspa (1987) Mouse keratinocytes derived from initiated skin or papillomas are resistant to DNA strand breakage by benzoyl peroxide: A possible mechanism for tumor promotion mediated by benzoyl peroxide. Carcinogenesis pp. 1827-1830.

17. Hennings, H., R. Shores, M.L. Wenk, E.F. Spangler, R. Tarone, and S.H. Yuspa (1983) Malignant conversion of mouse skin tumors is increased by tumor initiators and unaffected by tumor promoters. Nature 304:67-69.

18. Hennings, H., E.F. Spangler, R. Shores, P. Mitchell, D. Devor, A.K.M. Shamsuddin, K.M. Elgjo, and S.H. Yuspa (1986) Malignant conversion and metastasis of mouse skin tumors: A comparison of SENCAR and CD-1 mice. Env. Health Perspec. 68:69-74.

19. Hennings, H., R.A. Shores, M.C. Poirier, E. Reed, R.E. Tarone, and S.H. Yuspa (1990) Enhanced malignant conversion of benign mouse skin tumors by cisplatin. J. Natl. Cancer Inst. 82:836-840.

20. Ketterer, B. (1988) Protective role of glutathione and glutathione transferases in mutagenesis and carcinogenesis. Mutat. Res. 202:343-361.

21. Klein-Szanto, A.J.P., K.G. Nelson, Y. Shah, and T.J. Slaga (1983) Simultaneous appearence of keratin modifications and g-glutamyl transferase activity as indicators of tumor progression in mouse skin papillomas. J. Natl. Cancer Inst. 70:161-168.

22. O'Conneil, J.F., A.J.P. Klein-Szanto, D.M. DiGiovanni, J.W. Fries, and T.J. Slaga (1986) Malignant progression of mouse skin papillomas treated with ethylnitrosourea, N-methyl-N'-nitro-N-nitrosoguanidine, or 12-O-tetradecanoyl-phorbol-13-acetate. Cancer Lett. 30:269-274.

23. O'Connell, J.F., A.J.P. Klein-Szanto, D.M. DiGiovanni, J.W. Fries, and T.J. Slaga (1986) Enhanced malignant progression of mouse skin tumors by the free-radical generator benzoyl peroxide. Cancer Res. 46:2863-2865.

24. Reiners, J.J., personal communication.

25. Rotstein, J.B., J.F. O'Connell, and T.J. Slaga (1986) The enhanced progression of papillomas to carcinomas by peroxides in the 2-stage mouse skin model. Proc. Am. Assoc. Cancer Res. 27:143.

26. Rotstein, J.B., and T.J. Slaga (1988) Acetic acid, a potent agent of tumor progression in the multistage mouse skin model for chemical carcinogenesis. Cancer Lett. 42:87-90.

27. Saladina, A.J., J.C. Wiley, J.F. Lechner, R.C. Grafstrom, M. LaVeck, and C.C. Harris (1985) Effects of formaldehyde, acetaldehyde, benzoyl peroxide, and hydrogen peroxide on cultured normal human bronchial epithelial cells. Cancer Res. 45:2522-2526.

28. Shirai, T., S. Fujishima, M. Oshima, A. Masuda, and N. Ito (1984) Effects of butylated hydroxyanisole, butylated hydroxytoluene and NaCl on gastric carcinogenesis initiated with N-methyl-N'-nitro-N-nitrosoguanidine in F344 rats. J. Natl. Cancer Inst. 72:1189-1198.

29. Slaga, T.J., G.T. Bowden, and R.K. Boutwell (1975) Acetic acid, a potent stimulator of mouse epidermal macromolecular synthesis and hyperplasia but with weak tumor-promoting activity. J. Natl. Cancer Inst. 55:983-987.

30. Slaga, T.J., A.J.P. Klein-Szanto, L.L. Triplett, and L.P. Yotti (1981) Skin tumor-promoting activity of benzoyl peroxide, a widely used free radical-generating compound. Science 213: 1023-1025.

31. Slaga, T.J. (1983) Mechanisms involved in two-stage carcinogenesis in mouse skin In Mechanisms of Tumor Promotion. Vol. I. T.J. Slaga, ed. CRC Press, Boca Raton, Florida, pp. 1-16.

32. Slaga, T.J. (1989) Cellular and molecular mechanisms involved in multistage skin carcinogenesis. In Carcinogenesis: A Comprehensive Survey. Vol. II, C.J. Conti, T.J. Slaga, and A.J.P. Klein-Szanto, eds. Raven Press, New York, pp 1-18.

33. Zemaitis, M.A., and F.E. Greene (1976) Impairment of hepatic microsomal drug metabolism in the rat during daily disulfuram administration. Biochem. Pharmacol. 23:1355-1360.

HISTOPATHOLOGY OF HUMAN INTRAEPITHELIAL NEOPLASIA

WITH IMPLICATIONS FOR CHEMOPREVENTION STRATEGY

Charles W. Boone, Gary J. Kelloff, and Vernon E. Steele
Division of Cancer Prevention and Control
National Cancer Institute
Bethesda, Maryland 20892 USA

INTRODUCTION

Every epithelial cancer (i.e., invasive neoplasm) is preceded by a preinvasive stage of intraepithelial neoplasia that typically lasts for years. A better understanding of human intraepithelial neoplasia is of critical importance to investigators in the chemoprevention field; it can assist in the more rational design of drugs that will slow or stop the development of intraepithelial neoplasia and, therefore, the potential for subsequent development of invasive cancer, and also assist in the development of early biomarkers that are modulated by chemopreventive drugs.

MORPHOLOGICAL CHARACTERISTICS OF INTRAEPITHELIAL NEOPLASIA

Fortunately, the early development and widespread use of tissue biopsies and cytologic screening of exfoliated cells from the uterine cervix have produced a system of nomenclature and descriptive terminology that can be applied to other epithelia. In particular, the equivalence of the terms "dysplasia" and "intraepithelial neoplasia" is especially appropriate and is widely accepted by gynecologic cytopathologists who use the term "cervical intraepithelial neoplasia," or CIN (49).

From a survey of the literature, there appear to be seven defining morphological criteria of intraepithelial neoplasia, or dysplasia. These are: 1. Nucleus, increased size; 2. Nucleus, altered shape; 3. Nucleus, increased stain uptake; 4. Nucleus, pleomorphism (increased variation in size, shape, and stain uptake); 5. Mitoses, increased; 6. Mitoses, abnormal; 7. Maturation, disordered. Each of these criteria has been specifically listed for the term "dysplasia" in articles on uterine cervix (2,10,11,22,29,48,49), oral leukoplakia (6,46,53,55), larynx (34), lung (4), esophagus (52), colon (15,27,28,36,50), urinary bladder (9,17,20,37,38), and skin (25). Figure 1 presents an example of dysplastic change in a histological section of human vocal cord epithelium (Figs. 1 and 2 below are drawings derived from selected photographs

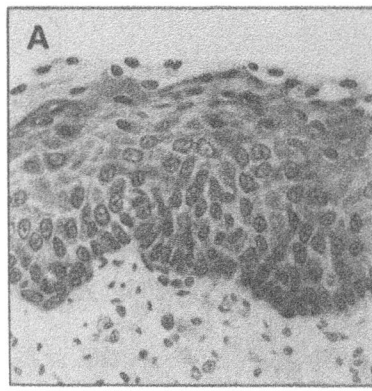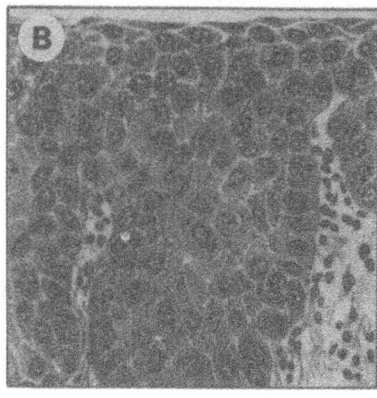

Figure 1. Human vocal cord squamous epithelium. A. Normal epithelium, showing maturation of basal cells into intermediate and superficial cells. B. Dysplastic epithelium. The nuclei are larger, vary more in size and shape (pleomorphism), an abnormal 3-group mitosis is present, and there is loss of normal maturation. Reprinted with permission of CRC Press, Inc. (58).

in the literature (5 and 34, respectively). Each of the criteria of epithelial dysplasia, except increased mitoses, is shown in panel B of Fig. 1 and is contrasted with the normal epithelium in panel A. The normal maturational stratification of basal, intermediate, and superficial cells seen in the normal epithelium are absent in the severely dysplastic epithelium.

ESTIMATING THE SEVERITY OF INTRAEPITHELIAL NEOPLASIA (DYSPLASIA)

The severity of intraepithelial neoplasia is estimated from the extent of the lesion as well as the degree of deviation from normal morphology. The initial clonal focus of proliferating neoplastic cells located on the basement membrane enlarges by expanding upward and laterally within the epithelium, replacing normal epithelial cells as it does so (57). Thus, mild dysplasia extends to the junction of lower and middle thirds of the epithelium, moderate dysplasia extends to the junction of middle and upper thirds, and severe dysplasia extends to the full thickness of the epithelium. In the case of cervical intraepithelial neoplasia, or CIN, grades 1, 2, and 3 are used to correspond to mild, moderate, and severe dysplasia (49).

In severe dysplasia of cervical and other epithelia, where there is complete absence of normal epithelial cells in the topmost layers, the term "carcinoma *in situ*" *is* sometimes used. There is now much well-reviewed evidence that "severe dysplasia" and "carcinoma *in situ*" are virtually identical and form one single spectrum of neoplastic disease (2,10). Unfortunately, the term "carcinoma *in situ*" connotes the false concept of an entity that is distinct from dysplasia; this term has misled some into assuming that the neoplastic process actually starts with the carcinoma *in situ*, thereby causing the preceding morphologic changes of dysplasia to be considered

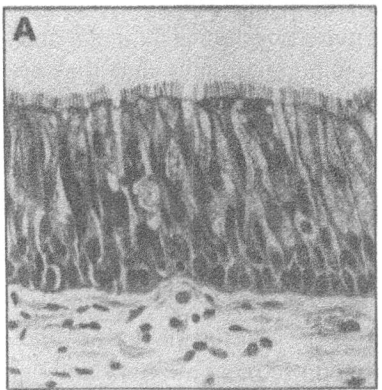

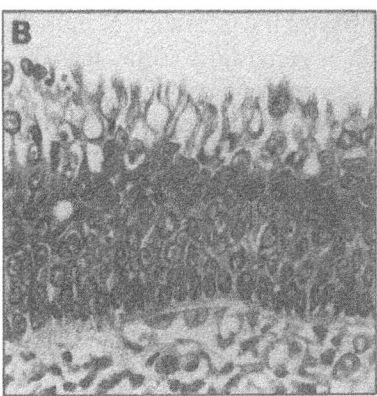
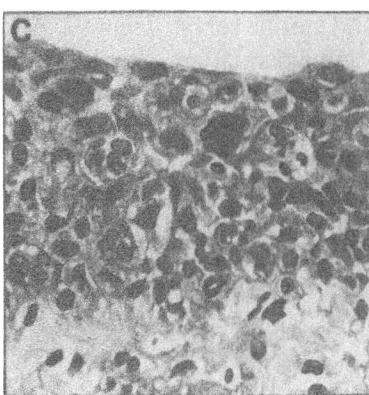

Figure 2. Human laryngeal epithelium. A. Normal respiratory epithelium of the larynx, showing goblet cells and ciliated cells. B. The figure shows "squamous metaplasia" in process, which is a response to chronic irritation, e.g., smoking. The respiratory epithelium is being replaced by proliferating squamous epithelium, which pushes up from below until the respiratory epithelium degenerates and is cast off. C. The subsequent squamous epithelium that forms has the potential of developing severe dysplasia, as shown here. Reprinted with permission CRC Press, Inc. (59).

"preneoplastic." On the contrary, the neoplastic process is a single indivisible continuum that begins within the epithelium as mild dysplasia and progresses through dysplasia of increasing severity until finally invasion across the basement membrane occurs, at which time, by definition, the terms "malignant," "cancer," or "carcinoma" apply. (It is a curious convention of cancer nomenclature that the same morphological changes identified as "severe dysplasia" before invasion are called "anaplasia" after invasion [51]). Thus, the neoplastic continuum of intraepithelial neoplasia, or dysplasia, may be called preinvasive, premalignant, or precancerous, but it is an error to call it preneoplastic. Foulds had this opinion on the subject: "The most frustrating gap in the terminology of pathology of tumors in man is a lack of a satisfactory name for the so-called pre-cancerous lesion. Reasons were given earlier for rejecting 'precancerous.' 'Preneoplastic,' sometimes now applied to lesions of the same general kind, is even more objectionable; it can only mean that the lesions are not neoplastic, whereas I maintain strongly that they are neoplastic and that this should be recognized in their designation" (18). The name that Foulds might have accepted is "intraepithelial neoplasia." The terms "dysplasia" and "intraepithelial neoplasia" will be considered biologically equivalent in this paper.

INTRAEPITHELIAL NEOPLASIA (DYSPLASIA) IN HUMAN EPITHELIA

The seven morphologic criteria of intraepithelial neoplasia (dysplasia) described above are the same in most of the nonglandular human epithelia that are prone to develop cancer, such as skin, oral epithelium, larynx, esophagus, lung, colon, bladder, and uterine cervix. Figure 2, A-C, illustrates the formation of intraepithelial neoplasia in human larynx. It is characteristic of respiratory tissue, including lung, that before dysplastic changes are seen, the respiratory epithelium is replaced by proliferating basal "reserve" cells that form a squamous epithelium resembling vocal cord epithelium. This change typically occurs in response to chronic irritation, such as that associated with tobacco smoking. Figure 2B shows this process, which is termed "squamous metaplasia." The respiratory epithelium is lifted up by the proliferating squamous cells, undergoes degeneration, and exfoliates. The squamous cells are then at risk to develop the morphological changes of dysplasia, which is shown in Fig. 2C.

INTRAEPITHELIAL NEOPLASIA (DYSPLASIA) IN THE DMBA/TPA MOUSE SKIN MODEL

The progression of the morphological changes of intraepithelial neoplasia (dysplasia) in the epithelium of skin papillomas of mice, induced by skin painting once with 7,12-dimethylbenz(α)anthracene (DMBA) and twice weekly with 12-0-tetra-decanoylphorbol-13-acetate (TPA), appears to be virtually identical to the analogous changes seen in human epithelia (1).

An initial period of hyperplasia is followed by dysplasia of increasing severity that is seen in 90% of the papillomas by the 20th week of skin painting with TPA (1). In addition, the epithelium of the papillomas becomes progressively more aneuploid, reaching 100% aneuploidy after 30 weeks of skin painting with TPA (12). The occurrence of aneuploidy during the course of intraepithelial neoplasia and prior to invasion is also seen in human epithelia, as described below.

Suitability of the Mouse Skin Model for Experimental Studies in Chemoprevention

The mouse skin model appears to lend itself very well to studying the effects of chemopreventive agents on the histomorphological progression of intraepithelial neoplasia. Each papilloma is typically monoclonal, being derived from a single stem cell of the so-called epidermal proliferative unit (33). A papilloma on the skin of a treated mouse represents a convenient visual marker for a focus of developing intraepithelial neoplasia that can be observed, treated, and biopsied, analogous to the human case of an adenomatous polyp of the colon which represents a visual marker for a focus of progressing intraepithelial dysplasia (28).

More importantly, the mouse skin model provides an opportunity to study mechanisms of <u>regression</u> of intraepithelial neoplasia at the histological and biochemical levels, both spontaneous regression and regression as modulated by chemopreventive drugs. There appear to be no reported histological studies specifically directed at documenting the mechanisms of mouse skin papilloma regression.

THE OCCURRENCE OF ANEUPLOIDY DURING INTRAEPITHELIAL NEOPLASIA

Aneuploidy, or an abnormal number of chromosomes per cell, is part of the neoplastic process and typically begins during the intraepithelial stage. Recent technological advances in quantitative flow and image cytospectrophotometry (30), including as source material paraffin-embedded archival material as well as fresh tissue from surgical specimens, biopsies, fine needle aspirates, and exfoliated cells, have revealed that aneuploidy is present in a very large percentage of cases of severe dysplasia, in a moderate percentage of cases of moderate dysplasia, and minimally or not at all in mild dysplasia, as applied to the cervix (26,48), skin (40), oral leukoplakia (23), larynx (8,13), lung (39), esophagus (47), stomach (32), and colorectum (24,45). After invasion, about 90% of solid tissue cancers produce an aneuploid peak by flow cytometry (21).

Causes of Aneuploidy

Aneuploidy is the result of unequal segregation of chromosomes to daughter cells during cell division. This is typically due to abnormalities in the structure and function of the mitotic apparatus, including: 1. abnormal centriolar replication or lack of movement to opposite poles (resulting in uni-, tri-, or multipolar mitoses); 2. abnormal spindle fiber polymerization or inadequate attachment to the kinetochore/-centromere complex (resulting in failure of chromosomes to orient at the metaphase plate, with collection around one or both poles to create "two-group" or "three-group" mitoses); 3. failure of chromatids to detach at the centromere (resulting in nondisjunction). Abnormalities of the mitotic apparatus can be caused by nongenotoxic drugs, including orthoquinones, catechol estrogens, stilbestrol, and colchicine (16,54), or by mutations. In the latter case, a permanent abnormality of the mitotic apparatus resulting from a mutation could generate aneuploid cells at each division.

Clonal Evolution During the Development of Intraepithelial Neoplasia

The concept and documentation of the clonal evolution in hematopoietic and lymphopoietic neoplasms has been described by Nowell (41,42). Clonal evolution is the continuous process within a tumor cell population of karyotypic, and therefore phenotypic, variation, with selection of clones most fit to survive and grow more rapidly under the prevailing set of selection pressures. Nowell (42) has noted that "clonal evolution might result from enhanced genetic instability within the tumor cell population, which increases the probability of further genetic alterations and their subsequent selection," and also that "the continued presence of multiple subpopulations within the tumor provides the basis for the heterogeneity that is also typically observed." When aneuploidy develops during the higher grades of human intraepithelial neoplasia, as documented above, with its associated expansion of karyotypic heterogeneity, conditions are established for the development of clonal evolution. Thus, clonal evolution may begin during the intraepithelial stage of neoplastic development prior to invasive cancer formation.

Foulds, in his description of "Intermediate Phase B" neoplastic lesions, which include mild, moderate, and severe dysplasia and carcinoma *in situ*, discusses the neoplastic "progression" of these lesions in terms of the stepwise acquisition of permanent new changes in morphology or function (19). This "intraepithelial neoplastic progression" could be an expression of the intraepithelial clonal evolution discussed above (42).

Aneuploidy as One Possible Basis for the Morphological Changes of Dysplasia

When aneuploidy is present in a neoplastic cell population, an increased number of chromosomes per cell could account for nuclear enlargement (35) and the associated karyotypic and, therefore, phenotypic heterogeneity could account for the pleomorphism, for example, wide variation in nuclear size, shape, and stain uptake, as well as the disordered maturation. Cells with abnormal mitoses are the hallmark of aneuploidy (22), including 2- and 3-group mitoses, lagging chromosomes, nondisjunction, and multipolar mitoses. Thus, aneuploidy *per se* can explain some of the morphological changes characteristic of dysplasia/intraepithelial neoplasia.

CHEMOPREVENTION STRATEGIES RELATED TO INTRAEPITHELLAL NEOPLASIA

From the extensive data of Barron and Richart (7), any degree of dysplasia of the uterine cervix is expected to progress to severe dysplasia/carcinoma *in situ* in about 80% of cases during a ten-year period. For mild dysplasia, the progression may be projected at 60% of cases over ten years. Once the stage of severe dysplasia/carcinoma *in situ* is reached, further progression to invasive cancer occurs in about 20% of cases within five years (43).

Since roughly the same time scale may apply to the rate of progression of intraepithelial neoplasia in other epithelia, the goal of cancer chemoprevention strategy during this years-long period should be to suppress or stop the process of intraepithelial neoplasia with chemopreventive drugs.

Two major factors have been shown to influence the rate of progression of intraepithelial neoplasia. The first is a sustained increase in proliferation rate. Preston-Martin *et al.* (44) have reviewed the numerous examples that show that a sustained increase in proliferative rate of an epithelium increases the cancer risk in that epithelium. The second is chronic exposure to carcinogens. Druckery (14) has shown, for a number of carcinogens, cancer types, and animal species, that the higher the daily dose of carcinogen over a prolonged period of exposure, the shorter the tumor latent period, or, said another way, the faster the progression of intraepithelial neoplasia to invasiveness. In consideration of the above two factors, chemoprevention strategy should include the use of drugs to suppress hyperproliferation and to block the effects of carcinogens. Examples of drugs which would act against hyperproliferation are difluoromethylornithine and, in the case of the hyperproliferation associated with familial polyposis, calcium salts (31).

With regard to blocking the effects of carcinogens, humans may be chronically exposed to carcinogens in the diet, atmosphere, personal habits (tobacco especially), and workplace. Wattenberg (56) has reviewed the mechanisms by which chemopreventive agents can block the activation of carcinogens, accelerate their biotransformation and elimination, or trap electrophilic intermediates. Wattenberg's Type A blocking compounds, which induce Phase II conjugating enzymes, are more desirable than compounds which modulate the Phase I cytochrome P450 mixed function oxidases because the latter enzymes carry the risk, however small, of activating rather than detoxifying ingested procarcinogens. Examples of desirable anticarcinogenic compounds are N-acetyl-l-cysteine and dithiolthiones such as oltipraz (3). The long-term administration of an anticarcinogenic drug should be considered in cases where prolonged exposure to a carcinogen is likely, for example, in chronic tobacco smokers who have failed cessation efforts.

REFERENCES

1. Aldaz, C.M., C.J. Conti, A.J.P. Klein-Szanto, and T.J. Slaga (1987) Progressive dysplasia and aneuploidy are hallmarks of mouse skin papillomas: Relevance to malignancy. Proc. Natl. Acad. Sci., USA 84:2029-2032.
2. Anderson, M.C. (1987) Premalignant and malignant diseases of the cervix. In Haines and Taylor Obstetrical and Gynaecological Pathology, 3rd ed., M. Haines, C.W. Taylor, and H. Fox, eds. Churchill Livingstone, New York, pp. 225-277.
3. Ansher, S.S., P. Dolan, and E. Bueding (1986) Biochemical effects of dithiolthiones. Food Chem. Toxicol. 24:405-415.
4. Auerbach, O., J.B. Gere, J.B. Forman, T.G. Petrick, H.J. Smolin, G.E. Muehsam, D.Y. Kassouny, and P.A. Stout (1957) Changes in the bronchial epithelium in relation to smoking and cancer of the lung. N. Engl. J. Med. 256:97-104.
5. Auerbach, O., E.C. Hammond, and L. Garfinkel (1970) Histological changes in the larynx in relation to smoking habits. Cancer 25:92-104.
6. Banoczy, J. (1982) Oral Leukoplakia. Martinus Nijhoff Publishers, Boston, pp. 64-74.
7. Barron, B.A., and R.M. Richart (1968) A statistical model of the natural history of cervical carcinoma based on a prospective study of 557 cases. J. Natl. Cancer Inst. 41:1343-1353.
8. Bjeldenkrantz, K., J. Lundgren, and J. Olofsson (1983) Single-cell DNA measurements in hyperplastic, dysplastic, and carcinomatous laryngeal epithelia with special reference to the occurrence of hypertetraploid cell nuclei. An. Quant. Cytol. 5:184-188.
9. Brawn, P.N. (1984) Interpretation of Bladder Biopsies. Raven Press, New York.
10. Buckley, C.H., E.B. Butler, and H. Fox (1982) Cervical intraepithelial neoplasia. J. Clin. Pathol. 35:1-13.

11. Burghardt, E. (1973) Early histological diagnosis of cervical cancer. In <u>Major Problems in Obstetrics and Gynecology</u>, Vol. 6, E.A. Friedman, ed. W.B. Saunders Company, Philadelphia, Chap. 2.

12. Conti, C.J., C.M. Aldaz, J. O'Connell, A.J.P. Klein-Szanto, and T.J. Slaga (1986) Aneuploidy, an early event in mouse skin tumor development. <u>Carcinogenesis</u> 7:1845-1848.

13. Crissman, J.D., and Y.S. Fu (1986) Intraepithelial neoplasia of the larynx. A clinicopathologic study of six cases with DNA analysis. <u>Arch. Otolaryngol. Head Neck Surg.</u> 112:522-528.

14. Druckrey, H. (1967) Quantitative aspects in chemical carcinogenesis. In <u>Potential Carcinogenic Hazards from Drugs: Evaluation of Risks</u>, UICC Monograph Series, Vol. 7, R. Truhaut, ed. Springer-Verlag, New York.

15. Elias, H. (1985) <u>Colonic Neoplasms</u>. Piccin Nuova Libraria, Padua, Italy, p. 45.

16. Epe, B., J. Hegler, and M. Metzler (1987) Site-specific covalent binding of stilbene-type and steroidal estrogens to tubulin following metabolic activation in vitro. <u>Carcinogenesis</u> 8:1271-1275.

17. Farrow, G.M., H. Barlebo, M. Enjoji, G. Chisolm, G.H. Friedell, G. Jacse, T. Kakizoe, L.G. Koss, T. Kotake, and W. Vahlensieck (1986) Transitional cell carcinoma in situ. In <u>Developments in Bladder Cancer</u> from the Proceedings of the First International Consensus Development Conference on Guidelines for Clinical Research in Bladder Cancer, Antwerp, Belgium, June 20-22, 1985, L. Denis, ed. Alan R. Liss, Philadelphia, pp. 85-96.

18. Foulds, L. (1969) <u>Neoplastic Development</u>, Vol. 1. Academic Press, New York, p. 92.

19. Foulds, L. (1975) <u>Neoplastic Development</u>, Vol. 2. Academic Press, New York, pp. 9-10.

20. Friedell, G.H., G.C. Parija, G.K. Nagy, and E.A. Soto (1980) The pathology of human bladder cancer. <u>Cancer</u> 45 (Suppl. 7):1823-1831.

21. Friedlander, M.L., D.W. Hedleu, and I.W. Taylor (1984) Clinical and biological significance of aneuploidy in human tumours. <u>J. Clin. Pathol.</u> 37: 961-974.

22. Fu, Y.S., J.W. Reagan, and R.M. Richart (1981) Definition of precursors. <u>Gynecol. Oncol.</u> 12:S220-S231.

23. Grassel-Pietrusky, R., E. Deinlein, and O.P. Hornstein (1982) DNA-aneuploidy rates in oral leukoplakias determined by flow-cytometry. <u>J. Oral Pathol.</u> 11:434-438.

24. Hammarberg, C., P. Slezak, and B. Tribukait (1984) Early detection of malignancy in ulcerative colitis. A flow-cytometric DNA study. <u>Cancer</u> 53:291-295.

25. Hashimoto, K., and A.H. Mehregan (1990) <u>Tumors of the Epidermis</u>. Butterworths, Boston.

26. Jacobsen, A., P.B. Kristensen, and H.K. Poulson (1983) Flow cytometric classification of biopsy specimens from cervical intraepithelial neoplasia. <u>Cytometry</u> 4:166-169.

27. Jass, J.R., and L.H. Sobin (1989) <u>Histological Typing of Intestinal Tumours</u>, 2nd ed. Springer-Verlag, New York.

28. Konishi, F., and B.C. Morson (1982) Pathology of colorectal adenomas: A colonoscopic survey. <u>J. Clin. Pathol.</u> 35:830-841.

29. Koss, L.G. (1979) <u>Diagnostic Cytology and Its Histopathologic Bases</u>, 3rd ed., 2 vols. J.B. Lippincott Company, Philadelphia.

30. Koss, L.G., B. Czerniak, B. Herz, and R.P. Wersto (1989) Flow cytometric measurements of DNA and other cell components in human tumors: A critical appraisal. Human Pathol. 20:528-548.

31. Lipkin, M., and H. Newmark (1985) Effect of added dietary calcium on colonic epithelial-cell proliferation in subjects at high risk for familial colonic cancer. N. Engl. J. Med. 313:1381-1384.

32. Macartney, J.C., and R.S. Camplejohn (1986) DNA flow cytometry of histological material from dysplastic lesions of human gastric mucosa. J. Pathol. 150:113-118.

33. Mackenzie, I.C. (1975) Ordered structure of the epidermis. J. Invest. Dermatol. 65:45-51.

34. Michaels, L. (1984) Pathology of the Larynx. Springer-Verlag, New York.

35. Miles, C.P., and L.G. Koss (1966) Diagnostic traits of interphase human cancer cells with known chromosome patterns. Acta Cytol. 10:21-25.

36. Morson, B.C., and L.S. Pang (1967) Rectal biopsy as an aid to cancer control in ulcerative colitis. Gut 8:423-434.

37. Murphy, W.M., and M.S. Soloway (1982) Urothelial dysplasia. J. Urol. 127: 849854.

38. Nagy, G.K., W.J. Frable, and W.M. Murphy (1982) Classification of premalignant urothelial abnormalities. Pathol. Ann. 17:219-233.

39. Nasiell, M., H. Kato, G. Auer, A. Zetterberg, V. Roger, and L. Karlen (1978) Cytomorphological grading and Feulgen DNA-analysis of metaplastic and neoplastic bronchial cells. Cancer 41:1511-1521.

40. Newton, J.A., R.S. Camplejohn, and D.H. McGibbon (1986) Aneuploidy in Bowen's Disease. Br. J. Dermatol. 114:691-694.

41. Nowell, P.C. (1976) The clonal evolution of tumor cell populations. Science 194:23-28.

42. Nowell, P.C. (1986) Mechanisms of tumor progression. Cancer Res. 46:2203-2207.

43. Peterson, O. (1956) Spontaneous course of cervical precancerous conditions. Am. J. Obstet. Gynecol. 72:1063-1071.

44. Preston-Martin, S., M.C. Pike, R.K. Ross, P.A. Jones, and B.E. Henderson (1990) Increased cell division as a cause of human cancer. Cancer Res. 50:7415-7421.

45. Quirke, P., J.B. Fozard, M.F. Dixon, J.E.D Dyson, G.R. Giles, and C.C. Bird (1986) DNA ploidy in colorectal adenomas. Br. J. Cancer 53:477-481.

46. Regezi, J.A., and J.J. Sciubba (1989) Oral Pathology. W.B. Saunders Company, Philadelphia, pp. 94-97.

47. Reid, B.J., R.C. Haggitt, C.E. Rubin, and P.S Rabinovitch (1987) Barrett's esophagus. Correlation between flow cytometry and histology in detection of patients at risk for adenocarcinoma. Gastroenterology 93:1-11.

48. Reid, R., and Y.S. Fu (1986) Is there a morphologic spectrum linking condyloma to cervical cancer? In Viral Etiology of Cervical Cancer, R. Peto and H. Zur Hausen, eds. Cold Spring Harbor Laboratory, Cold Spring Harbor, New York, pp. 91-113.

49. Richart, R. M. (1973) Cervical intraepithelial neoplasia. Pathol. Ann. 8:301-328.

50. Riddell, R.H., H. Goldman, D.F. Ransohoff, H.D. Appelman, C.M. Fenoglio, R.C. Haggitt, C. Ahren, P. Correa, S.R. Hamilton, B.C. Morson, S.C. Sommers, and J.H. Yardley (1983) Dysplasia in inflammatory bowel disease: Standardized classification with provisional clinical applications. Human

Pathol. 14:931-968.

51. Robbins, S.L., R.S. Cotran, and V. Kumar (1984) Pathological Basis of Disease, 3rd ed. W.B. Saunders Company, Philadelphia, p. 219.

52. Shu, I.-C., and L. Koss (1985) The Cytopathology of Esophageal Carcinoma: Precancerous Lesions and Early Cancer. Masson Publishing USA, Inc., New York.

53. Soames, J.V., and J.C. Southam (1985) Oral Pathology. Oxford University Press, Toronto, Canada, pp. 116-126.

54. Tsutsui, T., N. Suzuki, H. Maizumi, and J.C. Barrett (1990) Aneuploidy induction in human fibroblasts: Comparison with results in Syrian hamster fibroblasts. Mutat. Res. 240:241-249.

55. Waldron, C.A., and W.G. Shafer (1975) Leukoplakia revisited. A clinicopathologic study of 3,256 oral leukoplakias. Cancer 36:1386-1392.

56. Wattenberg, L.W. (1985) Chemoprevention of cancer. Cancer Res. 45:1-8.

57. Woodruff, M.F.A. (1988) Tumor clonality and its biological significance. Adv. Cancer Res. 50:197-229.

58. Kelloff, G.J., C.W. Boone, W. Malone, and V. Steele (1992) Recent results in preclinical and clinical drug development of chemopreventive agents at the National Cancer Institute. In Cancer Chemoprevention, L. Wattenberg, M. Lipkin, C.W. Boone, and G. Kelloff, eds. CRC Press, Inc., Boca Raton, Florida.

59. Boone, C.W., G.J. Kelloff, and V.E. Steele (1992) Histopathology of human intraepithelial neolasia with implications for chemoprevention strategy. In Cancer Chemoprevention, L. Wattenberg, M. Lipkin, C.W. Boone, and G. Kelloff, eds. CRC Press, Inc., Boca Raton, Florida.

POTENTIAL MECHANISMS OF ANTITUMORIGENESIS

BY PROTEASE INHIBITORS

Ann R. Kennedy
Department of Radiation Oncology
University of Pennsylvania School of Medicine
Philadelphia, Pennsylvania 19104-6072 USA

ABSTRACT

The mechanisms by which protease inhibitors suppress carcinogenesis are unknown. From our studies, we believe the first event in carcinogenesis is a high frequency epigenetic event and that a later event (presumably genetic) leads to the malignant state. Protease inhibitors appear to be capable of reversing the "initiating" event in carcinogenesis, even at long times after carcinogen exposure. Protease inhibitors are thought to stop an ongoing process begun by carcinogen exposure. Effects of protease inhibitors on the following phenomena are thought to be related to their anticarcinogenic activity: 1) ability to effect the expression of certain oncogenes, and 2) ability to return carcinogen-increased levels of certain proteolytic activities (e.g., Boc-val-pro-arg-MCA hydrolyzing activity) to normal levels.

Other effects of anticarcinogenic protease inhibitors have also been observed; for example, we have reported that they can bring carcinogen-induced, elevated levels of gene amplification to nearly normal levels. The mechanism(s) by which protease inhibitors suppress carcinogenesis will not be determined with certainty until the mechanisms involved in cancer induction are known.

INTRODUCTION

It is now clear that protease inhibitors are very effective as anticarcinogenic agents in both in vitro and in vivo model systems, as has been reviewed recently (23,24).   Much of our own work on the protease inhibitor suppression of carcinogenesis has been performed with the Bowman-Birk inhibitor (BBI) from soybeans. Thus far, we have shown that BBI prevents or suppresses radiation and chemically induced transformation in vitro (1,17,20,41,42) as well as carcinogenesis in vivo in the following systems: dimethylhydrazine-induced colon (7,36,38) and liver

(36) carcinogenesis; 7,12-dimethylbenz(a)anthracene-induced oral carcinogenesis (33); 3-methylcholanthrene-induced lung carcinogenesis (40); and methylbenzylnitrosamine-induced esophageal carcinogenesis (37). In each of the carcinogenesis systems we have utilized, various types of studies have been performed that are aimed at determination of the mechanism of action of the protease inhibitors in terms of their ability to suppress the carcinogenic process. These mechanistically oriented studies will be reviewed here.

MATERIALS AND METHODS

The most direct method of determining the anticarcinogenic mechanism of protease inhibitors is to identify and characterize the proteases with which they interact. Only a few proteases have been observed to interact with the anticarcinogenic protease inhibitors, and the characteristics of these proteases have been described in detail elsewhere (2,4,8,10,23,29). Proteases have been identified by both substrate hydrolysis and affinity chromatography. Utilizing substrate hydrolysis, we examined the ability of cell homogenates to cleave specific substrates and we then determined the ability of various protease inhibitors to affect that hydrolyzing ability. Utilizing affinity chromatography, we identified specific proteases directly interacting with the BBI on a BBI affinity column. The Boc-val-pro-arg-MCA hydrolyzing activity (2,8,23,29) and the Suc-ala-ala-pro-phe-AMC hydrolyzing activity (10) were identified by substrate hydrolysis and a 43 kDa protease has been identified by affinity chromatography (6). The functions of these proteases are not known. As the Boc-val-pro-arg-MCA hydrolyzing activity has characteristics similar to specific proteases known to be involved in growth factor processing, it has led us to speculate that it has a similar function (8, 29). Many types of studies are being performed in the laboratory now to determine the exact identities and functions of the proteases that are target enzymes for the anticarcinogenic protease inhibitors.

Another method of determining the mechanism of action of anticarcinogenic protease inhibitors is to study the characteristics of those protease inhibitors that are able to suppress carcinogenesis. Several different types of protease inhibitors have been shown to have anticarcinogenic activity, as has recently been reviewed (23,24). We have performed many studies to evaluate the ability of different protease inhibitors to affect radiation and chemically induced malignant transformations in vitro (23). Our work has suggested that at least two different types of proteolytic activities are involved in the protease inhibitor suppression of carcinogenesis (5).

Another approach in mechanistically oriented studies involves looking at the specific processes affected by the anticarcinogenic protease inhibitors. For example, we have observed that BBI inhibits the processing of endocytosed protein that is involved in transcytosis (6,34). A specialized cell culture system, involving Madin-Darby canine kidney cells, was necessary for the investigations on transcytosis. Other endpoints shown to be affected by anticarcinogenic protease inhibitors in cells include the expression patterns of specific oncogenes (such as myc and fos) (9,11-14,35) and gene amplification (16). The specific methods utilized to study these endpoints are described in detail in the individual publications cited above.

DISCUSSION

When studied by themselves at the anticarcinogenic levels utilized in our animal model systems, protease inhibitors have not produced any measurable effects (3,33,36-38,40). In animals exposed to cancer-causing agents, protease inhibitors suppress carcinogenesis without deleterious side effects. We believe that protease inhibitors suppress carcinogenesis by reversing critical carcinogen-induced changes. We have observed several different carcinogen-induced alterations that are affected by the anticarcinogenic protease inhibitors. Carcinogens are capable of producing elevated levels of both c-myc expression and certain types of proteolytic activity, such as the Boc-val-pro-arg-MCA hydrolyzing activity. In the presence of anticarcinogenic protease inhibitors, carcinogen-elevated levels of both Boc-val-pro-arg-MCA hydrolyzing activity (33) and c-myc expression (35) are reduced to normal levels (i.e., the levels observed in control, untreated animals). Similarly, carcinogen-elevated levels of gene amplification can be reduced to nearly normal levels by protease inhibitors (16). Which, if any, of these phenomena are the ones directly related to anticarcinogenic activity will not be known until the cause(s) of cancer become known. It is known that cancer-causing agents produce many different types of changes in cells. It is assumed that the ability to reverse the critical change produced by carcinogens to a normal state is responsible for the anticarcinogenic activity of protease inhibitors.

From other work of ours, we believe that the first event in carcinogenesis is a high frequency epigenetic event, and that a later event (or events) leads directly to the malignant state (18,19,26-28,30,32). Protease inhibitors appear to be capable of reversing the initiating, high frequency event, even when given at long times after carcinogen exposure. It is believed that protease inhibitors block an ongoing process begun by cellular exposure to cancer-causing agents (17,20,23). One example of an ongoing process begun by carcinogen exposure is radiation-induced recombination in yeast (15). This process occurs for long time periods and many cell divisions after radiation exposure (15) and is a process suppressed by protease inhibitors (39). We have hypothesized that such a process induced by carcinogens could eventually lead to the production of the cancer genotype unless the system is turned off by (anticarcinogenic) protease inhibitor treatment (17,20,23). It is likely that another example of an ongoing process is that present in the cells of patients with Bloom's Syndrome, a disease characterized by genetic instability. New chromosomal abnormalities are constantly being generated in cells from Bloom's Syndrome patients. In the presence of anticarcinogenic protease inhibitors, the levels of chromosome abnormalities become like those observed in normal human diploid cells, suggesting that protease inhibitors are capable of turning off the process leading to genetic instability (31). Our mechanistic studies on anticarcinogenic protease inhibitors are reviewed in greater detail elsewhere (12,18,21-25,29).

As discussed above, there are several potential mechanisms of action for the anticarcinogenic protease inhibitors. All of the effects we have observed, however, are related to the cellular alterations brought about by carcinogen exposure. Protease inhibitor treatment reverses the carcinogen-induced changes without affecting the normal levels of any of the phenomena studied (e.g., c-myc expression in the colon [35], Boc-val-pro-arg-MCA hydrolyzing activity in the oral epithelium

[33], gene amplification [16], etc.). While the endpoints that were measured may not be the critical one induced by carcinogens that leads to cancer, it is assumed that the anticarcinogenic protease inhibitors are having a similar effect on the carcinogen-produced alteration (or alterations) that is directly responsible for carcinogenesis.

ACKNOWLEDGEMENTS

The work on protease inhibitors discussed here has been supported by National Institutes of Health grants CA 22704 and CA 46496.

REFERENCES

1. Baturay, N.Z., and A.R. Kennedy (1986) Pyrene acts as a cocarcinogen with the carcinogens, benzo(a)pyrene, β-propiolactone and radiation in the induction of malignant transformation of cultured mouse fibroblasts; soybean extract containing the Bowman-Birk inhibitor acts as an anticarcinogen. Cell Biol. Toxicol. 2:21-32.

2. Billings, P.C., J.M. Habres, and A.R. Kennedy (1990) Inhibition of radiation-induced transformation of C3H/10T1/2 cells by specific protease substrates. Carcinogenesis 11:329-332.

3. Billings, P.C., P. Newberne, and A.R. Kennedy (1990) Protease inhibitor suppression of colon and anal gland carcinogenesis induced by dimethylhydrazine. Carcinogenesis 11:1083-1086.

4. Billings, P.C., W. St. Clair, A.J. Owen, and A.R. Kennedy (1988) Potential intracellular target proteins of the anticarcinogenic Bowman-Birk protease inhibitor identified by affinity chromatography. Cancer Res. 48:1798-1802.

5. Billings, P.C., A.R. Morrow, C.A. Ryan, and A.R. Kennedy (1989) Inhibition of radiation-induced transformation of C3H/10T1/2 cells by carboxypeptidase Inhibitor I and Inhibitor II from potatoes. Carcinogenesis 10:687-691.

6. Billings, P.C., J.M. Habres, W.C. Shen, and A.R. Kennedy (1990) Inhibition of a 45 kilodalton protease in C3H/10T1/2 cells by the anticarcinogenic Bowman-Birk protease inhibitor. Proc. Am. Assoc. Cancer Res. (Abstract #934) 31:157.

7. Billings, P.C., J.M. Habres, D.C. Liao, and S.W. Tuttle (in press) A protease activity in human fibroblasts which is inhibited by the anticarcinogenic Bowman-Birk protease inhibitor. Proc. Natl. Acad. Sci., USA.

8   Billings, P.C., J.A. Carew, C.E. Keller-McGandy, A. Goldberg, and A.R. Kennedy (1987) A serine protease activity in C3H/10T1/2 cells that is inhibited by anticarcinogenic protease inhibitors. Proc. Natl. Acad. Sci., USA 84:4801-4805.

9. Caggana, M., and A.R. Kennedy (1989) C-fos mRNA levels are reduced in the presence of antipain and the Bowman-Birk inhibitor. Carcinogenesis 10:2145-2148.

10. Carew, J.A., and A.R. Kennedy (1990) Identification of a proteolytic activity which responds to anticarcinogenic protease inhibitors in C3H/10T1/2 cells. Cancer Lett. 49:153-163.

11.   Chang, J.D., and A.R. Kennedy (1988) Cell cycle progression of C3H/10T1/2 and 3T3 cells in the absence of a transient increase in c-myc RNA levels. Carcinogenesis 9:17-20.12.

12.   Chang, J.D., and A.R. Kennedy (in press) Suppression of c-myc by anticarcinogenic protease inhibitors. In Protease Inhibitors as Cancer Chemopreventive Agents, Walter Troll and Ann R. Kennedy, eds. Plenum press, New York.

13.   Chang, J.D., P. Billings, and A.R. Kennedy (1985) C-myc expression is reduced in antipain-treated proliferating C3H/10T1/2 cells. Biochem Biophys. Res. Comm. 133:830-835.

14.   Chang, J.D., J.-H. Li, P.C. Billings, and A.R. Kennedy (1990) Effects of protease inhibitors on c-myc expression in normal and transformed C3H/10T1/2 cells. Molec. Carcin. 3:226-232.

15.   Fabre, F., and H. Roman (1977) Genetic evidence for inducibility of recombination competence in yeast. Proc. Natl. Acad. Sci., USA 74:1667-1671.

16.   Flick, M.B., and A.R. Kennedy (1991) Effect of protease inhibitors on DNA amplification in SV40-transformed Chinese hamster embryo cells. Cancer Lett. 56:102-108.

17.   Kennedy, A.R. (1982) Antipain, but not cycloheximide, suppresses radiation transformation when present for only one day at five days post-irradiation. Carcinogenesis 3:1093-1095.

18.   Kennedy, A.R. (1984) Promotion and other interactions between agents in the induction of transformation in vitro in fibroblasts. In Mechanisms of Tumor Promotion. Vol. III. "Tumor Promotion and Carcinogenesis In Vitro," T.J. Slaga, ed. CRC Press, pp. 13 -55.

19.   Kennedy, A.R. (1985) Evidence that the first step leading to carcinogen-induced malignant transformation is a high frequency, common event. In Carcinogenesis: A Comprehensive Survey, Volume 9: Mammalian Cell Transformation: Mechanisms of Carcinogenesis and Assays for Carcinogens, J.C. Barrett and R.W. Tennant, eds. Raven Press, New York, pp. 355-364.

20.   Kennedy, A.R. (1985) The conditions for the modification of radiation transformation in vitro by a tumor promoter and protease inhibitors. Carcinogenesis 6:1441-1446.

21.   Kennedy, A.R. (1988) Implications for mechanisms of tumor promotion and its inhibition by various agents from studies of in vitro transformation. In Tumor Promoters, Biological Approaches for Mechanistic Studies and Assay Systems, R. Langenbach, J.C. Barrett, and E. Elmore, eds. Raven Press, New York, pp. 201-212.

22.   Kennedy, A.R. (1990) Effects of protease inhibitors and vitamin E in the prevention of cancer. In Nutrients and Cancer Prevention, K.N. Prasad and F.L. Meyskens, Jr., eds. The Humana Press, Inc., Clifton, New Jersey, pp. 79-98.

23.   Kennedy, A.R. (in press) In vitro studies of anticarcinogenic protease inhibitors. In Protease Inhibitors as Cancer Chemopreventive Agents, Walter Troll and Ann R. Kennedy, eds. Plenum Press, New York.

24. Kennedy, A.R. (in press) Overview: Anticarcinogenic activity of protease inhibitors. In Protease Inhibitors as Cancer Chemopreventive Agents, Walter Troll and Ann R. Kennedy, eds. Plenum Press, New York.

25. Kennedy, A.R. (in press) Is there a critical target gene for the first step in carcinogenesis? Env. Health Persp.

26. Kennedy, A.R., and J.B. Little (1980) Investigation of the mechanism for the enhancement of radiation transformation in vitro by TPA. Carcinogenesis 1:1039-1047.

27. Kennedy, A.R., and J.B. Little (1981) High efficiency, kinetics and numerology of transformation by radiation in vitro. In Cancer: Achievements, Challenges and Prospects for the 1980's, Vol. 1, J.H. Burchenal and J.F. Oettgen, eds. Grune and Stratton, pp. 491-500.

28. Kennedy, A.R., and J.B. Little (1984) Evidence that a second event in X-ray-induced oncogenic transformation in vitro occurs during cellular proliferation. Rad. Res. 99:228-248.

29. Kennedy, A.R., and P.C. Billings (1987) Anticarcinogenic actions of protease inhibitors. In Anticarcinogenesis and Radiation Protection, P. A. Cerutti, O.F. Nygaard, and M.G. Simic, eds. Plenum Press, New York, pp. 285-295.

30. Kennedy, A.R., J. Cairns, and J.B. Little (1984) Timing of the steps in transformation of C3H/10T1/2 cells by X-irradiation. Nature 307:85-86.

31. Kennedy, A.R., B. Radner, and H. Nagasawa (1984) Protease inhibitors reduce the frequency of spontaneous chromosome abnormalities in cells from patients with Bloom syndrome. Proc. Natl. Acad. Sci., USA 81:1827-1830.

32. Kennedy, A.R., M. Fox, G. Murphy, and J.B. Little (1980) Relationship between X-ray exposure and malignant transformation in C3H 10Tl/2 cells. Proc. Natl. Acad. Sci., USA 77:7262-7266. (This article has been reproduced in: Readings in Mammalian Cell Culture, 2nd Edition, R. Pollack, ed., pp. 235-239, Cold Spring Harbor Laboratory, 1981.)

33. Messadi, P.V., P. Billings, G. Shklar, and A.R. Kennedy (1986) Inhibition of oral carcinogenesis by a protease inhibitor. J. Natl. Cancer Inst. 76:447-452.

34. Shen, W.-C., J. Wan, and D. Shen (1990) Proteolytic processing in a non-lysosomal compartment is required for transcytosis of protein-polylysine conjugates in cultured Madin-Darby canine kidney cells. Biochem. Biophys. Res. Comm. 166:316-323.

35. St. Clair, W.H., P.C. Billings, and A.R. Kennedy (1990) The effects of the Bowman-Birk protease inhibitor on c-myc expression and cell proliferation in the unirradiated and irradiated mouse colon. Cancer Lett. 52:145-152.

36. St. Clair, W., P. Billings, J. Carew, C. Keller-McGandy, P. Newberne, and A.R. Kennedy (1990) Suppression of DMH-induced carcinogenesis in mice by dietary addition of the Bowman-Birk protease inhibitor. Cancer Res. 50:580-586.

37. Von Hofe, E., P.M. Newberne, and A.R. Kennedy (in press). Inhibition of N-nitrosomethylbenzylamine-induced esophageal neoplasms by the Bowman-Birk protease inhibitor. Carcinogenesis.

38. Weed, H., R.B. McGandy, and A.R. Kennedy (1985) Protection against dimethylhydrazine-induced adenomatous tumors of the mouse colon by the dietary addition of an extract of soybeans containing the Bowman-Birk protease inhibitor. Carcinogenesis 6:1239-1241.

39.   Wintersberger, U. (1984) The selective advantage of cancer cells: A consequence of genome mobilization in the course of the induction of DNA repair processes?   (Model studies of yeast.)   In <u>Advances in Enzyme Regulation Volume 22</u>, Pergamon, Oxford, pp. 311-323.

40.   Witschi, H., and A.R. Kennedy (1989) Modulation of lung tumor development in mice with the soybean-derived Bowman-Birk protease inhibitor. <u>Carcinogenesis</u> 10:2275-2277.

41.   Yavelow, J., T.H. Finlay, A.R. Kennedy, and W. Troll (1983) Bowman-Birk soybean protease inhibitor as an anticarcinogen. <u>Cancer Res.</u> 43:2454-2459.

42.   Yavelow, J., M. Collins, Y. Birk, W. Troll, and A.R. Kennedy (1985) Nanomolar concentrations of Bowman-Birk soybean protease inhibitor suppress X-ray-induced transformation <u>in vitro</u>. <u>Proc Natl. Acad. Sci., USA</u> 82:5395-5399.

cDNA CLONING OF SERINE/THREONINE PHOSPHATASE

CATALYTIC SUBUNITS AND REVERSION OF THE MALIGNANT

PHENOTYPE TO THE NORMAL PHENOTYPE BY OKADAIC ACID,

A PROTEIN PHOSPHATASE INHIBITOR

Minako Nagao, Hiroshi Shima, Kazunori Sasaki, Yukihito Ishizaka,
Michie Nakayasu, and Takashi Sugimura

Carcinogenesis Division
National Cancer Center Research Institute
1-1, Tsukiji 5-chome, Chuo-ku, Tokyo 104, JAPAN

INTRODUCTION

Most oncogene products are protein kinases, and protein phosphatases are also considered to be involved in carcinogenesis. Phosphotyrosine phosphatases (PTPases) and serine/threonine phosphatases (PPases) seem to be involved in signal transduction for malignant transformation. Both PTPases and PPases are fairly large families of enzymes. PTPases can be classified into receptor types and nonreceptor types (6). However, so far, there are no receptor type PPases. PPases can be classified into PP-1 and PP-2 types, depending on their sensitivities to the protein inhibitors I-1 and I-2; PP-1 types being mainly present in the particulate fractions of cells, and PP-2 types in the soluble fraction (3).

Growth factors mainly activate receptor-type tryosine kinases that phosphorylate themselves, and their signals are transduced to downstream extracellular signal-regulated kinases (ERKs) and then serine/threonine kinases (2). Serine/threonine phosphatases also seem to be involved in the downstream signal transduction. We focused attention on PP-1 and PP-2A, a subtype of PP-2 that does not require any metal ion for activity. PP-1 and PP-2A are both composed of a catalytic subunit and regulatory subunits. We have cloned cDNAs for four different PP-1 catalytic subunits (13) and two different PP-2A catalytic subunits (8,9), and analyzed the levels of expression of the mRNAs for these catalytic subunits.

Okadaic acid is a specific and very potent inhibitor of PP-1 and PP-2A (1), and has been found to be a promoter in mouse skin carcinogenesis (14). Therefore, we used it to study the role of protein phosphatases in malignant transformation. We expected that okadaic acid would promote cells to be more malignant, however, the results were the opposite.

RESULTS

cDNA Cloning of PPase Catalytic Subunits

We have screened rat liver and testis cDNA libraries for PP-1 catalytic subunits using *dis2m*1 and *dis2m*2 mouse cDNAs (10) encoding two different PP-1 catalytic subunits as probes, and we have cloned four cDNAs for PP-1 catalytic subunits, under conditions of high and low stringencies (13). These cDNAs were named *PP-1α, PP-1γ1, PP-1γ2*, and *PP-1δ. PP-1γ2* was isolated from the testis library and the other cDNAs were from the liver library. We found that the protein encoded by *PP-1α* was identical with that of rabbit *PP-1α*, the *PP-1γ2* protein was identical with the mouse *dis2m*1 protein, and the *PP-1δ* protein was identical with the mouse *dis2m*2 protein. *PP-1γ1* and *PP-1γ2* mRNAs were transcribed from the same gene by alternative splicing and the carboxy-terminal regions of the proteins encoded by these cDNAs were different (13). The amino acid sequences encoded by these four cDNAs are shown in Fig. 1. Each member of this family has a consecutive unique amino acid sequence in the carboxy-terminal regions of the proteins encoded by these cDNAs were different (13). The amino acid sequences encoded by these four cDNAs are shown in Fig. 1. Each member of this family has a consecutive unique amino acid sequence in the carboxy-terminal region, suggesting that the carboxy-terminal region of each catalytic subunit defines the specificity by forming a complex with specific regulatory subunits that confine substrates by a targeting mechanism (3).

cDNAs for two PP-2A catalytic subunits were cloned from a rat liver cDNA library and named *PP-2Aα* (9) and *PP-2Aβ* (8). Both cDNAs encode 309 amino acid peptides and these peptides differ in only eight amino acids scattered in their N-terminal halves.

Expression in Normal Tissues and Tumors
Induced by Food-borne Carcinogenesis

Expression levels of mRNA for each of the PP-1 and PP-2A catalytic subunits were determined (8,13). We found that *PP-1α, PP-1γ1, PP-1γ2, PP-1δ, PP-2Aα,* and *PP-2Aβ* were expressed in all the organs we examined, but the expression level of *PP-2Aβ* was low in bone marrow, and that of *PP-1γ2* was exceptionally high in testis and very low in other organs. In rat hepatic tumors induced by the food-borne carcinogens 2-amino-3-methylimidazo[4,5-*f*]quinoline (IQ) and 2-amino-3,8-dimethyl-imidazo[4,5-*f*]quinoxaline (MeIQx), the expression of *PP-2Aα* was very much increased as compared with the normal portions of the livers of those tumor-bearing rats, or the livers of untreated rats (Fig. 2). The expression levels of *PP-1α* and *PP-1δ* were also increased but not as much as *PP-2Aα*.

```
rat PP-1α   M S D S E K L N L D S I I G R L L E V Q G S R P G K N V Q L T E N E I R G L C L K S R E I F L S Q P I L L E L E A P L K I C G D I H G Q Y Y D L L R L F E Y G G
rat PP-1γ1  M A D I D K L N I D S I I Q R L L E V R G S K P G K N V Q L Q E N E I R G L C L K S R E I F L S Q P I L L E L E A P L K I C G D I H G Q Y Y D L L R L F E Y G G
rat PP-1γ2  M A D I D K L N I D S I I Q R L L E V R G S K P G K N V Q L Q E N E I R G L C L K S R E I F L S Q P I L L E L E A P L K I C G D I H G Q Y Y D L L R L F E Y G G
rat PP-1δ   M A D G E - L N V D S L I T R L L E V R G C R P G K I V Q M T E A E V R G L C I K S R E I F L S Q P I L L E L E A P L K I C G D I H G Q Y T D L L R L F E Y G G

rat PP-1α   F P P E S N Y L F L G D Y V D R G K Q S L E T I C L L L A Y K I K Y P E N F F L L R G N H E C A S I N R I Y G F Y D E C K R R Y N I K L W K T F T D C F N C L P
rat PP-1γ1  F P P E S N Y L F L G D Y V D R G K Q S L E T I C L L L A Y K I K Y P E N F F L L R G N H E C A S I N R I Y G F Y D E C K R R Y N I K L W K T F T D C F N C L P
rat PP-1γ2  F P P E S N Y L F L G D Y V D R G K Q S L E T I C L L L A Y K I K Y P E N F F L L R G N H E C A S I N R I Y G F Y D E C K R R Y N I K L W K T F T D C F N C L P
rat PP-1δ   F P P E A N Y L F L G D Y V D R G K Q S L E T I C L L L A Y K I K Y P E N F F L L R G N H E C A S I N R I Y G F Y D E C K R R F N I K L W K T F T D C F N C L P

rat PP-1α   I A A I V D E K I F C C H G G L S P D L Q S M E Q I R R I M R P T D V P D Q G L L C D L L W S D P D K D V Q G W G E N D R G V S F T F G A E V V A K F L H K H D
rat PP-1γ1  I A A I V D E K I F C C H G G L S P D L Q S M E Q I R R I M R P T D V P D Q G L L C D L L W S D P D K D V L G W G E N D R G V S F T F G A E V V A K F L H K H D
rat PP-1γ2  I A A I V D E K I F C C H G G L S P D L Q S M E Q I R R I M R P T D V P D Q G L L C D L L W S D P D K D V L G W G E N D R G V S F T F G A E V V A K F L H K H D
rat PP-1δ   I A A I V D E K I F C C H G G L S P D L Q S M E Q I R R I M R P T D V P D T G L L C D L L W S D P D K D V Q G W G E N D R G V S F T F G A D V V S K F L N R H D

rat PP-1α   L D L I C R A H Q V V E D G Y E F F A K R Q L V T L F S A P N Y C G E F D N A G A M M S V D E T L M C S F Q I L K P A D K N K G K Y G Q F S G L N P G G R P I T
rat PP-1γ1  L D L I C R A H Q V V E D G Y E F F A K R Q L V T L F S A P N Y C G E F D N A G A M M S V D E T L M C S F Q I L K P A E K K K P - - - - - - - - N A T - R P V T
rat PP-1γ2  L D L I C R A H Q V V E D G Y E F F A K R Q L V T L F S A P N Y C G E F D N A G A M M S V D E T L M C S F Q I L K P A E K K K P - - - - - - - - N A T - R P V T
rat PP-1δ   L D L I C R A H Q V V E D G Y E F F A K R Q L V T L F S A P N Y C G E F D N A G G M M S V D E T L M C S F Q I L K P S E K - K A K Y - Q Y G G L N - S G R P V T

rat PP-1α   P P R - - N S A K - A K K
rat PP-1γ1  P P R G M I T - K Q A K K
rat PP-1γ2  P P R V G S G L N P S I Q K A S N Y R N N T V L Y E
rat PP-1δ   P P R - - - T A N P P K K R
```

Figure 1. Amino acid sequences encoded by four cDNAs of the rat PP-1 catalytic subunit family. Amino acids that are conserved in all four are boxed. A unique sequence of each isotype is present in the C-terminal region.

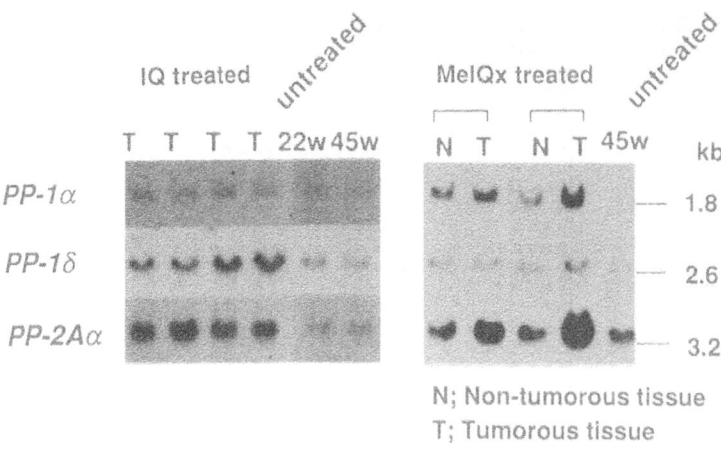

Figure 2. Expressions of *PP-1α*, *PP-1γ*, and *PP-2Aα* mRNAs in F344 rat liver tumors induced by IQ or MeIQx. Northern blot analysis was performed using 10 kg of total RNA. The levels of *PP-2Aα* expression were higher than in untreated rat liver, or nontumorous portions of the tumor-bearing livers.

Involvement of PP-2A in Malignant Cell Transformation

The role protein phosphatase 2A in malignant cell transformation was analyzed using NIH3T3 transformants. Transformants obtained by introduction of activated c-*raf* that encodes a serine/threonine kinase and of *ret*-II that encodes tyrosine kinase were cultured in the presence of 8 ng/ml of okadaic acid. On treatment, these cells became flattened, and showed clear contact inhibition (12), as seen in Fig. 3. The concentrations of okadaic acid used (less than 10 nM) did not affect the enzyme activity of PP-1 but inhibited that of PP-2A, as shown in Fig. 4. The IC_{50} of okadaic acid for PP-2A varies depending upon the PP-2A concentrations, but that for PP-1 is not affected by enzyme concentrations. Okadaic acid did not inhibit cell growth of the transformants during the exponential growth phase, but the growth reached plateau levels at lower cell densities with okadaic acid than without okadaic acid (12).

There are recent reports that polyoma middle T forms a complex with phosphatidylinositol 3-kinase, its regulatory subunit, PP-2A catalytic and regulatory subunits, and PP60[c-src] (4,11,16), and that the middle T antigens of host range mutants that cannot transform cells do not form stable complexes with PP-2A (15), suggesting that PP-2A plays a critical role in malignant transformation. We analyzed the affect of okadaic acid on a polyoma middle T gene (Py*MT*) transformant of NIH3T3 cells. When the transformant was cultured in the presence of 8 ng/ml of okadaic acid, the cells became flattened and show ever clear contact inhibition. In contrast, no contact inhibition was apparent with Ha-*ras* and Ki-*ras* transformants, although they showed slight morphological changes in the presence of 8 ng/ml of okadaic acid. With higher concentrations of okadaic acid, the cells died (12).

Table 1. Affect of okadaic acid on anchorage independent growth.

Cells	Oncogene	Number of Colonies/Plate		
		Okadaic Acid (ng/ml)		
		0	4	8
SIC2	*ret* II	556	63	1
2C-1	*raf*	246	28	0
MT3-4	Py*MT*	69	1	0
al-1	Ha-*ras*	329	270	174
Kr-6	Ki-*ras*	280	224	151
NIH3T3		0	0	0

We also examined the effect of okadaic acid on anchorage-independent growth of these transformants. As shown in Tab. 1, *ret*-II, *raf*, and Py*MT* transformants did not grow in soft agar containing 4 or 8 ng/ml okadaic acid, although in the presence of 8 ng/ml of okadaic acid, Ha-*ras* and Ki-*ras* transformants formed colonies with efficiencies of about 50% of those of untreated transformants.

Down regulation of fibronectin expression is frequently associated with cell transformation. Fibronectin mRNA expression was down regulated in *ret*-II and PY*MT* transformants, but the fibronectin mRNA levels of these transformants were markedly increased 2 and 3 days after adding okadaic acid to the medium. Cell surface fibronectin was also increased in the *ret*-II flat-revertants induced by okadaic acid treatment, as shown by incubating the cells with an antibody against human fibronectin, and fluorescein-labeled secondary antibody (12).

We also analyzed the tyrosine phosphorylation state of a ret^{TPC} transformant (7) before and after cell flattening, using a monoclonal antibody against phosphotyrosine. As expected, the phosphorylation state of the ret^{TPC} transformant was much higher than that of control NIH3T3 cells. However, the phosphorylation state of the flattened cells was also high and no difference was found in the phosphorylation states before and after treatment with okadaic acid. This result suggests that PP-2A is not involved in the tyrosine phosphorylation state. Thus, it seems to be involved downstream of tyrosine kinases.

DISCUSSION

The roles of protein phosphatases in glycogen metabolism have been studied extensively, but cDNA cloning has revealed that the roles of phosphatases are very complicated. The finding that okadaic acid is a potent and specific inhibitor of PP-1 and PP-2A led to the discovery that PP-1 and PP-2A are both involved in various biological activities. Using okadaic acid, we found that PP-2A is involved in malignant cell transformation. Like 10-*O*-tetradecanoyl phorbol 12-acetate (TPA),

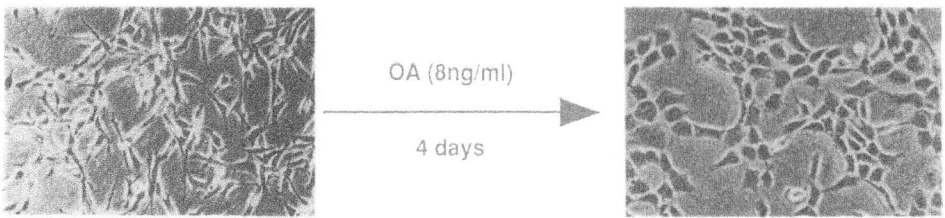

Figure 3. Morphological changes in *ret*-II transformant by treatment with okadaic acid.

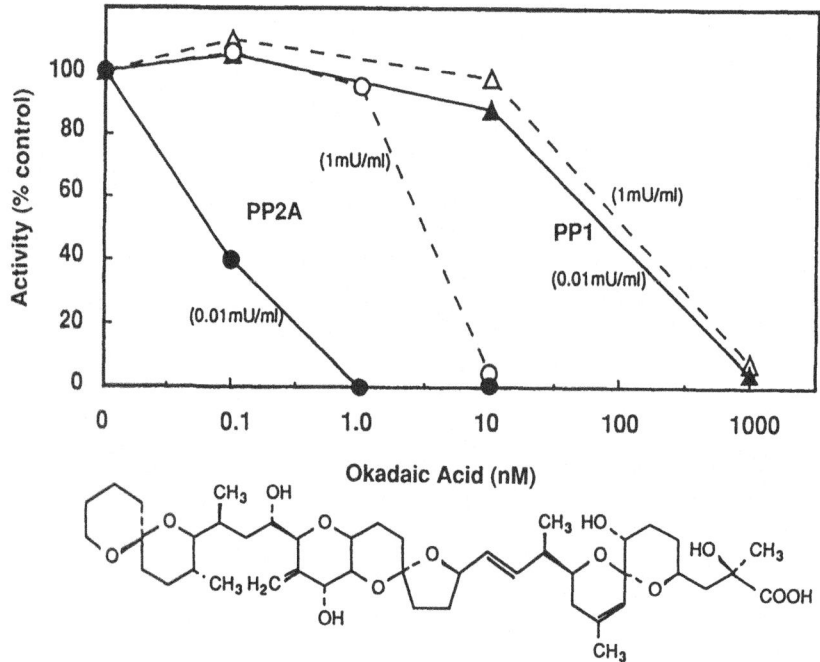

Figure 4. Effect of okadaic acid on purified PP-1 and PP-2A catalytic subunits (upper) and the structure of okadaic acid (below). Catalytic subunits of PP-1 and PP-2A were kindly purified by Drs. Tsuiki and Kikuchi (Tohoku University). Triangles (△ ▲) indicate PP-1 and circles (○ ●) indicate PP-2A. Open symbols indicate 1 mU/ml of the enzyme with phosphorylase *a* as substrate, and closed symbols indicate 0.01 mU/ml.

okadaic acid is a promoter of carcinogenesis in mouse skin, but these two compounds had quite different effects on NIH3T3 cell transformants. TPA changed the cell morphology of the *ret*-II transformants to a more malignant form (unpublished result). Moreover, unlike TPA, okadaic acid did not promote transformation of mouse fibroblast C3H10T1/2 cells in vitro (5).

The reasons for the difference in the effects of okadaic acid on in vitro cell transformation and in vivo skin carcinogenesis would be as follows: 1.) The affect of okadaic acid differs depending on the type of cells. 2.) The role of PP-2A differs depending on the oncogenes involved in malignant transformation. 3.) The application of high-doses of okadaic acid in vivo results in cell death and compensatory cell growth might occur as the consequence of the promoting affect of okadaic acid. In NIH3T3 transformants, PP-2A was suggested to be involved in signal transduction from activated *raf*, *ret*-II, *ret*^{TPC}, and PY*MT*. The effect of okadaic acid was reversible, and on removal of okadaic acid from the culture medium after 7-10 days, the flat cells changed to their transformed morphology within 7-10 days. The in vivo tumor forming ability of the flattened cell remains to be investigated. the effect of prolonged treatment with okadaic acid on the reversibility of its effect is now being investigated.

REFERENCES

1. Bialojan, C., and A. Takai (1988) Inhibitory effect of a marine-sponge toxin, okadaic acid, on protein phosphatases. *Biochem. J.* 256:283-290.
2. Boulton, T.G., S.H. Nye, D.J. Robbins, N.Y. Ip, E. Radziejewska, S.D. Morgenbesser, R.A. DePinho, N. Panayotatos, M.H. Cobb, and G.D. Yancopoulos (1991) ERKS: A family of protein-serine/threonine kinases that are activated and tyrosine phosphorylated in response to insulin and NGF. *Cell* 65:663-675.
3. Cohen, P. (1989) The structure and regulation of protein phosphatases. *Ann. Rev. Biochem.* 58:453-508.
4. Escobedo, J.A., S. Navankasattusas, W.M. Kavanaugh, D. Milfay, V.A. Fried, and L.T. Williams (1991) cDNA cloning of a novel 85 kd protein that has SH2 domains and regulates binding of PI3-kinase to the PDGF β-receptor. *Cell* 65:75-82.
5. Herschman, H.R., R.W. Lim, D.W. Brankow, and H. Fujiki (1989) The tumor promoters 12-*O*-tetradecanoylphorbol-13-acetate and okadaic acid differ in toxicity, mitogenic activity and induction of gene expression. *Carcinogenesis* 10:1495-1498.
6. Hunter, T. (1989) Protein-tyrosine phosphatases: The other side of the coin. *Cell* 58:1013-1016.
7. Ishizaka, Y., T. Ushijima, T. Sugimura, and M. Nagao (1990) cDNA cloning and characterization of *ret* activated in a human papillary thyroid carcinoma cell line. *Biochem. Biophys. Res. Comm.* 168:402-408.
8. Kitagawa, Y., R. Sakai, T. Tahira, H. Tsuda, N. Ito, T. Sugimura, and M. Nagao (1988) Molecular cloning of rat phosphoprotein phosphatase 2Aβ cDNA and increased expressions of phosphatase 2Aα and 2Aβ in rat liver tumors. *Biochem. Biophys. Res. Comm.* 157:821-827.
9. Kitagawa, Y., T. Taira, I. Ikeda, K. Kikuchi, S. Tsuiki, T. Sugimura, and M. Nagao (1988) Molecular cloning of cDNA for the catalytic subunit of rat liver

type 2A protein phosphatase, and detection of high levels of expression of the gene in normal and cancer cells. *Biochim. Biophys. Acta* 951:123-129.

10. Ohkura, H., N. Kinoshita, S. Miyatani, T. Toda, and M. Yanagida (1989) The fission yeast *dis2*[+] gene required for chromosome disjoining encodes one of two putative type 1 protein phosphatases. *Cell* 57:997-1007.

11. Pallas, D.C., K.K. Shahrik, B.L. Martin, S. Jaspers, T.B. Miller, D.L. Brautigan, and T.M. Roberts (1990) Polyoma small and middle T antigens and SV40 small t antigen form stable complexes with protein phosphatase 2A. *Cell* 60:167-176.

12. Sakai, R., I. Ikeda, H. Kitani, H. Fujiki, F. Takaku, U. Rapp, T. Sugimura, and M. Nagao (1989) Flat reversion by okadaic acid of *raf* and *ret*-II transformants. *Proc. Natl. Acad. Sci., USA* 86:9946-9950.

13. Sasaki K., H. Shima, Y. Kitagawa, S. Irino, T. Sugimura, and M. Nagao (1990) Identification of members of the protein phosphatase 1 gene family in the rat and enhanced expression of protein phosphatase 1*a* gene in rat hepatocellular carcinomas. *Japan. J. Cancer Res.* 81:1272-1280.

14. Suganuma, M. H. Fujiki, H. Suguri, S. Yoshizawa, M. Hirota, M. Nakayasu, M. Ojika, K. Wakamatsu, K. Yamada, and T. Sugimura (1988) Okadaic acid: An additional nonphorbol-12-O-tetradecanoate-13-acetate-type tumor promoter. *Proc. Natl. Acad. Sci., USA* 85:1768-1771.

15. Walter, G., A. Carbone, and W.J. Welch (1987) Medium tumor antigen of polyomavirus transformation-defective mutant NG59 is associated with 73-kilodalton heat shock protein. *J. Virol.* 61:405-410.

16. Walter, G., R. Ruediger, C. Slaughter, and M. Mumby (1990) Association of protein phosphatase 2A with polyoma virus medium tumor antigen. *Proc. Natl. Acad. Sci., USA* 87:2521-2525.

MECHANISMS OF ACTION OF NEW ANTITUMOR PROMOTERS

M. Suganuma,[1] S. Yoshizawa,[1] J. Yatsunami,[1] S. Nishiwaki,[1]
H. Furuya,[1]   S. Okabe,[1] R. Nishiwaki-Matsushima,[1] K. Frenkel,[2]
W. Troll,[2] A.K. Verma,[3] and H. Fujiki[1]

[1]Cancer Prevention Division
National Cancer Center Research Institute
Tsukiji 5-1-1, Chuo-ku, Tokyo 104, Japan

[2]New York University Medical Center
New York, New York 10016 USA

[3]University of Wisconsin
Madison, Wisconsin 53792 USA

INTRODUCTION

Sarcophytol A, isolated from the soft coral *Sarcophyton glaucum*, and (-)-epigallocatechin gallate (EGCG), isolated from Japanese green tea leaves, had inhibitory effects on tumor promotion by teleocidin, a tumor promoter of the 12-0-tetra-decanoylphorbol-13-acetate (TPA)-type on mouse skin initiated with 7,12-dimethylbenz(a)anthracene (DMBA) in two-stage carcinogenesis experiments (4,5,17). We called them antitumor promoters. Later it turned out that these antitumor promoters also have inhibitory effects on chemical carcinogenesis in various organs (9,13). Moreover, sarcophytol A and EGCG inhibited tumor promotion by okadaic acid, a tumor promoter of the non-TPA type, on mouse skin (7,8). As for the mechanisms of action of tumor promoters, teleocidin activates protein kinase C, whereas okadaic acid inhibits activities of protein phosphatases 1 and 2A (3,6). Since these two antitumor promoters similarly inhibited tumor promotion of the TPA and non-TPA types, it is of interest to study the mechanisms of action of these compounds at the biochemical level.

In addition, we have recently demonstrated that cryptoporic acid E inhibited tumor promotion of okadaic acid on mouse skin (12). However, a derivative of cryptoporic acid E, cryptoporic acid D, did not inhibit tumor promotion, although both compounds are potent inhibitors of superoxide anion radical release (12).

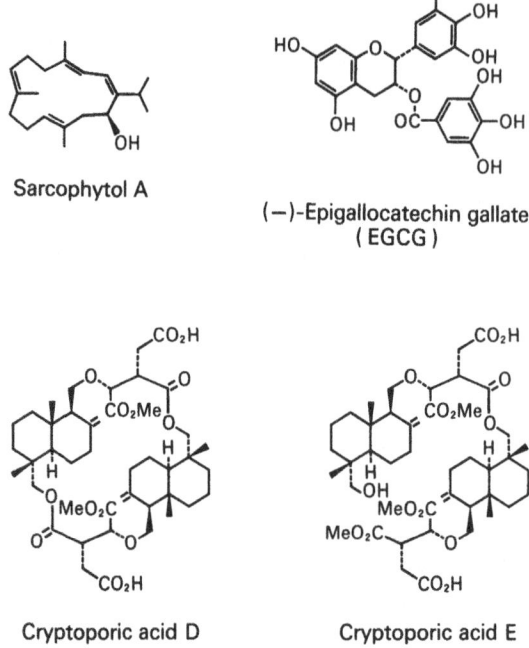

Sarcophytol A

(−)-Epigallocatechin gallate
(EGCG)

Cryptoporic acid D   Cryptoporic acid E

Figure 1. Structures of sarcophytol A, EGCG, and cryptoporic acids D and E.

In this paper, the studies on sarcophytol A and EGCG are reviewed with emphasis on mechanisms of their action. The differential effects of both cryptoporic acids D and E on tumor promotion are also discussed.

SARCOPHYTOL A

Sarcophytol A, chemically a member of the cembrane-type diterpenes (Fig. 1) (11), is a potent inhibitor of tumor promotion by teleocidin on mouse skin. This inhibition was achieved by application of 1.6 μg (5.5 nmol) of sarcophytol A in a ratio of 1:1 to a tumor promoter, teleocidin, as described previously (4). Sarcophytol A, within a range of micrograms, was also able to inhibit tumor promotion by okadaic acid. In this experiment, initiation with a single application of 100 μg DMBA to the skin of the backs of CD-1 mice was followed by repeated applications of 1 μg (1.2 nmol) of okadaic acid until week 20. Treatment with sarcophytol A at a dose of 3.5 μg (12.0 nmol) per application, prior to okadaic acid, reduced percentages of tumor-bearing mice from 80.0% to 46.7% in week 20 and the average numbers of tumors per mouse from 2.5 to 1.5. Sarcophytol A delayed the onset of tumor formation for three weeks. Therefore, the inhibitory effect of sarcophytol A was apparent in the early stages of tumor development induced by okadaic acid as well as teleocidin (7).

In order to obtain a more effective inhibition of tumor promotion, we thought that some other compounds could be used with sarcophytol A. Angiogenesis is induced in a tumor with a diameter of a few millimeters, and growth of the tumor is partly dependent on the angiogenesis factor (1). We studied the effect of co-treatment of sarcophytol A with an angiogenesis inhibitor, medroxyprogesterone

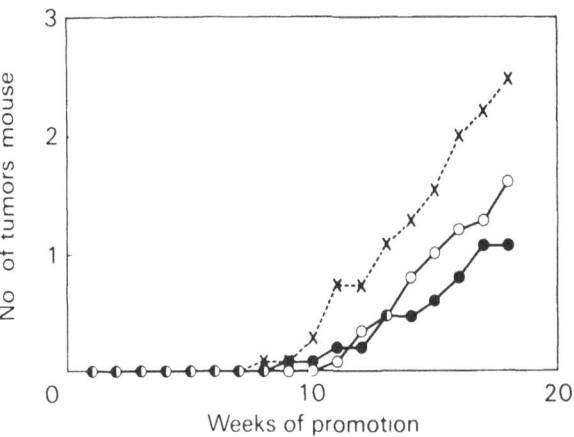

Figure 2. Inhibitory effects of co-treatment of sarcophytol A with MPA on tumor promotion by okadaic acid. Groups treated with DMBA and okadaic acid (×); with DMBA, okadaic acid, and sarcophytol A (o); and with DMBA, okadaic acid, and sarcophytol A plus MPA (•).

acetate (MPA), an angiostatic steroid (15), on tumor promotion of okadaic acid. Doses of 3.5 μg (12 nmol) sarcophytol A and 46.4 μg (120 nmol) MPA were applied simultaneously 15 min before each treatment with 1 μg (1.2 nmol) of okacdaic acid. Co-treatment was slightly more effective in inhibition of tumor promotion than without MPA (Fig. 2). The group treated with MPA alone did not show any significant inhibition (data not shown). Although the inhibition by sarcophytol A and MPA was not dramatically stronger, an increasing inhibition with co-treatment was observed. This result shows the possibility of obtaining an enhanced effect using other angiogenesis inhibitors.

Since sarcophytol A within a microgram range had inhibited tumor promotion by teleocidin, we thought sarcophytol A might inhibit the specific binding of teleocidin to its receptor. However, sarcophytol A did not share the receptor with teleocidin. Rather, we found that [^{3}H]sarcophytol A binds to various proteins, such as albumin, with various affinities (Verma et al., manuscript in preparation). Taken together, sarcophytol A directly enters the cells, but not through its own receptor, binds to various proteins in the cells, and then probably inhibits important mechanisms of tumor promotion through an undefined pathway, which should be further studied. Sarcophytol A is the first possible cancer chemopreventive agent derived from marine organisms.

EGCG

EGCG is the main constituent of polyphenolic compounds of Japanese green tea leaves (Fig. 1). As we reported previously, EGCG inhibits tumor promotion of teleocidin on mouse skin and chemical carcinogenesis in various organs (5,8,9,17).

Recently we found that treatment with 5 mg EGCG per application, prior to okadaic acid, completely inhibited tumor promotion of okadaic acid in a two-stage carcinogenesis experiment on mouse skin (Yoshizawa, et al., manuscript in preparation). The inhibitory effect of EGCG was more potent on tumor promotion by okadaic acid than that by teleocidin.

EGCG inhibited tumor promotion by both okadaic acid and teleocidin, which act by different mechanisms, on mouse skin. It is important to study how EGCG inhibits tumor promotion on mouse skin. We examined the direct interaction of EGCG with two different receptors of tumor promoters. Application of EGCG to mouse skin immediately caused a reduction of the specific binding of [^{3}H]okadaic acid to a particulate fraction as well as that of [^{3}H]TPA (Ref. 17, Yoshizawa et al., manuscript in preparation). Their specific bindings returned to normal levels after several hours. The evidence indicates that the tumor promoters okadaic acid and teleocidin had been applied to the skin when their specific binding was most reduced. The reduction of specific binding of the tumor promoters to their receptors might be induced by the astringency of EGCG, which we call "sealing of the cell membrane by EGCG" (8). In addition to the sealing effect of the cell membrane, we think that EGCG might have additional biochemical effects that inhibit carcinogenesis of the duodenum and the liver (Ref. 9, Muto et al., manuscript in preparation) by interrupting the interaction of various growth factors and hormones with their receptors in the membrane, subsequently resulting in an inhibition of tumor growth.

Recently, we (K.F. and W.T.) found that EGCG is a potent inhibitor of H_2O_2 formation by TPA-activated human polymorphonuclear leukocytes (manuscript in preparation). These results indicated that EGCG might partly act by suppressing oxyradical formation, in addition to the sealing of the cell membrane.

CRYPTOPORIC ACID E

Cryptoporic acid E is a dimer of drimane sesquiterpene isocitric acid, isolated from the fungus *Cryptoporus volvatus* (10). Cryptoporic acid E lacks macrocyclic conformation but has a hydroxyl group, whereas cryptoporic acid D is a macrocyclic compound (Fig. 1). The inhibition of the release of the superoxide anion radical from guinea-pig peritoneal macrophages induced by N-formyl-methionyl-leucyl-phenylalanine by cryptoporic acids D and E was studied. They were potent inhibitors (10). We first found that pretreatment with 10 mg cryptoporic acids D or E completely inhibited induction of ornithine decarboxylase (ODC) by 20 µg okadaic acid in mouse skin. Since formation of the superoxide anion radical and induction of ODC are thought to be closely associated with tumor promotion (2,14,16), the inhibitory effects of cryptoporic acids on tumor promotion in mouse skin were studied.

One week after initiation with 100 µg DMBA, 1 µg (1.2 nmol) of okadaic acid was applied to the skin of mice, twice a week until week 20. Five mg of either cryptoporic acid D or E were applied 10 min before each application of okadaic acid. Cryptoporic acid E inhibited tumor promotion; the percentages of tumor-bearing mice were reduced from 73.3% to 20.0% and the average numbers of tumors per mouse from 4.2 to 0.5 (Fig. 3A). However, treatment with cryptoporic acid D slightly enhanced rather than inhibited tumor promotion (Fig. 3B), and resulted in 93.3% tumor-bearing mice and 3.9 as the average number of tumors per mouse in week 20

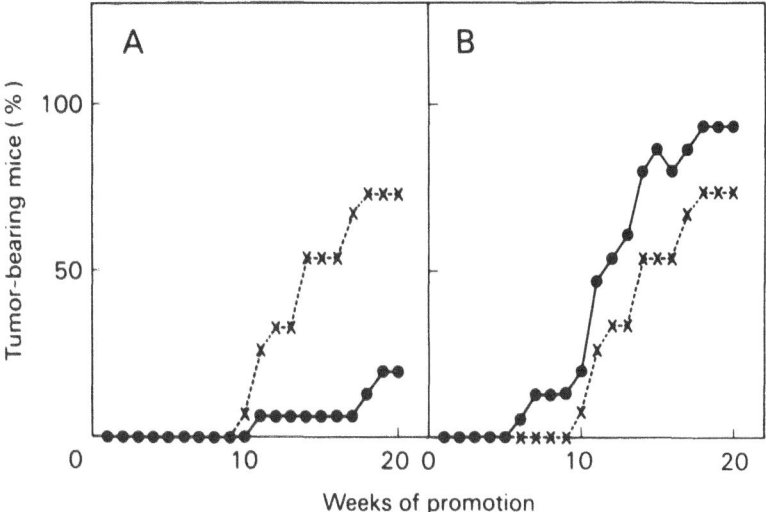

Figure 3. Inhibitory effect of cryptoporic acid E (A) compared with enhancing effect of cryptoporic acid D (B) on tumor promotion by okadaic acid. (A) Groups treated with DMBA and okadaic acid (X) and DMBA and okadaic acid plus cryptoporic acid E (●). (B) Groups treated with DMBA and okadaic acid (X), and DMBA and okadaic acid plus cryptoporic acid D (●).

(12). Thus, cryptoporic acid E is a new antitumor promoter. Although both compounds inhibited superoxide anion radical release and induction of ODC, cryptoporic acids D and E showed differential effects on tumor promotion of okadaic acid.

Since the reasons for this difference were not clear, we studied the effects of cryptoporic acids D and E on activation of protein kinases. Protein kinase C was eluted with 0.07 M NaCl, and another fraction of protein kinases was eluted with 0.2 M NaCl from DEAE-cellulose column chromatography of a mouse brain cytosolic fraction. Cryptoporic acid D activated protein kinase C in the presence of phosphatidylserine and Ca++ ion at a concentration of 30 μM, the potency of which is comparable to that of 100 nM teleocidin, and, moreover, stimulated phosphorylation of histone H1 by other protein kinases at a concentration of 100 μM. Thus, cryptoporic acid D directly activated protein kinase C as well as other protein kinases (12). In contrast, cryptoporic acid E did not significantly activate protein kinase C or other protein kinases. We now understand that cryptoporic acid D, due to its macrocyclic nature, has additional biochemical activities which are not present in cryptoporic acid E. From these results, cryptoporic acid D per se might have a tumor promoting activity. Cryptoporic acids provide evidence that the mechanisms of antitumor promotion are complicated. Further study on cryptoporic acids is anticipated.

CONCLUSION

At present, various compounds are found to inhibit tumor promotion. The antitumor promoters are structurally diverse and their mechanisms of action are heterogeneous. This article reviewed the study of sarcophytol A, EGCG, and cryptoporic acid E, which were identified as new antitumor promoters at our laboratory in Tokyo. The study of antitumor promoters is significant to cancer research for understanding the mechanism of tumor promotion by inhibitors, and for developing agents of cancer chemoprevention in humans. Collaborative studies with clinicians would seem to be very important.

ACKNOWLEDGEMENTS

This work was supported in part by Grants-in-Aid for Cancer Research from the Ministry of Education, Science and Culture, a grant for the Program for a Comprehensive 10-Year Strategy for Cancer Control from the Ministry of Health and Welfare of Japan, and by grants from the Foundation for Promotion of Cancer Research, the Uehara Memorial Life Science Foundation, and the Princess Takamatsu Cancer Research Fund. J. Yatsunami and A.K. Verma thank the Foundation for Promotion of Cancer Research, Japan, for support of their work at the National Cancer Center Research Institute, Tokyo. We would like to thank Dr. T. Sugimura, National Cancer Center, for his encouragement of the work.

REFERENCES

1. Folkman, J. (1985) Tumor angiogenesis. Adv. Cancer Res. 43:175-203.
2. Fujiki, H., and T. Sugimura (1987) New classes of tumor promoters: Teleocidin, aplysiatoxin, and palytoxin. Adv. Cancer Res. 49:223-264.
3. Fujiki, H., M. Suganuma, and T. Sugimura (1989) Significance of new environmental tumor promoters. Env. Carc. Revs. (J. Env. Sci. H.) C7(1):1-51.
4. Fujiki, H., M. Suganuma, H. Suguri, S. Yoshizawa, K. Takagi, and M. Kobayashi (1989) Sarcophytols A and B inhibit tumor promotion by teleocidin in two-stage carcinogenesis in mouse skin. J. Cancer Res. Clin. Oncol. 115:25-28.
5. Fujiki, H., M. Suganuma, H. Suguri, K. Takagi, S. Yoshizawa, A. Ootsuyama, H. Tanooka, T. Okuda, M. Kobayashi, and T. Sugimura (1990) New antitumor promoters: (-)-Epigallocatechin gallate and sarcophytols A and B. In Antimutagenesis and Anticarcinogenesis Mechanisms II, Y. Kuroda, D.M. Shankel, and M.D. Waters, eds. Plenum Publishing Corp., pp. 205-212.
6. Fujiki, H., M. Suganuma, S. Nishiwaki, S. Yoshizawa, J. Yatsunami, R. Matsushima, H. Furuya, S. Okabe, S. Matsunaga, and T. Sugimura (1992) Specific mechanistic aspects of animal tumor promoters: The okadaic acid pathway. In Relevance of Animal Studies to Evaluation of Human Cancer Risk, R. D'Amato, T.J. Slaga, W. Farland, and C. Henry, eds. John Wiley & Sons, Inc., New York, pp. 337-350.
7. Fujiki, H., M. Suganuma, S. Yoshizawa, J. Yatsunami, S. Nishiwaki, H. Furuya, S. Okabe, R.N. Matsushima, S. Matsunaga, Y. Muto, T. Okuda, and T. Sugimura (1992) Sarcophytol A and (-)-epigallocatechin gallate (EGCG), non-toxic inhibitors of cancer development. In Workshop on Cancer Chemoprevention.
8. Fujiki, H., S. Yoshizawa, T. Horiuchi, M. Suganuma, J. Yatsunami, S. Nishiwaki, S. Okabe, R.N. Matsushima, T. Okuda, and T. Sugimura (1992) Anticarcinogenic effects of (-)-epigallocatechin gallate (EGCG). Preventive Medicine.

9. Fujita, Y., T. Yamane, M. Tanaka, K. Kuwata, J. Okuzumi, T. Takahashi, H. Fujiki, and T. Okuda (1989) Inhibitory effect of (-)-epigallocatechin gallate on carcinogenesis with N-ethyl-N'-nitro-N-nitrosoguanidine in mouse duodenum. Jpn. J. Cancer Res. 80:503-505.

10. Hashimoto, T., M. Tori, Y. Mizuno, Y. Asakawa, and Y. Fukazawa (1989) The superoxide release inhibitors, cryptoporic acids C, D, and E: Dimeric drimane sesquiterpenoid ethers of isocitric acid from the fungus *Cryptoporus volvatus*. J. Chem. Soc. Chem. Comm. 4:258-259.

11. Kobayashi, M., T. Nakagawa, and H. Mitsuhashi (1979) Marine terpenes and terpenoids. I. Structures of four cembrane-type diterpenes: sarcophytol-A, sarcophytol-A acetate, sarcophytol-B, and sarcophytonin-A, from the soft coral, *Sarcophyton glaucum*. Chem. Pharm. Bull. (Tokyo) 27:2382-2387.

12. Matsunaga, S., H.F. Suguri, S. Nishiwaki, S. Yoshizawa, M. Suganuma, T. Hashimoto, Y. Asakawa, and H. Fujiki (1991) Differential effects of cryptoporic acids D and E, inhibitors of superoxide anion radical release, on tumor promotion of okadaic acid in mouse skin. Carcinogenesis 12:1129-1131.

13. Narisawa, T., M. Takahashi, M. Niwa, Y. Fukaura, and H. Fujiki (1989) Inhibition of methylnitrosourea-induced large bowel cancer development in rats by sarcophytol A, a product from a marine soft coral *Sarcophyton glaucum*. Cancer Res. 49:3287-3289.

14. O'Brien, T.G., R.C. Simsiman, and R.K. Boutwell (1975) Induction of the polyamine-biosynthetic enzymes in mouse epidermis by tumor-promoting agents. Cancer Res. 35:1662-1670.

15. Oikawa, T., A. Hiragun, Y. Yoshida, H.A. Fuse, T. Tominaga, and T. Iwaguchi (1988) Angiogenic activity of rat mammary carcinomas induced by 7,12-dimethylbenz(a)anthracene and its inhibition by medroxyprogesterone acetate: Possible involvement of antiangiogenic action of medroxyprogesterone acetate in its tumor growth inhibition. Cancer Lett. 43:85-92.

16. Troll, W., K. Frenkel, and G. Teebor (1984) Free oxygen radicals: necessary contributors to tumor promotion and cocarcinogenesis. In Cellular Interaction by Environmental Tumor Promoters, H. Fujiki, E. Hecker, R.E. Moore, T. Sugimura, and I.B. Weinstein, eds. Jpn. Sci. Soc. Press, Tokyo/VNU Science Press, Utrecht, pp. 207-217.

17. Yoshizawa, S., T. Horiuchi, H. Fujiki, T. Yoshida, T. Okuda, and T. Sugimura (1987) Antitumor promoting activity of (-)-epigallocatechin gallate, the main constituent of "tannin" in green tea. Phytotherapy Res. 1:44-47.

Fujita, Y., T. Yamane, M. Tanaka, K. Kuwata, ..., and T. Okada (1980) Inhibitory effect ...

MURINE α/β INTERFERONS INHIBIT BENZO(A)PYRENE ACTIVATION

AND MUTAGENESIS IN MICE

P. Hrelia, M. Morotti, F. Vigagni, F. Maffei, M. Paolini,
and G. Cantelli Forti

Department of Pharmacology
University of Bologna
Bologna, ITALY

ABSTRACT

In addition to their antiviral and immune regulatory properties, interferons (IFNs) are known to depress hepatic cytochrome P450-dependent metabolism. As many chemical mutagens and carcinogens require bioactivation by the mixed-function monooxygenase (MFO) system in order to be genotoxic, a combined genetic and biochemical approach was used to establish whether IFNs could inhibit the activation of benzo(a)pyrene (BaP) to the ultimate clastogenic metabolite(s) in vivo.

Treatment of mice with murine IFN-α/β depressed cytochrome P450 content, as well as ethoxyresorufin O-deethylase activity (EROD), as a probe of class IA1 P450 isozymes, for 24 hrs and delayed the attainment of normal levels to approximately 30 hrs. After IFNs plus BaP treatment, EROD activity showed a reduction up to 70% after 24 hrs with an enhancement in activity at 30 hrs. A positive correlation exists between the rate of inhibition of oxidative BaP hepatic metabolism and inhibition of clastogenic effects in vivo, as scored in the bone marrow chromosome aberration assay.

INTRODUCTION

Interferons (IFNs) are inducible glycoproteins which share antiviral, immunomodulatory, and antiproliferative effects in eukaryotic cells. Although discovered over 30 years ago, it is only recently that, by use of molecular techniques, several IFNs have been produced in sufficient amounts and with adequate purity to permit extensive human clinical trials. Widespread interest in IFNs as potential therapeutic agents has been stimulated by the observation that IFNs elicit antitumor activity in animals and humans. The mechanism of this antitumor effect is not fully understood (2). IFN preparations have also been demonstrated to impair drug metabolism by depressing the level of hepatic microsomal cytochrome P450 (5,9,14).

Since the rate of metabolism and the stereospecificity of metabolism are dependent on the types and amounts of P450s available, IFNs might drastically alter the metabolic activation of some carcinogens. Decreases in P450 content and associated catalytic activities paralleled an in vitro suppression of liver-dependent mutagenesis of certain promutagens/procarcinogens, such as N-acetylaminofluorene, benzo(a)pyrene, aflatoxin B_1 and dimethylbenzo(a)anthracene (12,13). So far as we know, no evidence exists to demonstrate depression of hepatic promutagen activation and mutagenesis in vivo. An integrated approach utilizing both cytogenetic and enzymatic endpoints was used to evaluate whether altered P450 content could be correlated with changes in the ability of mouse liver to metabolically activate BaP into clastogenic intermediates in vivo.

MATERIALS AND METHODS

Animals and Chemicals

Male Swiss albino mice CD1 strain (Nossan S.r.l., Correzzana, Milan, Italy), aged 7-8 wks (29 ± 3 g body wt), were maintained under conditions of controlled temperature, humidity, and lighting with free access to water and standard laboratory diet prior to use. Purchased animals were allowed 1 wk acclimation period before beginning treatment.

Benzo(a)pyrene (98% purity) was obtained from Aldrich (Milan, Italy); murine IFN-α/β (4.7 x 10^5 IU/mg proteins); colcemid, 7-ethoxyresorufin and NADPH were purchased from Sigma Chemical Co. (St. Louis, MO, USA); all other chemicals and solvents were of analytical grade purity. BaP was dissolved in corn oil and IFNs were reconstituted with distilled water immediately before use. All solutions were administered to mice i.p. at a level of 0.1 ml/kg body wt.

Treatment

Groups of mice (3 mice per group) were assigned as negative/solvent control or experimental groups. Mice in the latter experimental groups were administered either murine IFN-α/β (50,000 IU IFNs daily for 2 days), BaP (80 mg/Kg--one injection), or IFNs plus BaP (BaP on the 2nd day of IFN treatment). Mice were sacrificed at 24 hrs and 30 hrs after the last injection.

Bone Marrow Chromosome Aberration Analysis

One hour prior to sacrifice, control and treated animals were injected i.p. with a dose of 4 mg/Kg of colcemid to arrest cells at metaphase. For cytological preparations of bone marrow cells, the conventional hypotonic-giemsa schedule was followed. Scoring was done by a single observer with randomized slides. Classification of aberrations was done according to Savage (15). Wherever possible, at least 100 metaphases from each animal were examined for chromosomal aberrations.

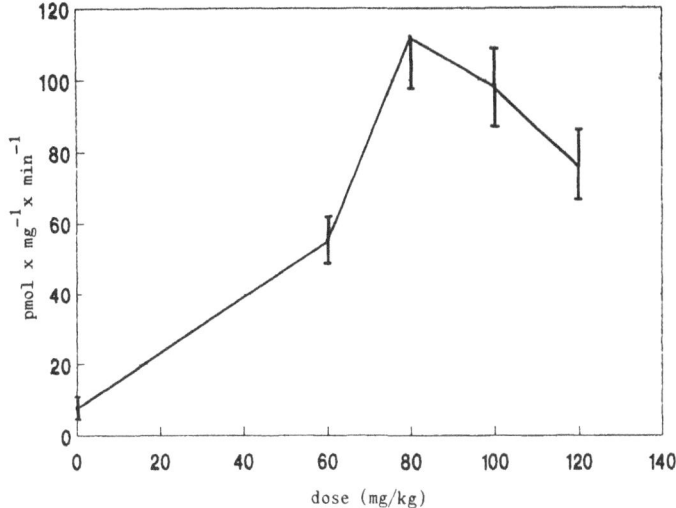

Figure 1. Expression of ethoxyresorufin O-deethylase (EROD) activity in hepatic S9 fraction from benzo(a)pyrene treated mice.

Preparation of S9 Fraction

Livers from treated animals were aseptically removed and homogenized with a Potter homogenizer fitted with a Teflon pestle in 4 ml/g of 0.05 M Tris-HCl buffer (pH 7.4) containing KCl at 1.15% (w/v) for isotonicity. The homogenate thus obtained was centrifuged at 9,000 g for 20 min. The S9 fractions were immediately processed for determination of specific enzyme activities.

P450 Content

P450 was determined by the difference in the spectrum between the reduced form plus carbon monoxide, and the oxidized form at 450-490 nm (10).

Ethoxyresorufin O-Deethylase (EROD) Activity

EROD activity was determined at 37°C by monitoring resorufin formation using a Hitachi spectrophotofluorimeter (mod. 650/40), according to Lubet et al. (8). Reaction mixture consisted of 2.0 ml of 0.05 M Tris-HCl buffer (pH 7.4), 0.025 M $MgCl_2$, 1.7 μM of ethoxyresorufin, and 50-100 μl of sample. Reactions were initiated by the addition of 130 mM NADPH to the cuvette. The rate of formation of resorufin was calculated by comparing the rate of increase in relative fluorescence to the fluorescence of known amounts of resorufin.

Determination of Protein Concentration

Protein concentration was determined according to the method described by Lowry (7) and Bailey (1), using bovine serum albumin as the standard. The samples were diluted 1,000 times to provide a suitable protein concentration.

Table 1. Incidence of structural chromosome aberrations induced in bone marrow cells of mice treated with murine α/β-interferons, benzo(a)pyrene, and α/β-interferons plus benzo(a)pyrene

Treatment	Sampling time	Metaphases scored/animal	Break-type aberrations[a]						Gaps	Total aberrations[b]
			chromatid	iso-chromatid	fragment	ring	exchange	Total		
Control	-	300/3	7	-	4	-	-	3.67±1.20	11	3.00±1.00
α/β interferon	24	300/3	4	-	10	-	-	4.67±1.67	6	4.63±0.34
Benzo(a)pyrene	24	300/3	23	-	9	-	-	10.67±1.76[c]	27	8.67±0.33
	30	300/3	42	2	28	-	-	24.67±5.24[c]	26	18.67±2.96
Benzo(a)pyrene + α/βinterferon	24	300/3	3	-	19	-	-	7.33±1.20	10	5.00±0.58
	30	300/3	18	1	18	-	-	12.67±3.84[d]	28	10.33±2.60

a) Isochromatid breaks, rings and exchanges were each counted as 2 breaks.
b) Values as mean per 100 cells ± S.E.
c) p < 0.01 with respect to control, CHI square test.
d) p < 0.01 with respect to respective time B(a)P treated mice, CHI square test.

Table 2. Effect of murine α/β-interferon (INF-α/β) on cytochrome P450 levels and oxidative benzo(a)pyrene (BaP) hepatic metabolism in mice

TREATMENT	SAMPLING TIME (h)	7-ETHOXYRESORUFIN O-DEETHYLASE (class IA1 P450) pmol × mg⁻¹ × min⁻¹	CYTOCHROME P450 nmol × mg⁻¹
Control (corn oil)		11.03 ± 0.60	0.12 ± 0.03
IFN-α/β (50.000 UIx2)	24	6.53 ± 0.55 (40.8)[a]**	0.05 ± 0.02 (58.3)[a]**
	30	10.07 ± 0.42 (8.70)[a]	0.13 ± 0.03 (0)[a]
B(a)P (80 mg/kg)	24	111.8 ± 13.9	0.17 ± 0.04
	30	157.5 ± 19.1	0.17 ± 0.04
IFN-α/β (50.000 UIx2) + B(a)P (80 mg/Kg)	24	33.2 ± 3.99 (70.3)[b]**	0.10 ± 0.02 (41.2)[b]**
	30	83.9 ± 10.2 (46.7)[b]**	0.16. ± 0.04 (6.0)[b]

Values are the mean ± S.D. of 3-4 animals.
In parentheses the percentage loss with respect to controls(a) and respective time B(a)P (b) treated mice.
* P < 0.05.
** P < 0.01.

Statistical Analysis

Results of enzyme analysis were submitted to statistical analysis using Wilcoxon's rank method as reported by Box and Hunter (3). CHI-square test was used to compare experimental and control data in cytogenetic analyses.

The combined data for each animal group was further converted to a square root scale to homogenize the intersubject variation within groups. Statistical significance (t after significant F) was determined by analysis of variance (AOV) on the converted scale (19). In this analysis, exposure-related trends and differences among groups were considered. Mean ± S.E. for the untransformed data are presented.

RESULTS

BaP is a well-known inducer of the MFO system. Preliminary experiments have elucidated the appropriate dose to be tested. The inductive profile shows a 15-fold increase over control in EROD (P450IA1-linked) activity in BaP-treated mice at 80 mg/kg body wt, with a decrease in activity at higher doses. Because, as shown in Fig. 1, there is a toxic effect of BaP on EROD activity above 80 mg/Kg, and as toxicity may be a confounding parameter, the BaP dose of 80 mg/Kg was used in subsequent studies in the elucidation of the combined effect of BaP and IFNs.

Comparisons of aberration frequencies obtained from bone marrow metaphase analysis were used to determine the effects of murine INF-α/β upon the clastogenic potency of BaP in vivo. The structural chromosome aberrations encountered were mainly of the chromatid type and comprised chromatid breaks, fragments (paired and unpaired), and isochromatid breaks, of which the first two categories were more frequent. All these aberrations together are referred to here as break-type aberrations. Because of their unknown significance, gaps were considered separately. Total aberrations included both break-type aberrations and gap-type aberrations. Table 1 summarizes the kind and frequency of chromosome aberrations scored in bone marrow cells after the different treatments.

Cytogenetic analysis showed a time-dependent increase in the frequency of chromosome aberrations in BaP-treated animals. The maximal response in chromosome aberrations occurred 30 hrs after BaP treatment ($p < 0.01$). In fact, a decrease in aberrations was recorded at a 48-hr interval (data not shown). The drop in aberration maxima indicates some cytotoxicity of the treatment. In fact, the mitotic profile shows a reduction of 45% in the mitotic index at 30 hrs (data not shown).

Concomitant treatment with IFN-α/β significantly reduced the BaP-dependent increases in mutagenesis. Aberration frequencies were lowered by 31.3% and 48.7% ($p < 0.01$) at 24 and 30 hrs, respectively. Analysis of variance shows a treatment dependency ($p < 0.05$).

Table 2 shows the effect of murine IFN-α/β on P450 levels and oxidative metabolism of mice. As the overall activity of cytochrome P450-dependent monooxygenases measured in microsomal samples is the summation of the individual contribution of many isozymic P450 forms, the deethylation of 7-ethoxyresorufin was chosen as a probe of class IA1 P450, which typically metabolizes polycyclic aromatic hydrocarbons. A positive correlation was found between the rate of inhibition of critical enzyme activity for metabolism of BaP and of inhibition of clastogenicity in vivo.

The administration of 50,000 IU of IFN-α/β resulted in a 58.3% loss of cytochrome P450 in S9 fractions prepared from the liver of mice treated daily for 2 days, and sacrificed 24 hrs after the last injection of IFNs. This loss was accompanied by a 40.8% decrease in EROD activity. The kinetic analysis indicates that the depression of P450 and EROD activity is a relatively short-lived effect, since P450 and in vivo oxidative reactions were restored to near-normal levels by 30 hrs.

In mice receiving IFN-α/β plus BaP, which strongly induces the P450 function, a 70.3% loss in EROD activity was demonstrated within 24 hrs (Tab. 1). However, this inhibition was temporary, since an enhancement of 61% in activity compared to the level expressed at 24 hrs was reached within 30 hrs.

DISCUSSION

The present study describes for the first time the effect of treatment with murine IFN-α/β on promutagen/procarcinogen activation in vivo. The extent to which hepatic microsomal P450 content and P450-dependent catalytic activity, such as ethoxyresorufin O-deethylase (P450IA1-linked), were depressed in mice treated for 2 consecutive days with 50,000 UI IFN-α/β correlated with the inhibitory activity of this IFN preparation B(a)P clastogenicity in vivo.

In this study we used murine IFN-α/β as it was demonstrated that the effects of IFN are species specific (4) and that autologous but not heterologous IFNs confer antiviral activity and impair drug metabolism (6,11). The kinetics of depression and recovery of murine hepatic P450 levels and oxidative reactions after treatment with IFN-α/β confirmed that the effect on drug metabolism is relatively short-lived, as previously reported by other authors. The latter demonstrated decreased hepatic P450 content and decreased oxidative activities in both uninduced and induced animals, due to pure IFN preparations and IFN inducers (5,9,14).

Induction or depression of the P450 system can alter markedly the therapeutic effectiveness or toxicity of drugs and other xenobiotics. It has been reported that recombinant human leucocyte IFNs decreased the rate of metabolism in vivo of different substrates, such as hexobarbital, 7-ethoxycoumarin, benzphetamine, and zoxazolamine (11,14). In particular, BaP metabolism was decreased up to 40% (14).

Previous studies have been designed to determine whether IFN preparations could inhibit the activation of chemical promutagens to ultimate mutagens. The in vivo induction of IFN-dependent effects seems to be specific. For mice sensitized with *Mycobacterium bovis* there was a preferential suppression of the ability of liver homogenate from treated animals to metabolically activate BaP compared with the promutagen N-acetylaminofluorene and aflatoxin B_1 in the Ames test (12). Moreover, Reiners (13) demonstrated that the preincubation of keratinocytes with IFN-β partially inhibited 7,12-dimethylbenzo(a)anthracene and BaP dependent mutagenesis on V79 cells.

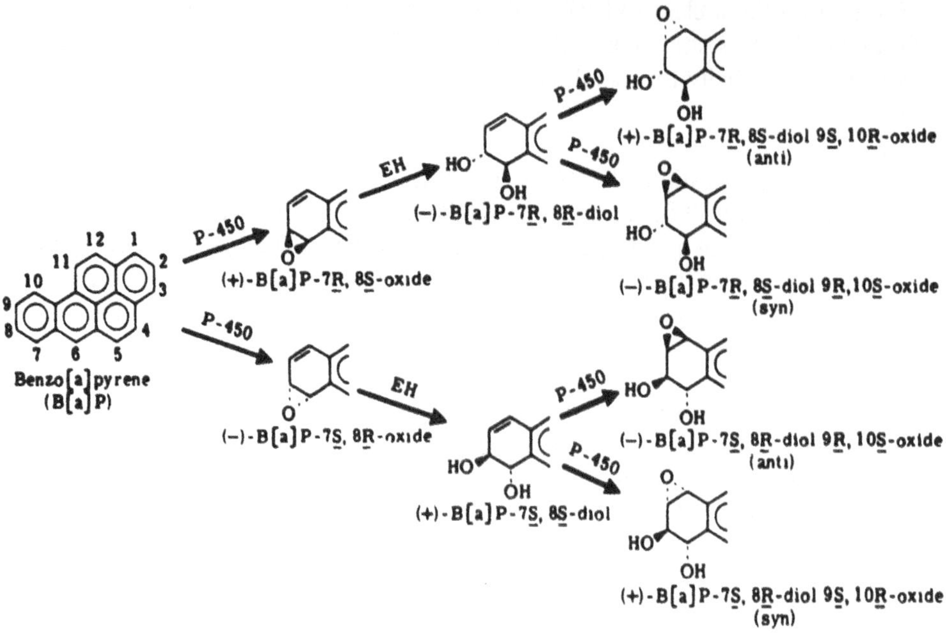

Figure 2. Stereochemistry of the metabolic activation of benzo(a)pyrene. P450, microsomal cytochrome P450-linked monooxygenase. EH, microsomal epoxide hydrolase.

This study has shown an in vivo inhibition of procarcinogen activation and inhibition of clastogenic effects. It has been well established that BaP is an MFO inducer and requires metabolic activation in order to express its genotoxic properties. P450IA1-lined enzymes are involved in the activation of BaP to epoxide and diol-epoxides (Fig. 2).

The increase of the P450 enzyme activity by BaP is dose-dependent and time-dependent. Induction of oxidative metabolism parallels an increase in chromosome aberrations, especially of break-type. Concomitant treatment with IFN-α/β causes a significant reduction (up to 70%) of the specific oxidative metabolism of BaP, which parallels a decrease in the cytogenetic response in vivo (up to 48%), probably due to a reduced formation of the putative ultimate carcinogen, 7,8-diol-9,10-oxide.

A critical question regarding the mechanism of IFN-α/β is whether the detected effects upon mutation induction are imputable only to inhibition of liver-dependent promutagen metabolism by the interferon. The maximum reduction in BaP metabolism appears within 24 hrs, and so it is likely to contribute to reduction in the activation of BaP and chromosome aberration formation.

Of course, the contribution of other IFN-sensitive processes cannot be excluded. In fact, genotoxic effects depend more on other factors, such as DNA repair systems, inhibition or enhancement of cell proliferation, and host immunocompetence, than they do on metabolism. A large antiproliferative effect was reported for IFN preparations. How much of the protective effect shown in this study was due to inhibition of biotransformation of BaP and how much to an antiproliferative effect of IFN-α/β on early formation of aberrant cells has not been determined.

Depression of bone marrow cell proliferation was demonstrated in vitro (17), but this effect has not been demonstrated in vivo when administered to adult animals in acute experiments. Inhibition of cell cycle progression by IFNs was demonstrated by in vitro studies with cell lines and strains (16). Studies on the combination of IFN with cytotoxic agents focused on the biochemical and cell kinetic effects of IFNs and suggested that IFN may be acting as a biochemical modulator (2,18).

In this study, mitotic index data (not shown) on the combination of BaP and IFNs indicated a slight effect on cell cycle, but it does not appear to be significant, compared with BaP effects. Further studies are needed to elucidate the effects of IFNs on BaP clastogenicity in vivo.

ACKNOWLEDGEMENTS

We are grateful to Prof. Norman N. Trieff, Department of Preventive Medicine and Community Health, The University of Texas Medical Branch at Galveston (USA), for critical comments on the manuscript and helpful advice. This investigation was supported by Ministry of University and Scientific and Technological Research (MURST) of Italy and by CNR (National Research Council of Italy), special "Bilateral Project," contract No. 90.01435.CT04.

REFERENCES

1. Bailey, Y.L. (1967) Techniques in Protein Chemistry. Elsevier, Amsterdam, pp. 340-341.
2. Balkwill, F.R., S. Mowshowitz, S.S. Seilman, E.M. Moodie, D.B. Griffin, K.H. Fantes, and C.R. Wolf (1984) Positive interactions between interferon and chemotherapy due to direct tumor action than effects on host drug-metabolizing enzymes. Cancer Res. 44:5249-5255.
3. Box, G.E.P., and W.G. Hunter (1978) Statistics for experiments. Wiley, New York, pp. 80-82.
4. Craig, P.I., S.J. Williams, E. Cantrill, and G.C. Farrell (1989) Rat but not human interferons suppress hepatic oxidative drug metabolism in rats. Gastroenterology 97:999-1004.
5. Crowe, D.O., J.J. Reiners, Jr., D.E. Nerland, and G. Sonnenfeld (1986) Kinetics of depression and recovery of murine hepatic cytochrome P450 levels after treatment with the interferon inducer polyriboinosinic-polyribocytidylic acid. JNCI 76(5):879-883.
6. Franklin, M.R., and B.S. Finkle (1985) Effect of murine gamma interferon on the mouse liver and its drug-metabolizing enzymes: Comparison with human hybrid alpha-interferon. J. Interferon Res. 5:265-272.
7. Lowry, O.H., H.J. Rosenbrough, A.L. Farr, and R.J. Randall (1951) Protein measurement with the Folin phenol reagent. J. Biol. Chem. 193:265-275.
8. Lubet, R.A., R.T. Mayer, J.W. Cameron, W.N. Raymond, D.H. Burke, T. Wolff, and F.P. Guengerich (1985) Dealkylation of pentoxyresorufin: A rapid

and sensitive assay for measuring induction of cytochrome P450 by phenobarbital and other xenobiotics in the rat. <u>Arch. Biochem. Biophys.</u> 238:43-48.

9. Mannering, G.J., K.W. Renton, R. Azhary, and L.B. Deloria (1980) Effects of interferon-inducing agents on hepatic cytochrome P450 drug metabolizing systems. <u>Ann. NY Acad. Sci.</u> 350:314-331.

10. Omura, T., and R. Sato (1964) The carbon monoxide-binding pigment of liver microsomes. I. Evidence for its hemoprotein nature. <u>J. Biol. Chem.</u> 239:2370-2378.

11. Parkinson, A., J. Lasker, and M.J. Kramer (1982) Effect of three recombinant human leukocyte interferons on drug metabolism in mice. <u>Drug Metab. Dispos.</u> 10:579-585.

12. Reiners, J.J., Jr., D. Crowe, C. McKeown, D.E. Nerland, and G. Sonnenfeld (1984) Gamma interferon induction depresses murine hepatic promutagen/procarcinogen activation. <u>Carcinogenesis</u> 5:125-128.

13. Reiners, J.J., Jr. (1985) Interferon-beta inhibits 7,12dimethylbenzo(a)-anthracene-dependent mutagenesis in a keratinocyte cell-mediated mutation assay. <u>Carcinogenesis</u> 6:1781-1785.

14. Renton, K.W., and J. Mannering (1976) Depression of the hepatic cytochrome P450 monooxygenase system by administered Tilorone (2,7-bis[2-(diethyl-amino)ethoxy] fluoren-9-one dihydrochloride). <u>Drug Metab. Dispos.</u> 4:223-231.

15. Savage, J.R.K. (1976) Annotation: Classification and relationships of induced chromosomal structural changes. <u>J. Med. Gen.</u> 13:103-122.

16. Tamm, I., B.R. Jasny, W.J. Pledger, and I. Tamm (1987) Antiproliferative action of interferons. In <u>Mechanisms of Interferon Actions,</u> Vol. 2, L.M. Pfeiffer, ed. Boca Reton, CRC Press, FL, pp. 25-28.

17. Vant'Hull, S., H. Schellekens, B. Lowenberg, and M.J. de Vries (1978) Influence of interferon preparations on the proliferative capacity of human and mouse bone marrow cells in vitro. <u>Cancer Res.</u> 38:911-914.

18. Wadler, S., and E.L. Schwartz (1990) Antineoplastic activity of the combination of interferon and cytotoxic agents against experimental and human malignancies: A. Review. <u>Cancer Res.</u> 50:3473-3486.

19. Whorton, E.B. (1985) Some experimental design and analysis considerations for cytogenetics studies. <u>Env. Muta.</u> 7:9-15.

INHIBITION OF MALIGNANT TUMOR CELL INVASION:

AN APPROACH TO ANTI-PROGRESSION

Adriana Albini and Annamaria Colacci

Istituto Nazionale per la Ricerca sul Cancro
16132 Genova, ITALY
and Istituto Nazionale per la Ricerca sul Cancro
Biotechnology Satellite Unit
40126 Bologna, ITALY

INTRODUCTION

Considerable attention should be given to primary prevention of neoplastic transformation. The best approaches to anticarcinogenesis are preventative measures to reduce exposure to environmental pollutants and improving life styles that may be responsible for genetic and epigenetic cell damage. Once neoplastic transformation has occurred, however, the major target of the oncologist becomes hindering tumor progression. It is difficult to define the onset of tumor progression and whether it involves a continous evolution or a step-like process. It is easier to define the consequence of tumor progression: malignant lesions with the ability to invade and to metastasize to distant organs. While benign tumors can often be cured simply by surgical removal, metastases remain the most frequent cause of cancer mortality.

Metastasis is a complex phenomenon in which several biological parameters of the cell are altered (49). Fig. 1 indicates some of the essential events in the transition from a normal cell to a malignant tumor. Our review will focus on one of the characteristics of all metastatic cells: the acquisition of the capability to cross extracellular matrix compartments in order to translocate through the body (45). This phenomenon, which requires the steps sketched in Fig. 2, is referred to as invasiveness. Since metastasis derives from the gain of numerous new properties--all of which are necessary, although none is sufficient--the prevention of any one of these properties should be enough to "short-circuit" the onset of the malignant phenotype. Here we describe the invasive phenomenon and the most frequently used methods to assess invasiveness in vivo and in vitro. Focusing on in vitro studies, we will present some of the drugs and molecules that, as efficient inhibitors of invasion, are possible candidates for use as antimetastatic agents.

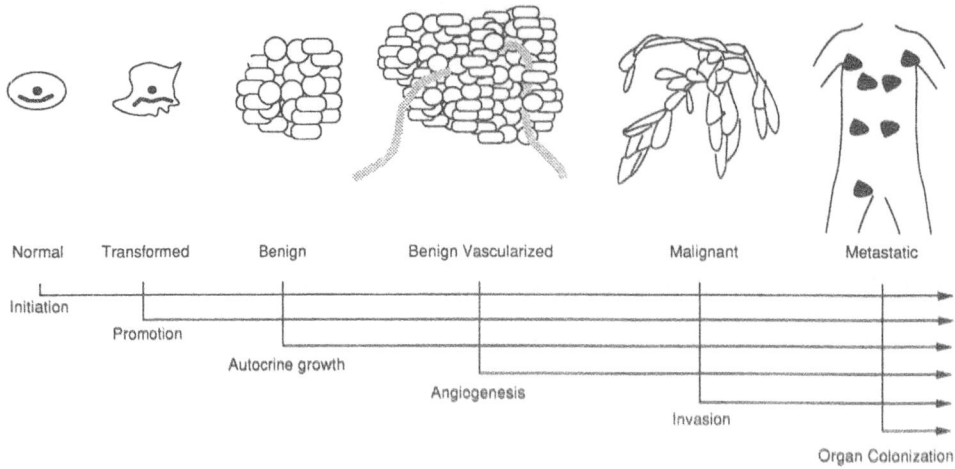

Normal	Transformed	Benign	Benign Vascularized	Malignant	Metastatic

Initiation

Promotion

Autocrine growth

Angiogenesis

Invasion

Organ Colonization

Figure 1. Illustration of the transition from normal cells to metastatic tumor cells. The last four steps sketched represent tumor progression events.

MECHANISMS OF INVASION

Malignant tumor cells must cross extracellular matrix compartments in order to metastasize (for reviews see Refs. 45, 49). Since most cells metastasize via the vascular system, they have to cross vascular basement membranes (Fig. 2), which are generally impermeable to normal cells (44,82). However, invasion can occur naturally in the adult and during development. Leukocytes are normally invasive, and tropho- blasts of the first trimester human placenta are also able to invade basement membranes in situ (94). Invasion of basement membranes by malignant cells occurs as a result of several steps. These include binding to basement membrane compo- nents, activation of proteolytic enzyme production, digestion of extracellular matrix, and migratory response to chemotactic or chemokinetic stimuli (49).

Laminin is the major biologically active component of basement membranes (36). Metastatic and invasive cells adhere preferentially to laminin (84). Exposure of tumor cells to laminin enhances their capability to form experimental metastases and also to attach to basement membranes (Terranova, et al., 1984). Adhesion to laminin also triggers the production of type IV collagenase (89) and of plasminogen activator. Laminin, therefore, has both direct and indirect effects on tumor cell invasion.

Metastatic tumor cells produce high amounts of type IV collagenase and other metalloproteases able to digest basement membrane type IV collagen (49). Meta- static cell lines produce plasminogen activator (87), which converts the plasminogen to plasmin. Plasmin-mediated tissue degradation is involved in basement membrane digestion and tumor cell invasion (55,63). Plasmin activates collagenase type IV as

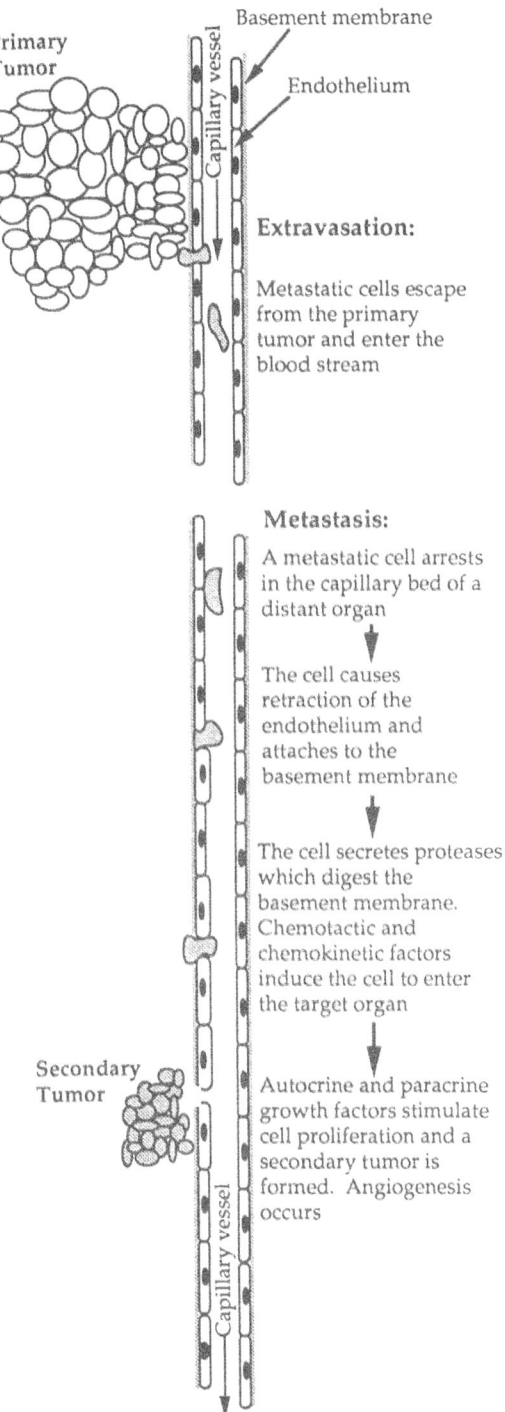

Primary
Tumor

Basement membrane

Capillary vessel

Endothelium

Extravasation:

Metastatic cells escape
from the primary
tumor and enter the
blood stream

Metastasis:

A metastatic cell arrests
in the capillary bed of a
distant organ

The cell causes
retraction of the
endothelium and
attaches to the
basement membrane

The cell secretes proteases
which digest the
basement membrane.
Chemotactic and
chemokinetic factors
induce the cell to enter
the target organ

Secondary
Tumor

Autocrine and paracrine
growth factors stimulate
cell proliferation and a
secondary tumor is
formed. Angiogenesis
occurs

Capillary vessel

Figure 2. Tumor cell extravasation into a blood vessel and colonization of a distant organ, during the process of metastasis. Crossing of extracellular matrices, in particular basement membranes, is referred to as "invasion."

well (65). The plasminogen activator system is modulated by matrix components in breast carcinoma cells (64). Cathepsin D is also involved in matrix degradation phenomena (11,67) and Cathepsin B has a role in tumor malignancy (75).

Motility is a requirement for translocation through basement membranes, and several factors have been implicated in the induction of migratory properties. The major classes of chemoattractants are extracellular matrix proteins and their proteolytic fragments, blood cell products, and growth factors (2,91). Autocrine motility factors, produced by the malignant cells themselves, have been reported (47). All these interactions are mediated through specific chemoattractant receptors on the cell surface. How the chemotactic signal is transduced in order to make invasion possible is under debate.

With the aim of understanding some of the mechanisms of A375M melanoma cell invasion, Welch et al. (92) tried several agents in a matrigel invasion assay (see later). Drugs that cause microtubule disruption (such as vincristine, vinblastine, colchicine, colcemid) inhibited invasion, as previously shown for fibroblastic cell chemotaxis (2). Dibutyril cAMP, a cAMP agonist, was inhibitory, while 8-Br-cAMP was stimulatory. Cholera toxin and forskolin, stimulators of adenyl cyclase, also inhibited invasion. In contrast, pertussis toxin, a G protein inhibitor, had no effect on invasion. Pertussis toxin is also ineffective on insulin peptide-mediated melanoma cell chemotaxis (81); however, it affects autocrine motility factor-induced migration (39). It is, therefore, not clear which mechanisms of signal transduction are essential for fibroblast and tumor cell chemotaxis and tumor cell invasion. There are indications that different pathways of signal transduction are activated in response to different agents. It is possible that, as for other eukaryotic systems, there are redundant mechanisms for achieving a response.

MODELS FOR THE STUDY OF MALIGNANT CELL INVASIVENESS

Invasion and the metastatic phenotype have been principally studied in "in vivo" models, most frequently mice (17). Murine tumors can be studied in most cases in the syngeneic animal; for human tumors there is a vast literature on the use of athymic nude mice (25). The most common routes of injection are subcutaneously, in the test called "spontaneous metastasis assay," or in the tail vein (experimental metastasis). In the latter, only the final steps of the metastatic process are represented. More recently, injection into the spleen has been suggested as a valuable assay (26). Also, a tendency to inject malignant cells orthotopically (i.e., in the colon for colon carcinoma, in the eye for retinoblastoma) has arisen (18,19,93). The orthotopic route of transplantation seems to furnish more reliable models of metastatic colonization from a given site.

The use of animal studies is quite appropriate; however, large-scale testing of new potential antimetastatic compounds, as well as rapid tests of malignancy, cannot be achieved with the mouse system. Even the use of the nude mouse has not completely overcome problems of immune response when using cells of nonmurine origin. The chick chorioallantoic membrane has been employed as an invasion-metastasis test in an immune-deficient host (42,72). Furthermore, several assays have been developed to study tumor cell invasion in vitro (for a review, see Ref. 82). For instance, entire organs have been cultured and the invasion assessed histologically in the "organotypical" tests of invasiveness (50). Cultured cells, fibroblasts, smooth muscle, and endothelial cells (62), have also been exposed to malignant cells and the ability to penetrate the monolayer has been correlated to the cells' metastatic potential in vivo. Alternatively, natural extracellular matrix barriers, such as the amnion, the bladder wall, the lens or others have been used to assess tumor cell invasion (30,46,76). More artificial systems have been proposed in recent years. Basement membrane components, laminin and type IV collagen, compacted together at high pressure, form dense disks that act as a barrier to the passage of nonmetastatic cells (83).

Based on these observations, we have introduced the use of a gel-like reconstituted basement membrane, matrigel, in a quantitative assay (5) that measures both the chemotactic and invasive abilities of cells and is, therefore, termed "chemoinvasion."

THE CHEMOINVASION ASSAY

Several studies have indicated that the components of basement membranes will bind to one another and are able to reconstitute into a gel under physiological conditions. The gels can be formed either by purified type IV collagen, laminin, and heparin sulphate proteoglycan (29) or by a crude extract of authentic basement membranes, such as the placenta, or a basement membrane-producing tumor, the EHS sarcoma (38). The gel-like extract of the EHS tumor has been characterized and named "matrigel."

Analysis of matrigel shows that it contains all major basement membrane components (38), even if the proportions might be altered with respect to natural basement membranes. Matrigel is a biologically active substrate and can induce the differentiation of epithelial, endothelial, neural cells, and others (37,51). When plated on or in matrigel (Fig. 3, upper panel), nonmetastatic cells remain as single cells or small clumps on the surface, while malignant cells are able to grow and penetrate the gel, showing an invasive morphology (4,5,8,56). Electron microscopical analysis has revealed that malignant cells migrating through the gel leave "tunnels" behind (40).

In the chemoinvasion assay (5), a thin layer of matrigel (25 μg/13 mm filter is generally used) is coated onto a polycarbonate chemotaxis filter. The filter is then assembled in a Boyden chamber migration device, separating two distinct chambers (Fig. 3, lower panel). In the lower one is placed the chemoattractant, often fibroblast-conditioned medium, which contains potent migratory stimuli for tumor cells (58). In the upper compartment a cell suspension is introduced in serum-free medium. Potential invasion inhibitors can be added to the assay (see later). Highly invasive cells adhere to the matrigel coating, traverse the barrier, and appear on the lower surface of the coated filter. Noninvasive cells also bind to matrigel, but are not able to degrade it and, therefore, remain in the upper compartment of the chamber. The number of migrated cells is evaluated after 6 hrs of incubation at 37°C in 5% CO_2. A thicker coating lengthens the time necessary for penetration allowing, the migration of fewer cells over the same period of time. Stringency of the conditions can, therefore, be modulated.

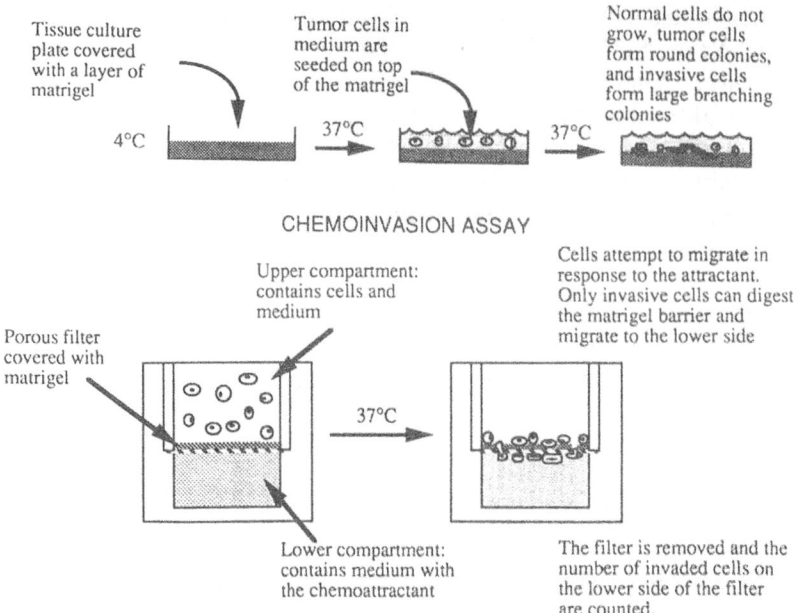

Figure 3. Illustration of two assays employing the reconstituted basement membrane matrigel for the in vitro assessment of malignant cell invasion.

We and others have used the chemoinvasion assay to study a large number of cells and systems. The assay and its variations have also been referred to as "matrigel invasion assay" or "Boyden chamber invasion assay" or "membrane invasion culture system" (16,66,92). A very good correlation between invasion in vitro and metastatic potential in vivo has been found with this model.

USE OF THE CHEMOINVASION ASSAY TO SCREEN FOR POTENTIAL ANTIMETASTATIC DRUGS

All of the events involved in invasion can be specific targets for agents and drugs with antimetastatic potential. Metastatic tumor cells attach preferentially to laminin (LN), the major component of basement membranes (84). This interaction might provide the initial adhesion to that membrane. Malignant cells have high numbers of available laminin receptors, belonging to two major classes: the laminin binding proteins, having high affinity (with the 67 kDa receptor being the one known the longest) (85), and low affinity ones, the integrins (24,41). Integrins are heterodimeric calcium-dependent receptors (1,54), composed of an α and a β subunit, both of which have a long extracellular domain, a transmembrane, and a short (with the exception of β4) cytoplasmic domain. LN, like fibronectin, vitronectin, thrombospondin, and fibrinogen, has an RGD sequence (71). Treatment of melanoma cells with laminin increases metastasia (86). Interference with integrins and the integrin-binding sequence RGD inhibits invasion (23) and metastasis of tumor cells. Anti-bodies to the β1 integrin subunit inhibit invasion (95). RGD is an inhibitor of adhesion (68) and chemotaxis (3), and both these events might be involved in the metastatic

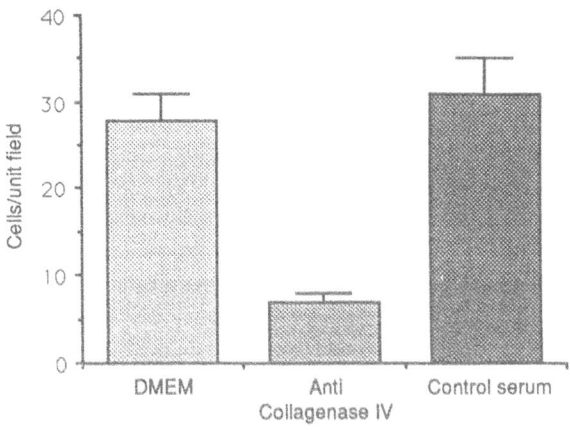

Figure 4. Inhibition of invasion by antibodies specific for collagenase type IV/gelatinase of 72 kDa. Rabbit antisera or nonimmune serum, in the same concentrations, were added to the upper compartment of Boyden chambers in the "chemoinvasion assay" (Fig.3). Fibroblast conditioned medium was used to stimulate migration. Target cells were HT-1080, derived from a malignant human fibrosarcoma. An inhibition of about 70% of invasion was noticed when using the collagenase antibodies, while rabbit serum itself had no effect. (Adapted from Albini et al., Ref. 7)

colonization. The laminin peptide YIGSR, which has cell binding properties (28), is an inhibitor of invasion in the chemoinvasion assay, and is also capable of inhibiting experimental metastasis of melanoma cells in vivo (33). On the contrary, another LN peptide, IKVAV, enhances experimental metastasis (34). This could be related to this peptide's angiogenic activity.

Adhesion to LN stimulates production of collagenase IV (89), an enzyme associated with the malignant phenotype (78). Two major forms of collagenase IV are known, of 72 and 92 kDa. These enzymes are secreted in a latent form, and activation occurs mainly through the removal of an aminoterminal propeptide of circa 80 amino acids (80). The destruction of the interaction between a sulfydryl group contained in this region and the metal atom of the enzyme is crucial for the activation of proteolytic activity, a mechanism common to several metalloproteases (90). A collagenase inhibitor from Searle, SC44463, mimics a fragment of the collagen helix and binds to the active site of collagenase. A hydroxamic group chelates the Zn atom on the protein, necessary for the lytic activity. This compound is an efficient inhibitor of invasion for malignant cells (65) and Kaposi's sarcoma cells (88). It has also been demonstrated to be a good inhibitor of metastasis (65).

We have used a different approach for inhibition of collagenase IV activity, also based on its activation by peptide cleavage. Stetler-Stevenson et al. (77) have shown that a synthetic peptide encompassing a highly conserved region of the aminopropeptide of collagenase IV is a good inhibitor of collagenolytic activity in an enzymatic assay. This peptide is able to repress invasion in the chemoinvasion assay (6).

Another mechanism of inactivation of collagenase IV is complex formation with TIMPs, the tissue inhibitors of metalloproteases. These are specific proteins that bind to and inhibit the action of certain metalloproteinases. It has been shown that exogenous TIMP-1 can control invasion (74), and that antisense suppression of TIMP-1 synthesis is able to make 3T3 cells invasive and metastatic (35).

TIMP-2 is a member of this class of metalloproteinase-inhibitors that binds preferentially to the 72 kd collagenase IV form (27,79). We have recently shown that TIMP-2 is an excellent inhibitor of invasion of basement membrane, at doses of approximately 10 µg/ml (7). Furthermore, recombinant TIMP-2 inhibits invasion of smooth muscle cell layers (14). The direct involvement of collagenase IV in the invasive process is also demonstrated by the fact that antibodies inhibiting this molecule's activity were also able to decrease the penetration of A2058 melanoma (31) and HT1080 fibrosarcoma cells (7) through a reconstituted basement membrane. The effect of anti-type IV collagenase antibodies on HT1080 cell invasion (7) is shown in Fig. 4. Differentiative agents, such as retinoic acid or butyric acid, often affect the collagenolytic activity of metastatic cells, acting as anti-invasive molecules (53,61).

Other proteases are involved in tumor cell invasion of basement membranes, among which are plasminogen activators (PA) (87). Antibodies to PA are able to inhibit invasiveness of melanoma cells through fibrin gels and matrigel (55). Matrigel invasion of HCT116 cells is dependent on plasmin production mediated by urokinase (73). In this system, a specific urokinase receptor peptide antagonist exerted a dose-dependent reduction of matrigel invasion. Reich et al. have shown that the generation of active collagenase IV enzyme can be accomplished via plasminogen activator and plasminogen (65).

Interesting substances with protease inhibitor activities that have been shown to inhibit invasion through matrigel-coated filters include bestatin, an immune-modulator (70), and tetra-p-amidino-phenoxy-neopentane, an aromatic polyamidine (22).

The third target for inhibition of invasion is motility, since binding to matrix and enzyme production does not lead to metastasis in the absence of cell migration. The interference could target either the chemoattractant receptor or the chemoattractant itself. However, while there are several peptides and drugs that effect adhesion to or degradation of basement membranes that have potential in vivo applications, there are little data available on the approach of inhibiting chemotaxis. The best-studied molecules are matrix-derived chemoattractants, the inhibition of which can be achieved by antibodies or peptides, as already described in the section dedicated to cell adhesion. We know that general inhibitors of cytoskeletal function and cell motility, as well as blockers of signal transduction pathways (43,92), abolish invasion. However, this kind of treatment would impair the entire function of the body.

ANTI-INVASION: A MATTER OF SELECTIVITY

It is important to emphasize that, while with invasion assays in vitro we study the direct interaction of a cell with basement membranes, the in vivo situation is more complex. We have to consider drugs that are not toxic and have selective specificity for the invasive process. Since adhesion to basement membrane and its degradation occurs in physiological conditions--mostly in inflammatory responses--in very specific times and places, both matrix proteins and their receptors as well as collagenase IV inhibitors are good candidates for in vivo approaches. Animal studies so far performed have shown that the antimetastatic treatment with such agents does not affect the health of the host.

A WORD ON ANGIOGENESIS

An essential factor in metastatic spread and growth is tumor neoangiogenesis (20,48). The fact that the tumor mass becomes vascularized is a requirement not only for the growth but also for the transport of cells shed into the circulation. Once the metastatic "seed" has reached the target organ, proper blood support is necessary for its development into a secondary lesion. Even though angiogenesis is secondary to tumor invasion, the similarity of the two processes makes it worthwhile to mention it in this chapter.

During the process of tumor neoangiogenesis, endothelial cells are stimulated to migrate from a vessel to a new site by specific angiogenic factors. In order to translocate they have to cross their own basement membrane. Supernatants of Kaposi's sarcoma (KS) cells, a tumor which has been postulated to recruit host vascular cells, are able to make endothelial and smooth muscle cells invasive in vitro (88). This could be due to the presence in the conditioned media of high concentrations of certain growth factors able to stimulate invasion, such as FGF (59) or PDGF (52), which induces collagenase secretion (12).

It appears that endothelial cell invasion through basement membrane occurs through the same mechanisms as malignant cell invasion of a blood vessel wall. Adhesion to basement membrane and proteolytic digestion by the endothelial cell are necessary as well as migration in response to chemotactic or chemokinetic stimuli. Studies from our laboratory show that laminin antibodies and a collagenase inhibitor, common targets for malignant cell invasion, also prevent tumor cell-stimulated endothelial cell invasion of matrigel (88). Since angiogenesis occurs sporadically in the adult human body, it represents another potential target for antimetastatic intervention.

CHEMICAL CARCINOGENS AND INVASION

Most of the readers of this book are involved in the study of chemically induced carcinogenesis. The current in vivo models for multistep carcinogenesis, hepato- and skin carcinogenesis, allow for the study of tumor progression of preneoplastic forms or benign tumors as a consequence of chemical exposure. However, they provide no information on metastatic spread. Recent work has shown that cell treatment with chemical carcinogens can induce the invasive phenotype in vitro. Pretreatment of a murine squamous carcinoma cell line with the initiating and promoting agent benzoyl peroxide in the culture medium increases the capability of the cells to migrate through a porous filter coated with matrigel (9). Phorbol esters are inducers of invasion for certain cells (21), and they increase invasiveness of transformed cells (10). Cells transformed with the initiating agent MNNG (methyl-nitro-nitroso-guanidine) express markers typical of invasive cells, such as specific enhancement of laminin and collagen receptor complexes (15). Furthermore, tumor promoters stimulate type IV collagenase/gelatinase in normal and malignant cells (60).

Our own work shows that in vitro transformation of Balbc/3T3 cells with initiating agents (BaP) or initiating-promoting agents (1,1,2,2-tetrachloro-ethane) accelerates malignant progression, giving rise to cell populations able to invade matrigel in vitro and to metastasize in the animal (13,57).

These studies confirm a link between chemical carcinogens and tumor progression and substantiate that the induction of the invasive phenotype is a process generally involved in malignancy. Chemoprevention, therefore, is an example of "better than never," since in case of an existing neoplastic lesion it could nevertheless slow down progression.

ACKNOWLEDGEMENTS

The work was financed by grants of CNR, BTBS N.91.01269.PF70 and AIRC. We are grateful to Dr. Douglas Noonan and Dr. Remy Adatia (Istituto Nazionale per la Ricerca sul Cancro, Genova, Italy) for revision of the manuscript.

REFERENCES

1. Albelda, S.M., and C.A. Buck (1990) Integrins and other cell adhesion molecules. FASEB J. 4:2868-2880.
2. Albini, A., B. Adelmann-Grill, and P.K. Muller (1985) Fibroblast chemotaxis. Collagen Rel. Res. 5:283-296.
3. Albini, A., G. Allavena, A. Melchiori, F. Giancotti, H. Richter, G.R. Martin, and G. Tarone (1987) Chemotaxis of 3T3 and SV3T3 cells to fibronectin is mediated through the cell attachment site in fibronectin and a fibronectin cell surface receptor. J. Cell. Biol. 105:1867-1872.
4. Albini, A., S.L. Aukerman, D.M. Noonan, R.C. Ogle, R. Fridman, G.R. Martin, and I.J. Fidler (1989) The in vitro invasiveness and interactions with laminin of K-1735 melanoma cells. J. Clin. Exp. Met. 7:437-451.
5. Albini, A., Y. Iwamoto, H.K. Kleinman, G.R. Martin, S.A. Aaronson, J.M. Kozlowski, and R.N. McEwan (1987) A rapid in vitro assay for quantitating the invasive potential of tumor cells. Cancer Res 47:3239-3245.
6. Albini, A., A. Melchiori, S. Parodi, L. Santi, L.A. Liotta, and W.G. Stetler-Stevenson (1990) TIMP-2 and synthetic peptides derived from the profragment of collagenase type IV inhibit invasion in the "chemoinvasion assay." Clin. Exper. Metastasis. 8 suppl 1:66.
7. Albini, A., A. Melchiori, L. Santi, L.A. Liotta, P. Brown, and W.G. Stetler-Stevenson (1991) Tumor cell invasion inhibited by TIMP-2. J. Natl. Cancer Inst. 83:775-779.
8. Allavena, G., A. Melchiori, O. Aresu, L. Ferreri-Santi, R. McEwan, G. Kitten, S. Parodi, and A. Albini (1988) In vitro models for studying the malignant phenotype: Chemotaxis and chemoinvasion. In Cancer Metastasis. Biological and Biochemical Mechanisms and Clinical Aspects. G. Prodi, L.A. Liotta, P.L. Lollini, S. Garbisa, S. Gorini, and K. Hellman, eds. Plenum Press, New York, pp. 215-226.
9. Bonfil, R., S. Momiki, C. Conti, and A. Klein-Szanto (1989) Benzoyl peroxide enhances the invasive ability of a mouse epidermal carcinoma cell line. Int. J. Cancer 44:165-169.

10. Bonfil, R., S. Momiki, R. Fridman, R. Reich, R. Reddel, C. Harris, and A. Klein-Szanto (1989) Enhancement of the invasive ability of a transformed human bronchial epithelial cell line by 12-0-tetradecanoyl-phorbol-13-acetate and diacylglycerol. Carcinogenesis 10:2335-2338.

11. Briozzo, P., M. Morisset, F. Capony, C. Rougeot, and H. Rochefort (1988) In vitro degradation of extracellular matrix with Mr 52,000 cathepsin D secreted by breast cancer cells. Cancer Res. 48:3688-3692.

12. Chua, C.C., D.E. Geiman, G.H. Keller, and R.L. Ladda (1985) Induction of collagenase secretion in human fibroblast cultures by growth promoting, factors. J. Biol. Chem. 260:5213-5216.

13. Colacci, A., A. Albini. A. Melchiori, P. Nanni, G. Nicoletti, S. Parodi, N. Bobola, and S. Grilli (1992) Role of chlorocompounds in multistep carcinogenesis: Acquisition of tumorigenic, metastatic and invasive potential by BALB/c 3T3 cells treated with 1,1,2,2-tetrachloroethane. Int. J. Cancer (Submitted)

14. DeClerk, Y.A., T.D. Yean, D. Chan, H. Shimada, and K.E. Langley (1991) Inhibition of tumor invasion of smooth muscle cell layers by recombinant human metalloproteinase inhibitor. Cancer Res. 51:2151-2157.

15. Dedhar, S., and R. Saulnier (1990) Alterations in integrin receptor expression on chemically transformed human cells: Specific enhancement of laminin and collagen receptor complexes. J. Cell Biol. 110:481-489.

16. Erkell, L.J., and V. Schirrmacher (1988) Quantitative in-vitro assay for tumor cell invasion through extracellular matrix or into protein gels. Cancer Res. 48:6933-6937.

17. Fidler, I.J. (1990) Critical factors in the biology of human cancer metastases: 28th G.H. Clowes Memorial Award Lecture. Cancer Res. 50:613-618.

18. Fidler, I.J. (1991) Orthotopic implantation of human colon carcinoma into nude mice provides a valuable model for the biology and therapy of metastasis. Cancer Met. Rev. 10:229-243.

19. Fidler, I.J., S. Naito, and S. Pathak (1990) Orthotopic implantation is essential for the selection, growth and metastasis of human renal cell cancer in nude mice. Cancer Metas. Rev. 9:149-165.

20. Folkman, J., and M. Klagsbrun (1987) Angiogenic factors. Science 235:442-447.

21. Fridman, R., J. Lacal, R. Reich, D. Bonfil, and C. Ahn (1990) Differential effects of phorbol ester on the in vitro invasiveness of malignant and non-malignant human fibroblast cells. J. Cell Physiol. 142:55-60.

22. Gambari, R., R. Barbieri, G. Feriotto, D. Spandidos, and C. Nastruzzi (1990) Effects of the proteinase inhibitor tetra-p-amidino-phenoxy-neopentane on in vitro adhesion and invasiveness of tumor cells. Anticancer Res. 10:259-263.

23. Gehlsen, K.R., W.S. Argraves, M.D. Pierschbacher, and E. Ruoslahti (1988) Inhibition of in vitro tumor cell invasion by Arg-Gly-Asp-containing synthetic peptides. J. Cell Biol. 106:925-930.

24. Gehlsen, K.R., L. Dillner, E. Engvall, and E. Ruoslahti (1988) The human laminin receptor is a member of the integrin family of cell adhesion receptors. Science 241:1228-1229.

25. Giavazzi, R. (1991) The nude mouse in oncology research. In Metastatic Models, E. Boven and B. Winograd, eds. CRC Press, Inc., London, pp. 117-132.

26. Giavazzi, R., J.M. Jessup, D.E. Campbell, S.M. Walker, and I.J.F. Fidler (1986) Experimental nude mouse model of human colorectal cancer liver metastases. J. Natl. Cancer Inst. 77:1303-1308.

27. Goldberg, G.I., B.L. Marmer, G.A. Grant, A.Z. Eisen, S. Wilhelm, and C. He (1989) Human 72-kilodalton type IV collagenase forms a complex with a tissue

inhibitor of metalloproteases designated TIMP-2. <u>Proc. Natl. Acad. Sci., USA</u> 86:8207-8211.

28. Graf, J., R.C. Ogle, F.A. Robey, M. Sasaki, G.R. Martin, Y. Yamada, and H.K. Kleinman (1987) A pentapeptide from the laminin B1 chain mediates cell adhesion and binds the 67 000 laminin receptor. <u>Biochem.</u> 26:6896-6900.

29. Grant, D.S., C.P. Leblond, H.K. Kleinman, S. Inoue, and J.R. Hassel (1989) The incubation of laminin, collagen IV and heparin sulfate proteoglycan at 35°C yields basement-membrane-like structures. <u>J. Cell Biol.</u> 108:1567-1574.

30. Hart, I.R., and I.J. Fidler (1978) An in vitro quantitative assay for tumor cell invasion. <u>Cancer Res.</u> 38:3218-3224.

31. Hoyhtya, M., E. Hujanen, T. Turpeenniemi-Hujanen, U. Thorgeirsson, L. Liotta, and K. Tryggvason (1990) Modulation of type-IV collagenase activity and invasive behavior of metastatic human melanoma (A2058) cells in vitro by monoclonal antibodies to type-IV collagenase. <u>Int. J. Cancer</u> 46:282-286.

32. Humphries, M., Y. Yasuda, K. Olden, and K. Yamada (1988) The cell interaction sites of fibronectin in tumour metastasis. <u>Ciba Found. Symp.</u> 141:75-93.

33. Iwamoto, Y., F.A. Robey, J. Graf, M. Sasaki, H.K. Kleinman, Y. Yamada, and G.R. Martin (1987) YIGSR, a synthetic laminin pentapeptide, inhibits experimental metastasis formation. <u>Science</u> 238:1132-1134.

34. Kanemoto, T., R. Reich, L. Royce, D. Greatorex, S.H. Adler, N. Shiraishi, G.R. Martin, Y. Yamada and H.K. Kleinman (1990) Identification of an amino acid sequence from the laminin A chain that stimulates metastasis and collagenase IV production. <u>Proc. Natl. Acad. Sci., USA</u> 87:2279-2283.

35. Khokha, R., P. Waterhouse, S. Yagel, P.K. Lala, C.M. Overall, G. Norton, and D.T. Denhardt (1989) Antisense RNA-induced reduction in murine TIMP levels confers oncogenicity on Swiss 3T3 cells. <u>Science</u> 243:947-950.

36. Kleinman, H.K., F.B. Cannon, G.W. Laurie, J.R. Hassel, M. Aumailley, V.P. Terranova, G.R. Martin, and M. Dubois-Dalcq (1985) Biological activities of laminin. <u>J Cell Biochem.</u> 27:317.

37. Kleinman, H.K., J. Graf, Y. Iwamoto, G. Kitten, R. Ogle, M. Sasaki, Y. Yamada, G.R. Martin, and L. Luckenbill-Edds (1987) Role of basement membranes in cell differentiation. <u>Ann. N.Y. Acad. Sci.</u> 513:134-145.

38. Kleinman, H.K., M.L. McGarvey, J.R. Hassell, V.L. Star, F. Cannon, G.W. Laurie, and G.R. Martin (1986) Basement membranes complexes with biological activity. <u>Biochemistry</u> 25:312-318.

39. Kohn, E., M. Stracke, C. Freter, M. Lippman, E. Schiffman, and L. Liotta (1988) Autocrine motility factor AMF stimulates phosphatidylinositol turnover in human melanoma cells. <u>Proc. Am. Assoc. Cancer Res. Ann. Meet.</u> 29:65.

40. Kramer, R.H., K.G. Bensch, and J. Wong (1986) Invasion of reconstituted basement membrane matrix by metastatic human tumor cells. <u>Cancer Res.</u> 46:1980-1989.

41. Kramer, R.H., K.A. McDonald, E. Crowley, D.M. Ramos, and C. Damsky (1989) Melanoma cell adhesion to basement membrane mediated by Integrin-related Complexes. <u>Cancer Res.</u> 49:393-402.

42. Leighton, J. (1964) Invasion, and metastasis of hetreologous tumors in the chick embryo. <u>Prog. Exp. Tumor Res.</u> 4:98-125.

43. Lester, B., J. McCarthy, Z. Sun, R. Smith, L. Furcht, and A. Spiegel (1989) G-protein involvement in matrix-mediated motility and invasion of low experimental metastatic B16 melanoma clones. <u>Cancer. Res.</u> 49:5940-5948.

44. Liotta, L.A. (1984) Tumor invasion and metastases: The role of basement membrane. Am. J. Pathol. 117:339-348.

45. Liotta, L.A. (1986) Tumor invasion and metastases: Role of the extracellular matrix: Rhoads memorial award lecture. Cancer Res. 46:1-7.

46. Liotta, L.A., W.C. Lee, and D.J. Morakis (1980) New method for preparing large surfaces of intact basement membrane for tumor invasion studies. Cancer Lett 11:141-147.

47. Liotta, L.A., R. Mandler, G. Murano, D.A. Katz, R.K. Gordon, P.K. Chiang, and E. Schiffman (1986) Tumor cell autocrine motility factor. Proc. Natl. Acad. Sci., USA 83:3302-3306.

48. Liotta, L., P. Steeg, and W. Stetler-Stevenson (1991) Cancer metastasis and angiogenesis: An imbalance of positive and negative regulation. Cell 64:327-336.

49. Liotta, L., U. Wewer, N. Rao, E. Schiffmann, M. Stracke, R. Guirguis, U. Thorgeirsson, R. Muschel, and M. Sobel (1988) Biochemical mechanisms of tumor invasion and metastases. Prog. Clin. Biol. Res. 256:3-16.

50. Mareel, M., G.K. DeBruyré, F. Vandesante, and C. Dragonetti (1981) Immunohistochemical study of embryonic chick heart invaded by malignant cells in three-dimensional culture. Invas. Metast. 1:195-204.

51. Martin, G.R., and R. Timpl (1987) Laminin and other basement membrane components. Ann. Rev. Cell Biol. 3:57-85.

52. Maslow, D.E, and V.P. Terranova (1988) Modulation of cancer cell motility and invasiveness by epidermal and platelet-derived growth factors. Proc. Am. Assoc. Cancer Res. Ann. Meet. 29:67.

53. McGarvey, T., S. Siberman, and B. Persky (1990) The effect of butyric acid and retinoic acid on invasion and experimental metastasis of murine melanoma cells. Clin. Exp. Metastasis 8:433-448.

54. Mecham, R.P. (1991) Receptors for laminin on mammalian cells. FASEB J. 5:2538-2546.

55. Meissauer, A., M. Kramer, M. Hofmann, L. Erkell, E. Jacob, V. Schirrmacher, and G. Brunner (1991) Urokinase-type and tissue-type plasminogen activators are essential for in vitro invasion of human melanoma cells. Exp. Cell Res. 192:453-459.

56. Melchiori, A., S. Carlone, G. Allavena, O. Aresu, S. Parodi, S.A. Aaronson, and A. Albini (1990) Invasiveness and chemotactic activity of oncogene transformed NIH/3T3 cells. Anticancer Res. 10:37-44.

57. Melchiori, A., A. Colacci, P.L. Lollini, C. De Giovanni, S. Carlone, S. Grilli, S. Parodi, and A. Albini (1992) Induction of invasive and experimental metastasis potential in Balb/c3T3 cells by benzo(a)pyrene transformation. (Submitted for publication.)

58. Mensing, H., A. Albini, T. Krieg, B.F. Pontz, and P.K. Muller (1984) Enhanced chemotaxis of tumor-derived and virus-transformed cells to fibronectin and fibroblast conditioned medium. Int. J. Cancer 33:43-48.

59. Mignatti, P., R. Tsuboi, E. Robbins, and D.B. Rifkin (1989) In vitro angiogenesis on the human amniotic membrane: Requirement for basic fibroblast growth factor-induced proteinases. J. Cell Biol. 108:671-682

60. Moll, U., G. Youngleib, K. Rosinski, and J. Quigley (1990) Tumor promoter-stimulated Mr 92,000 gelatinase secreted by normal and malignant human cells: Isolation and characterization of the enzyme from HT1080 tumor cells. Cancer Res. 50:6162-6170

61. Nakajima, M., D. Lotan, M.M. Baig, R.M. Carralero, W.R. Wood, M.J.C. Hendrix, and R. Lotan (1989) Inhibition by retinoic acid of type IV collagenolysis

and invasion through reconstituted basement membrane by metastatic rat mammary adenocarcinoma cells. Cancer Res. 49:1698-1706.

62. Nicolson, G.L. (1982) Metastatic tumor cell attachment and invasion assay utilizing vascular endothelial cell monolayers. J. Histochem. Cytoch. 30:214-220.

63. Ossowski, S. (1988) In vivo invasion of modified chorioallantoic membrane by tumor cells: The role of cell surface-bound urokinase. J. Cell Biol. 197:2437-2445.

64. Pourreau-Schneider, N., P. Delori, B. Boutiere, D. Arnoux, F. George, J. Sampol, and P.M. Martin (1989) Modulation of plasminogen activator systems by matrix components in two breast cancer cell lines MCF-7 and MDA-MB-231. J. Natl. Cancer Inst. 81:259-266.

65. Reich, R., E.W. Thompson, Y. Iwamoto, G.M. Martin, J.R. Deason, G.C. Fuller, and R. Miskin (1988) Effects of inhibitors of plasminogen activator, serine protinases, and collagenase IV on the invasion of basement membranes by metastatic cells. Cancer Res. 48:3307-3312.

66. Repesh, L. (1989) A new in vitro assay for quantitating tumor cell invasion. Invasion-Metastasis 9:192-208.

67. Rochefort, H., P. Augereau, P. Briozzo, F. Capony, V. Cavailles, G. Freiss, M. Garcia, T. Maudelonde, M. Morisset, and F. Vignon (1988) Structure, function, regulation and clinical significance of the 52K pro-cathepsin D secreted by breast cancer cells. Biochimie 70:943-949.

68. Ruoslahti, E., and M.D. Pierschbacher (1987) New perspectives in cell adhesion: RGD and integrins. Science 38:491-497.

69. Saiki, I., J. Murata, T. Makabe, Y. Matsumoto, Y. Ohdate, Y. Kawase, Y. Taguchi, T. Shimojo, F. Kimizuka, and I. Kato (1990) Inhibition of lung metastasis by synthetic and recombinant fragments of human fibronectin with functional domains. Jpn. J. Cancer Res. 81:1003-1011.

70. Saiki, I., J. Murata, K. Watanabe, H. Fujii, F. Abe, and I. Azuma (1989) Inhibition of tumor cell invasion by ubenimex (bestatin) in vitro. Jpn. J. Cancer Res. 80:873-878.

71. Sasaki, M., H.K. Kleinman, H. Huber, R. Deutzmann, and Y. Yamada (1988) Laminin, a multidomain protein. The A chain has a unique globular domain and homology with the basement membrane proteoglycan and the laminin B chains. J. Biol. Chem. 263:16536-16544.

72. Scher, C.D., C. Haudenschild, and M. Klagsbrun (1976) The chick chorioallantoic membrane as a model system for the study of tissue invasion by viral transformed cells. Cell 8:373-382.

73. Schlechte, W., M. Brattain, and D. Boyd (1990) Invasion of extracellular matrix by cultured colon cancer cells: Dependence on urokinase receptor display. Cancer Comm. 2:173-179.

74. Shultz, R.M., S. Silberman, B. Persky, A.S. Bajowski, and D.F. Carmichael (1988) Inhibition of human recombinant tissue inhibitor of metalloproteinases of human amnion invasion and lung colonization by murine B16-F10 melanoma cells. Cancer Res. 48:5539-5545.

75. SIoane, B.F., K. Moin, E. Kzepela, and J. Rozhin (1990) Cathepsin B and its enogenous inhibitors: The role in tumor malignancy. Cancer Metastasis Rev. 9:333-352.

76. Starkey, J.R., H.L. Hosick, D.R. Stanford, and H.D. Liggett (1984) Interaction of metastatic tumor cells with bovine lens capsule basement membrane. Cancer Res. 44:1585-1594.

77. Stetler-Stevenson, W.G., J. Talano, M. Gallagher, H.C. Krutzsch, and L.A. Liotta (1991) Inhibition of human type IV collagenase by a highly conserved peptide sequence derived from its prosegment. American J. Medic. Scien. 302:163-170.

78. Stetler-Stevenson, W.G. (1990) Type IV collagenases in tumor invasion and metastasis. Cancer Metastasis Rev. 9:289-303.

79. Stetler-Stevenson, W.G., H. Krutzsch, and L.A. Liotta (1989) Tissue inhibitor of metalloproteinase-2 (TIMP-2): A new member of the metalloproteinase inhibitor family. J. Biol. Chem. 264:17374-17378.

80. Stetler-Stevenson, W.G., H. Krutzsch, M. Wacher, M.K. Margulies, and L.A. Liotta (1989) The activation of human type IV collagenase proenzyme. J. Biol. Chem. 264:1353-1356.

81. Stracke, M.L., E. Kohn, S. Aznavoorian, L. Wilson, D. Salomon, H. Krutzsch, L.A. Liotta, and E. Schiffmann (1988) Insulin-like growth factors stimulate chemotaxis in human melanoma cells. Biochem. Biophys. Res. Comm. 153:1076-1083.

82. Terranova, P., E.S. Hujanen, and G.R. Martin (1986) Basement membranes and the invasive activity of metastatic tumor cells. J. Natl. Cancer Inst. 77:311-316.

83. Terranova, V.P., E.S. Hujanen, D.M. Loeb, G.R. Martin, L. Thornburg, and V. Glushko (1986) Use of a reconstituted basement membrane to measure cell invasiveness and select for highly invasive tumor cells. Proc. Natl. Acad. Sci., USA 83:465-469.

84. Terranova, V.P., L.A. Liotta, R.G. Russo, and G.R. Martin (1982) Role of laminin in the attachment and metastasis of murine tumor cells. Cancer Res. 42:2265-2269.

85. Terranova, V.P., C.N. Rao, T. Kalebic, I.M. Margulies, and L.A. Liotta (1983) Laminin receptors on human breast carcinoma cells. Proc. Natl. Acad. Sci., USA 80:444-448.

86. Terranova, V.P., J.E. Williams, L.A. Liotta, and G.R. Martin (1984) Modulation of the metastatic activity of melanoma cells by laminin and fibronectin. Science 226:982-984.

87. Testa, J.E., and J.O. Quigley (1990) The role of urokinase-type plasminogen activator in aggressive tumor cell behaviour. Cancer Metastasis Rev. 9:353-367.

88. Thompson, E., S. Nakamura, T. Shima, A. Melchiori, G. Martin, S. Salahuddin, R. Gallo, and A. Albini (1991) Supernatants of acquired immunodeficiency syndrome-related Kaposi's sarcoma cells induce endothelial cell chemotaxis and invasiveness. Cancer Res. 51:2670-2671.

89. Turpeenniemi-Hujianen, T., U.P. Thorgeirrson, C.N. Rao, and L.A. Liotta (1986) Laminin increases the release of type IV collagenase from malignant cells. J. Biol. Chem. 261:1883-1889.

90. Van Wart, H.E, and H. Birkedal-Hansen (1990) The cysteine switch: A principle of regulation of metalloproteinase activity with potential applicability to the entire matrix metalloproteinase gene family. Proc. Natl. Acad. Sci., USA 87:5578-5582.

91. Varani, J., and P.A. Ward (1982) Tumor cell chemotaxis. In The biological basis of metastasis, I. Hart and L.A. Liotta, eds. Martinus Nijhoff, The Hague, pp. 99-112.

92. Welch, D.A., T.J. Lobl, E.A. Seftor, P.J. Wack, P.A. Aeed, K.H. Yohem, R.E.B. Seftor, and M.J.C. Hendrix (1989) Use of the membrane invasion culture system (MICS) as a screen for anti-invasive agents. Int. J. Cancer 43:449-457.

93. Xu, H., J. Sumegi, S.X. Hu, A. Namerjee, A., E. Uzvolgyi, and W.F. Benedict (1990) Intraocular tumor formation of RB reconstituted retinoblastoma cells.

Cancer Res. 51:4481-4485.

94. Yagel, S., R.S. Parher, J.J. Jeffrey, and P.K. Lala (1988) Normal nonmetastatic human trophoblast cells share *in vitro* invasive properties of malignant cells. J. Cell Physiol. 136:455-462.

95. Yamada, K., D. Kennedy, S. Yamada, H. Gralnick, W. Chen, and S. Akiyama (1990) Monoclonal antibody and synthetic peptide inhibitors of human tumor cell migration. Cancer Res. 50:4485-4496.

DIRECT INTERCEPTION OF MUTAGENS

AND CARCINOGENS BY BIOMOLECULES

Philip E. Hartman and Zlata Hartman
Department of Biology
The Johns Hopkins University
Baltimore, Maryland 21218 USA

ABSTRACT

Five points are emphasized: 1. Chemical interception and mere physical exclusion of mutagens and carcinogens constitute the major means by which mutations in cellular DNA are prevented. DNA repair processes comprise critical, but relatively minor, modes of genetic protection. 2. Disruption of a mutagen-interception defense mechanism can lead to substantial increases in mutagenesis and can preordain sites to eventual tumor formation. 3. Quantitation of the relative contributions of various blocking molecules is often simplified by the fact that protection can be calculated merely through knowledge of the measured concentration of the antimutagen and its rate of reaction with specific mutagens as measured in straightforward in vitro tests. 4. Two recently recognized defensive molecules, carnosine and ergothioneine, are put forward as examples of interesting chemical interceptor molecules. 5. Essentially all antimutagens are in fact "double-edged swords." Situations can be artificially constructed that can lead to generation of toxic species from molecules that are normally antimutagens; in isolated cases some of these interactions can be pictured as having deleterious consequences in vivo. This may be one reason why a number of important antimutagens are often sequestered, either in different tissues or by binding to dispensable macromolecules.

INTRODUCTION

The International Conference on Mechanisms of Antimutagenesis and Anticarcinogenesis-I (ICMAA-I) (72,73) and ICMAA-II (53) cover multiple facets involved in protection of cellular DNA from what otherwise would be an avalanche of mutagenic/ carcinogenic damage. New information accumulated since the first conference has served to consolidate some views expressed at that meeting (34) and to point out others. This paper will stress five points that seem under-appreciated at the present time. While a number of antimutagens do not act through a single mechanism, for the sake of brevity we will stress interception mechanisms, realizing that other effects may be operating as adjunct antagonizers of mutagens under particular circumstances.

INTERCEPTION MECHANISMS ARE EXTREMELY IMPORTANT

As pointed out elsewhere (43), the overwhelming percentage of exogenous and endogenous potentially mutagenic agents is blocked from ever encountering cellular DNA. Much of this quenching of mutagenic activity occurs through mechanisms preventing entry into critical cells (for example, stem cells) or by interception intracellularly of exogenous and endogenous mutagens (Fig. 1). DNA repair processes, while certainly essential and impressively precise, are of much lesser total importance in the preservation of genetic integrity. The numbers in Fig. 1 are somewhat arbitrary and are merely educated guesses of averages for all mutagens. Nevertheless, even with substantial adjustments, the overall picture would change little. While it is said that "An ounce of prevention is worth a pound of cure," what cells and tissues actually use is a pound of prevention and an ounce of cure.

DISRUPTION OF AN INTERCEPTING DEVICE
LEADS TO AN INCREASE IN GENETIC DAMAGE

If chemical interception of mutagens/carcinogens is important, we should see increased mutagenesis/carcinogenesis when one of these quenching mechanisms is disrupted or eliminated. A few examples of particular interest to the authors will be briefly mentioned here.

Ames Salmonella Strains

The collection of Ames Salmonella mutagen-tester strains lack the majority of their outer cell walls due to *rfa* mutations (4). Although not examined directly, these strains probably also lack the contents of their periplasms. The periplasm of *Salmonella* has been calculated to comprise about one-third of the total cell volume (80) and is packed with a wide variety of molecules (62,65). The molecules may include reduced glutathione, since this molecule is exported by laboratory bacterial strains (66) in quantities sufficient to be protective against micromolar concentrations of N-methyl-N'-nitro-N-nitrosoguanidine (MNNG) and other agents reactive with aliphatic thiols (67).

The relative impact of disrupting the outer cellular barrier of *Salmonella* varies greatly with the mutagen under test. With some small "direct-acting" compounds such as N-methyl-N-nitrosourea and N-methyl-N-nitrosoacetamide, there is no significant effect of cell wall structure on mutagenicity (13), but many other agents are decidedly more mutagenic for *rfa* strains than for their wild-type counterparts. In many cases the increase in mutagenicity due to a *rfa* genetic background is larger than for a DNA excision repair-defective (*uvr*) genetic background (4). An example is shown in Table 1. One component of this increased mutagenicity in some cases could merely be a generalized increase in cell permeability. Another component in some cases could be loss of quenching mechanisms that disallow entry of particular mutagens, for example electrophiles that react with one or more components of the periplasm.

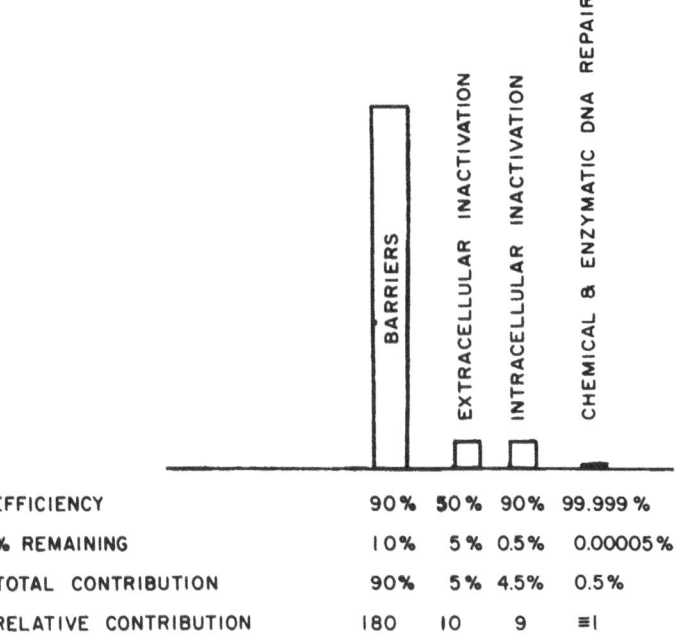

	BARRIERS	EXTRACELLULAR INACTIVATION	INTRACELLULAR INACTIVATION	CHEMICAL & ENZYMATIC DNA REPAIR
EFFICIENCY	90%	50%	90%	99.999%
% REMAINING	10%	5%	0.5%	0.00005%
TOTAL CONTRIBUTION	90%	5%	4.5%	0.5%
RELATIVE CONTRIBUTION	180	10	9	≡1

Figure 1. Relative roles in decreasing mutagenic damage in cellular DNA played by tissue and cellular organization ("barriers," see text), extra- and intracellular quenching of mutagens, and by DNA repair mechanisms. While relatively inefficient, the various modes of mutagen interception collectively comprise the predominant means of protection of cellular DNA fidelity. Adapted from Ref. 43.

Intestinal Metaplasia

Small focal lesions that are probably mutagen-induced and are extremely frequent accumulate and slowly expand during the lifetime of mammals, including man (33,42). Prominent among such lesions is "intestinal metaplasia" of the stomach. Expanding clusters of cells with many properties of small or of large intestinal epithelial cells form sites of mixed surface epithelium in the stomach mucosa. Such sites, containing cells "transdetermined" to act structurally and functionally as intestinal cells, disrupt multiple barriers to mutagen penetration (34,42). These intestinalized foci also perhaps comprise microenvironments conducive to increased mutagen and mitogen production localized at these specific sites (16). Foci of intestinalized mucosa may be "preinitiated" in the sense that no oncogenes or tumor suppressor genes have been mutated. It is only a critical stage in terminal differentiation that seems to have been permanently altered in the switch to metaplasia. The altered cells and their immediate progeny are "self" and thus are relatively free from immunological surveillance. At the same time, they exhibit an increased level of DNA replication and cell turnover characteristic of the intestine as opposed to the slower cell turnover typical of gastric epithelium, and the topology of cells synthesizing DNA is shifted (46,54,88). Such mixed epithelia result in disruption of multiple defense mechanisms and have been termed "barrier breakdown." Thus, these sites often are precursors of peptic ulcers (48) and gastric cancers (31,42). That is, the sites are primed to progress through stages that, on rare occasion, eventually

Table 1. Number of His$^+$ mutations per plat induced by 100 μg per plate of hycanthone furoate in *Salmonella* strains competent and defective in excision repair (*uvrB*) and in outer cell wall lipopolysaccharide side-chain biosynthesis (*rfa*). Adapted from Hartman et al. (37). For a review of hycanthone mutagenicity, see Refs. 35 and 52A.

Strain	Induced Reversions Per Plate	Relative No. of Induced Reversions
hisC3076	19	≡ 1
TA1952 (hisC3076 uvrB)	63	3
TA1537 (hisC3076 uvrB rfa)	2145	112
hisD3052	10	≡ 1
TA1534 (hisD3052 uvrB)	165	17
TA1538 (hisD3052 uvrB rfa)	2420	242
TA1978 (hisD3052 rfa)	379	38

gain the multiple mutations requisite for gastric carcinoma formation (16,34,42,54). A number of studies, including two from our laboratory, have pinpointed sites of dysplasia and very early gastric carcinomas as arising in regions of intestinal metaplasia (42,57,58). Prior intestinalization is certainly not required for gastric carcinoma formation, but intestinal metaplasia does predominate as a step in the carcinoma sequence in the elderly in whom gastric cancer most often arises (42). It is clearly possible to induce gastric cancers at unintestinalized sites, and these "undifferentiated" tumors often eventually develop transdetermined intestinal foci as they grow (85-87). But one step in the major pathway of gastric carcinoma formation in man involves what is probably a simple genetic change leading to intestinal cells embedded in the gastric mucosa.

Ability to induce mitogenesis has been proposed as one very important factor in the carcinogenicity of chemicals (2,3). The increased mitoses found in intestinal metaplasia, as noted above, could well introduce an important mitogenic factor in the gastric carcinoma sequence. Even a small shift in steady state probability of "self maintenance" (i.e., the ratio at cell division of stem versus differentiating progeny cells; see Ref. 68) of the altered metaplastic stem cells could have an important effect. In addition, gastric carcinogens themselves induce gastric mucosal damage and, thus, probably are mitogenic, whereas nongastric carcinogens do not induce mucosal damage (81,82).

Sodium chloride acts as a gastric cocarcinogen by damaging gastric mucosa and serving as a mitogen. Sodium chloride increases the number of cells in S phase and induces ornithine decarboxylase activity (15,26,64) but does not appear to enhance mutagen penetration into the total mucosa (15). (It still remains possible that stem cell susceptibility is altered.) These data are in keeping with earlier observations that sodium chloride acts predominantly as a gastric cocarcinogen and not merely as a tumor promoter; salt does not act after the carcinogenic process has been initiated (74,83,84). High salt intake has been implicated as one factor in induction of stomach cancer in man, and decreased salt intake has been flagged as one component in the declining incidence of gastric cancer (32,47,49,50). Data are consistent with the idea that salt intake enhances induction of "undifferentiated" gastric carcinomas directly and "differentiated" carcinomas indirectly via induction of sites of intestinal metaplasia.

Glutathione Status and Ataxia Telangiectasia

Ataxia telangiectasia (AT) is a pleiotropic human autosomal recessive disorder whose manifestations include a predisposition to cancer and a cellular sensitivity to ionizing radiation damage. The spontaneous mutation frequency is elevated in AT cells (6). Defects in DNA repair by AT cells have been searched for extensively with only modestly suggestive results supporting the DNA repair hypothesis (25,69). It now appears that in at least some cases of AT the cellular effects are due to impaired cysteine uptake and, thus, slow intracellular regeneration of glutathione; AT homozygous cells require 24 hr to reconstitute about 30% of their normal intra-cellular glutathione content as opposed to 100% repletion in 6 hr by wild-type cells (56).

The bias of the scientific community toward examination for DNA repair rather than other causal factors not only has delayed pinpointing of the primary defect in glutathione status of AT cells but also appears, at the present time, to hinder general acceptance of the glutathione explanation. Antimutagens are of extreme importance in the carcinogenic process and should be seriously considered for investigation. Over 300 different cancer-prone genetic disorders in man are listed by McKusick (55), and most of these disorders are pleiotropic. It seems quite likely that a number of these will turn out to affect the production and/or the localization of antimutagens.

CHEMICAL REACTIVITIES CAN BE MEASURED BY SIMPLE ASSAYS

Nonenzymatic mutagen-inactivation reactions are relatively simple to estimate. They involve in vitro determinations of reaction rates of mutagen-antimutagen interactions and determination of the measured concentrations of the interceptor molecules in, for example, particular fluids of biological origin. Although exogenous singlet oxygen is not mutagenic for bacteria (19), the proposed approach is nicely exemplified by the quantitative studies of Kanofsky (51) on the quenching of singlet oxygen by human plasma. The contributions of various suspect biomolecules to the total quenching were calculated from their plasma concentrations and their quenching constants when tested individually. Thus, each biomolecule could be assigned a percentage contribution of total plasma quenching (51). Interestingly, albumin was the most significant quencher; albumin has been pointed out as a significant blocker of toxic species (28-30,43). Additionally, the sum of the quenching activities of the individual molecules tested was not significantly different from that for plasma (51). This indicates that all major singlet oxygen antagonists present in plasma were

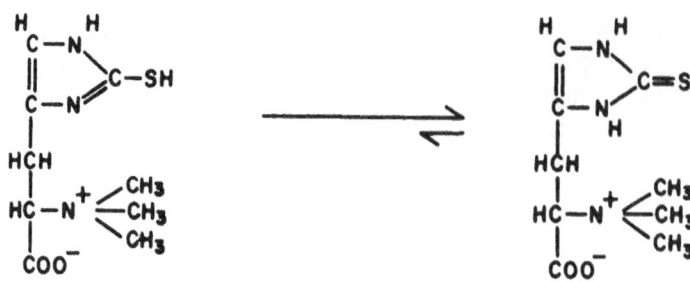

Figure 2. Ergothioneine (2-thiol-L-histidine betaine) depicted as a thiol (left) and as the predominant thione (right). After Ref. 36.

detected and, thus, there were not other important quenchers to be searched for and characterized. Similarly, although slightly more complicated, analyses could be performed for other defensive biomolecules acting against arrays of chemical mutagens, as has already been illustrated in the case of glutathione (43). The compilation of antimutagenic profiles (Ref. 94 and M. Waters). Conference presentation is an important step in this direction. However, supplementary "chemical destruction profiles" would be extremely easy and fast to run; a vast amount of predictive data could be accumulated in a short amount of time via HPLC analyses, etc. The field of antimutagenesis, however, needs to recruit additional competent chemists to its ranks in order to achieve this swing away from complicated and cumbersome biological assays. In the case of glutathione, the world literature on reaction rates (with and without glutathionyl-S-transferases) with various xenobiotics still awaits quantitative compilation.

In another approach to the measurement of relative protective activities of various putative blocking agents in plasma, Frei et al. (24) examined the temporal disappearance of antioxidants in response to reactive agents. Information on the sequential depletion of protector biomolecules was supplemented by investigation of the effects of addition of a particular antimutagen and by gel fractionation of plasma in order to further pinpoint its most active constituents (22).

One thing that would greatly facilitate assessment of biomolecules that interrupt flow of mutagens to DNA would be a reliable catalogue of the molecular components in critical biological fluids. Generally, antimutagens that intercept in the absence of enzyme intervention will be present in the micromolar to millimolar range to be effective. Yet most of the literature values on their contents, in fluids as diverse as seminal fluid to the aqueous humor of the human lens, are from older assays that have been replaced by advanced and much more precise assays developed only in recent years. Quantitative reassessments are called for. In the human, aside from the external cavities where an enormous variety of ingested foreign molecules may be present, there is not going to be found an unmanageable array of different classes of natural antimutagens/anticarcinogens.

TWO INTERESTING CANDIDATES AS SIGNIFICANT INTERCEPTORS: CARNOSINE AND ERGOTHIONEINE

The dipeptide L-carnosine (β-alanyl-L-histidine) is found in millimolar concentrations in the striated muscles of mammals (references in Ref. 43). Carnosine appears both to act as a buffer and to serve as an antioxidant of significance (7-10,17,20,40,41, 52). Although some of these results (7-10,17) have been questioned as based on the ability of a carnosine preparation to inhibit a commonly used version of the thiobarbituric acid test (5), the questioned results do coincide with data obtained by other independent methodologies. Carnosine-metal complexes can generate active oxygen species (see below, "Unusual Cellular Circumstances") and, thus, differing transition metal content may underlie some of the discordant results.

Carnosine is not effective in blocking mutagenesis in *Salmonella* by *N*-methyl-*N'*-nitro-*N*-nitrosoguanidine and 4-nitroquinoline-*N*-oxide but is active against t-butyl hydroperoxide (45); it remains untested against other chemical mutagens (43).

Ergothioneine (Fig. 2) is an extremely interesting antioxidant and antimutagen (1,36,41,43,45). Synthesized by fungi, ergothioneine is taken up by the roots of plants. Mammals assimilate a very high percentage of ingested ergothioneine and conserve it in particular tissues where it serves as an adjunct to the differently reactive aliphatic thiols, such as glutathione. Thus, antioxidant/antimutagenic defenses in some tissues and body fluids can be readily modulated by dietary supplementation with this naturally occurring compound that seems treasured and to which mammals have been exposed through evolution. The very limited experimental work available suggests that ergothioneine can inhibit mutagenicity of a variety of compounds and has a different spectrum of activity from glutathione. Ergothioneine also may assist in counteracting oxidative damage through its strong chelating ability for divalent cations, especially copper (1,60,61). This chelating ability lessens formation of hydroxyl radicals from hydrogen peroxide and perhaps also blocks deleterious hydroxylation reactions (1,36,59).

"UNUSUAL" CELLULAR CIRCUMSTANCES CAN CONVERT ANTIMUTAGENS TO TOXIC SPECIES

War against the enemy requires constant coordination and assessment; otherwise, a defensive force can end up killing some of its own troops. The same is found in nature's systems. Cells possess dynamic compensatory processes, but these can occasionally go awry or become manipulated and, thus, create damaging species from otherwise protective molecules.

Ergothioneine

What largely comprises a defense in daily survival against deleterious insults may be turned around into a situation producing cellular damage. Hundreds of quinones occur in nature, with the great majority accumulated as secondary metabolic products by fungi and by plants (89). It turns out that some of these quinones produce singlet oxygen when illuminated alone. In the presence of reducing agents such as ergothioneine, urate, or ascorbate the quinones are reduced to semiquinones which, in turn, are reoxidized with the production of a second toxic form of active oxygen, namely, superoxide (21,38,39,44). Such "redox cycling" (76) may actually be used by plant pathogenic fungi to produce superoxide radicals as one weapon in their attack on host cells (Fig. 3). This would allow a barrage of two different toxic oxygen species whose alternate generation could well circumvent specific plant defenses (Legend, Fig. 3). The main point to be made here, however, is not to conjecture how

Figure 3. Plant invasion by certain pathogenic fungi producing α-hydroxy quinones is proposed as being facilitated by a repeated, stepwise production of active oxygen species synchronized by light. 1.) Fungal mycelia contain inactive or water soluble glycosides of reduced α-hydroxyl quinones <u>Sequestered in vacuoles</u>. 2.) Fungi growing in the extracellular spaces of the plant host export the quinones. 3.) The nontoxic precursors are activated int he extracellular space. 4.) The activated quinones emit singlet oxygen when illuminated in daytime in the absence of extracellular reducing agents, damaging plant cell membranes. 5.) Reducing agents (ASC = ascorbate, GSH = reduced glutathione, etc.) leak out of damaged, proximal plant cells. 6.) Redox cycling (76) begins in the presence of reducing agents with the generation of superoxide, which is a different damaging active oxygen species than the singlet oxygen originally produced. Reducing agents become oxidized. 7.) At night, photomediated singlet oxygen and superoxide formation ceases. Extracellular reducing agents and α-hydroxy quinones assume their completely oxidized forms. The system is now primed for light-mediated singlet oxygen production, cycling back to 4.). The process is repeated: 4.) → 7.) and 1.) → 7.), just as day follows night. This hypothetical sequence was constructed from data in Refs. 21, 38, 39, and 44.

fungi may switch host defenses to fight plants (and how plant photodynamic quinones may be similarly used to fight insect predators). Rather, it is important to realize that molecules that are truly beneficial under many circumstances can be turned into cell and DNA-damaging agents in special circumstances, not just in plants but also in people. For example, free radicals generated during redox cycling of quinones have been shown to be mutagenic (18,76); yet in normal circumstances lipid-soluble, natural benzoquinones such as coenzyme Q (ubiquinones) are thought to serve as physiologically significant antioxidants (23).

Glutathione

While reduced glutathione (GSH) plays a major role in interrupting the transit of mutagens to DNA, GSH can have unfavorable consequences under particular circumstances (reviewed in Ref. 43). These include: (a) intracellular activation to mutagenic species of dihaloalkanes and nitrosamides, (b) formation of active oxygen species and ultimately DNA-damaging hydroxyl radicals during auto-oxidation, a process that is enhanced if GSH is converted to cysteinylglycine by the action of γ-glutamyltranspeptidase (77,78), and (c) oxidation to thionyl radicals. Thus, GSH can be protective or damaging, depending on its concentration (41,67,70) and how well it is protected against conversion to more damaging derivatives.

Ascorbate

It has long been recognized that ascorbic acid can produce active oxygen species during auto-oxidation. Cell and DNA damage are greatly accelerated when transition metal ions such as Cu^{2+} are present together with ascorbate (14,79; reviewed in Refs. 30,43). The interplay of protector molecules, for example, ascorbate with other cellular reducing agents, can be of importance in blocking deleterious effects under normal circumstances (e.g., 63,71,95).

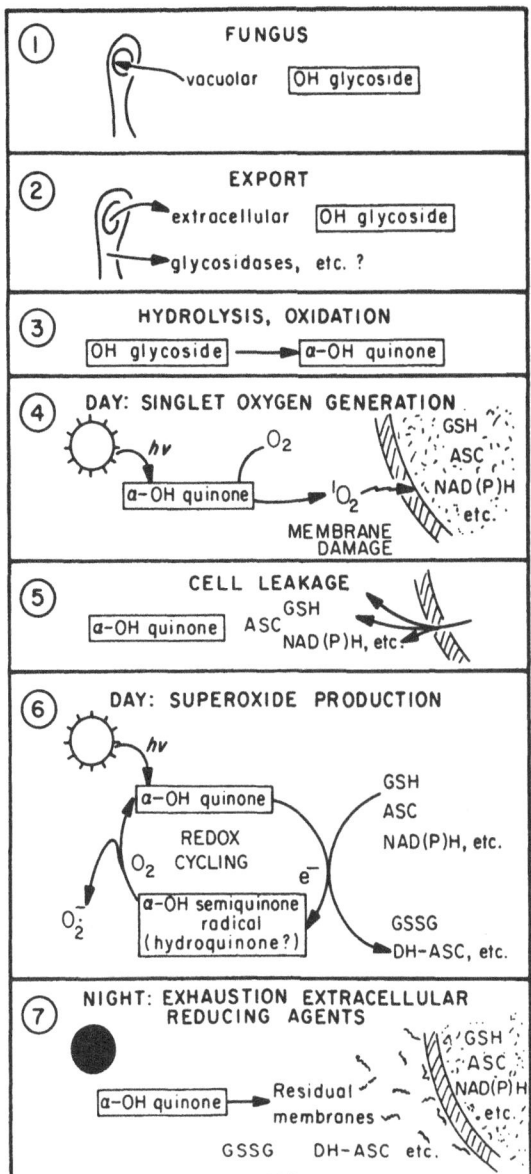

① FUNGUS

vacuolar |OH glycoside|

② EXPORT

extracellular |OH glycoside|

glycosidases, etc. ?

③ HYDROLYSIS, OXIDATION

|OH glycoside| ⟶ |α-OH quinone|

④ DAY: SINGLET OXYGEN GENERATION

hν   O₂   GSH
                              ASC
|α-OH quinone| ⟶ ¹O₂   NAD(P)H
                              etc.
MEMBRANE
DAMAGE

⑤ CELL LEAKAGE
                    GSH
|α-OH quinone| ASC
                    NAD(P)H, etc.

⑥ DAY: SUPEROXIDE PRODUCTION

hν

|α-OH quinone| GSH
                          ASC
REDOX   NAD(P)H, etc.
CYCLING
O₂   e⁻

O₂⁻
|α-OH semiquinone
radical
(hydroquinone?)| GSSG
                          DH-ASC, etc.

⑦ NIGHT: EXHAUSTION EXTRACELLULAR
REDUCING AGENTS
                              GSH
                              ASC
                              NAD(P)H
|α-OH quinone| ⟶ Residual   etc.
                    membranes
        GSSG   DH-ASC etc.

Carnosine and Carnosine-Ergothioneine Mixtures

In experiments examining the prevention by ergothioneine of singlet oxygen production by illuminated dyes and the quenching of singlet oxygen by carnosine (20,40), we have observed a slow production of active oxygen species in phosphate buffer solutions of L-carnosine preparations (Z. Hartman and P.E. Hartman, unpublished observations). This oxygen-requiring activity includes the production of H_2O_2, as measured by the ability of catalase to retard the activity. Activity does not

Table 2. Some major antioxidants in the normal and cataractous human eye lens.

Compound	Normal Lens	Cataracts	Reference
Glutathione	6.5 mM	0.9 mM	(9)
Carnosine	24 μM	5.2 μM	(9)
Ergothioneine	5 mM	2.9 mM	(75)
Ascorbate	1.3 – 2 mM	1.3 – 2 mM[a]	(90–92)
Pyruvate	400 μM	?	(93)

[a]However, ascorbate content of aqueous humor decreased about 50%.

require light or the addition of a photosensitizer. Strikingly, activity is enhanced 7- to 14-fold in the presence of equimolar ergothioneine or any of a number of 2-mercapto-imidazoles, and activity is increased about two-fold by thiourea or dimethylthiourea. Activity in mixtures is linearly proportional to carnosine concentration and to mercaptoimidazole concentration. Activity is slightly enhanced by addition of micromolar concentrations of copper or cobalt ions but not by addition of ferrous or ferric ions. Chelex pretreatment of phosphate buffer and deionized water solutions (14) used in the experiments greatly reduces activity, indicating that carnosine-transition metal complexes are responsible for the activity. The reaction appears very carnosine-specific in the sense that a variety of closely related imidazoles are completely inactive. L-histidine and glycyl-L-histidine remain inactive even when transition metal ions have been added. Addition of superoxide dismutase has little effect on the reaction. Our current hypothesis is that a metal-carnosine complex, particularly in the presence of a mercaptoimidazole, generates superoxide. It is known that histidine peptide-metal chelates can serve as superoxide dismutases, producing H_2O_2 from superoxide (11,27). The metal ion also could play a role in reduction of H_2O_2 to hydroxyl radicals (see Ref. 30). Formation of hydroxyl radicals could account for the assay reaction measured, namely, bleaching of N,N-dimethyl-4-nitrosoaniline (assay method described in Ref. 40).

Sequestered Molecules

Regardless of precise mechanism, it is clear from the above that each of two molecules normally active in retarding cell damage can have quite the opposite effect when mixed. This may be one reason why some of the native biomolecules most active in antimutagenesis are sequestered. Separation and protection may occur by localization in different tissues, in different cellular compartments, or by specific binding to macromolecules, such as albumin when circulating. Besides molecules such as ascorbate, bilirubin, and retinoic acid, albumin binds carnosine (12). Such site-specific binding to albumin of useful yet potentially deleterious molecules may withdraw them from damaging activation or, at least, target damage to the dispensable carrier protein itself.

In a search of the literature, only one tissue in man has been located where both carnosine and ergothioneine have been reported as being present at concentrations that could give rise to significant active oxygen species in the absence of any enzymatic catalysis. This tissue is the eye lens (Tab. 2). Perhaps the presence of other antioxidants impedes oxidative damage to lens proteins. Such protein damage is apparently of importance in formation of senile cataracts (92) just as DNA damage is of importance in carcinogenesis. Or, it is possible that even in the eye lens carnosine and ergothioneine are sequestered so that each could individually have beneficial impacts and be prevented from interacting. In either event, it is of interest from the point of view of carcinogenesis to consider what roles cellular disruption and abnormal partitioning of antimutagens might contribute to the sequential process. Could conversion of such "double-edged swords" in vitro be mimicked on occasion in vivo?

ACKNOWLEDGEMENTS

We thank Dr. Saul Roseman for his support during our studies of anti-mutagens and Ms. Donna Bragg for her expertise on the word processor. Contribution No. 1467 of the Department of Biology, The Johns Hopkins University.

REFERENCES

1. Akanmu, D., R. Cecchini, O.I. Aruoma, and B. Halliwell (1991) The antioxidant action of ergothioneine. Arch. Biochem. Biophys. 288:10-16.
2. Ames, B.N., and L.S. Gold (1990) Too many rodent carcinogens: Mitogenesis increases mutagenesis. Science 249:970-971.
3. Ames, B.N., and L.S. Gold (1990) Chemical carcinogenesis: Too many rodent carcinogens. Proc. Natl. Acad. Sci., USA 87:7772-7776.
4. Ames, B.N., F.D. Lee, and W.E. Durston (1973) An improved bacterial test system for the detection and classification of mutagens and carcinogens. Proc. Natl. Acad. Sci., USA 70:782-786.
5. Aruoma, O.I., M.J. Laughton, and B. Halliwell (1989) Carnosine, homocarnosine and anserine: Could they act as antioxidants in vivo? Biochem. J. 264:863-869.
6. Bigbee, W.L., R.G. Langlois, M. Swift, and R.H. Jensen (1989) Evidence for an elevated frequency of in vivo somatic cell mutations in ataxia telangiectasia. Amer. J. Hum. Genet. 44:402-408.
7. Boldyrev, A.A. (1990) Retrospectives and perspectives on the biological activity of histidine-containing dipeptides. Intl J. Biochem. 22:129-132.
8. Boldyrev, A.A., A.M. Dupin, M.A. Batrukova, N.I. Bavykina, G.A. Korshunova, and Y.P. Shvachkin (1989) A comparative study of synthetic carnosine analogs as antioxidants. Comp. Biochem. Physiol. 94B:237-240.
9. Boldyrev, A.A., A.M. Dupin, A.Y. Bunin, M.A. Babizhaev, and S.E. Severin (1987) The antioxidative properties of carnosine, a natural histidine-containing dipeptide. Biochem. Intl. 15:1105-1113.
10. Boldyrev, A.A., A.M. Dupin, E.V. Pindel, and S.E. Severin (1988) Antioxidative properties of histidine-containing dipeptides from skeletal muscles of vertebrates. Comp. Biochem. Physiol. 89B:245-250.

11. Brigelius, R., R. Spöttl, W. Bors, E. Lengfelder, M. Saran, and U. Weser (1974) Superoxide dismutase activity of low molecular weight Cu^{2+}-chelates studied by pulse radiolysis. FEBS Lett. 47:72-75.

12. Brown, C.E., and W.E. Antholine (1979) Chelation chemistry of carnosine. Evidence that mixed complexes may occur in vivo. J. PhYs. Chem. 83:3314-3319.

13. Brundrett, R.B., M. Colvin, E.H. White, J. McKee, P.E. Hartman, and D.L. Brown (1979) Comparison of mutagenicity, antitumor activity, and chemical properties of selected nitrosoureas and nitrosamides. Cancer Res. 39:1328-1333.

14. Buettner, G.R. (1988) In the absence of catalytic metals ascorbate does not autooxidize at pH 7: Ascorbate as a test for catalytic metals. J. Biochem. Biophys. Meth. 16:27-40.

15. Charnley, G., and S.R. Tannenbaum (1985) Flow cytometric analysis of the effect of sodium chloride on gastric cancer risk in the rat. Cancer Res. 45:5608-5616.

16. Charnley, G., S.R. Tannenbaum, and P. Correa (1982) Gastric cancer: An etiologic model. In Nitrosamines and Human Cancer, Banbury Report 12, P.N. Magee, ed. Cold Spring Harbor Laboratory, New York, pp. 503-522.

17. Chasovnikova, L.V., V.E. Formazyuk, V.I. Sergienko, A.A. Boldyrev, and S.E. Severin (1990) The antioxidative properties of carnosine and other drugs. Biochem. Intl. 20:1097-1103.

18. Chesis, P.L., D.E. Levin, M.T. Smith, L. Ernster, and B.N. Ames (1984) Mutagenicity of quinones: Pathways of metabolic activation and detoxification. Proc. Natl. Acad. Sci.,USA 81:1696-1700.

19. Dahl, T.A., W.R. Midden, and P.E. Hartman (1988) Pure exogenous singlet oxygen: Nonmutagenicity in bacteria. Mutat. Res. 201:127-136.

20. Dahl, T.A., W.R. Midden, and P.E. Hartman (1988) Some prevalent biomolecules as defenses against singlet oxygen damage. Photochem. Photobiol. 47:357-362.

21. Daub, M.E. (1987) The fungal photosensitizer cercosporin and its role in plant disease. In Light-Activated Pesticides, J.R. Heitz and K.R. Downum, eds. Amer. Chem. Soc., Washington, pp. 271-280.

22. Frei, B., L. England, and B.N. Ames (1989) Ascorbate is an outstanding antioxidant in human blood plasma. Proc. Natl. Acad. Sci., USA 86:6377-6381.

23. Frei, B., M.C. Kim, and B.N. Ames (1990) Ubiquinol-10 is an effective lipid-soluble antioxidant at physiological concentrations. Proc. Natl. Acad. Sci., USA 87:4879-4883.

24. Frei, B., R. Stocker, and B.N. Ames (1988) Antioxidant defenses and lipid peroxidation in human blood plasma. Proc. Natl. Acad. Sci., USA 85:9748-9752.

25. Friedberg, E.C., U.K. Ehmann, and J.I. Williams (1979) Human diseases associated with defective DNA repair. Adv. Radiat. Biol. 8:85-174.

26. Furihata, C., Y. Sato, M. Hosaka, T. Matsushima, F. Furukawa, and M. Takahashi (1984) NaCl induced ornithine decarboxylase and DNA synthesis in rat stomach mucosa. Biochem. Biophys. Res. Commun. 121:1027-1032.

27. Gulyaeva, N.V. (1988) Superoxide-scavenging activity of carnosine in the presence of copper and zinc ions. Biochemistry (Biokhimiya) 52:1051-1054.

28. Gutteridge, J.M.C. (1986) Antioxidant properties of the proteins caeruloplasmin, albumin and transferrin. Biochim. Biophys. Acta 869:119-127.

29. Halliwell, B. (1988) Albumin -- An important extracellular antioxidant? Biochem. Pharmacol. 37:569-571.

30. Halliwell, B., and J.M.C. Gutteridge (1987) Free Radicals in Biology and Medicine. Clarendon Press, Oxford, 346 pp.

31. Hartman, P.E. (1982) Nitrates and nitrites: Ingestion, pharmacodynamics and toxicology. In Chemical Mutagens: Principles and Methods for Their Detection, Vol. 7, F.J. de Serres and A. Hollaender, eds. Plenum Press, New York, pp. 211-294.

32. Hartman, P.E. (1982) Nitrite load in the upper gastrointestinal tract -- Past, present, and future. In Nitrosamines and Human Cancer, Banbury Report 12, P.N. Magee, ed. Cold Spring Harbor Laboratory, New York, pp. 415-431.

33. Hartman, P.E. (1983) Mutagens: Some possible health impacts beyond carcinogenesis. Environ. Mutag. 5:139-152.

34. Hartman, P.E. (1986) Interception of toxic agents/mutagens/carcinogens: Some of nature's novel strategies. In Antimutagenesis and Anticarcinogenesis: Mechanisms, D.M. Shankel, P.E. Hartman, T. Kada, and A. Hollaender, eds. Plenum Press, New York, pp. 169-179.

35. Hartman, P.E. (1989) Early years of the Salmonella mutagen tester strains: Lessons from hycanthone. Environ. Molec. Mutag. 14 (Suppl. 16):39-45.

36. Hartman, P.E. (1990) Ergothioneine as antioxidant. Meth. Enzymol. 186:310-318.

37. Hartman, P.E., H. Berger, and Z. Hartman (1973) Comparison of hycanthone ("etrenol"), some hycanthone analogs, myxin and 4-nitroquinoline-1-oxide as frameshift mutagens. J. Pharmacol. Exp. Ther. 186:390-398.

38. Hartman, P.E., W.J. Dixon, T.A. Dahl, and M.E. Daub (1988) Multiple modes of photodynamic action by cercosporin. Photochem. Photobiol. 47:699-703.

39. Hartman, P.E., and M.A. Goldstein (1989) Superoxide generation by photomediated redox cycling of anthraquinones. Env. Molec. Mutag. 14:42-47.

40. Hartman, P.E., Z. Hartman, and K.T. Ault (1990) Scavenging of singlet molecular oxygen by imidazole compounds: High and sustained activities of carboxyl terminal histidine dipeptides and exceptional activity of imidazole-4-acetic acid. Photochem. Photobiol. 51:59-66.

41. Hartman, P.E., Z. Hartman, and M.J. Citardi (1988) Ergothioneine, histidine, and two naturally occurring histidine dipeptides as radioprotectors against γ--irradiation inactivation of bacteriophages T4 and P22. Radiat. Res. 114:319-330.

42. Hartman, P.E., and R.W. Morgan (1985) Mutagen-induced focal lesions as key factors in aging: A review. In Molecular Biology of Aging: Gene Stability and Gene Expression, R.S. Sohal, L.S. Birnbaum, and R.G. Cutler, eds. Raven Press, New York, pp. 93-135.

43. Hartman, P.E., and D.M. Shankel (1990) Antimutagens and anticarcinogens: A survey of putative interceptor molecules. Env. Molec. Mutag. 15:145-182.

44. Hartman, P.E., C.K. Suzuki, and M.E. Stack (1989) Photodynamic production of superoxide in vitro by altertoxins in the presence of reducing agents. Appl. Env. Micro. 55:7-14.

45. Hartman, Z., and P.E. Hartman (1987) Interception of some direct-acting mutagens by ergothioneine. Env. Molec. Mutag. 10:3-15.

46. Hattori, T., and S. Fujita (1979) Tritiated thymidine autoradiographic study on histogenesis and spreading of intestinal metaplasia in human stomach. Path. Res. Pract. 164:224-237.

47. Hirayama, T. (1984) Epidemiology of stomach cancer in Japan, with special reference to the strategy for the primary prevention. Japan J. Clin. Oncol. 14:159-168.

48. Holt, K.M., and J.I. Isenberg (1985) Peptic ulcer disease: Physiology and pathophysiology. Hospital Practice (#1):89-93, 96-101, 104-106.

49. Joossens, J.V., and J. Geboers (1983) Epidemiology of gastric cancer: A clue to etiology. In Precancerous Lesions of the Gastrointestinal Tract, P. Sherlock, B.C. Morson, L. Barbara, and U. Veronesi, eds. Raven Press, New York, pp. 97-113.

50. Joossens, J.V., and J. Geboers (1987) Dietary salt and risks to health. Amer. J. Clin. Nutr. 45:1277-1288.

51. Kanofsky, J.R. (1990) Quenching of singlet oxygen by human plasma. Photochem. Photobiol. 51:299-303.

52. Kohen, R., Y. Yamamoto, K.C. Cundy, and B.N. Ames (1988) Antioxidant activity of carnosine, homocarnosine, and anserine present in muscle and brain. Proc. Natl. Acad. Sci., USA 85:3175-3179.

52A. Kramers, P.G.N., J.M. Gentile, B.J.A.M. Gryseels, P. Jordan, N. Katz, K.E. Mott, J.J. Mulvihill, J.L. Seed, and H. Frohberg (1991) Review of the genotoxicity and carcinogenicity of antischistosomal drugs: Is there a case for a study of mutation epidemiology? Mutat. Res. 257:49-89.

53. Kuroda, Y., D.M. Shankel, and M.D. Waters, eds. (1990) Antimutagenesis and Anticarcinogenesis: Mechanisms II, Plenum Press, New York, 485 pp.

54. Lipkin, M. (1988) Biomarkers of increased susceptibility to gastrointestinal cancer: New application to studies of cancer prevention in human subjects. Cancer Res. 48:235-245.

55. McKusick, V.A. (1988) Mendelian Inheritance in Man. 8th Edition. The Johns Hopkins University Press, Baltimore, Maryland.

56. Meredith, M.J., and M.L. Dodson (1987) Impaired glutathione biosynthesis in cultured human ataxia-telangiectasia cells. Cancer Res. 47:4576-4581.

57. Morgan, R.W., J.M. Ward, and P.E. Hartman (1981) Aroclor 1254-induced intestinal metaplasia and adenocarcinoma in the glandular stomach of F344 rats. Cancer Res. 41:5052-5059.

58. Morgan, R.W., J.M. Ward, and P.E. Hartman (1981) Detection of mutagens-carcinogens: Carcinogen-induced lesions pinpointed by alkaline phosphatase activity in fixed gastric specimens from rats. JNCI 66:941-945.

59. Motohashi, N., and I. Mori (1986) Thiol-induced hydroxyl radical formation and scavenger effect of thiocarbamides on hydroxyl radicals. J. Inorgan. Biochem. 26:205-212.

60. Motohashi, N., I. Mori, and Y. Sugiura (1976) Complexing of copper ion by ergothioneine. Chem. Pharm. Bull. (Tokyo) 24:2364-2368.

61. Motohashi, N., I. Mori, Y. Sugiura, and H. Tanaka (1974) Metal complexes of ergothioneine. Chem. Pharm. Bull. (Tokyo) 22:654-657.

62. Nikaido, H., and M. Vaara (1987) Outer membrane. In Escherichia coli and Salmonella typhimurium. Cellular and Molecular Biology, F.C. Neidhardt, ed. Amer. Soc. Microbiol., Washington, pp. 7-22.

63. Niki, E. (1987) Interaction of ascorbate and α-tocopherol. Ann. N.Y. Acad. Sci. 498:186-198.

64. Ohgaki, H., Z. Szentirmay, M. Take, and T. Sugimura (1989) Effects of 4-week treatment with gastric carcinogens and enhancing agents on proliferation of gastric mucosa cells in rats. Cancer Lett. 46:117-122.

65. Oliver, D.B. (1987) Periplasm and protein secretion. In Escherichia coli and Salmonella typhimurium. Cellular and Molecular Biology, F.D. Neidhardt, ed. Amer. Soc. Microbiol., Washington, pp. 56-69.

66. Owens, R.A., and P.E. Hartman (1986) Export of glutathione by some widely used Salmonella typhimurium and Escherichia coli strains. J. Bact. 168:109-114.

67. Owens, R.A., and P.E. Hartman (1986) Glutathione: A protective agent in Salmonella typhimurium and Escherichia coli as measured by mutagenicity and by growth delay assays. Env. Mutag. 8:659-673.

68. Potten, C.S. (1984) Clonogenic, stem and carcinogen-target cells in small intestine. Scandinavian J. Gastroenterol. 19(Suppl.):3-14.

69. Rainbow, A.J. (1991) Host cell reactivation of sunlamp-exposed adenovirus in fibroblasts from patients with Bloom's syndrome, ataxia telangiectasia, and Huntington's disease. Env. Molec. Mutag. 17:98-103.

70. Rowley, D.A., and B. Halliwell (1982) Superoxide-dependent formation of hydroxyl radicals in the presence of thiol compounds. FEBS Lett. 138:33-36.

71. Sevanian, A., K.J.A. Davies, and P. Hochstein (1985) Conservation of vitamin C by uric acid in blood. J. Free Radic. Biol. Med. 1:117-124.

72. Shankel, D.M., P.E. Hartman, T. Kada, and A. Hollaender, eds. (1986) Antimutagenesis and Anticarcinogenesis: Mechanisms, Plenum Press, New York, 605 pp.

73. Shankel, D.M., P.E. Hartman, T. Kada, and A. Hollaender (1987) Synopsis of the First International Conference on Antimutagenesis and Anticarcinogenesis: Mechanisms Env. Mutag. 9:87-103.

74. Shirai, T., K. Imaida, S. Fukushima, R. Hasegawa, M. Tatematsu, and N. Ito (1982) Effects of NaCl, Tween 60 and a low dose of N-ethyl-N'-nitro-N-nitrosoguanidine on gastric carcinogenesis of rat given a single dose of N-methyl-N'nitro-N-nitrosoguanidine. Carcinogenesis 3:1419-1422.

75. Shukla, Y., O.P. Kulshrestha, and K.P. Khuteta (1981) Ergothioneine content in normal and senile human cataractous lenses. Indian J. Med. Res. 73: 472-473.

76. Smith, M.T., C.G. Evans, H. Thor, and S. Orrenius (1985) Quinone-induced oxidative injury to cells and tissues. In Oxidative Stress, H. Sies, ed. Academic Press, New York, pp. 91-113.

77. Stark, A.-A., A. Arad, S. Siskindovich, D.A. Pagano, and E. Zeiger (1989) Effect of pH on mutagenesis by thiols in Salmonella typhimurium TA102. Mutat. Res. 224:89-94.

78. Stark, A.-A., E. Zeiger, and D.A. Pagano (1988) Glutathione mutagenesis in Salmonella typhimurium is a γ-glutamyl-transpeptidase-enhanced process involving active oxygen species. Carcinogenesis 9:771-777.

79. Stich, H.F., L. Wei, and R.F. Whiting (1979) Enhancement of the chromosome-damaging action of ascorbate by transition metals. Cancer Res. 39:4145-4151.

80. Stock, J.B., B. Rauch, and S. Roseman (1977) Periplasmic space in Salmonella typhimurium and Escherichia coli. J. Biol. Chem. 252:7850-7861.

81. Tabuchi, Y., T. Mitsuno, and T. Sugiyama (1975) Mucosal damage induced by various gastric carcinogens in the glandular stomach of the rat. J. Natl. Cancer Inst. 55:1395-1401.

82. Tabuchi, Y., T. Ogino, T. Mitsuno, and T. Sugiyama (1974) Possible role of mucosal damage in stomach carcinogenesis with N-methyl-N'-nitro-N-nitrosoguanidine in the rat. J. Natl. Cancer Inst. 52:1589-1594.

83.	Takahashi, M., T. Kokubo, F. Furukawa, Y. Kurokawa, and Y. Hayashi (1984) Effects of sodium chloride, saccharin, phenobarbital and aspirin on gastric carcinogenesis in rats after initiation with N-methyl-N'-nitro-N-nitrosoguanidine. Gann 75:494-501.

84.	Takahashi, M., T. Kokubo, F. Furukawa, Y. Kurokawa, M. Tatematsu, and Y. Hayashi (1983) Effect of high salt diet on rat gastric carcinogenesis induced by N-methyl-N'-nitro-N-nitrosoguanidine. Gann 74:28-34.

85.	Tatematsu, M., C. Furihata, T. Katsuyama, R. Hasegawa, J. Nakanowatari, D. Saito, M. Takahashi, T. Matsushima, and N. Ito (1983) Independent induction of intestinal metaplasia and gastric cancer in rats treated with N-methyl-N'-nitro-Nnitrosoguanidine. Cancer Res. 43:1335-1341.

86.	Tatematsu, M., C. Furihata, T. Katsuyama, K. Miki, H. Honda, Y. Konishi, and N. Ito (1986) Gastric and intestinal phenotypic expression of human signet ring cell carcinomas revealed by their biochemistry, mucin histochemistry, and ultrastructure. Cancer Res. 46:4866-4872.

87.	Tatematsu, M., T. Katsuyama, C. Furihata, H. Tsuda, and N. Ito (1984) Stable intestinal phenotypic expression of gastric and small intestinal tumor cells induced by N methyl-N'-nitro-N-nitrosoguanidine or methylnitrosourea in rats. Gann 75:957-965.

88.	Teir, H. (1963) Mitotic rate and cell differentiation in the diseased human gastric mucosa. Gastroenterology 44:536-539.

89.	Thomson, R.H. (1987) Naturally Occurring Quinones III. Recent Advances, Chapman and Hall, London, 732 pp.

90.	Varma, S.D. (1981) Superoxide and lens of the eye: A new theory of cataractogenesis. Intl. J. Quant. Chem. 20:479-484.

91.	Varma, S.D. (1987) Ascorbic acid and the eye with special reference to the lens. Ann. N.Y. Acad. Sci. 498:280-305.

92.	Varma, S.D. (1991) Scientific basis for medical therapy of cataracts by antioxidants. Amer. J. Clin. Nutr. 53:335S-345S.

93.	Varma, S.D., and S.M. Morris (1988) Peroxide damage to the eye lens in vitro. Prevention by pyruvate. Free Rad. Res. Commun. 4:283-290.

94.	Waters, M.D., A.L. Brady, H.F. Stack, and H.E. Brockman (1990) Antimutagenicity profiles for some model compounds. Mutat. Res. 238:57-85.

95.	Winkler, B.S. (1987) In vitro oxidation of ascorbic acid and its prevention by GSH. Biochim. Biophys. Acta 925:258-264.

SPHINGOLIPIDS INHIBIT MULTISTAGE CARCINOGENESIS

AND PROTEIN KINASE C

Carmia Borek and Alfred H. Merrill, Jr.

New England Medical Center
Department of Radiation Oncology
Division of Radiation and Cancer Biology
Boston, Massachusetts 02111 USA
and the Department of Biochemistry
Emory University Medical School
Atlanta, Georgia 30322 USA

INTRODUCTION

Sphingolipids have long been associated with cell surface phenomena relevant to signal transduction (reviewed in Ref. 1). They are found primarily on the external surface of the plasma membrane, and affect the properties of cell surface receptors. they undergo changes with cell growth differentiation and neoplastic transformation; bind cytoskeleton elements and lectins; participate in cell to cell communication; act as cell surface antigens; and alter the behavior of cellular protein kinases. The underlying mechanisms of action for sphingolipids are poorly understood. Recent findings have suggested that complex sphingolipids, such as gangliosides, may interact directly with receptors to modulate their functions, and the sphingolipid turnover may remove one type of bioactive compound to produce sphingosine lyses, sphingolipids, and ceramides that serve as lipid second messengers (4).

SPHINGOLIPIDS IN MULTISTAGE CARCINOGENESIS

Transformation of normal cells to malignancy proceeds through several discernable stages. Two early states of transformation have been characterized as initiation and promotion. Initiation is achieved by exposing cells to a single dose of physical or chemical carcinogens. Promotion is induced by repetitive treatments with noncarcinogenic tumor promoters, such as phorbol-12-myristate-13-acetate (PMA), and leads to an augmented outgrowth of initiated cells and enhanced transformation rates (reviewed in Ref. 2).

Group	Irradiation	Week: 1	2	3	4	5	6	# Transformed foci /cells at risk	Transformation frequency (x 10^{-4} ± SE)
1	None							0/52100	< 0.01
2	Day 0							48/78360	6.1 ± 0.9
3	Day 0	So						16/58500	2.7 ± 0.6
4	Day 0		So					14/51280	2.7 ± 0.7
5	Day 0		PMA					65/56555	11.5 ± 1.5
6	Day 0		PMA + So					14/50196	2.8 ± 0.6
7	Day 0		PMA + So	So				16/51096	3.1 ± 0.9
8	Day 0		PMA + So	PMA				37/43475	8.5 ± 1.4
9	Day 0	So	PMA					33/58290	5.7 ± 1.2
10	Day 0	Sa						8/27772	2.9 ± 0.4
11	Day 0		Sa					9/27210	3.3 ± 0.7
12	Day 0		PMA + Sa					4/17844	2.2 ± 0.8
13	Day 0		PMA + Sa	Sa				8/25886	3.1 ± 0.7
14	Day 0		PMA + Sa	PMA				18/26376	6.8 ± 1.2
15	Day 0	Sa	PMA					10/20475	4.9 ± 1.1

Figure 1. The effects of sphingosine (So) and sphinganine (Sa) in suppressing radiogenic transformation and its enhancement by the tumor promoter PMA. C3H 10T1/2 cells were exposed to gamma rays and treated with various combinations of 0.16 μM PMA and/or 3 μM sphingosine or sphinganine, which were added daily as the 1:1 complex with bovine serum albumin. The frequency of transformations is defined as number of transformed foci types II and III per 10^4 cells at risk. No transformation was observed in control (unirradiated) C3H 10T1/2 cells treated with PMA, the long-chain bases alone, or the carriers bovine serum albumin and DMSO under the same conditions. The data are given as the mean values of at least three experiments ± standard errors (S.E.) (4).

The major cellular receptor for phorbol ester tumor promoters (and potentially others) is a family of Ca^{++} and phospholipid-dependent protein kinases termed protein kinase C, which bind phorbol esters with high affinity (3). Protein kinase C is known to play a central role in signal transduction and growth control; however, its function in multistage carcinogenesis is poorly understood. The identification of agents that can inhibit protein kinase C as well as tumor promotion is of importance for elucidating the role of protein kinase C in carcinogenesis. In addition, such agents may act as endogenous antipromoters and dietary antirisk factors and, therefore, play an important role in cancer prevention.

Sphingosine and sphinganine have been shown to inhibit protein kinase C in cell-free systems (1). Our work is based on the hypothesis that these long-chain bases may affect tumor promotion and in parallel affect protein kinase C. If this is the case, these bases could influence carcinogenesis since they are the backbone of moieties of complex sphingolipids and are natural constituents of cells and food products (1). We, therefore, studied whether the long-chain sphingoid bases will inhibit radiation-induced transformation in C3H 10T1/2 fibroblasts as well as the promotion of this transformation process by PMA in radiation-initiated cells. We correlated the effects of these compounds on initiation and promotion with their ability to modulate protein kinase C (4).

Mouse C3H 10T1/2 cells were exposed to gamma rays and treated with various combinations of PMA and/or long-chain bases as seen in Fig. 1. The cultures were fixed and stained and transformation was scored as the number of types II and III foci expressed as the frequency of transformation for surviving cells. Unirradiated cells or cells treated with PMA alone showed nontransformed foci; radiation alone significantly increased the number of foci, and irradiation followed by treatment of the cells with PMA during weeks 2 and 3 postirradiation caused a highly reproducible doubling in transformation frequency.

Both sphingosine and sphinganine reduced transformation frequency to half of that observed in cells exposed to radiation alone. Moreover, they completely blocked the PMA-induced increase in transformed foci when added during weeks 2 and 3 after radiation.

The inhibition of transformation was partially reversed by removing the long-chain bases and treating the cells continuously with PMA during weeks 4 through 6, whereas transformation frequency remains low and only sphingosine or sphingonine was present during the later period. The inhibition of transformation by long-chain bases and their reversal of inhibition by PMA, in the later period, support the hypothesis that long-chain bases and phorbol esters act competitively at a single site.

The addition of N-acetylsphingosine, a ceramide analog, did not affect transformation nor did it have any effect on the promotion of transformation in radiation-initiated cells. Thus, it appears that the free amino group is required for inhibition of multistage carcinogenesis.

Sphingosine and sphinganine inhibit transformation when added during the first week after radiation. This indicates that long-chain bases have the ability to suppress early events in radiogenic transformation. This, too, was partially reversed when PMA was added in weeks 2 and 3, but not to the level observed in cells exposed to PMA alone.

THE EFFECT OF LONG-CHAIN BASES AND PMA ON PROTEIN KINASE C

An important mechanism for the inhibitory action of long-chain bases on transformation may be by inhibiting the activation of protein kinase C. Protein kinase C activities were measured in irradiated and unirradiated C3H 10T1/2 cells exposed to single or combined treatments with PMA and sphinganine as well as in untreated controls. The protocols used were the same ones used for transformation experiments.

The phorbol ester caused a major loss of protein kinase C activity (4), a typical occurrence in cells exposed continuously to phorbol ester over extended periods of time. The remaining activity was mainly in the particulate fraction, which is consistent with the translocation of protein kinase C to membranes in response to PMA. The radiation of the cells also caused some loss of protein kinase C, suggesting that exposure to cells to X-rays may produce an activation and a later down-regulation of protein kinase C. The loss was exacerbated when gamma ray treatment was combined with a subsequent exposure of the cells to PMA (4).

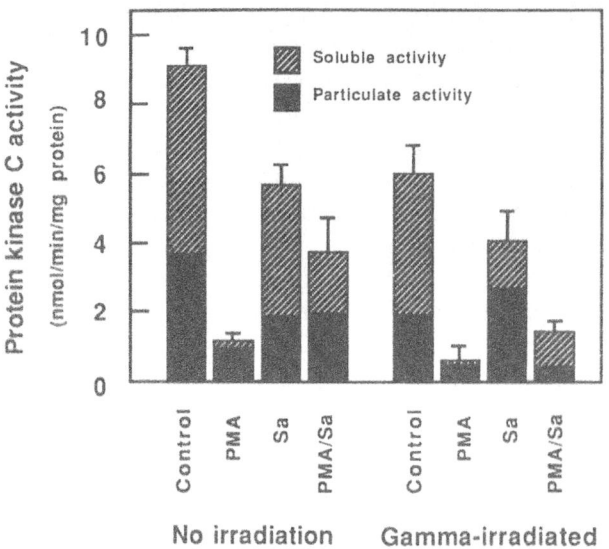

Figure 2. Protein kinase C activities in soluble and particulate fractions from C3H 10T1/2 cells. The cells were exposed to gamma-rays and treated with PMA and/or sphinganine (Sa) for 48 hr in essentially the same manner as for Fig. 1. The cells were scraped from the dishes, disrupted, centrifuged to separate soluble and particulate fractions, and assayed for protein kinase C activity with histone IIIs as the phosphate acceptor. The data are given as the mean values of at least three experiments ± standard error (S.E.) (4).

The most striking effect of sphinganine on protein kinase C activity was seen in groups that received PMA and sphinganine. In both control and irradiation cells, the total and particulate activities of protein kinase C were much higher in cells exposed to PMA and sphinganine than in those treated with PMA alone. This suggests that the inhibition of the protein kinase C by long-chain bases may reduce the PMA-induced down regulation of the enzyme.

DISCUSSION

Our data show that sphingosine and sphinganine inhibit multistage carcinogenesis. They support the hypothesis that long-chain bases, as potent inhibitors of protein kinase C, can block tumor promotion by phorbol esters. While it should be borne in mind that long-chain bases may have additional targets in cells, our findings with C3H 10T1/2 cells provide evidence for an involvement of protein kinase C in transformation.

Long-chain bases also partially protected protein kinase C from down-regulation by phorbol esters. This raises the question whether the determining event in promotion is the activation of protein kinase C by phorbol esters, or the reduction in activity that occurs during the prolonged exposure to PMA, required for tumor promotion. Partial down-regulation of protein kinase C appears to play a role in the multistep process of transformation by chemical carcinogens, radiation, and by dominant transforming genes. Hence, the ability of long-chain bases to inhibit transformation might be related to the protection of a portion of the protein kinase C from down-regulation rather than a reduction in the activity of the enzyme per se.

The suppression of radiation-induced transformation by long-chain bases in groups that were not treated with PMA may represent the effects of these compounds on a variety of cellular and molecular processes. It may reflect an inhibition of protein kinase C activation by endogenous factors such as diacylglycerols and unsaturated fatty acids. Other agents such as hormones and various naturally-occurring compounds may act as endogenous tumor promoters under the appropriate conditions. Transformation of C3H 10T1/2 cells by radiation has been associated with the activation of cellular oncogenes (5), and one mechanism by which oncogenes transform cells has been postulated to be via constitutive activation of protein kinase C.

While a variety of agents have been reported to act as anticarcinogens, or to inhibit protein kinase C, long-chain bases are particularly interesting in their combined action as inhibitors of transformation and modulators of protein kinase C. Sphingolipids are natural constituents of cells and tissues. Thus, the turnover of endogenous sphingolipids to long-chain bases may be one of the factors that influence cell susceptibility to neoplastic conversion.

Sphingolipids are constituents of foods and may act as protective factors and anticarcinogens and play an important role in cancer prevention (4).

ACKNOWLEDGEMENTS

This work was supported in part by a contract from the National Dairy Council and grants from the National Science Foundation (NSF-DCB-871083) and National Institutes of Health (GM33369).

REFERENCES

1. Merrill, A.H., Jr., and D.D. Jones (1990) *Biochem. Biophys. Acta* 1044:1-12.
2. Borek, C. (1985) *J. Pharm. Therapeut.* 27:99-142.
3. Nishizuka, Y. (1986) *Science* 233:305-312.
4. Borek, C., A. Ong, V.L. Stevens, E. Wang, and A.H. Merrill, Jr. (1991) Long-chain (sphingoid) bases inhibit multistage carcinogenesis in mouse C3H 10T1/2 cells treated with radiation and phorbol-12-myristate-13-acetate. 88:1953-1957.
5. Borek, C., A. Ong, and H. Mason (1987) *Proc. Natl. Acad. Sci., USA* 84:794-798.

RECENT RESULTS IN PRECLINICAL AND CLINICAL DRUG

DEVELOPMENT OF CHEMOPREVENTIVE AGENTS AT THE

NATIONAL CANCER INSTITUTE

Gary J. Kelloff, Charles W. Boone, Winfred Malone,
and Vernon Steele

Chemoprevention Branch
Division of Cancer Prevention and Control
National Institutes of Health
Bethesda, Maryland 20892 USA

INTRODUCTION

A primary goal of the Chemoprevention Program of the Division of Cancer Prevention and Control (DCPC) at the National Cancer Institute (NCI) is the development of safe and effective chemopreventive agents. This NCI effort is organized as an applied drug development science program, with clinical trials as the final endpoint. The objective of this paper is to establish a perspective for the applied research in drug development that is sponsored by the Chemoprevention Program, and to present recent results and progress in the development of the most promising chemopreventive agents.

BACKGROUND

The Chemoprevention Program began preclinical chemopreventive agent testing about five years ago with the intention of supplying successful candidate drugs for clinical testing. At that time, there were already three classes of agents on which significant drug development in chemoprevention had been made. These were the retinoids, β-carotene, and calcium compounds.

With regard to the retinoids, that is, compounds structurally and functionally related to retinol (vitamin A), a number of animal efficacy studies had been completed (30,31). From their biological activity, there seemed little doubt that these agents held high promise for chemoprevention. In the Chemoprevention Program, efforts have been undertaken with retinoids to help clinical investigators obtain grants, compounds, and adequate dosage forms. The Chemoprevention Program has also reviewed and given advice on clinical protocols and assisted the investigators with INDs and other interactions with the Food and Drug Administration (FDA).

At present, to help assure the future development of the retinoids, the Chemoprevention Program is sponsoring extensive preclinical toxicology testing, including long-term carcinogenicity bioassays. Preclinical efficacy testing is also being done with the most promising retinoid, all-*trans*-N-(4-hydroxyphenyl)retinamide (4-HPR).

Five years ago, the scientific rationale for the development of β-carotene as a chemopreventive agent evolved quite differently from that for the retinoids. It was case-control epidemiology data from lung cancer patients (3,4,13,19,20,22,25,28,35,45), a chemical structure indicating ability to scavenge free radicals, and bioconversion to vitamin A that made β-carotene seem a potential chemopreventive compound of interest. In contrast to the retinoids, there was little concern about toxicity. On the other hand, there were no animal efficacy data. A few clinical trials were being started. Because both NCI and the FDA considered the agent to be safe, the Chemoprevention Program was able to initiate additional clinical trials, even with well subjects.

At present, the results of these clinical trials are still being awaited, with the exception that there has been one negative result from testing the chemopreventive effect of β-carotene in skin cancer (15). Although this outcome appears to be conclusive, the dosage used may have been lower than the effective dose for skin cancer. Meanwhile, the Chemoprevention Program is looking at the activity of β-carotene in animal efficacy experiments to help provide insight for interpreting the clinical data. Recently, β-carotene proved to be efficacious against animal lung tumors.

The chemopreventive potential of calcium was shown by the protective effect of calcium against proliferation of cancer cells in the colon of patients at high risk for cancer (6,16,24). Calcium has shown chemopreventive activity at the cellular level (6,23,24,50), in animals (1,33,51), and in clinical tests (23). The challenge now is to conduct clinical trials with adequate drug formulations; with adequate control of the confounding variables of vitamin D, phosphate, and fat levels; and with identification of valid endpoints.

These three prototypic classes of chemopreventive agents (retinoids, β-carotene, and calcium) serve to illustrate the diversity and the multidisciplinary nature of chemopreventive agent drug development.

SIX STEPS IN THE NCI CHEMOPREVENTIVE DRUG DEVELOPMENT PROCESS

Agent Selection and Data Analysis

Literature search and database development. First, a comprehensive search of the published literature relevant to chemoprevention was carried out. Agents and associated test results and other biological activities related to chemoprevention were compiled into a set of databases known collectively as the "Master List of Chemopreventive Agents." These databases are updated continually and are a major source of new candidates for testing by NCI. Progress and results of testing in the Chemoprevention Branch Preclinical Testing Program were also compiled and incorporated into a database known as the "Desktop Database." The information in this database is useful in analyzing test data and in identifying and selecting appropriate additional testing for promising agents already in the Preclinical Testing Program.

Identification of candidate agents. Next, candidate agents were identified from the "Master List" based on one or more factors related to chemopreventive efficacy. Such factors included previously demonstrated inhibitory activity of a cancer endpoint *in vivo, in vitro* inhibition of cell transformation, and epidemiological data suggesting chemopreventive activity. Agents were also identified because they had other activities related to chemopreventive efficacy, such as antimutagenicity, inhibition of enzymes associated with cell proliferation (e.g., ornithine decarboxylase and those involved in arachidonic acid metabolism), or stimulation of enzymes involved in carcinogen detoxification (e.g., glutathione-S-transferases). Finally, experts in chemoprevention in academia, the pharmaceutical industry, and government were asked for their help and advice in selecting new and interesting compounds.

Evaluation, prioritization, and selection of agents. The candidate agents were next evaluated and ranked for further testing. Ranking was based on criteria related to their potential as clinical chemopreventives and the value of the information that their testing would provide to the Chemoprevention Program. These criteria were based on factors such as the potency of the activity observed; the species, models, and target organs against which chemopreventive activity was observed; postulated mechanism(s) of action; the quality of previous testing; and comparisons with other agents in the same chemical and biological classes that were already on test.

Data on the toxicity of the candidate agents were also summarized and analyzed. These data came from documents prepared by the pharmaceutical industry, as well as from the general literature. Agents with the lowest chronic toxicity and those best studied usually received higher priorities for testing.

Additional factors were considered in ranking the candidate agents, including bioavailability, half-life, and other pharmacokinetic parameters; cost, commercial availability, and patent status; likely clinical cohorts; and dosage forms. Highest priority went to agents that were bioavailable, biologically stable, commercially obtainable, administered orally, and likely to be applicable to relatively large or high-risk clinical cohorts.

Once the chemicals had been ranked, a panel of experts was asked to provide their opinions on the candidates and available data. Although a major aspect of the agent selection process was the identification of new agents for testing, equally important was the evaluation of agents that had already had significant previous testing elsewhere. For the previously tested agents, it became critical to determine what further testing the agent must have to move it to clinical trials. In some cases, all early efficacy testing was unnecessary because sufficient data were already available, and the agents could go directly to toxicology studies.

Table 1. In vitro screens

Cell Substrate	Carcinogen	Promoter	Endpoint: Inhibition of
Human Lung Tumor Cells (A427)	None	None	Anchorage Independent Growth
Mouse Epidermal Cells (JB6)	None	TPA	Anchorage Independent Growth
Rat Tracheal Epithelial Cells (RTE)	B(a)P	None	Transformed Foci or Colonies
Mouse Mammary Organ Culture (MMOC)	DMBA	TPA	Hyperplastic Alveolar Nodules

Abbreviations: B(a)P=Benzo(a)pyrene, DMBA=7,12-Dimethylbenz(a)anthracene, TPA=12-O-Tetradecanoylphorbol-13-acetate

In Vitro Efficacy Screening

Testing *in vitro* has been under development for only 2-3 years, since the Chemoprevention Program first wanted to develop an *in vivo* database to be used in validating *in vitro* assays. The strengths of *in vitro* testing are: 1.) efficiency in terms of cost and time in evaluation of a given agent, 2.) sensitivity and ease of quantitation, 3.) controlled test conditions, and 4.) most important, potential use of human cells.

The primary limitation of *in vitro* screening is the inability to determine with any certainty the relevance of the observed activity to potential efficacy (or toxicity) in an intact organism. Thus, any promising agents identified by *in vitro* techniques must be evaluated further by *in vivo* screening. *In vivo,* such factors as bioavailability, tissue distribution, metabolism, and species variability can be evaluated. In choosing the *in vitro* test systems, it was thought to be important to use organ culture, epithelial cells, primary cells (as opposed to cell lines), and, where possible, human cells. The assays selected are listed in Tab. 1.

The human lung tumor A427 cell line was the only tumor cell line used. It has been surprisingly effective in measuring antiproliferative activity, especially of agents that are arachidonic acid metabolism inhibitors. JB6 mouse epidermal cells were used primarily to detect antipromoters.

The rat tracheal epithelial (RTE) cell assay is useful in measuring anti-mutagens, and it appears to be accurate in predicting *in vivo* results. The mouse mammary gland organ culture assay, or MMOC assay, is useful in that test conditions may be set so that they can measure antimutagens, antipromoters, or both effects.

Table 2 compares the activity of selected chemopreventive agents in the RTE, A427, and MMOC assays with their chemopreventive activity *in vivo.* The agents are classified as antimutagens, antiproliferatives, or both, based on their anticipated mechanisms of action and previously demonstrated biological activities. As described above, the RTE assay detected antimutagens, the A427 assay detected antiprolifera-

Table 2. Chemopreventive activities of antimutagens and antiproliferatives

	In Vitro			In Vivo				
	RTE	A427	MMOC	Skin	Mammary	Lung	Colon	Bladder
Antimutagens								
NAC	+	NE	+		+	+		
Oltipraz	+	NE	+	+	+	+		+
Mesna	+	NE	+				NE	+
Bismuthiol I	+	NE						+
Antiproliferatives								
DFMO	NE	NE	+		+		+	+
Ibuprofen	NE	+	+		+	NE		
Molybdenum	NE	+	+			NE	NE	NE
Piroxicam	NE	+	NE				+	
Antimutagens and Antiproliferatives								
DHEA	+	+	+		+		+	
4-HPR	+	+	+	+	+	+		

Abbreviations: + = Chemopreventive activity observed; significant at p < 0.05. NE = No significant chemopreventive effect observed.
DFMO=2-Difluoromethylornithine, DHEA=Dehydroepiandrosterone, 4-HPR=all-*trans*-N-(4-Hydroxy)phenylretinamide, Mesna=2-Mercaptoethane sulfonate, NAC=N-Acetyl-*l*-cysteine, A427=Human Lung Tumor A427 Cell Assay, MMOC=Mouse Mammary Gland Organ Culture Assay, RTE=Rat Tracheal Epithelial Cell Focus Assay

tives, and the MMOC assay detected both. At this time, the available data are insufficient to estimate the concordance of the *in vitro* versus the *in vivo* results. However, current trends are promising.

In Vivo Efficacy Screening

About five years ago, several criteria were used to select a battery of *in vivo* screening models that would be relevant to the Chemoprevention Program. These criteria are as follows: 1.) The models must represent target organs that are applicable to the human cancer problem. Thus, models of lung, mammary glands, colon, bladder, prostate, and skin were chosen. Since many chemopreventive agents appear to be tissue specific, testing in multiple target tissues is critical. 2.) The assays used must be rapid. Currently, the tests run from 90-180 days. 3.) The models involve induction of tumors with both direct-acting and metabolically-activated carcinogens. Although it is not clear that chemical carcinogenesis allows an ideal extrapolation to human cancers, models employing spontaneous tumors are impractical.

The animal models now used in the Preclinical Testing Program are listed in Table 3. Note that both pulmonary adenocarcinoma and tracheal squamous cell carcinoma models are used for lung. Both the rat and the mouse colon models have advantages. In the mammary gland, a direct-acting carcinogen (MNU) is used in one model, and a metabolically-activated carcinogen (DMBA) is used in the second model. Hormonal dependence, metastatic behavior, and the known molecular biology of the *ras* oncogene with respect to MNU make the MNU model desirable

(26,31,52). The OH-BBN bladder model had already been used extensively for retinoid evaluation (30,31). Finally, the skin model was chosen because it is quantitative, rapid, and well studied (5,8,17).

Testing with the *in vivo* models began in September, 1985. Results are now available in at least one assay for 65 agents. Another 40 or so agents are on test *in vivo*. The testing status of 12 of the most promising agents to date is summarized in Table 4, where they are listed in roughly the rank order of their expected promise as clinical chemopreventive agents. Before each agent was tested in a given model, a maximum tolerated dose or MTD was determined in terms of the maximum dose that would not produce weight suppression. Chemoprevention testing was done at doses of 0.4 MTD and 0.8 MTD. The chemopreventive was given from before the administration of the carcinogen to the end of the experiment. In Table 4, a positive result signifies that the tumor incidence in the treated group was less than that in the control group at a significance level of $p=0.05$.

From Table 4, the 12 most promising chemopreventive agents based on efficacy in animal screening are as follows.

All-*trans*-N-(4-Hydroxyphenyl)retinamide (4-HPR). 4-HPR was chemopreventive in the lung, mammary, and skin models. 4-HPR has previously been shown to be positive in the bladder (30-32). The negative result in Table 4 was possibly because, at 0.4 mM in the diet, 4-HPR was at a level well under the usual effective dose of 1.0-2.0 mM (30).

2-Difluoromethylornithine (DFMO). DFMO appears to have broad chemopreventive efficacy, being positive in every model tested except the lung-MNU (i.e., trachea) model. DFMO has also been effective in combination with other agents,

Table 3. In vivo screens

Organ Model	Species	Carcinogen	Endpoint: Inhibition of
Lung	Hamster	DEN	Adenocarcinoma
Trachea	Hamster	MNU	Squamous Cell Carcinoma
Colon	Mouse	MAM	Adenocarcinoma
	Rat	AOM	Adenocarcinoma
Mammary	Rat	MNU	Adenocarcinoma
	Rat	DMBA	Adenocarcinoma
Bladder	Mouse	OH-BBN	Transitional Cell Carcinoma
Skin	Mouse	DMBA/TPA	Papilloma/Carcinoma

Abbreviations: AOM=Azoxymethane, DEN=*N,N*-Diethylnitrosamine, DMBA=7,12-Dimethylbenz(*a*)anthracene, MAM=Methylazoxymethanol, MNU=*N*-Methyl-*N'*-nitrosourea, OH-BBN=*N*-Butyl-*N*-(4-hydroxybutyl)nitrosamine, TPA=12-*O*-Tetradecanoylphorbol-13-acetate

including piroxicam (37). One-year chronic toxicity studies have been completed in rats and dogs. DFMO is currently in both Phase I and Phase II clinical trials.

Piroxicam. The first positive results with piroxicam, a long-acting nonsteroidal antiinflammatory agent, were obtained in the colon (36). Later, it was also found to be very active in bladder and skin. An 18-month preclinical toxicology study has been completed. Because piroxicam is a well-known pharmaceutical used in the treatment of arthritis, its clinical safety for chronic use has already been established. A Chemoprevention Program-sponsored Phase I clinical study has been completed.

Oltipraz. Oltipraz, a dithiolthione and Phase II metabolic enzyme inducer (e.g., glutathione-S-transferases), was positive in every model in which it was tested. It is currently being evaluated in one-year chronic toxicity studies in rats and dogs, and two Phase I clinical trials are in progress.

Calcium. Three Phase II clinical studies with calcium salts are nearing completion. Since glucaric acid and its derivatives have demonstrated chemopreventive activity in assays in mammary glands and colon carried out by the Chemoprevention Program and in several studies reported in the literature (9-11, 47-49), calcium glucarate is being studied in case the glucarate anion should provide some additional efficacy over that of the calcium ion.

N-Acetyl-l-cysteine (NAC). Table 4 shows that NAC has chemopreventive activity in the trachea and MNU mammary models. NAC has been marketed as the pharmaceutical Mucomyst® for the last 20 years, with no reports of significant toxicity (29). It is currently in Phase I clinical testing.

β-Carotene. The historical epidemiological data associating lowered risk of cancer (particularly lung cancer) with increased levels of, β-carotene was described above. In Table 4, the positive results obtained in the lung and mammary models are especially noteworthy. Unlike primates, rodents do not absorb β-carotene after oral administration (21), so the agent was administered to hamsters and rats by subcutaneous injection three times weekly in order to obtain adequate blood levels. When detectable blood levels were obtained, chemopreventive activity was observed.

Ibuprofen. Ibuprofen is a non-steroidal antiinflammatory pharmaceutical with known safety (39). Positive results were obtained in the bladder model. It is currently in Phase I testing.

Dehydroepiandrosterone (DHEA) and DHEA analog 8354. As is well-known, DHEA is a natural metabolite produced in man in the testosterone/estrogen synthetic pathway. In men over 50 years of age, DHEA serum levels have been positively correlated with longevity and inversely correlated with cardiovascular disease risk (2). Besides the chemopreventive activity shown in Table 2, the chemopreventive efficacy of both DHEA and its fluorinated analog 8354 has been shown in numerous animal models (34,40-44). Analog 8354 appears to have excellent promise for chemoprevention because, unlike DHEA, it is not converted to hormonally-active compounds (43). In addition, the analog produces less of the weight loss effects that have confounded some of the animal tumor studies of DHEA. In general, in both preclinical efficacy and toxicity studies conducted by the Chemoprevention Program, the analog appears to have little toxicity.

Table 4. Promising chemopreventive agents

	Lung DEN	Trachea MNU	Colon Mouse	Colon Rat	Mammary DMBA	Mammary MNU	Bladder	Skin	Toxicology	Clinical Trials
4-HPR	+	NE	NE		+	+	NE	+	√	√
[Related Compounds: Vitamin A (Retinol), 13-*cis*-Retinoic Acid]									√	√
DFMO	OT	NE	OT	+	+	+	+	OT	√	√
Piroxicam	OT			+		OT	+	+	√	√
Oltipraz	OT	+	+	+	+	+	+	+	√	√
Calcium Glucarate			+			NE				√
[Other Calcium Salts]										√
NAC		+		OT		+	OT	OT		√
β-Carotene	OT	NE				+				√
Ibuprofen	OT	NE		OT		OT	+		√	√
DHEA Analog 8354				+		+			√	
DHEA	OT		+			+	NE	+		
Tamoxifen						+				
Glycyrrhetinic Acid		NE	+	OT		NE	OT	+	√	√

Abbreviations: + = Chemopreventive activity observed; significant at p < 0.05. NE = No significant chemopreventive effect observed. OT = Agent is currently on test. √ = Agent has been or is being evaluated. Agents: DFMO=2-Difluoromethylornithine, DHEA=Dehydroepiandrosterone, 4-HPR=all-*trans*-N-(4-Hydroxyphenyl)retinamide, NAC=N-Acetyl-l-cysteine. Carcinogens used to induce tumors: DEN=N,N-Diethylnitrosamine, DMBA=7,12-Dimethylbenz(a)anthracene, MNU=N-Methyl-N'-nitrosourea

Tamoxifen. Tamoxifen is a well-known antiestrogen that is already being used clinically in the adjuvant treatment of breast cancer (7,14). Since 4-HPR and tamoxifen have been shown to bind to two different estrogen receptor sites (12,27), it is possible that the two agents given together will have a synergistic effect.

Glycyrrhetinic acid. Glycyrrhetinic acid is found in licorice root and is an antiinflammatory with weak antilipoxygenase activity (18). It is used commercially as a flavoring agent in candy and tobacco products. In Europe it is used for treatment of ulcers (38). It is currently in Phase I clinical trials.

Another important mode of *in vivo* efficacy evaluation is the testing of agents in combination. As noted above, chronic toxicity is a major problem in the development of chemopreventive agents. One approach to reducing the risk of toxicity is administering the chemopreventive agents at lower doses. To this end, additive and synergistic chemopreventive effects may play a significant role. That is, two or more agents administered together may provide the same or better efficacy at lower (and, presumably, less toxic) doses than any of the agents singly. Unfortunately, in terms of drug development, combinations of agents also greatly complicate the IND approval process.

Thus far, the Chemoprevention Branch has sponsored the testing of numerous agent combinations. Results of testing with some of the most promising combinations are shown in Table 5. Particularly interesting are the synergistic effects observed in the lung with, β-carotene and 4-HPR or vitamin A; in the bladder with DFMO and oltipraz or 4-HPR; in the colon with DFMO and piroxicam; and in mammary glands with 4-HPR and tamoxifen or DHEA.

Table 5. Agent combinations showing significant
enhancement of chemopreventive activity

	Lung	Colon	Mammary	Bladder
β-Carotene + 4-HPR	+S			
β-Carotene + vitamin A	+S			
DFMO + 4-HPR	+A		+I	+S
DFMO + oltipraz		+	+	+S
DFMO + piroxicam		+S		+
4-HPR + tamoxifen			+S	
4-HPR + DHEA			+S	

Abbreviations: + = Chemopreventive activity observed; significant at p < 0.05,
but not greater than the best of the individual agents. +A = Chemopreventive
activity observed; the combination was more efficacious than any of the
individual agents (*i.e.*, the effect was additive). +S = Chemopreventive activity
observed; the combination was more efficacious than the sum of the efficacies
of the individual agents (*i.e.*, the effect was synergistic). +I =
Chemopreventive activity observed; the combination was less efficacious than
the best individual agent. Agents tested: DFMO=2-Difluoromethylornithine,
DHEA=Dehydroepiandrosterone, 4-HPR=all-*trans*-N-(4-Hydroxy)phenylretin-
amide.

Drug Procurement

Drug procurement is a very critical part of chemoprevention drug development. Assuring adequate supplies and formulations is one of the most difficult and
time-consuming aspects of moving promising agents to clinical trials. The methods
and policies involved in drug procurement are left to be the subject of another paper.

Toxicology

In the preclinical toxicology testing of chemopreventive agents, the tests
required are generally not different from those for other noncancer, chronic use
drugs, and the tests follow standardized protocols. They include acute, 30-day, 90-
day, and one-year toxicity studies along with two-year carcinogenicity studies.
Administration of the agent, physical examination, laboratory testing, and necropsy
and histology are generally the same as for other drugs.

As in efficacy testing, it is sometimes necessary to go beyond the standard
protocols to adequately evaluate the potential toxicity of a given chemopreventive
agent. Judgment must be applied in interpreting results and in choosing appropriate
dose ranges for testing. For instance, the interpretation of tolerable toxicity is based
on the human cohort that will be treated with the agent. Different criteria apply to
treatments aimed at the general population than to those aimed at groups at high risk
for cancer, those with premalignant lesions, and those with previously treated cancers.

Table 6. Clinical studies

	Phase I	Phase II	Phase III
Retinoids			
4-HPR	√		√
13-*cis*-Retinoic Acid		√ (2)	√
Retinol (Vitamin A)	√	√ (2)	√ (2)
all-*trans*-Retinoic Acid			√
β-Carotene	√	√ (5)	√ (2)
Calcium		√ (3)	
Piroxicam	√	√	
DFMO	√ (3)	√	
Oltipraz	√ (2)		
Glycyrrhetinic Acid	√		
NAC	√		
Ibuprofen	√		
Folic Acid		√	
Vitamins C and E		√	

Abbreviations: DFMO=2-Difluoromethylornithine, 4-HPR=all-*trans*-N-(4-Hydroxy)phenylretinamide, NAC=N-Acetyl-*l*-cysteine, √=Clinical trial in progress or completed, (No.)=Number of clinical trials, if more than one

Often the available toxicity data on an agent entering the drug development program are very scanty, especially if the agent is not a pharmaceutical. Because of the cost and time involved in carrying out the required preclinical toxicity testing, the Chemoprevention Program thus far has given priority to developing agents that are already FDA-approved pharmaceuticals. The Chemoprevention Program is exploring the possibility of doing early toxicity screening in cultured human cells. Even though the FDA has not yet established criteria for interpreting such *in vitro* tests, they may help in the early prioritization of agents.

It is worth emphasizing once again that minimizing chronic toxicity in applied chemoprevention is paramount. The public health impact of a moderately effective chemopreventive that is non-toxic (for example, calcium) tends to far outweigh one that is very effective, but is associated with some degree of toxicity. However, as stated above, the human risk group in which the drug is to be used is an important factor that will determine the level of toxicity that can be tolerated.

Clinical Trials

Clinical trials are categorized as Phase I, II, and III. Phase I trials are designed to determine the dose-related safety and toxicity of the proposed chemopreventive agent. In these trials, a method for estimating the concentration of the compound in blood is sought. Also, pharmacokinetics are assessed, including the usual "ADME" parameters of absorption, distribution, metabolism, and excretion. Both Phase II and Phase III trials have to do with the determination of efficacy. Phase II trials are small scale. Usually, modulation of intermediate endpoints of cancer is measured. Phase III trials are much larger, and usually cancer incidence reduction is the endpoint.

Table 6 summarizes the 35 clinical studies that have been or are being sponsored by the Chemoprevention Program. As is evident from this table, the three agents on which significant progress had been made prior to efforts of the Chemoprevention Program are now quite far along in clinical trials. Eleven clinical efficacy studies are underway for the retinoids, eight for β-carotene, and three for calcium. In addition, six active compounds are either in Phase I trials or have finished them. These compounds are piroxicam, DFMO, oltipraz, glycyrrhetinic acid, NAC, and ibuprofen. Two compounds, piroxicam and DFMO, are also in Phase II trials.

FUTURE DIRECTIONS

These studies indicate that it is not unduly optimistic to assume that clinical chemoprevention research will begin to yield practical applications for the reduction of cancer incidence. The time required to carry out a full clinical evaluation of a chemopreventive agent in a cancer incidence reduction study is of great concern, and so the Chemoprevention Program will be expanding the Phase II clinical studies. More experimentation related to the validation of intermediate biomarkers that can be modulated with chemopreventive agents is also being added to the Chemoprevention Program. These studies with validated markers could significantly shorten the clinical efficacy evaluation of promising agents.

REFERENCES

1. Appleton, G.V.N., P.W. Davies, J.B. Bristol, and R.C.N. Williamson (1987) Inhibition of intestinal carcinogenesis by dietary supplementation with calcium. Br. J. Surg. 74:523-525.
2. Barrett-Connor, E., K.T. Khaw, and S.C. Yen (1986) A prospective study of dehydroepiandrosterone sulfate, mortality, and cardiovascular disease. N.Engl. J.Med. 315:1519-1524.
3. Bjelke, E. (1975) Dietary vitamin A and human lung cancer. Int. J. Cancer 15:561-565.
4. Bond, G.G., F.E. Thompson, and R.R. Cook (1987) Dietary vitamin A and lung cancer: Results of a case-control study among chemical workers. Nutr. Cancer 9:109-121.
5. Boutwell, R.K. (1989) Model systems for defining initiation, promotion, and progression of skin neoplasms. In Skin Carcinogenesis. Mechanisms and Human Relevance. Vol. 298, T.J. Slaga, A.J.P. Klein-Szanto, R.K. Boutwell, D.E. Stevenson, H.L. Spitzer, and B. D'Motto, eds. CRC Press, Inc., Boca Raton, Florida, pp. 3-15.
6. Buset, M., M. Lipkin, S. Winawer, S. Swaroop, and E. Friedman (1986) Inhibition of human colonic epithelial cell proliferation in vivo and in vitro by calcium. Cancer Res. 46:5426-5430.
7. Cuzick, J., and M. Baum (1986) Tamoxifen and contralateral breast cancer. Lancet 2:282.
8. DiGiovanni, J., T.J. Slaga, D.L. Berry, and M.R. Juchau (1980) Inhibitory effects of environmental chemicals on polycyclic aromatic hydrocarbon carcinogenesis. In Carcinogenesis; A Comprehensive Survey, Volume 5.

Modifiers of Chemical Carcinogenesis: An Approach to the Biochemical Mechanism and Cancer Prevention, T.J. Slaga, ed. Raven Press, New York, pp. 145-167.

9. Dwivedi, C., A.A. Downie, and T.E. Webb (1988) Effect of dietary calcium glucarate on 7,12-dimethylbenz(a)anthracene-induced skin tumorigenesis in CD-1 mice. Cleve. Clin. J. Med. 55:561-564.

10. Dwivedi, C., A.A. Downie, and T.E. Webb (1989) Modulation of chemically initiated and promoted skin tumorigenesis in CD-1 mice by dietary glucarate. J. Env. Pathol. Toxicol. Oncol. 9:253-259.

11. Dwivedi, C., O.A. Oredipe, R.F. Barth, A.A. Downie, and T.E. Webb (1989) Effects of the experimental chemopreventative agent, glucarate, on intestinal carcinogenesis in rats. Carcinogenesis 10:1539-1541.

12. Fontana, J.A. (1987) Interaction of retinoids and tamoxifen on the inhibition of human mammary carcinoma cell proliferation. Exp. Cell Biol. 55:136-144.

13. Fontham, E.T., L.W. Pickle, W. Haenszel, P. Correa, Y.P. Lin, and R.T. Falk (1988) Dietary vitamins A and C and lung cancer risk in Louisiana. Cancer 62:2267-2273.

14. Furr, B.J.A., and V.C. Jordan (1984) The pharmacology and clinical uses of tamoxifen. Pharmacotherapy 25:127-205.

15. Greenberg, E.R., J.A. Baron, T.A. Stukel, M.M. Stevens, J.S. Mandel, S.K. Spencer, P.M. Elias, N. Lowe, D.W. Nierenberg, G. Bayrd, J.C. Vance, D.H. Freeman, Jr., W.E. Clendenning, T. Kwan, and the Skin Cancer Prevention Study Group (1990) A clinical trial of β-carotene to prevent basal-cell and squamous-cell cancers of the skin. N. Engl. J. Med. 323:789-795.

16. Gregoire, R.C., H.S. Stern, K.S. Yeung, J. Stadler, S. Langley, R. Furrer, and W.R. Bruce (1989) Effect of calcium supplementation on mucosal cell proliferation in high risk patients for colon cancer. Gut 30:376-382.

17. Hennings, H. (1987) Tumor promotion and progression in mouse skin. In Mechanisms of Carcinogenesis, Vol. II. Multistep Models of Carcinogenesis, J.C. Barrett, ed. CRC Press, Inc., Boca Raton, Florida, pp. 59-71.

18. Hile, J.P. (1985) GRAS status of licorice, (glycyrrhiza), ammoniated glycyrrhizin, and monoammonium glycyrrhizinate. Fed. Reg. 50:21043.

19. Hirayama, T. (1979) Diet and cancer. Nutr. Cancer 1:67-81.

20. Kolonel, L.N., M.W. Hinds, A.M.Y. Nomura, J.H. Hankin, and J. Lee (1985) Relationship of dietary vitamin A and ascorbic acid intake to the risk for cancers of the lung, bladder, and prostate in Hawaii. NCI Monograph 69:137-142.

21. Krinsky, N.I., M.M. Mathews-Roth, S. Welankiwar, P.K. Sehgal, N.C.G. Lausen, and M. Russett (1990) The metabolism of [^{14}C]β-carotene and the presence of other carotenoids in rats and monkeys. J. Nutr. 120:81-87.

22. Kvale, G., E. Bjelke, and J.J. Gart (1983) Dietary habits and lung cancer risk. Int. J. Cancer 31:397-405.

23. Lipkin, M., E. Friedman, S.J. Winawer, and H. Newmark (1989) Colonic epithelial cell proliferation in responders and nonresponders to supplemental dietary calcium. Cancer Res. 49:248-254.

24. Lipkin, M., and H. Newmark (1985) Effect of dietary calcium on colonic epithelial-cell proliferation in subjects at high risk for familial colonic cancer. N. Engl. J. Med. 313:1381-1384.

25. MacLennan, R., J. Da Costa, N.E. Day, C.H. Law, Y.K. Ng, and K. Shanmugaratnam (1977) Risk factors for lung cancer in Singapore Chinese, a population with high female incidence rates. Int. J. Cancer 20:854-860.

26. McCormick, D.L., R.G. Mehta, C.A. Thompson, N. Dinger, J.A. Caldwell, and R.C. Moon (1982) Enhanced inhibition of mammary carcinogenesis by combined treatment with N-(4-hydroxyphenyl)retinamide and ovariectomy. Cancer Res. 42:508-512.
27. McCormick, D.L., and R.C. Moon (1986) Retinoid-tamoxifen interaction in mammary cancer prevention. Carcinogenesis 7:193-196.
28. Menkes, M.S., G.W. Comstock, J.P. Vuillemier, K.J. Helsing, A.A. Rider, and R. Brookmeyer (1986) Serum β-carotene, vitamins A and E, selenium, and the risk of lung cancer. N. Eng. J. Med. 315:1250-1254.
29. Miller, L.F., and B.H. Rumack (1983) Clinical safety of high oral doses of acetylcysteine. Semin. Oncol. 10:76-85.
30. Moon, R.C., D.L. McCormick, P.J. Beci, Y.F. Shealy, F. Frickel, J. Paust, and M.B. Sporn (1982) Influence of 15 retinoic acid amides on urinary bladder carcinogenesis in the mouse. Carcinogenesis 3:1469-1472.
31. Moon, R.C., D.L. McCormick, and R.G. Mehta (1983) Inhibition of carcinogenesis by retinoids. Cancer Res. 43:2469s-2475s.
32. Moon, R.C., and R.G. Mehta (1989) Chemoprevention of experimental carcinogenesis in animals. Prev. Med. 18:576-591.
33. Pamukcu, A.M., S. Yalciner, and G.T. Bryan (1977) Inhibition of carcinogenic effect of bracken fern (Pteridium aquilinum) by various chemicals. Cancer 40:2450-2454.
34. Pashko, L.L., R.J. Rovito, J.R. Williams, E.L. Sobel, and A.G. Schwartz (1984) Dehydroepiandrosterone (DHEA) and 3β-methylandrost-5-en-17-one: Inhibitors of 7,12-dimethylbenz(a)anthracene (DMBA)-initiated and 12-O-tetradecanoyl-phorbol-13-acetate (TPA)-promoted skin papilloma formation in mice. Carcinogenesis 5:463-466.
35. Peto, R., R. Doll, J.D. Buckley, and M.B. Sporn (1981) Can dietary β-carotene materially reduce human cancer rates? Nature 290:201-208.
36. Reddy, B.S., H. Maruyama, and G. Kelloff (1987) Dose-related inhibition of colon carcinogenesis by dietary piroxicam, a nonsteroidal antiinflammatory drug, during different stages of rat colon tumor development. Cancer Res. 47:5340-5346.
37. Reddy, B.S., J. Nayini, K. Tokumo, J. Rigotty, E. Zang, and G. Kelloff (1990) Chemoprevention of colon carcinogenesis by concurrent administration of piroxicam, a nonsteroidal antiinflammatory drug with d,l-a-difluoro-methylornithine, an ornithine decarboxylase inhibitor, in diet. Cancer Res. 50:2562-2568.
38. Reynolds, J.E.F., and A.B. Prasad, eds. (1982) Martindale: The Extra Pharmacopoeia 28th ed. The Pharmaceutical Press, London, pp. 691-692.
39. Royer, G.L., C.E. Seckman, and I.R. Welshman (1984) Safety profile: Fifteen years of clinical experience with ibuprofen. Am. J. Med. 77:25-34.
40. Schwartz, A.G. (1979) Inhibition of spontaneous breast cancer formation in female C3H (A vy/a) mice by long-term treatment with dehydroepi-androsterone. Cancer Res. 39:1129-1132.
41. Schwartz, A.G., D.K. Fairman, M. Polansky, M.L. Lewbart, and L.L. Pashko (1989) Inhibition of 7,12,-dimethylbenz(a)anthracene-initiated and 12-O-tetradecanoylphorbol-13-acetate-promoted skin papilloma formation in mice by dehydroepiandrosterone and two synthetic analogs. Carcinogenesis 10:1809-1813.

42. Schwartz, A.G., G.C. Hard, L.L. Pashko, M. Abou-Gharbia, and D. Swern (1981) Dehydroepiandrosterone: An anti-obesity and anti-carcinogenic agent. Nutr. Cancer 3:46-53.

43. Schwartz, A.G., M.L. Lewbart, and L.L. Pashko (1988) Novel dehydroepiandrosterone analogues with enhanced biological activity and reduced side effects in mice and rats. Cancer Res. 48:4817-4822.

44. Schwartz, A.G., and R.H. Tannen (1981) Inhibition of 7,12-dimethylbenz(a)anthracene- and urethane-induced lung tumor formation in A/J mice by long-term treatment with dehydroepiandrosterone. Carcinogenesis 2:1335-1337.

45. Shekelle, R.B., S. Liu, W.J. Raynor, M. Lepper, C. Maliza, A.H. Rossof, O. Paul, A.M. Shryock, and J. Stamler (1981) Dietary vitamin A and risk of cancer in the Western Electric Study. Lancet 2:1185-1189.

46. Walaszek, Z., A.K. Adams, and E. Flores (1989) Inhibition of 7,12 dimethylbenz(a)anthracene (DMBA)-induced rat mammary carcinogenesis by glucarate. Proc. Am. Assoc. Cancer Res. 30:170 abstract #672.

47. Walaszek, Z., and M. Hanausek-Walaszek (1987) Dietary glucarate inhibits rat mammary tumorigenesis induced by N-methyl-N-nitrosourea. Proc. Am. Assoc. Cancer Res. 28:153 abstract #607.

48. Walaszek, Z., M. Hanausek-Walaszek, J.P. Minton, and T.E. Webb (1986) Dietary glucarate as anti-promoter of 7,12-dimethylbenz(a)anthracene-induced mammary tumorigenesis. Carcinogenesis 7:1463-1466.

49. Walaszek, Z., M. Hanausek-Walaszek, and T.E. Webb (1986) Dietary glucaratemediated reduction of sensitivity of murine strains to chemical carcinogenesis. Cancer Lett. 33:25-32.

50. Wargovich, M.J., V.W.S. Eng, H.L. Newmark, and W.R. Bruce (1983) Calcium ameliorates the toxic effect of deoxycholic acid on colonic epithelium. Carcinogenesis 4:1205-1207.

51. Wargovich, M.J., L.C. Stephens, and K.N. Gray (1989) Effect of two human nutrient density levels of calcium on promotional phase of colon tumorigenesis in the F344 rat. Proc. Am. Assoc. Cancer Res. 30:196 abstract #777.

52. Zarbl, H., S. Sukumar, A.V. Arthur, D. Martin-Zanca, and M. Barbacid (1985) Direct mutagenesis of Ha-ras-1 oncogenes by N-nitroso-N-methylurea during initiation of mammary carcinogenesis in rats. Nature 315:382-385.

DIETARY INHIBITORS AGAINST

MUTAGENESIS AND CARCINOGENESIS

Hikoya Hayatsu, Tomoe Negishi, and Sakae Arimoto

Faculty of Pharmaceutical Sciences
Okayama University
Tsushima, Okayama 700, JAPAN

INTRODUCTION

It has become increasingly clear that mutagenesis plays a central role in the multiple steps of carcinogenesis (42,51,155). Finding inhibitors in the diet and studying the mechanisms of their actions are important in providing humans a preventive measure against cancer (170). Mutagenesis- and carcinogenesis-inhibitors have been extensively studied, although the mechanisms of the inhibitions are, in many cases, not well understood.

LISTS OF INHIBITORS

Since there are many previously published reviews available on this subject (10,22,36,43,45,46,70,104,131,136), we would like to focus our attention here on recent findings. Tables 1 and 2 are summaries of the inhibitions, as studied in experimental systems, that have been reported during these five years, Table 1 for anticarcinogenesis and Table 2 for antimutagenesis. In these tables, we list only those inhibitors that may be present in food, with selected man-made additives included, such as those currently under trial for humans in the United States (84).

In this paper, we would like to describe in some detail our own work related to the subject of dietary inhibitors of mutagenesis. We found that porphyrins, notably hemin and chlorophyll, can trap mutagens having polycyclic structures. The resulting effect is inhibition of their mutagenicity in organisms that range from bacteria to insects.

We recently investigated the mechanistic aspect of the antimutagenic action of a green-tea component, epigallocatechin gallate. Studies on this inhibitor may be of primary importance because green tea has long been ingested by people in many areas of the world without any apparent adverse effect (108).

Antimutagenesis and Anticarcinogenesis Mechanisms III,
Edited by G. Bronzetti *et al.*, Plenum Press, New York, 1993

Table 1. Inhibitors against induced cancers in animals (1986-1991)

Carcinogen	Inhibitor	Target tissue and animal[a]	Reference
Aflatoxin B1	Methionine, Choline	Liver (m)	103
	Indole-3-carbinol	Rainbow trout (*Salmo gairdneri*)	19
Azaserine	Neurotensin	Pancreas (r)	144
	Selenium, Retinoid	Pancreas, Liver (r)	17
Azoxymethane	Benzyl selenocyanate	Colon (r)	26
	Cysteamine	Colon (r)	146
	Calorie restriction	Colon (r)	69
	Ethanol	Colon (r)	40
	Glucarate	Intestine (r)	24
	Neostigmine	Colon (r)	147
	Inositol hexaphosphate	Large intestine (r)	151
	Oltipraz	Intestine (r)	123
	n-6 Polyunsaturated fats	Colon (r)	105
	Putrescine	Colon (r)	145
Benzo[a]pyrene (B[a]p)	Soy sauce	Forestomach (m)	11
B[a]p diol epoxide	Tannic acid, green tea extract	skin (m)	62
N-Bis(2-hydroxypropyl-nitrosamine	Antioxidants	Lung (r)	44

Carcinogen	Agent	Organ (species)	Ref.
7,12-Dimethylbenz[a]-anthracene	Calorie restriction	Mammary gland (r)	63
	β-Carotene	Mouth (h)	49,137
	β-Carotene	Skin (m)	77
	Nordihydroguaiaretic acid, Diallyl sulfide	Skin (m)	8
	Garlic (*Allium sativum*)	(m)	122
	Garlic extract	Mouth (h)	86
	Garlic oil, Onion oil	Skin (m)	117
	Glycyrrhetinic acid	Skin (m)	161
	Glycyrrhizin	Skin (m)	2
	Green tea polyphenols	Skin (m)	163
	Ixora javanica extract	Skin (m)	96
	Monoterpenoids	Mammary gland (r)	127
	Rosemary extract	Mammary gland	133
	Tamoxifen-PSK	Mammary gland (r)	53
	Selenium	Mammary gland(r)	57
	Vitamin E	Mouth (h)	150
	Vitamin A-deficiency	Skin (m)	23
	6 modulators	Mammary gland (r)	121
1,2-Dimethylhydrazine	N-Acetylcysteine	Colon (r)	172
	Bowman-Birk protease inhibitor	Gastrointestinal tract, Liver (m)	135
	Bowman-Birk protease inhibitor	Colon (m)	12
	Chymostatin	Anal gland (m)	12
	Milk protein	Colon (m)	115

Tab. 1. Continued

Carcinogen	Inhibitor	Organ	Ref.
	Diallyl sulfide	Colon (m)	164
	Psyllium husk, Cellulose	Colon (r)	125
	13-*cis*-Retinoic acid	Colon (r)	107
	Potassium chloride	Small intestine (r)	58
Estrogen	Vitamin C	Kidney (h)	78
	Dicumarol, Butylated hydroxyanisole	Kidney (h)	126
Methylazoxymethanol acetate	Chlorogenic acid	Large intestine, Liver (h)	89
3-Methylchoranthrene	Bowman-Birk protease inhibitor (soy bean)	Lung (m)	173
	Green tea polyphenols	Skin (m)	163
N-Methyl-*N*''-nitro-*N*-nitrosoguanidine	γ-Amino-*N*-butyric acid, Baclofen	Gastrointestine (r)	141
	Cysteamine	Stomach (r)	142
	Phenylalanine	Stomach (r)	54
	Potassium chloride	Stomach (r)	143
4-(Methylnitrosamino)-1-(3-pyridyl)-1-butanone (NNK)	Arylalkyl isothiocyanates	Lung (m)	48, 90
	Ellagic acid, Butylated hydroxyanisole	Lung (m)	116
	D-Limonene, Citrus fruit oils	Lung, Forestomach (m)	169

N-Methyl *N*-nitrosourea	Dietary fiber, fat	Mammary gland	16
	Sarcophytol A	Large bowel (r)	97
	Retinoic acid	Mammary gland (r)	74
	Retinoid-Tamoxifen	Mammary gland (r)	85
Nitrite and Dibutylamine	Soybean	Liver, Bladder (m)	88
N-Nitrosobenzylmethylamine	Ellagic acid	Esophagous (r)	82
	Riboflavin, Nicotinic acid, Minerals	Esophagous (r)	153
N-Nitroso(2-hydroxypropyl)-(2-oxopropyl)amine	Casein-restriction	Pancreas (h)	66
N-Nitrosodiethylamine	Organosulfur compounds, Monoterpenes	Lung, Forestomach (m)	168
N-Nitrosomorpholine	Neostigmine	Liver (r)	148
Ovarian hormone	Caffeine	Mammary gland (m)	152
Transplanted tumors	Oleic acid, Linoleic acid, their methyl esters	(m)	180
	Eicosapentaenoic acid	(m)	149
UV irradiation	Green tea polyphenol	Skin (m)	160

a) r, rat; m, mouse; h, hamster

HEMIN AND ITS DERIVATIVES

Hemin, the isolated form of the blood pigment heme (Fig. 1), can inhibit the mutagenic activity of heterocyclic amines as tested with the Salmonella reversion assay (Tab. 3). As these heterocyclic amines are indirect mutagens, requiring for their mutagenicity an S9-mediated conversion into reactive hydroxyamino compounds, the inhibition by hemin can either be on the S9 reaction or on the mutagenic action of the activated compounds. We have found that both of these processes are inhibited by hemin. Thus, in the S9-mediated conversion of Trp-P-2 into Trp-P-2 (NHOH), the presence of hemin was inhibitory for the reaction. The yield of Trp-P-2 (NHOH) in this reaction decreased with the increasing dose of hemin in the incubation mixture (5). Furthermore, hemin inhibited the direct-acting mutagenicity of every preactivated heterocyclic amine examined (Tab. 3).

Protoporphyrin, a compound lacking the iron atom in the center of the hemin-structure, was also inhibitory against these heterocyclic amines either in their indirect forms or in their direct-acting forms (Tab. 3). The inhibition, however, was less powerful than that of hemin. Likewise, biliverdin, the ring-opened metabolite of protoporphyrin, was inhibitory toward these mutagens, but again the effect was smaller than by hemin.

We have spectroscopic evidence that hemin can form complexes with both Trp-P-2 and Trp-P-2 (NHOH) in aqueous solutions (5). It is possible that such complex formation precludes the mutagen from being a substrate for the metabolizing enzymes and also abolishes the ability of the active mutagen to attack cellular genetic targets. We have also found that hemin mediates a rapid degradation of Trp-P-2 (NHOH), a phenomenon that may play a part in the inhibitory action of hemin (5). It should be noted that myoglobin and hemoglobin, but not globin, show inhibitory activities toward direct-acting mutagens Trp-P-2 (NHOH) and Glu-P-1 (NHOH) and that this inhibition is attributable to the oxidative degradation of the hydroxylamino compounds accelerated by these heme proteins (6).

Heme and heme proteins as components of food and also as constituents of the human body, therefore, deserve further studies for evaluating them as defense mechanisms against environmental mutagens.

CHLOROPHYLL AND CHLOROPHYLLIN

Chlorophyll, the green pigment in vegetables, and its hydrolysis product chlorophyllin are porphyrin derivatives (Fig. 1), and therefore may be expected to show inhibitory actions against the mutagenicity of heterocyclic amines. This was indeed so, as demonstrated with the Salmonella assay for chlorophyllin (Tab. 3). In addition to the bacterial experiments, we have explored the antimutagenic activity in Drosophila, with somatic cell mutations as the endpoint.

The Drosophila wing spot test, originally developed by Würgler and co-workers (174) and used by Yoo et al. (175) for the mutagenicity test of food-pyrolysate mutagens, was carried out according to the method described previously (99,101).

Table 2. Inhibitors against mutagenesis (1986-1991).

Mutagen	Inhibitor	Assay system[a]	Reference
2-Acetylaminofluorene	Isoflavones	A	83,158
Active oxygens	Thiols	A	21
Aflatoxin B₁	Caffeic acid, eugenol, ellagic acid	A	27
	Chlorophyllin	A	20,165,171
	Ellagic acid	A, DNA damage in cultured tissues	82
	Glycyrrhetinic acid	A	161
	Green tea polyphenols	A, V79 cells	162
AF-2	Butylhydroxytoluene	A, *E. coli hcr⁻*	176
9-Aminoacridine	Caffeine	*E. coli* K12	120
2-Aminoanthracene	Chlorophyllin	*Salmonella arabinose* resistance	165
	Flavonoides from *Psoralea corylifolia*	A	157
o-Aminoazotoluene	Retinol	A	154
2-Aminofluorene	Diacylmethane derivatives	A	159
	Glycyrrhetinic acid	A	161

Tab. 2. Continued

Benzo[a]pyrene (B[a]P)	Human saliva	Umu-test	110
	Isoflavones	A	83, 158
	Milk fermented by S. thermophilus & L. Bulgaricus	A	13
	Taurine	A	75
	Aquatic plants	A	29
	Catechin	A	93
	Chlorophyllin	Salmonella arabinose resistance	165
	Copper complex (SOD analog)	A	124
	Desert mushroom extract	A	41
	Dialyzates of vegetables and fruits	A	132
	Erythrosine (food dye)	A	76
	Furfural	A	67
	Glycyrrhetinic acid	A	161
	Green tea polyphenols	A, V79 cells	162
	Oolong tea	A, Chromosome aberrations in CHL cells	65
	Peony root extract	A	128
	Phenol carboxylic acids	A	129
	Wheat sprout extract	A	118
B[a]P diol epoxide	Ellagic acid analogs	A	60, 134
	Saponins	A	25

Compound	Modifier	Test system	Ref.
Bleomycin	Taurine	A	75
2-tert-Butyl-p-quinone from butylhydroxy-anisole plus nitrite	Vegetable & fruit juice	A	87
Cigarette smoke condensate	Catechin	A	94
Cyclophosphamide	Uric acid	Micronuclei in mouse	3
1,8-Dinitropyrene	Microsomal lipid	A	130
Dibenzo[a,c]fluoranthene	Norharman	A	177
Diiodohydroxyquinoline	Ascorbic acid	In vivo micronucleus	33
	L-Cysteine	In vivo micronucleus	32
7,12-Dimethylbenz[a]anthrachene	Catechin	A	93
	Desert mushroom extract	A	41
Doxorubicin	Taurine	A	75
Ethydium bromide	Chlorophyllin	S. cerevisiae	14

Tab. 2. Continued

Gingerol	Zingerone in ginger extract	A	92
Heterocyclic amines	Chlorophyllin	A	20
	Hemin	A	5
	Lactic acid bacteria	Binding to bacteria	178
	Methylxanthines	A	91
	Retinol	A	56
IQ	Tryptamine	A	1
Maleic hydrazide	Diethyldithiocarbamate	Tradescantia	38
4-(Methylnitrosamino)-1-(3-pyridyl)-1-butanone (NNK)	Catechin	Rat hepatocytes	79
N-Methyl-N'-nitro-N-nitrosoguanidine	p-Aminobenzoic acid	A	37
	Caffeic acid, eugenol, ellagic acid	A	27
	Chlorophyllin	Salmonella arabinose resistance	165
	Dialyzates of vegetables & fruits	A	132
	Garlic extract	A	64
	Instant coffee	Salmonella	106

	Tea extract	A, E. coli WP2	59
	protein pyrolyzate	E. coli	72
	Phenols	E. coli WP2	73
	o-Vanillin	E. coli	166
Methylnitrosourea (formation)	p-Aminobenzoic acid	A	35
	Curcumins	A	95
Mutagenic complexes (coal dust, airborne particles, fried beef)	Chlorophyllin, retinol, vitamins	A	112
N-Nitrosodimethylamine	Acetylcysteine	A	31
	Ellagic acid	A	80
N-Nitrosodiethylamine	Diethyldithiocarbamate	Tradescantia	38
4-Nitroquinoline 1-oxide (4NQO)	Butylhydroxytoluene	A, E. coli hcr$^-$	14
	Garlic	E. coli WP2	179
	Milk fermented by S. thermophilus & L. Bulgaricus	A	13
	Paeonol	E. coli	30
	Protein pyrolyzate	E. coli	72
	Sesquiterpene lactones	E. coli WP2	71

Tab. 2. Continued

Nitrofluorene	Aquatic plant extract	A	29
Nitrophenylenediamine	Citrus fruits	A	9
Pepper	Whole casein	Salmonella	52
Trp-P-1/Trp-P-2	Chlorophyll, Chlorophyllin	A, Drosophila	99
	Dialyzates of vegetables & fruits	A	132
	Furfural	A	67
	Glycyrrhiza extract	A	140
	Hemin	A	5
	Herb extracts	A	98
	Lactic acid bacteria	Binding to bacterial cells	178
	Oolong tea	A, Chromosome aberration in CHL cells	65
Tabacco-specific N-nitrosamines	Hydroxychavicol, betel leaf extract	A, in vivo micronucleus	4,114
UV	Vanillin	V79 cells	55
	p-Aminobenzoic acid	Salmonella	34

a) A, Ames test

In the wing spot test, genotoxicity is detected by phenotypic alterations of adult-fly wing hairs, alterations resulting from mutations in the somatic cell chromosomes of the larvae (39,174). An administration of Trp-P-2 at a dose of 2 mg per bottle induced both small single spots (one or two mutant hairs in a clone), with a frequency of 1.20 spots per wing (the control value, 0.29), and large single spots (three or more mutant hairs in a clone), with a frequency of 0.30 spots per wing (control, 0.03). Twin spots (the two different phenotype mutants co-existing in a clone) were also induced with a frequency of 0.07 per wing (control, 0.02). These values are in agreement with those previously reported by Yoo et al. (175) for the genotoxicity of Trp-P-2 in this system. When 200 mg chlorophyll was co-administered with Trp-P-2 to these larvae, the values of mutant clone numbers per wing decreased: small single spots to 0.28, large single spots to 0.15, and twin spots to 0.01. The decrease in small single spots represents a complete inhibition, because the frequency of spontaneous induction was 0.29. Chlorophyllin was also effective in reducing the frequencies of mutant clone formation. In Fig. 2, we show the dose dependence of these inhibitions. Both chlorophyll and chlorophyllin seem to be less effective in suppressing large single spot formation than in suppressing small singles. The reason for this difference is not clear.

A finding relevant to the mechanism of these inhibitions is the following: We obtained evidence showing that chlorophyll and chlorophyllin can make complexes with Trp-P-2. First, an equimolar mixture of chlorophyllin (chlorophyll) and Trp-P-2 showed an absorption spectrum different from that of a calculated sum of spectra for the individual components. This can be seen from the different spectra shown in Fig. 3. Second, chlorophyllin covalently linked to Sepharose can efficiently adsorb Trp-P-2, whereas Sepharose itself does not adsorb Trp-P-2 (data not shown). These observations suggest that chlorophyllin and chlorophyll can trap Trp-P-2 by forming a complex. We detected no degradation of Trp-P-2 in the presence of these pigments, a fact indicating that degradation of Trp-P-2 is unlikely to be the cause of the inhibition. Therefore, a straightforward mechanism of these inhibitions is that Trp-P-2 becomes no longer available to organisms due to formation of the chlorophyll complex. In a preliminary experiment using rats, we observed that chlorophyllin can accelerate the excretion of mutagenic components into the feces following an oral administration of Trp-P-2.

Green-yellow vegetables have been suggested to be anticarcinogenic, based on epidemiological evidence (119). Since chlorophyll is a main component of green vegetables (15), it is possible that chlorophyll makes a contribution, among other components of vegetables, to the beneficial effect of vegetables for human health. It should be noted that other workers also have reported antimutagenic activities of chlorophyll and chlorophyllin (for review, see Ref. 46). Especially remarkable in this regard are a series of recent experiments performed by Bailey and Dashwood. They reported the inhibitory property of chlorophyllin against aflatoxin B_1-DNA binding in rainbow trout (20) and against IQ-induced DNA-adduct formation in the liver of rats (18).

EPIGALLOCATECHIN GALLATE

Epidemiological studies have shown that the habit of drinking green tea lowers the risk for colon cancer (61). It is reported that (-)-epigallocatechin gallate

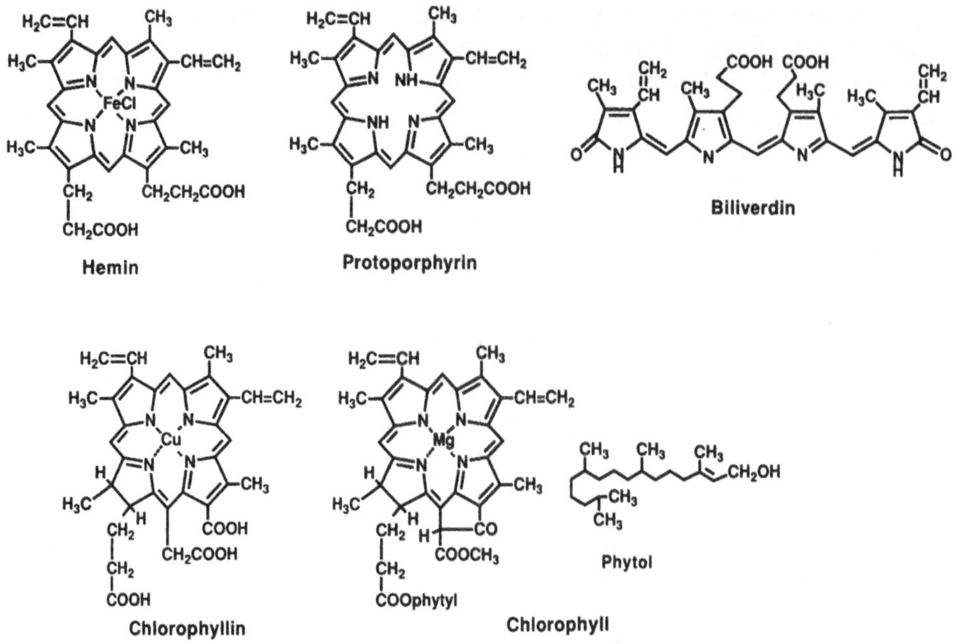

Figure 1. Structures of porphyrin derivatives.

(EGCG), a major polyphenol in tea, inhibits the promotion of induced cancers in animals (28,138). Also, there are reports dealing with the protective effects of tea components against mutagenesis and carcinogenesis (see Tab. 1 and 2). An earlier investigation in our laboratory has shown that tea polyphenols may be antimutagenic against several direct-acting mutagens, as tested in the Salmonella assay (111). Thus, we have investigated the inhibitory actions of EGCG against several carcinogenic mutagens, with a variety of genotoxic endpoints as the measure of the actions.

In studying the effect of EGCG against the activities of mutagens *in vitro,* we have avoided the use of metabolizing-enzyme systems (i.e., "S9 mix" in the Salmonella mutation assay) because plant polyphenols are known to interact with proteins (68). As Tab. 4 shows, EGCG effectively suppresses mutagenicity in *Salmonella typhimurium* TA98 of direct-acting mutagens, Trp-P-2 (NHOH) and Glu-P-1 (NHOH), the activated forms of food-derived mutagens Trp-P-2 and Glu-P-1, respectively. The suppressing effect of EGCG was weaker for B(a)P diol epoxide, and was undetectable for rice-nitrite mutagens (47) or for rice bran-nitrite mutagens (unpublished). Therefore, the inhibitory action of EGCG is not indiscriminating among mutagens, but rather is specific for certain mutagens.

The antimutagenic activity of EGCG was detectable not only with bacteria but also with mammalian cells in culture. Thus, as Tab. 5 shows, the mutation frequency

Table 3. Inhibitory effect of hemin and other porphyrin derivatives on the mutagenicity of heterocyclic amines.[a]

Mutagen	Mutagenicity in S. typhimurium TA98		Dose for 50% inhibition (nmole/plate)				Presence of S9[d]
	Amount of mutagen (nmole)	Number of revertants per plate	Hemin	Proto-porphyrin	Biliverdin	Chloro-phyllin	
Indirect acting							
Trp-P-1	1.8	12,350	3.0	50	200	100	+
Trp-P-2	0.1	7,750	20	100	500	200	+
Glu-P-1	1.7	25,600	75	n.i.[b]	300	100	+
Glu-P-2	9.1	6,010	40	n.i.[b]	200	150	+
AαC	80	6,580	30	n.i.[b]	500	150	+
MeAαC	250	3,940	25	50	500	200	+
Direct-acting							
Trp-P-2(NHOH)	0.05	15,400	0.5	2.2	7.4	1.8	−
Activated Trp-P-1	0.5[c]	2,300	1.0	1.0	1.0	1.0	−
Activated Glu-P-1	1[c]	6,360	15	350	350	200	−
Activated Glu-P-2	5[c]	1,230	25	150	80	90	−
Activated AαC	200[c]	2,090	30	7,500	n.i.[b]	250	−
Activated MeAαC	400[c]	1,350	7	25	70	10	−
Activated IQ	2[c]	2,040	15	25	100	10	−
Activated MeIQ	0.15[c]	3,570	8	3	25	2	−
Activated MeIQx	0.4[c]	2,570	25	20	250	30	−

a) Data taken from ref. 5 and ref. 7.
b) No inhibition.
c) Nanomole of heterocyclic amine subjected to the pre-activation with S9 (5).
d) 10 μl of S9/plate was used.

401

of FM3A cells induced by Trp-P-2(NHOH) can be decreased to about one-third of the original value by addition of a 17-fold excess EGCG. The activity of rice-bran-nitrite mutagens was not subject to inhibition by EGCG, as was the case observed with Salmonella.

Effect of EGCG against genotoxic actions of carcinogens in vivo was investigated by means of Drosophila genotoxicity tests. The larvae were grown on feeds containing indirectly-acting mutagens with or without added EGCG, and the adult flies that emerged from the larvae were examined for genetic damage using the wing-spot test (39,99,102) and the *in vivo* repair test (102). EGCG was effective in suppressing Trp-P-1-mediated wing spot formation which is a result of somatic cell mutation (Fig. 4). Both the small and the large single spots caused by Trp-P-1 were suppressed by EGCG, and the effect was dependent on the dose of EGCG (see the left half of the figure). EGCG was also effective in diminishing the lethal DNA damages caused by Trp-P-1 in Drosophila, as indicated by the significantly smaller decrease in the male/female ratios compared to those in the Trp-P-1-only flies (the right half of the figure). Figure 5 illustrates the effect of EGCG against the activity of other mutagens in these Drosophila tests. EGCG was strongly suppressive against the genotoxic actions of 4NQO, either in single spot formation or in causing repairable DNA damage. A weaker inhibition was found in the effect of EGCG against spot formation by NDMA, and only a marginal suppression was observed against spot formation by aflatoxin B_1 (note that in the case of aflatoxin B_1, its genotoxicity was weak compared to that of 4NQO and NDMA, and therefore the evaluation of the EGCG suppression was difficult).

Finally, we investigated whether EGCG can inhibit DNA single-strand breaks caused by active oxygen species derived from Glu-P-1 (NHOH) *in vitro*. We have previously shown that the N-hydroxyl forms of heterocyclic amines decompose in aqueous media by an oxidative process, generating active oxygen species, presumably hydroxyl radicals, and that these active oxygen species can cause DNA single strand breaks either *in vitro* (50,156) or *in vivo* (167). When EGCG was added to a mixture of Glu-P-1 (NHOH) and a double-stranded, super-coiled circular DNA, a strong suppression of the formation of strand breaks was observed (Tab. 6).

EGCG is the major constituent of green tea polyphenols. The amount in a cup of green tea ranges from 10-30 mg (Sakata, I. et al., unpublished data). Therefore, the intake of EGCG in people with a habit of drinking green tea is substantial. As shown above, EGCG exhibits efficient suppressing capability in a variety of assay systems against genotoxic effects of several different carcinogens.

It is important to elucidate the mechanisms underlying these effects. The inhibition of Glu-P-1 (NHOH)-mediated DNA strand breaks by EGCG is probably due to the antioxidant nature of this polyphenol (113). One possibility is that EGCG prevents the oxidative decomposition of Glu-P-1(NHOH), a process necessary for causing DNA strand breaks (50,156). This mode of action, however, seems unlikely because the decomposition of Glu-P-1 (NHOH) in a buffered aqueous solution at pH 6 was not hindered by addition of a ten-fold excess EGCG, as studied by HPLC analysis of the reaction mixture (data not shown). Therefore, the most likely

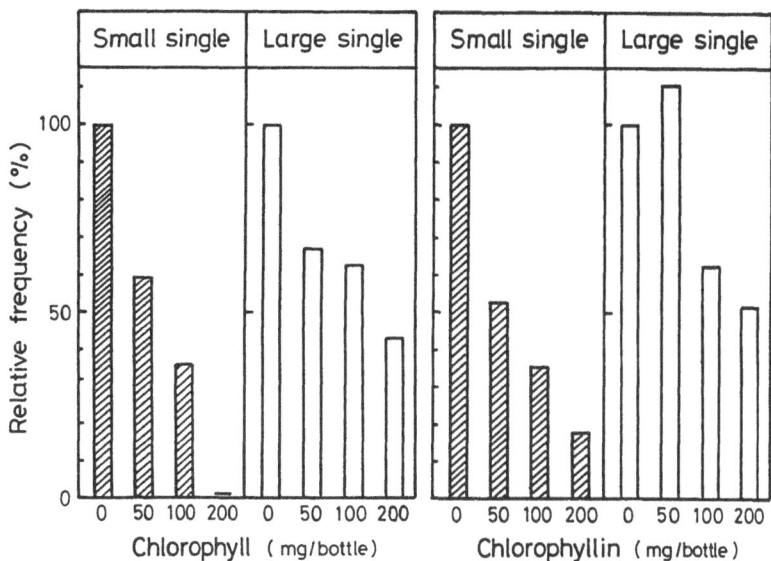

Figure 2. Inhibition of Trp-P-2 genotoxicity by chlorophyll and chlorophyllin. Relative frequency (RF) was calculated as follows:

$$RF (\%) = \frac{\text{spots/wing (Trp-P-2 + pigment) - spots/wing (solvent)}}{\text{spots/wing (Trp-P-2) - spots/wing (solvent)}} \times 100$$

Trp-P-2 dose/bottle was 2 mg. The solvent was distilled water.

mechanism is the scavenging of hydroxyl radicals, the putative attacking agent on the DNA chain, by EGCG.

To explore possible molecular interactions between EGCG and those heterocyclic amines, we measured absorption spectra of Trp-P-2 (NHOH) and Glu-P-1 (NHOH) in the presence of equimolar EGCG, expecting that some nonadditive spectra might result if there were such interactions as were found for porphyrins with these mutagens. No such spectral shifts, either in terms of the absorption intensities or in terms of the absorbance peak positions, were detected (data not shown). This finding has suggested the absence of direct interactions between EGCG and these carcinogens. To further confirm this concept, we have prepared a Sepharose-supported EGCG (40 μmol EGCG/g Sepharose) having a 12-carbon spacer (unpublished work), and explored whether this ligand-bearing Sepharose can adsorb Trp-P-2 and Glu-P-1 in aqueous media. No adsorption took place (data not shown).

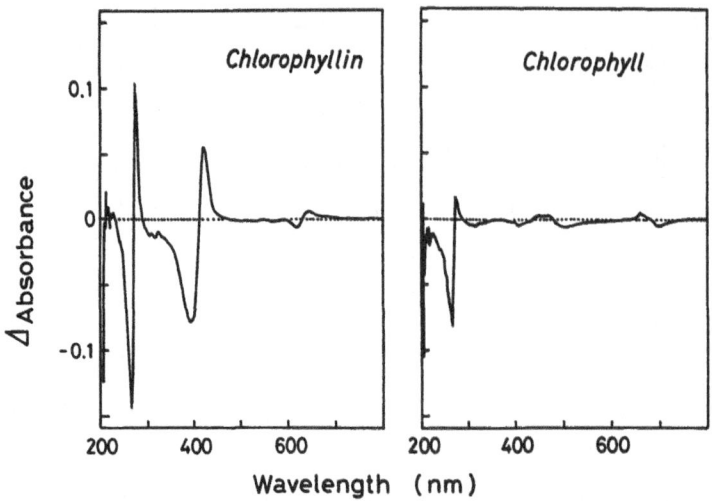

Figure 3. Difference spectra for mixtures of Trp-P-2 and chlorophyll/chlorophyllin. The solutions contained Trp-P-2 and/or pigment (20 μM each) in 0.1 *M* sodium phosphate buffer at pH 7.4.

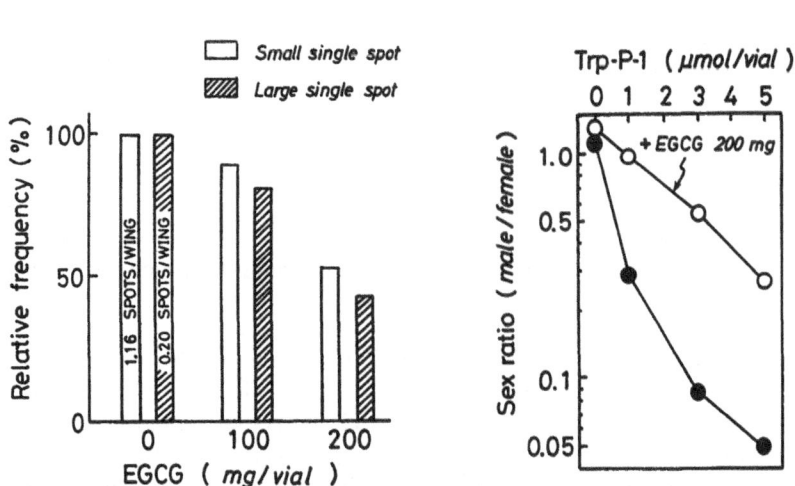

Figure 4. Inhibitory effect of EGCG on the Trp-P-1 genotoxicity in Drosophila. In the wing spot test, Trp-P-1 was used at 2.7 mg/vial. The total numbers of wings examined in the spot test were 205 for the EGCG-0 mg data, 124 for the 100 mg data, and 312 for the 200 mg data. In the DNA repair test, the standard protocol (102) was used.

Table 4. Effect of EGCG on the activity of mutgens in Salmonella.[a]

Mutagen	EGCG (nmol)	No. of revertants /plate	% activity
Trp-P-2 (NHOH)		Strain TA98	
(7.5 pmol)	0	968	100
	10	472	48
	100	26[b]	1[b]
Solvent only		22	–
Glu-P-1 (NHOH)		TA98	
(0.5 nmol)	0	769	100
	10	223	27
	100	128[b]	23[b]
Solvent only		22	–
B(a)P diol epoxide		TA100	
(0.2 nmol)	0	1418	100
	100	1530	109
	1000	359	18
Solvent only		131	–
Nitrite-treated		TA100	
rice	0	805	100
(10 g rice	100	819	102
equivalent)	1000	642	77
Solvent only		91	–
Nitrite-treated		TA100	
rice-bran	0	413	100
(0.5 g bran	100	425	105
equivalent)	1000	395	93
Solvent only		139	–

a) All assays were in the absence of S9 mix.
b) Partial killing of bacteria (~ 70% in the surviving fraction, as determined by plating on nutrient agar) was observed, and consequently corrections were made in the calculation of '% activity'.

The molecular mechanism underlying these antigenotoxic EGCG actions is obviously not a simple inhibitor-mutagen interaction. EGCG may thus influence the cellular mechanisms that are related to induced mutagenesis, e.g., DNA synthesis and repair processes. EGCG may also alter cellular ability for mutagen-intake; in this case, in such a way that only certain mutagens, but not others, are blocked from entering into cells. EGCG may also intercept mutagen actions by interacting with the mutagen-derived ultimate active principles. Such an example is the inhibition of Glu-P-1 (NHOH)-induced DNA strand breaks by EGCG. In this case, EGCG is probably acting not as a scavenger of the Glu-P-1 (NHOH) molecules themselves but as a scavenger of the reactive oxygen radicals generated by spontaneous degradation of the mutagen. There may be many possible modes of action by which EGCG interferes with process of mutagenesis, and further studies are needed for elucidating these mechanisms.

Table 5. Effect of EGCG on the mutagenesis in FM3A cells.

Compound[a]	No. of ouabain resistant mutants scored / total No. of original cells	Mutation frequency	Colony-forming ability (%) ± S.D. (n=15)
None	$2 / 1.5 \times 10^7$	1.5×10^{-7}	86.6 ± 9.7
EGCG	$3 / 1.5 \times 10^7$	2.2×10^{-7}	90.4 ± 11.8
Trp-P-2(NHOH)	$166 / 1.5 \times 10^7$	1.5×10^{-5}	74.4 ± 7.0
Trp-P-2(NHOH) + EGCG	$70 / 1.5 \times 10^7$	5.1×10^{-6}	91.7 ± 13.7
Rice bran-HNO$_2$	$21 / 1.5 \times 10^7$	1.6×10^{-6}	89.4 ± 9.3
Rice bran-HNO$_2$ + EGCG	$22 / 1.5 \times 10^7$	2.0×10^{-6}	72.9 ± 9.6

a) Concentration of compound : EGCG, 50 nmol/ml ; Trp-P-2(NHOH), 3 nmol/ml; Rice bran-HNO$_2$, 5 gE/ml.

Table 6. Inhiition of EGCG of Glu-P-1 (NHOH)-induced PM2 DNA single-strand breaks.

Agent	Concn of EGCG (mM)	Strand break (%)
Glu-P-1(NHOH) (1 mM) [a]		
+ EGCG	0	75
	0.2	20
	2	11
	20	5
	200	0

a) Incubation: 2 hr at pH 7.4 (0.1 M sodium phosphate buffer) and 37°C.

CONCLUDING REMARKS

Prevention of cancer by adopting appropriate dietary habits instead of administering some drug is a practical and safe way of achieving the goal. Since preventive measures require intake of some chemical material for one's entire lifetime, the material must be 100% guaranteed to be harmless. Because of this inherent requirement for the inhibitor, natural food components are the best candidates as ideal inhibitors. One such example is the green-tea component EGCG.

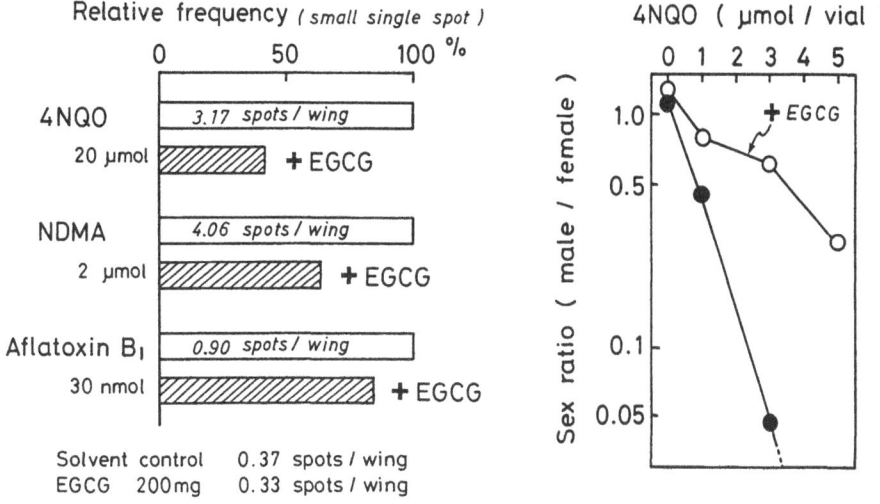

Mutagenicity: Wing spot test   **DNA damage**: *In vivo* DNA repair test

Relative frequency *(small single spot)*   4NQO (μmol / vial)

4NQO   3.17 spots / wing
20 μmol   + EGCG

NDMA   4.06 spots / wing
2 μmol   + EGCG

Aflatoxin B₁   0.90 spots / wing
30 nmol   + EGCG

Solvent control   0.37 spots / wing
EGCG   200mg   0.33 spots / wing

Figure 5. Effect of EGCG on the genotoxicity of mutagens. EGCG was added at 200 mg per vial. The numbers of wings examined were 116 for 4NQO, 154 for 4NQO + EGCG, 53 for NDMA, 72 for NDMA + EGCG, 80 for aflatoxin B$_1$, 82 for aflatoxin B$_1$ + EGCG, 128 for the control, and 126 for EGCG-only.

Once the mechanism of the inhibitory action has been elucidated, the usefulness of this compound as a tea component would become more convincing.

We hope that there will be more dietary inhibitors discovered by future studies. The progress of these studies will eventually enable us to establish a "preventive diet" with scientifically sound supporting evidence.

ACKNOWLEDGEMENTS

This work was supported by a Grant-in-Aid for Cancer Research (03151042) from the Ministry of Education, Science and Culture, Japan, and a Grant for Cancer Research from the Ministry of Health and Welfare, Japan.

REFERENCES

1.   Abu-Shakra, A., C. Ioannides, and R. Walker (1987) Effect of tryptamine on the mutagenic activity of 2-amino-3-methylimidazo [4,5-*f*]quinoline (IQ) and related azaarenes in the Ames test. <u>Mutagenesis</u> 2:51-56.
2.   Agarwal, R., Z.Y. Wang, and H. Mukhtar (1991) Inhibition of mouse skin tumor-initiating activity of DMBA by chronic oral feeding of glycyrrhizin in drinking water. <u>Nutr. Cancer</u> 15:187-193.

3. Al-Bekairi, A.M., S. Qureshi, M.S. Chaudhry, and A.H. Shah (1991) Uric acid as an inhibitor of cyclophosphamide-induced micronuclei in mice. Mutat. Res. 262:115-118.

4. Amonkar, A.J., P.R. Padma, and S.V. Bhide (1989) Protective effect of hydroxychavicol, a phenolic component of betel leaf, against the tobacco-specific carcinogens. Mutat. Res. 210:249-253.

5. Arimoto, S., and H. Hayatsu (1989) Role of hemin in the inhibition of mutagenic activity of 3-amino-1-methyl-5H-pyrido[4,3-b]indole (Trp-P-2) and other aminoazaarenes. Mutat. Res. 213:217-226.

6. Arimoto, S., Y. Ohara, K. Hiramoto, and H. Hayatsu (1987) Inhibitory effect of myoglobin and hemoglobin on the direct-acting mutagenicity of protein pyrolysate heterocyclic amine derivatives. Mutat. Res. 192:253-258.

7. Arimoto, S., Y. Ohara, T. Namba, T. Negishi, and H. Hayatsu (1980) Inhibition of the mutagenicity of amino acid pyrolysis products by hemin and other biological pyrrole pigments. Biochem. Biophys. Res. Comm. 92:662-668.

8. Athar, M., H. Raza, D.R. Bickers, and H. Mukhtar (1990) Inhibition of benzoyl peroxide-mediated tumor promotion in 7,12-dimethylbenz (a)anthracene-initiated skin of Sencar mice by antioxidants nordihydroguaiaretic acid and diallyl sulfide. J. Invest. Dermatol. 94:162-165.

9. Bara, S., and I.S. Grover (1989) Antimutagenicity of some citrus fruits in *Salmonella typhimurium.* Mutat. Res. 222:141-148.

10. Bartsch, H., I.K. O'Neill, and R. Schulte-Hermann, eds. (1987) Relevance of N-Nitroso Compounds to Human Cancer: Exposures and Mechanisms. IARC Scientific Publications No. 84, Lyon, France.

11. Benjamin, H., J. Storkson, A. Nagahara, and M.W. Pariza (1991) Inhibition of benzo(a)pyrene-induced mouse forestomach neoplasia by dietary soy sauce. Cancer Res. 51:2940-2942.

12. Billings, P.C., P.M. Newberne, and A.R. Kennedy (1990) Protease inhibitor suppression of colon and anal gland carcinogenesis induced by dimethylhydrazine. Carcinogenesis 11:1083-1086.

13. Bodana, A., and D.R. Rao (1990) Antimutagenic activity of milk fermented by *Streptococcus thermophilus* and *Lactobacillus bulgaricus.* J. Dairy Sci. 73:3378-3384.

14. Bronzetti, G., A. Galli, and C. Della Croce (1990) Antimutagenic effect of chlorophyllin. In Antimutagenesis and Anticarcinogenesis Mechanisms II, Y. Kuroda, D.M. Shankel, and M.D. Waters, eds. Plenum, New York, pp. 463-468.

15. Chipchase, M.I.H. (1961) Chemical components in plant tissues. In Biochemists' Handbook, C. Long, ed. E. Spon and F.N. Spon, Ltd., London, pp. 1032-1033.

16. Cohen, L.A., M.E. Kendall, E. Zang, C. Meschter, and D.P. Rose (1991) Modulation of N-nitrosomethylurea-induced mammary tumor promotion by dietary fiber and fat. J. Natl. Cancer Inst. 83:496-501.

17. Curphey, T.J., E.T. Kuhlmann, B.D. Roebuck, and D.S. Longnecker (1988) Inhibition of pancreatic and liver carcinogenesis in rats by retinoid- and selenium-supplemented diets. Pancreas 3:36-40.

18. Dashwood, R.H. (1992) Protection by chlorophyllin against the covalent binding of 2-amino-3-methylimidazo [4,5-f]quinoline (IQ) to rat liver DNA. Carcinogenesis 13:113-118.

19. Dashwood, R.H., D.N. Arbogast, A.T. Fong, C. Pereira, J.D. Hendrick, and C.S. Bailey (1989) Quantitative inter-relationships between aflatoxin B_1

carcinogen dose, indole-3-carbinol and carcinogen dose, target organ DNA adduction and final tumor response. Carcinogenesis 10:175-181.

20. Dashwood, R.H., V. Breinholt, and G. Bailey (1991) Chemopreventive properties of chlorophyllin: Inhibition of Aflatoxin B_1 (AFB_1) -DNA binding *in vivo* and anti-mutagenic activity against AFB_1 and two heterocyclic amines in the Salmonella mutagenicity assay. Carcinogenesis 12:939-942.

21. De Flora, S., C. Bennicelli, P. Zanacchi, and F. D'Agostini (1989) Mutagenicity of active oxygen species in bacteria and its enzymic or chemical inhibition. Mutat. Res. 214:153-158.

22. De Flora, S., P. Zanacchi, A. Izzotti, and H. Hayatsu (1991) Mechanisms of food-borne inhibitors of genotoxicity relevant to cancer prevention. In Mutagens in Food: Detection and Prevention, H. Hayatsu, ed. CRC Press, Boca Raton, Florida, pp. 158-180.

23. De Luca, L.M., R.L. Shores, E.F. Spangler, and M.L. Wenk (1989) Inhibition of initiator-promoter-induced skin tumorigenesis in female SENCAR mice fed a vitamin A-deficient diet and reappearance of tumors in mice fed a diet adequate in retinoid or β-carotene. Cancer Res. 49:5400-5406.

24. Dwivedi, C., O.A. Oredipe, R.F. Barth, A.A. Downie, and E. Thomas (1989) Effects of the experimental chemopreventative agent, glucarate, on intestinal carcinogenesis in rats. Carcinogenesis 10:1539-1541.

25. Elias, R., M. De Meo, E. Vidal-Ollivier, M. Laget, G. Balansard, and G. Dumenil (1990) Antimutagenic activity of some saponins isolated from *Calendula officinalis* L., *C. arvensis* L. and *Hedera helix* L. Mutagenesis 5:327-331.

26. Fiala, E.S., C. Joseph, O. Soon Sohn, K. El-Bayoumy, and B.S. Reddy (1991) Mechanism of benzylselenocyanate inhibition of azoxymethane-induced colon carcinogenesis in F344 rats. Cancer Res. 51:2826-2830.

27. Francis, A.R., T.K. Shetty, and R.K. Bhattacharya (1989) Modification of the mutagenicity of aflatoxin B_1 and N-methyl-N'-nitro-N-nitrosoguanidine by certain phenolic compounds. Cancer Lett. 4:177-182.

28. Fujiki, H., M. Suganuma, H. Suguri, K. Takagi, S. Yoshizawa, A. Ootsuyama, H. Tanooka, T. Okuda, M. Kobayashi, and T. Sugimura (1990) New antitumor promoters: (-)-Epigallocatechin gallate and sarcophytols A and B. In Antimutagenesis and Anticarcinogenesis Mechanisms II, Y. Kuroda, D.M. Shankel, and M.D. Waters, eds. Plenum, New York, pp. 205-212.

29. Fujimoto, T., Y. Ose, T. Sato, H. Matsuda, H. Nagase, and H. Kito (1987) Antimutagenic factors in aquatic plants. Mutat. Res. 178:211-216.

30. Fukuhara, Y., and D. Yoshida (1987) Paeonol: A bio-antimutagen isolated from a crude drug, moutan cortex. Agric. Biol. Chem. 51:1441-1442.

31. Garland, W.A., E.P. Norkus, W.A. Kuenzig, and M.L. Powell (1988) The effect of N-acetylcysteine on the toxicity induced by N-nitrosodimethylamine. Drug Metab. Dispos. 16:162-165.

32. Ghaskadbi, S., S.V. Pavaskar, and V.G. Vaidya (1987) Bioantimutagenic effect of L-cystein on diiodohydroxyquinoline-induced micronuclei in Swiss mice. Mutat. Res. 187:219-222.

33. Ghaskadbi, S., and V.G. Vaidya (1989) In vivo antimutagenic effect of ascorbic acid against mutagenicity of the common antiamebic drug diiodohydroxyquinoline. Mutat. Res. 222:219-222.

34. Gichner, T., I. Baburek, J. Veleminsky, and A. Kappas (1991) UV-irradiation potentiates the antimutagenicity of *p*-aminobenzoic and *p*-aminosalicylic acids in *Salmonella typhimurium*. Mutat. Res. 249:119-123.

35. Gichner, T., and J. Veleminsky (1988) Inhibitory effects of *p*-aminobenzoic acid on the formation and mutagenicity of *N*-nitroso compounds. Mutagenesis 3:329-331.

36. Gichner, T., and J. Veleminsky (1988) Inhibitors of N-nitroso compounds-induced mutagenicity. Mutat. Res. 195:21-43.

37. Gichner, T., J. Veleminsky, I.A. Rapoport, and S.V. Vasileva (1987) Antimutagenic effect of *p*-aminobenzoic acid on the mutagenicity of *N*-methyl-*N'*-nitro-*N*-nitrosoguanidine in *Salmonella typhimurium*. Mutat. Res. 192:95-98.

38. Gichner, T., J. Veleminsky, and R. Rieger (1988) Antimutagenic effects of diethyldithiocarbamate towards maleic hydrazine- and *N*-nitrosodietylamine-induced mutagenicity in the tradescantia mutagenicity assay. Biol. Plant 30:14-19.

39. Graf, U., F.E. Würgler, A.J. Katz, H. Frei, H. Juon, C.B. Hall, and P.G. Kale (1984) Somatic mutation and recombination test in *Drosophila melanogaster*. Env. Muta. 6:153-188.

40. Hamilton, S.R., O.S. Shon, and E.S. Fiala (1988) Inhibition by dietary ethanol of experimental colonic carcinogenesis induced by high-dose azoxymethane in F344 rats. Cancer Res. 48:3313-3318.

41. Hannan, M.A., A.A. Al-Dakan, H.Y. Aboul-Enein, and A.A. AlOthaimeen (1989) Mutagenic and antimutagenic factor(s) extracted from a desert mushroom using different solvents. Mutagenesis 4:111-114.

42. Harris, C.C. (1991) Chemical and physical carcinogenesis: Advances and perspectives for the 1990s. Cancer Res. (Suppl.) 51:5012s-5044s.

43. Hartman, P.E., and D.M. Shankel (1990) Antimutagens and anticarcinogens: A survey of putative interceptor molecules. Env. Molec. Mutag. 15:145-182.

44. Hasegawa, R., F. Furukawa, K. Toyoda, M. Takahashi, Y. Hayashi, M. Hirose, and N. Ito (1990) Inhibitory effect of antioxidants on *N*-bis (2-hydroxypropyl) nitrosamine-induced lung carcinogenesis in rats. Jpn. J. Cancer Res. 81:871-877.

45. Hayatsu, H., ed. (1991) Mutagens in Food: Detection and Prevention. CRC Press, Boca Raton, Florida.

46. Hayatsu, H., S. Arimoto, and T. Negishi (1988) Dietary inhibitors of mutagenesis and carcinogenesis. Mutat. Res. 202:429-446.

47. Hayatsu, H., and T. Hayatsu (1989) Mutagenicity arising from boiled rice on treatment with nitrous acid. Jpn. J. Cancer Res. 80:1021-1023.

48. Hecht, S.S., M.A. Morse, K.I. Eklind, and F.-L. Chung (1991) A/J mouse lung tumorigenesis by the tobacco-specific nitrosamine 4-(methylnitrosamino)-1-(3-pyridyl)-1-butanone and its inhibition by arylalkyl isothiocyanates. Exp. Lung Res. 17:501-511.

49. Hibino, T., K. Shimpo, K. Kawai, T. Chihara, K. Maruta, M. Arai, T. Nagatsu, and K. Fujita (1990) Polyamine levels of urine and erythrocytes on inhibition of DMBA-induced oral carcinogenesis by topical beta-carotene. Biogenic Amines 7:209-216.

50. Hiramoto, K., K. Negishi, T. Namba, T. Katsu, and H. Hayatsu (1988) Superoxide dismutase-mediated reversible conversion of 3-hydroxyamino-1-methyl-5*H*-pyrido[4,3-*b*]indole, the *N*-hydroxy derivative of Trp-P-2, into its nitroso derivative. Carcinogenesis 9:2003-2008.

51. Hollstein, M., D. Sidransky, B. Vogelstein, and C.C. Harris (1991) Mutations in human cancer. Science 253:49-53.

52. Hosono, A., K. Shashikanth, and H. Otani (1988) Antimutagenic activity of whole casein on the pepper-induced mutagenicity to streptomycin-dependent

strain SD 510 of *Salmonella typhimurium* TA98. J. Dairy Res. 55:435-442.

53. Iino, Y., M. Yoshida, M. Izuo, and H. Takikawa (1987) Effects of tamoxifen on the growth of DMBA-induced mammary tumors in rats during and after administration of PSK or picibanil. Horumon to Rinsho 35:1353-1357.

54. Iishi, H., M. Tatsuta, M. Baba, S. Okuda, and H. Taniguchi (1990) Protection by oral phenylalanine against gastric carcinogenesis induced by N-methyl-N'-nitro-N-nitrosoguanidine in Wistar rats. Br. J. Cancer 62:173-176.

55. Imanishi, H., Y.F. Sasaki, K. Matsumoto, M. Watanabe, T. Ohta, Y. Shirasu, and K. Tutikawa (1990) Suppression of 6-TG-resistant mutations in V79 cells and recessive spot formations in mice by vanillin. Mutat. Res. 243:151-158.

56. Ioannides, C., A.D. Ayrton, A. Keele, D.F.V. Lewis, P.R. Flatt, and R. Walker (1990) Mechanism of the in vitro antimutagenic action of retinol. Mutagenesis 5:257-262.

57. Ip, C., and C. Hayes (1989) Tissue selenium levels in selenium-supplemented rats and their relevance in mammary cancer protection. Carcinogenesis 10:921-925.

58. Jacobs, M.M. (1990) Potassium inhibition of DMH-induced small intestinal tumors in rats. Nutr. Cancer 14:95-101.

59. Jain, A.K., K. Shimoi, Y. Nakamura, T. Kada, Y. Hara, and I. Tomita (1989) Crude tea extracts decrease the mutagenic activity of N-methyl-N'-nitro-N-nitrosoguanidine in vitro and in intragastric tract of rats. Mutat. Res. 210:1-8.

60. Josephy, P.D., H.L. Lord, and V.A. Snieckus (1990) Inhibition of benzo[a]pyrene dihydrodiol epoxide mutagenicity by synthetic analogs of ellagic acid. Mutat. Res. 242:143-149.

61. Kato, I., S. Tominaga, A. Matsuura, Y. Yoshii, M. Shirai, and S. Kobayashi (1990) A comparative case-control study of colorectal cancer and adenoma. Jpn. J. Cancer Res. 81:1101-1108.

62. Khan, W.A., Z.Y. Wang, M. Athar, D.R. Bickers, and H. Mukhtar (1988) Inhibition of the skin tumorigenicity of (+)-7,8-a-dihydroxy-9a, 10a-epoxy-7,8,9,10-tetrahydrobenzo[a]pyrene by tannic acid, green tea polyphenols and quercetin in Sencar mice. Cancer Lett. 42:7-12.

63. Klurfeld, D.M., C.B. Welch, L.M. Lloyd, and D. Kritchevsky (1989) Inhibition of DMBA-induced mammary tumorigenesis by caloric restriction in rats fed high-fat diets. Int. J. Cancer 43:922-925.

64. Knasmuller, S., R. De Martin, G. Domjan, and A. Szakmary (1989) Studies on the antimutagenic activities of garlic extract. Env. Molec. Mutagen. 13:357-365.

65. Kojima, H., B. Miwa, M. Mori, M. Osaki, and H. Konishi (1989) Desmutagenic effect of oolong tea. Shokuhin Eiseigaku Zasshi 30:233-239.

66. Kokkinakis, D.M. (1990) Sex difference in the dietary modulation of pancreatic cancer in Syrian hamsters treated continuously with N-nitroso(2-hydroxypropyl)(2-oxopropyl)amine. Carcinogenesis 11:1909-1913.

67. Kong, Z.L., K. Shinohara, M. Mitsuiki, H. Murakami, and H. Omura (1989) Desmutagenicity of furan compounds towards some mutagens. Agric. Biol. Chem. 53:2073-2079.

68. Kumar, R., and M. Singh (1984) Tannins: Their adverse role in ruminant nutrition. J. Agric. Food Chem. 32:447-453.

69. Kumar, S.P., S.J. Roy, K. Tokuymo, and B.S. Reddy (1990) Effect of different levels of calorie restriction on azoxymethane-indued colon carcinogenesis in male F344 rats. Cancer Res. 50:5761-5766.

70. Kuroda, Y., D.M. Shankel, and M.D. Waters, eds. (1990) Antimutagenesis and Anticarcinogenesis Mechanisms II. Plenum, New York.

71. Kuroda, M., D. Yoshida, and H. Kodama (1987) Bio-antimutagenic effects of sesquiterpene lactones from costus root oil. Agric. Biol. Chem. 51:585-587.

72. Kushi, A., and D. Yoshida (1987) Antimutagenic effects of pyrolyzate of protein on chemically induced mutagenesis in Escherichia coli. Agric. Biol. Chem. 51:1435-1437.

73. Kushi, A., and D. Yoshida (1987) Antimutagenic effects of phenols on MNNG-induced mutagenesis in Escherichia coli. Agric. Biol. Chem. 51:1439-1440.

74. Lacroix, A., C. Doskas, and P.V. Bhat (1990) Inhibition of growth of established N-methyl-N-nitrosourea-induced mammary cancer in rats by retinoic acid and ovariectomy. Cancer Res. 50:5731-5734.

75. Laidlaw, S.A., M.F. Dietrich, M.P. Lamtenzan, H.I. Vargas, J.B. Block, and J.D. Kopple (1989) Antimutagenic effects of taurine in a bacterial assay system. Cancer Res. 49:6600-6604.

76. Lakdawalla, A.A., and M.S. Netrawali (1988) Mutagenicity, comutagenicity, and antimutagenicity of erythrosine (FD + C Red 3), a food dye, in the Ames/Salmonella assay. Mutat. Res. 204:131-139.

77. Lambert, L.A., W.H. Koch, W.G. Wamer, and A. Kornhouse (1990) Antitumor activity in skin of Skh and Sencar mice by two dietary β-carotene formulations. Nutr. Cancer 13:213-221.

78. Liehr, J.G., D. Roy, and A. Gladek (1989) Mechanism of inhibition of estrogen-induced renal carcinogenesis in male Syrian hamsters by vitamin C. Carcinogenesis 10:1983-1988.

79. Liu, L., and A. Castoguay (1991) Inhibition of the metabolism and genotoxicity of 4-(methylnitrosamino)-1-(3-pyridyl)-1-butanone (NKK) in rat hepatocytes by (+)-catechin. Carcinogenesis 12:1203-1208.

80. Lord, H.L., V.A. Snieckus, and P.D. Josephy (1989) Reevaluation of the effect of ellagic acid on dimethylnitrosamine mutagenicity. Mutagenesis 4:453-455.

81. Mandal, S., A. Ahuja, N.M. Shivapurkar, S. Cheng, J.D. Groopman, and G.D. Stoner (1987) Inhibition of aflatoxin B_1 mutagenesis in Salmonella typhimurium and DNA damage in cultured rat and human tracheobronchial tissues by ellagic acid. Carcinogenesis 8:1651-1656.

82. Mandal, S., and G.D. Stoner (1990) Inhibition of N-nitrosobenzylmethylamine-induced esophageal tumorigenesis in rats by ellagic acid. Carcinogenesis 11:55-61.

83. Manikumar, G., K. Gaetano, M.C. Wani, H. Taylor, T.J. Hughes, J. Warner, R. McGivney, and M.E. Wall (1989) Plant antimutagenic agents, 5. Isolation and structure of two new isoflavones, fremontin, and fremontone from Psorothamnus fremontii. J. Nat. Prod. 52:769-773.

84. Marx, J. (1991) Efforts to prevent cancer are on the increase. Science 253:613.

85. McCormick, D.L., and R.C. Moon (1986) Retinoid-tamoxifen interaction in mammary cancer chemoprevention. Carcinogenesis 7:193-196.

86. Meng, C.L., and K.W. Shyu (1990) Inhibition of experimental carcinogenesis by painting with garlic extract. Nutr. Cancer 14:207-217.

87. Mizuno, M., M. Toda, G. Danno, K. Kanazawa, and M. Natake (1988) Desmutagenic effects of sulfhydryl compounds on a mutagen formed from butylated hydroxyanisole reacted with sodium nitrite. Agric. Biol. Chem. 52:2843-2849.

88. Mokhtar, N.M., H.A. Ibrahim, N.B. El-Din, and N.Z. Moharram (1988) Effect of soybean feeding on experimental carcinogenesis. III. Carcinogenicity of nitrite and dibutylamine in mice: A histopathological study. Eur. J. Cancer Clin. Oncol. 24:403-411.

89. Mori, H., T. Tanaka, H. Shima, T. Kuniyasu, and M. Takahashi (1986) Inhibitory effect of chlorogenic acid on methylazoxymethanol acetate-induced carcinogenesis in large intestine and liver of hamsters. Cancer Lett. 30:49-54.

90. Morse, M.A., K.I. Eklind, S.S. Hecht, K.G. Jordan, C.-I. Choi, D.H. Desai, S.G. Amin, and F.-L. Chung (1991) Structure-activity relationships for inhibition of 4-(methylnitrosamino)-1-(3-pyridyl)-1-butanone lung tumorigenesis by arylalkyl isothiocyanates in A/J mice. Cancer Res. 51:1846-1850.

91. Murray, B.P., A.R. Boobis, R. De La Torre, J. Segura, and D.S. Davies (1988) Inhibition of dietary mutagen activation by methylxanthines. Biochem. Soc. Trans. 16:620-621.

92. Nagabhushan, M., A.J. Amonkar, and S.V. Bhide (1987) Mutagenicity of gingerol and shogaol and antimutagenicity of zingerone in Salmonella/microsome assay. Cancer Lett. 36:221-233.

93. Nagabhushan, M., A.J. Amonkar, U.J. Nair, U. Santhanam, N. Ammigan, A.V. D'Souza, and S.V. Bhide (1988) Catechin as an antimutagen: Its mode of action. J. Cancer Res. Clin. Oncol. 114:177-182.

94. Nagabhushan, M., and S.V. Bhide (1988) Anti-mutagenicity of catechin against environmental mutagens. Mutagenesis 3:293-296.

95. Nagabhushan, M., U.J. Nair, A.J. Amonkar, A.V. D'Souza, and S.V. Bhide (1988) Curcumins as inhibitors of nitrosation in vitro. Mutat. Res. 202 (1):163-169.

96. Nair, S.C., B. Panikkar, K.G. Akamanchi, and K.R. Panikkar (1991) Inhibitory effects of Ixora javanica extract on skin chemical carcinogenesis in mice and its antitumour activity. Cancer Lett. 60:253-258.

97. Narisawa, T., M. Takahashi, M. Niwa, Y. Fukaura, and H. Fujiki (1989) Inhibition of methylnitrosourea-induced large bowel cancer development in rats by sarcophytol A, a product from a marine soft coral Sarcophyton glaucum. Cancer Res. 49:3287-3289.

98. Natake, M., K. Kanazawa, M. Mizuno, N. Ueno, T. Kobayashi, G. Dnno, and S. Minamoto (1989) Herb water-extracts markedly suppress the mutagenicity of Trp-P-2. Agric. Biol. Chem. 53:1423-1425.

99. Negishi, T., S. Arimoto, C. Nishizaki, and H. Hayatsu (1989) Inhibitory effect of chlorophyll on the genotoxicity of 3-amino-1-methyl-5-H-pyrido[4,3-b]indole (Trp-P-2). Carcinogenesis 10:145-149.

100. Negishi, T., S. Arimoto, C. Nishizaki, and H. Hayatsu (1990) Inhibition of the genotoxicity of 3-amino-1-methyl-5-H-pyrido[4,3-b]indole (Trp-P-2) in Drosophila by chlorophyll. In Antimutagenesis and Anticarcinogenesis Mechanisms II, Y. Kuroda, D.M. Shankel, and M.D. Waters, eds. Plenum, New York, pp. 341-344.

101. Negishi, T., K. Negishi, H. Ryo, S. Kondo, and H. Hayatsu (1988) The genotoxicity of N^4-aminocytidine in the Drosophila wing spot test. Mutagenesis 3:11-13.

102. Negishi, T., T. Shiotani, K. Fujikawa, and H. Hayatsu (1991) The genotoxicities of N-nitrosamines in Drosophila melanogaster in vivo: The correlation of mutagenicity in the wing spot test with the DNA damages detected by the DNA-repair test. Mutat. Res. 252:119-128.

103. Newberne, P.M., V. Suphiphat, M. Locniskar, and J.L.V. De Camargo (1990) Inhibition of hepatocarcinogenesis in mice by dietary methyl donors methionine and choline. Nutr. Cancer 14:175-181.

104. Newmark, H. (1984) A hypothesis for dietary components as blocking agents of chemical carcinogenesis: Plant phenolics and pyrrole pigments. Nutr. Cancer 6:58-70.

105. Nicholson, M.L., J.P. Neoptolemos, H.A. Clayton, I.C. Talbot, and P.R.F. Bell (1990) Inhibition of experimental colorectal carcinogenesis by dietary n-6 polyunsaturated fats. Carcinogenesis 11:2191-2197.

106. Obana, H., and S. Nakamura (1989) Possible antimutagenic effect of an instant coffee. Chem. Express 4:581-584.

107. O'Dwyer, P.J., T.S. Ravikumar, D.P. McCabe, and G. Steele, Jr. (1987) Effect of 13-cis-retinoic acid on tumor prevention, tumor growth, and metastasis in experimental colon cancer. J. Surg. Res. 43:550-557.

108. Oguni, I., and M. Hara (1990) The Beneficial Effect of Tea. Chunichi Newspaper Press, Nagoya (in Japanese).

109. Ohgaki, H., S. Takayama, and T. Sugimura (1991) Short review of the carcinogenicities of mutagens in food pyrolysates. In Mutagens in Food: Detection and Prevention, H. Hayatsu, ed. CRC Press, Boca Raton, Florida, pp. 219-228.

110. Okaka, M., S. Nakamura, K. Miura, and K. Morimoto (1989) The antimutagenic effects of human saliva investigated by umu-test. I. Effects of filtration and storage at low temperatures. Nippon Eiseigaku Zasshi 44:1009-1013.

111. Okuda, T., K. Mori, and H. Hayatsu (1984) Inhibitory effect of tannins on direct-acting mutagens. Chem. Pharm. Bull. 32:3755-3758.

112. Ong, T., W.Z. Whong, J.D. Stewart, and H.E. Brockman (1989) Comparative antimutagenicity of 5 compounds against 5 mutagenic complex mixtures in Salmonella typhimurium strain TA98. Mutat. Res. 222:19-25.

113. Osawa, T., K. Namiki, and S. Kawakishi (1990) Role of dietary antioxidants in protection against oxidative damage. In Antimutagenesis, and Anticarcinogenesis Mechanisms II, Y. Kuroda, D.M. Shankel, and M.D. Waters, eds. Plenum, New York, pp. 139-153.

114. Padma, P.R., A.J. Amonkar, and S.V. Bhide (1989) Antimutagenic effects of betel leaf extract against the mutagenicity of two tobacco-specific N-nitrosamines. Mutagenesis 4:154-156.

115. Papenburg, R., G. Bounous, D. Fleiszer, and P. Gold (1990) Dietary milk protein inhibits the development of dimethylhydrazine induced malignancy. Tumor Biol. 11:129-136.

116. Pepin, P., G. Rossignol, and A. Castonguay (1990) Inhibition of NNK-induced lung tumorigenesis in A/J mice by ellagic acid and butylated hydroxyanisole. Cancer J. 3:266-273.

117. Perchellet, J.-P., E.M. Perchellet, and S. Belman (1990) Inhibition of DMBA-induced mouse skin tumorigenesis by garlic oil and inhibition of two tumor-promotion stages by garlic and onion oils. Nutr. Cancer 14:180-193.

118. Peryt, B., J. Miloszewska, B. Tudek, M. Zielenska, and T. Szymczyk (1988) Antimutagenic effects of several subfractions of extract from wheat sprout toward benzo[a]pyrene-induced mutagenicity in strain TA98 of Salmonella typhimurium. Mutat. Res. 206:221-225.

119. Phillips, R.L. (1975) Role of life-style and dietary habits in risk of cancer among Seventh-Day Adventists. Cancer Res. 35:3515-3522.

120. Pons, F.W., and P. Mueller (1990) Strong antimutagenic effect of caffeine on 9-aminoacridine-induced frameshift mutagenesis in *Escherichia coli* K12. Mutagenesis 5:363-366.

121. Rao, A.R., S.P. Hussain, J.N. Jannu, M.V.R Kumari, and A.S. Sadhana (1990) Modulatory influences of tamoxifen, tocopherol, retinyl acetate, aminoglutethimide, ergocryptine and selenium on DMBA-induced initiation of mammary carcinogenesis in rats. Indian J. Exp. Biol. 28:409-416.

122. Rao, A.R., A.S. Sadhana, and H.C. Goel (1990) Inhibition of skin tumors in DMBA-induced complete carcinogenesis system in mice by garlic (*Allium sativum*). Indian J. Exp. Biol. 28:405-408.

123. Rao, C.V., K. Tokomo, G. Kelloff, and B.S. Reddy (1991) Inhibition by dietary oltipraz of experimental intestinal carcinogenesis induced by azoxymethane in male F344 rats. Carcinogenesis 12:1051-1055.

124. Reiners, J.J., Jr., E. Brott, and J.R.J. Sorenson (1986) Reductase activity by copper complexes. Carcinogenesis 7:1729-1732.

125. Roberts-Andersen, J., T. Mehta, and R.B. Wilson (1987) Reduction of DMH-induced colon tumors in rats fed psyllium husk or cellulose. Nutr. Cancer 10:129-136.

126. Roy, P.M., K. Kazakoff, and K. Carlson (1990) Inhibition of estrogen-induced kidney carcinogenesis in Syrian hamsters by modulators of estrogen metabolism. Carcinogenesis 11:567-570.

127. Russin, W.A., J.D. Hoesly, C.E. Elson, M.A. Tanner, and M.N. Gould (1989) Inhibition of rat mammary carcinogenesis by monoterpenoids Carcinogenesis 10:2161-2164.

128. Sakai, Y., H. Nagase, Y. Ose, H. Kito, T. Sato, M. Kawai, and M. Mizuno (1990) Inhibitory action of peony root extract on the mutagenicity of benzo[a]pyrene. Mutat. Res. 244:129-134.

129. Sakai, Y., Y. Ose, H. Kitoh, T. Sato, H. Hasegawa, Y. Yoshioka, M. Kawai, M. Mizuno, and I. Moriguchi (1990) Studies on the quantitative structure activity relationship of antimutagenic phenol carboxylic acids to benzo[a]pyrene. Eisei Kagaku 36:304-313.

130. Shah, A.B., R.D. Combes, and I.R. Rowland (1991) Interaction with microsomal lipid as a major factor responsible for S9-mediated inhibition of 1,8-dinitropyrene mutagenicity. Mutat. Res. 249:93-104.

131. Shankel, D.M., P.E. Hartman, T. Kada, and A. Hollaender, eds. (1986) Antimutagenesis and Anticarcinogenesis Mechanisms. Plenum Press, New York.

132. Shinohara, K., S. Kuroki, M. Miwa, Z.L. Kong, and H. Hosoda (1988) Antimutagenicity of dialyzates of vegetables and fruits. Agric. Biol. Chem. 52:1369-1375.

133. Singletary, K.W., and J.M. Nelshoppen (1991) Inhibition of 7,12-dimethylbenz[a]anthracene(DMBA)-induced mammary tumorigenesis and of in vivo formation of mammary DMBA-DNA adducts by rosemary extract. Cancer Lett. 60:169-175.

134. Smart, R.C., M.T. Huang, R.L. Chang, J.M. Sayer, D.M. Jerina, and A.H. Conney (1986) Disposition of the naturally occurring antimutagenic plant phenol, ellagic acid, and its synthetic derivatives, 3-*O*-decylellagic acid and 3,3'-di-*O*-methylellagic acid in mice. Carcinogenesis 10:1663-1667.

135. St. Claire, W.H., P.C. Billings, J.A. Carew, C. Keller-Mcgandy, P. Newberne, and A. R. Kennedy (1990) Suppression of dimethylhydrazine-induced

carcinogenesis in mice by dietary addition of the Bowman-Birk protease inhibitor. <u>Cancer Res.</u> 50:580-586.

136. Stich, H.F., ed. (1983) <u>Carcinogens</u> and <u>Mutagens</u> <u>in</u> <u>the</u> <u>Environment</u>, Vols. I & II. CRC Press, Boca Raton.

137. Suda, D., J. Schwartz, and G. Shklar (1986) Inhibition of experimental oral carcinogenesis by topical beta carotene. <u>Carcinogenesis</u> 7:711-715.

138. Suganuma, M., S. Yoshizawa, J. Yatsunami, S. Nishiwaki, H. Furuya, S. Okabe, R. Nishiwaki-Matsushima, K. Frenkel, W. Troll, A.K. Verma, and H. Fujiki (1992) Mechanisms of action of new antitumor promoters. This volume, pp. 317-323.

139. Sugimura, T. (1985) Carcinogenicity of mutagenic heterocyclic amines formed during the cooking process. <u>Mutat. Res.</u> 150:33-41.

140. Tanaka, M., N. Mano, E. Akazai, Y. Narui, F. Kato, and Y. Koyama (1987) Inhibition of mutagenicity by glycyrrhiza extract and glycyrrhizin. <u>J. Pharmacobio-Dyn.</u> 10:685-688.

141. Tatsuta, M., H. Iishi, M. Baba, A. Nakaizumi, M. Ichii, and H. Taniguchi (1990) Inhibition by γ-amino-n-butyric acid and baclofen of gastric carcinogenesis induced by N-methyl-N'-nitro-N-nitrosoguanidine in Wistar rats. <u>Cancer Res.</u> 50:4931-4934.

142. Tatsuta, M., H. Iishi, M. Baba, M. Ichii, A. Nakaizumi, H. Uehara, and H. Taniguchi (1990) Attenuating effect of bromocriptine on cysteamine anticarcinogenesis of stomach cancers induced by N-methyl-N'-nitro-N-nitrosoguanidine. <u>Cancer Res.</u> 50:5308-5311.

143. Tatsuta, M., H. Iishi, M. Baba, M. Ichii, A. Nakaizumi, H. Uehara, and H. Taniguchi (1991) Protective effect by potassium chloride against gastric carcinogenesis induced by N-methyl-N'-nitro-N-nitrosoguanidine in spontaneously hypertensive rats. <u>Jpn. J. Cancer Res.</u> 82:280-285.

144. Tatsuta, M., H. Iishi, M. Baba, and A. Nakaizumi (1991) Inhibition by neurotensin of azaserine-induced carcinogenesis in rat pancreas. <u>Int. J. Cancer</u> 47:408-412.

145. Tatsuta, M., H. Iishi, M. Baba, M. Ichii, A. Nakaizumi, H. Uehara, and H. Taniguchi (1991) Inhibition by putrescine of experimental carcinogenesis in rat colon induced by azoxymethane. <u>Int. J. Cancer</u> 47:738-741.

146. Tatsuta, M., H. Iishi, M. Baba, and H. Taniguchi (1989) Tissue norepinephrine depletion as a mechanism for cysteamine inhibition of colon carcinogenesis induced by azoxymethane in Wistar rats. <u>Int. J. Cancer</u> 44:1008-1011.

147. Tatsuta, M., H. Iishi, H. Yamamura, M. Baba, and H. Taniguchi (1988) Inhibition by isoproterenol and neostigmine of experimental carcinogenesis [induced] in rat colon by azoxymethane. <u>Br. J. Cancer</u> 58:619-620.

148. Tatsuta, M., H. Iishi, M. Baba, H. Uehara, and A. Nakaizumi (1990) Inhibition by neostigmine of hepatocarcinogenesis induced by N-nitrosomorpholine in Sprague-Dawley rats. <u>Br. J. Cancer</u> 62:773-775.

149. Tisdale, M.J., and S.A. Beck (1991) Inhibition of tumor-induced lipolysis *in vitro* and cachexia and tumor growth *in vivo* by eicosapentaenoic acid. <u>Biochem. Pharmacol.</u> 41:103-107.

150. Trickler, D., and G. Shklar (1987) Prevention by vitamin E of experimental oral carcinogenesis. <u>J. Natl. Cancer Inst.</u> 78:165-169.

151. Ullah, A., and A.M. Shamsuddin (1990) Dose-dependent inhibition of large intestinal cancer by inositol hexaphosphate in F344 rats. <u>Carcinogenesis</u> 11:2219-2222.

152. VanderPloeg, L.C., and C.W. Welsch (1991) Inhibition by caffeine of ovarian hormone-induced mammary gland tumorigenesis in female GR mice. Cancer Lett. 56:245-250.

153. Van Rensburg, S.J., J.M. Hall, and P.S. Gathercole (1986) Inhibition of esophageal carcinogenesis in corn-fed rats by riboflavin, nicotinic acid, selenium, molybdenum, zinc, and magnesium. Nutr. Cancer 8:163-170.

154. Victorin, K., L. Busk, and U.G. Ahlborg (1987) Retinol (vitamin A) inhibits the mutagenicity of O-aminoazotoluene activated by liver microsomes from several species in the Ames test. Mutat. Res. 179:41-48.

155. Vogelstein, B., and K.W. Kinzler (1992) Carcinogens leave fingerprints. Nature 455:209.

156. Wakata, A., T. Oka, K. Hiramoto, A. Yoshioka, K. Negishi, Y. Wataya, and H. Hayatsu (1985) DNA strand cleavage *in vitro* by 3hydroxyamino-1-methyl-5-H-pyrido[4,3-b]indole, a direct-acting mutagen formed in the metabolism of carcinogenic 3-amino-1-methyl-5-H-pyrido[4,3-b]indole. Cancer Res. 45:5867-5871.

157. Wall, M.E., M.C. Wani, G. Manikumar, P. Abraham, H. Taylor, T.J. Hughes, J. Warner, and R. McGivney (1988) Plant antimutagenic agents. 2. Flavonoids. J. Nat. Prod. 51:1084-1091.

158. Wall, M.E., M.C. Wani, G. Manikumar, H. Taylor, and R. McGivney (1989) Plant antimutagens. 6. Intricatin and intricationol, new antimutagenic homoisoflavonoids from *Hoffmanosseggia intricata*. J. Nat. Prod. 52:774-778.

159. Wang, C.Y., M. Lee, H. Nagase, and K. Zukowski (1989) Inhibition by diacylmethane derivatives of mutagenicity and nucleic acid binding of 2-aminofluorene derivatives. J. Natl. Cancer Inst. 81:1743-1747.

160. Wang, Z.Y., R. Agarwal, D.R. Bickers, and H. Mukhtar (1991) Protection against ultraviolet B radiation-induced photocarcinogenesis in hairless mice by green tea polyphenols. Carcinogenesis 12:1527-1530.

161. Wang, Z.Y., R. Agarwal, Z.C. Zhou, D.R. Bickers, and H. Mukhtar (1991) Inhibition of mutagenicity in *Salmonella typhimurium* and skin tumour initiating and tumor promoting activities in SENCAR mice by glycyrrhetinic acid: Comparison of 18 α- and 18 β-stereoisomers. Carcinogenesis 12:187-192.

162. Wang, Z.Y., S.J. Cheng, Z.C. Zhou, M. Athar, W.A. Khan, D.R. Bickers, and H. Mukhtar (1989) Antimutagenic activity of green tea polyphenols. Mutat. Res. 223:273-285.

163. Wang, Z.Y., W.A. Khan, D.R. Bickers, and H. Kukhtar (1989) Protection against polycyclic aromatic hydrocarbon-induced skin tumor initiation in mice by green tea polyphenols. Carcinogenesis 10:411-415.

164. Wargovich, M.J. (1987) Diallyl sulfide, a flavor component of garlic (*Allium sativum*), inhibits dimethylhydrazine-induced colon cancer. Carcinogenesis 8:487-489.

165. Warner, J.R., J. Nath, and T.M. Ong (1991) Antimutagenicity studies of chlorophyllin using the salmonella arabinose-resistant assay system. Mutat. Res. 262:25-30.

166. Watanabe, K., T. Ohta, and Y. Shirasu (1989) Enhancement and inhibition of mutation by O-vanillin in *Escherichia coli*. Mutat. Res. 218:105-109.

167. Wataya, Y., K. Yamane, K. Hiramoto, Y. Ohtsuka, Y. Okubata, K. Negishi, and H. Hayatsu (1988) Generation of intracellular active oxygens in mouse FM3A cells by 3-hydroxyamino-1-methyl-5-H-pyrido[4,3-b]indole, the activated Trp-P-2. Jpn. J. Cancer Res. (Gann) 79:576-579.

168. Wattenberg, L.W., V.L. Sparnins, and G. Barany (1989) Inhibition of *N*-nitrosodiethylamine carcinogenesis in mice by naturally occurring organosulfur compounds and monoterpenes. Cancer Res. 49:2689-2692.

169. Wattenberg, L.W., and J.B. Coccia (1991) Inhibition of 4(methylnitrosamine)-1-(3-pyridyl)-1-butanone carcinogenesis in mice by D-limonene and citrus fruit oils. Carcinogenesis 12:115-117.

170. Weinstein, I.B. (1991) Cancer prevention: Recent progress and future opportunities. Cancer Res. (Suppl.) 51:5080s-5085s.

171. Whong, W.Z., J. Stewart, H. Brockman, and T.M. Ong (1988) Comparative antimutagenicity of chlorophyllin and five other agents against aflatoxin B_1-induced reversion in *Salmonella typhimurium* strain. Terat. Carcino. Muta. 8:215-224.

172. Wilpart, M., A. Spender, and M. Roberfroid (1986) Anti-initiation activity of *N*-acetylcysteine in experimental colonic carcinogenesis. Cancer Lett. 31:319-324.

173. Witschi, H., and A.R. Kennedy (1989) Modulation of lung tumor development in mice with the soybean-derived Bowman-Birk protease inhibitor. Carcinogenesis 10:2275-2277.

174. Würgler, F.E., and E.W. Vogel (1986) *In vivo* mutagenicity testing using somatic cells of *Drosophila melanogaster*. In Chemical Mutagens, Vol. 10, F.J. de Serres, ed. Plenum, New York, pp. 1-72.

175. Yoo, M.-A., H. Ryo, T. Todo, and S. Kondo (1985) Mutagenic potency of heterocyclic amines in the *Drosophila* wing spot test and its correlation to carcinogenic potency. Jpn. J. Cancer Res. (Gann) 76:468-473.

176. Yoshida, Y. (1990) Study on mutagenicity and antimutagenicity of BHT and its derivatives in a bacterial assay. Mutat. Res. 242:209-217.

177. Zajdela, F., O. Perin-Roussel, and S. Saguem (1987) Marked differences between mutagenicity in Salmonella and tumor-initiating activities of dibenzo*[a,f]*fluornathene proximate metabolites; initiation inhibiting activity of norharman. Carcinogenesis 8:461-464.

178. Zhang, X.-B., Y. Ohta, and A. Hosono (1990) Antimutagenicity and binding of lactic ancid bacteria from a chinese cheese to mutagenic pyrolysates. Dairy Sci. 73:2702-2710.

179. Zhang, Y., X. Chen, and Y. Yu (1989) Antimutagenic effect of garlic *(Allium sativum L.)* on 4NQO-induced mutagenesis in *Escherichia coli* WP2. Mutat. Res. 227:215-219.

180. Zhu, Y.-P., S.-W. Su, and C.-H. Li (1989) Growth-inhibition effects of oleic acid, linoleic acid, and their methyl esters on transplanted tumors in mice. J. Natl. Cancer Inst. 81:1302-1306.

OXIDANTS AND MITOGENESIS AS CAUSES OF

MUTATION AND CANCER: THE INFLUENCE OF DIET

Mark K. Shigenaga and Bruce N. Ames

Division of Biochemistry and Molecular Biology
Barker Hall
University of California
Berkeley, California 94720 USA

ABSTRACT

A very high level oxidative damage to DNA occurs during normal metabolism. In each rat cell, the steady-state level of this damage is estimated to be about 10^6 oxidative adducts, and about 10^5 new adducts are formed daily. This endogenous DNA damage appears to be a major contributor to cancer and aging. The oxidative damage rate in mammalian species with a high metabolic rate, short life span, and high age-specific cancer rate such as in rats is much higher than the rate in humans, long-lived mammals with a lower metabolic rate, and a lower age-specific cancer rate. It is argued that deficiency of micronutrients, that protect against oxidative DNA damage, is a major contributor to human cancer.

Epidemiological studies, a large body of experimental evidence, and theoretical work on the mechanisms of carcinogenesis point to mitogenesis as a major contributor to cancer. Dividing cells, compared to nondividing cells, are at an increased risk for mutations due to: 1.) conversion of DNA adducts to mutations; 2.) chance of mitotic recombination, gene conversion, and nondisjunction; and, 3.) increased exposure of DNA to mutagens. Mitogenesis also increases the probability of gene amplification and loss of 5-methylcytosine. Dietary interventions that lower mitogenesis, such as calorie restriction, decrease cancer incidence.

AGING, CANCER, AND ENDOGENOUS SOURCES OF DNA DAMAGE

A marked decrease in age-specific cancer rates has accompanied the marked increase in life span that has occurred in 60 million years of mammalian evolution; for example, at two years of age, cancer rates are high in rodents, but extremely low in humans. Cancer incidence increases with approximately the fifth power of age, both in short-lived species such as rats and mice and in long-lived species such as humans (34, 83). Thus, cancer is one of the degenerative diseases of old age, though exogenous factors can substantially increase it (e.g., cigarette smoking in humans) or decrease it (e.g., calorie restriction in rodents). One important factor in longevity appears to be basal metabolic rate (30), which is much lower in man than in rodents and could markedly affect the level of endogenous mutagens produced by normal metabolism (2,94).

Aging is thought to occur because nature selects for many genes that have immediate survival value, but that have long-term deleterious consequences (113,114). The burst of NO, O_2, H_2O_2, and HOC1 from white blood cells, for example, protects against bacterial and viral infections, but contributes to DNA damage and mutation (92, 117). It seems plausible that DNA damage is likely to be critical for both aging (44) and cancer. One view of the somatic mutation theory of aging is that the amount of maintenance of somatic tissues is always less than that required for indefinite survival because a considerable proportion of an animal's resources is devoted to reproduction at a cost to maintenance. Thus, some DNA damage that is induced in somatic cells by endogenous mutagens will accumulate with time and contribute to cancer.

Four endogenous processes leading to significant DNA damage are likely to be oxidation (3,47,100), methylation, deamination, and depurination (90). The importance of these processes is supported by the existence of specific DNA repair glycosylases for oxidative, methylated, and deaminated adducts, and a repair system for apurinic sites that are produced by spontaneous depurination (67). The measurement of DNA adducts by new methods shows that DNA damage produced by oxidation (see below) could be the most significant endogenous damage.

OXIDATIVE DNA DAMAGE

The impressive inverse correlation that exists between metabolic rate and longevity and the strong epidemiological findings indicating inverse correlations between consumption of dietary antioxidants and various types of cancers (12, 15) have led us to focus our research efforts on the damage produced by reactive by-products of O_2 and the possible dietary interventions that may protect DNA from this damage (40, 41).

Oxidants are produced as by-products of mitochondrial electron transport, various oxygen-utilizing enzyme systems, peroxisomes and other processes associated with normal aerobic metabolism, as well as by lipid peroxidation (Figs. 1 and 2). Oxidants that escape the numerous antioxidant defenses can damage cellular macro-molecules, including DNA, and such damage can lead to mutations and cancer.

Nonspecific DNA repair enzymes excise DNA adducts to release deoxynucleo-tides, and specific DNA repair glycosylases release free bases. Deoxynucleotides are enzymatically hydrolyzed to deoxynucleosides that are not usually further metabolized, and both these and the free bases may be recovered in the urine. Two products of oxidative damage to DNA are thymine glycol and 5-hydroxymethyluracil. A specific DNA glycosylase repair enzyme in mouse cells repairs 5-hydroxymethyluracil and

$$O_2 \xrightarrow{e^-} O_2^{\bar{}} \xrightarrow{e^-} H_2O_2 \xrightarrow{e^-} \cdot OH \xrightarrow{e^-} H_2O$$

$$O_2 \xrightarrow[\text{LIGHT}]{\text{DYE}} {}^1O_2$$

Figure 1. Oxidants from normal metabolism. The formation of superoxide, hydrogen peroxide, and hydroxyl radicals by successive additions of electrons to oxygen. Cytochrome oxidase adds four electrons fairly efficiently during energy generation in mitochondria, but, some of these toxic intermediates are inevitable by-products. The same oxidants are produced in copious quantities from phagocytic cells. Singlet oxygen is generated from oxygen by the absorption of energy from a dye activated by light.

differs from the specific DNA glycosylase repair enzyme for thymine glycol in mousecells (56). The existence of these *specific* repair enzymes points to the importance of this type of DNA damage *in vivo*.

The postulated importance of endogenously produced oxidative damage to DNA in aging and age-related degenerative pathologies such as cancer has prompted efforts to develop rapid methods that measure this damage (2,5,23,31,39,86,94). Endogenously produced oxidative damage to DNA has been assayed by measuring the urinary levels of the known radiation damage products thymine glycol, thymidine glycol, hydroxymethyluracil, and hydroxymethyldeoxyuridine by HPLC with UV detection (2,23,91). Our results indicate that normal humans excrete a total of about 100 nmol/day of the first three compounds. We have considerable evidence that most of this total is derived from repair of oxidized DNA, rather than from alternative sources, such as diet bacterial flora (6,23,91). This 100 nmol may, therefore, represent an average of about 10^3 oxidized thymine residues per day for each of the body's 6×10^{13} cells. Because these products are only three of ≈ 20 major products of oxidative damage of DNA (20,105), the total number of all types of oxidative hits to DNA per cell per day may be about 10^4 in man and about 10^5 in the rat.

A more easily assayed product of oxidative DNA damage is oxo^{8}dG (8-oxo-7,8-dihydrodeoxyguanosine; 8-hydroxy-2'-deoxyguanosine), which can be measured with high sensitivity by HPLC-EC (38). The oxo^{8}dG generates mutations by causing G to T transversions (93,116) and is formed in DNA by gamma-irradiation (33) and various carcinogens (37,62).

Immunoaffinity columns, prepared with polyclonal antibodies (31) or monoclonal antibodies that recognize oxo^{8}dG, facilitate the isolation of oxo^{8}dG, oxo^{8}Gua, and oxo^{8}G from urine, biological fluids, and spent media recovered from bacterial cultures. Estimates of the urinary oxo^{8}dG and oxo^{8}Gua in rats fed nucleic acid-free diets indicates that oxo^{8}Gua is excreted at a rate $\approx$10-fold higher than oxo^{8}dG. Similar results have been obtained upon analysis of spent media recovered from cultures of *Escherichia coli* (Park et al., unpublished data). These results suggest that the glycosylase activity, which has recently been isolated and characterized (25), may be quantitatively the more important of the two repair activities that have been proposed to catalyze the excision of this damage product from DNA. Measurement of the steady-state level of oxo^{8}dG lesions in liver DNA from 1-yr-old Fischer 344

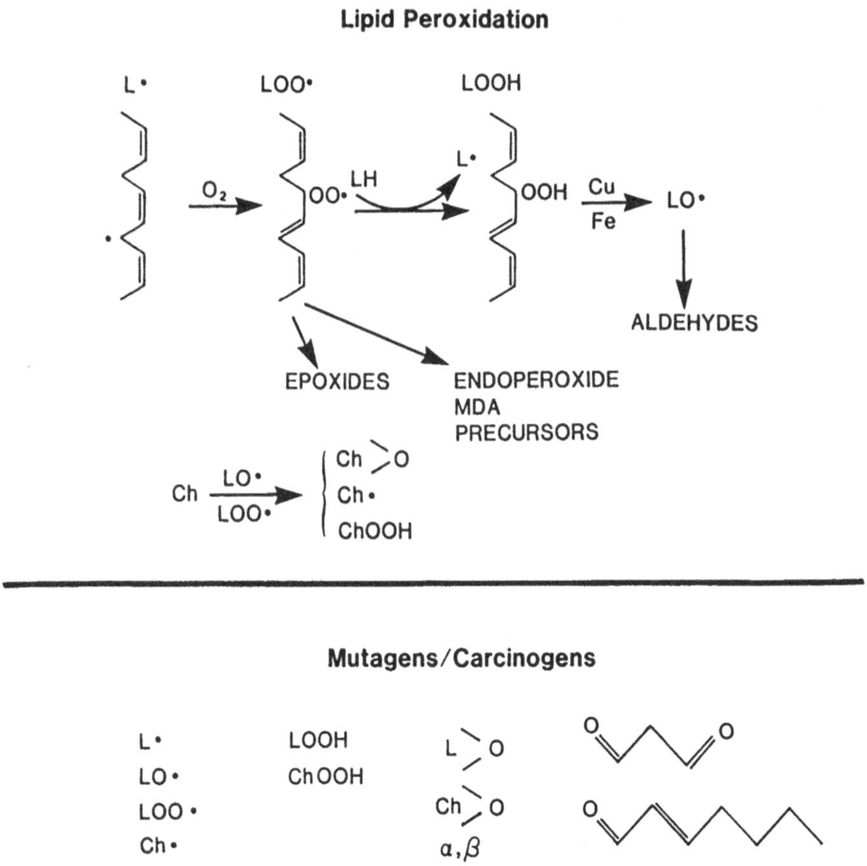

Lipid Peroxidation

Figure 2. Lipid peroxidation (4). L = lipid radical; LΣ = alkoxy lipid radical; LOOΣ = hydroperoxy lipid radical; LOOH = lipid hydroperoxide; Ch = cholesterol; Ch>O = cholesterol epoxide; L>O = lipid epoxide; MDA = malondialdehyde.

and oxo[8]Gua are excreted per cell per day (39). Taking into account that this lesion is only one of $\approx$20 major gamma-irradiation-induced DNA damage products (105), we estimate that each rat cell contains approximately 10^6 oxidatively damaged bases in its DNA (39,86) and that about 10^5 oxidative hits to the DNA occur per rat cell per day (23,39). This estimate based on oxo[8]dG is in agreement with the earlier estimate based on thymidine glycol. The repair rate almost equals the damage rate; however, we estimate that 80 oxo[8]dG residues accumulate per cell per day in DNA isolated from kidney of young and old rats (39). This accumulation of oxo[8]dG and other oxidative damage products may play an important role in the occurrence of spontaneous mutations and age-associated increases in tumor formation.

The oxidative DNA damage rate as measured by thymidine glycol excretion in urine for mouse, rat, monkey, and man is related to the metabolic rate and is inversely related to the age-specific cancer rate and life span (2). We have found a similar relation with urinary oxo[8]dG excretion (94).

DEFENSES AGAINST OXIDANTS: ANTIOXIDANTS AS ANTIMUTAGENS AND ANTICARCINOGENS

Many defense mechanisms within the organism have evolved to limit the levels of reactive oxidants and the damage they inflict. These effects vary and include the direct inhibition of oxidant-induced damage, inhibition of oxidant-induced mitogenesis, protection against oxidant-induced degradation of immune surveillance mechanisms, and, possibly, inhibition of metastases. Among the defenses are enzymes such as SOD, catalase, and GSH peroxidase as well as the dietary antioxidants ascorbate, α-tocopherols, β-carotene, glutathione, and ubiquinol (41). We have been particularly interested in dietary antioxidants because their level in humans can be altered readily.

Inhibition of Oxidative DNA Damage by a Dietary Antioxidant

A study of oxo^8dG in sperm DNA(40) indicates that oxidative DNA damage is correlated inversely with seminal fluid ascorbate concentrations that, in turn, are related to the amount of dietary ascorbate consumed. In individuals whose dietary intake of ascorbate was decreased from 250 to 5 mg/day, the levels of seminal fluid ascorbate levels were halved and the levels of oxo^8dG in sperm DNA were doubled; repletion of dietary ascorbate led to a decrease in oxo^8dG levels, to levels similar to those of the pretreatment controls. In a separate group consisting of 25 subjects, the steady-state levels of oxo^8dG were also found to correlate inversely with the content of seminal plasma ascorbate. Together, these results demonstrate the protective effect of dietary ascorbate against oxidative damage to human sperm DNA. This protective effect may be especially relevant to those individuals who have low ascorbate levels and whose progeny are at a higher risk for birth defects, such as smokers.

Oxidants, Antioxidants, and Mitogenesis

Mitogenesis could lead to oxidative DNA damage both from the increase in oxygen consumption (118) and exposure of DNA to endogenous oxidants. Surgical removal of two-thirds of a rat liver (partial hepatectomy) stimulates a rapid proliferation of cells that is accompanied by a two-fold increase in the oxo^8dG levels in DNA one day after surgery, followed the next day by a spike in the urinary excretion rate of oxo^8dG (46). The levels of oxo^8dG in the DNA isolated from liver tissue two days following the spike decrease to one-half of that measured prior to the operation, indicating that repair of oxo^8dG is efficient.

The intracellular antioxidant glutathione appears to play a critical role in protecting the cell from mitogenesis-induced oxidative damage (46). Depleting rats of intracellular glutathione by pretreatment with dl-buthionine-S,R-sulfoximine for three days prior to and one week following partial hepatectomy results in a five-fold increase in the levels of oxo^8dG in DNA and a two-fold increase in the urinary excretion rates when compared to the corresponding sample time points of untreated, partially hepatectomized rats. Taken together, these results indicate that oxidative DNA damage accompanies the proliferation of cells, and that under normal conditions antioxidant activities can effectively but not completely protect the cell from this damage. Under conditions of glutathione depletion, oxidative damage to DNA is even more extensive. Deficiencies in other antioxidants may have similar consequences.

Antioxidants and Immune Function

Protecting the immune system from excessive oxidant exposure, by consuming dietary antioxidants (74) or by calorie restriction (108), delays the decline in tumoricidal immune function that is associated with increasing age. The protective effects of these interventions may also be relevant to maintaining immune function despite increased tumor burden. For example, as tumor growth progresses, host immune defenses become less effective in suppressing its growth (71), a possible consequence of oxidant-induced decline in immune effector cell response. The oxidants produced by polymorphonuclear leucocytes and macrophages (10) in response to tumor invasion, as well as the oxidants that appear to be produced by the tumor cells themselves (98), could account for the decreased effectiveness of the tumoricidal effector cells that is associated with increased tumor burden. The proliferation of T and B cells, natural killer cells, and lymphokine-activated killer cells that are required to mount an effective defense against pathogens and tumor cells can be inhibited markedly upon exposure to oxidants. The effectiveness of the immune effector cells would be especially compromised during chronic inflammatory diseases in which high levels of oxidants are chronically released from activated PMNs and macrophages. Antioxidants could conceivably serve a dual protective role in such a case by increasing the responsiveness of the immune effector cells to transformed cells at the site of chronic inflammation while decreasing the mitogenesis stimulated by PMN and macrophage-derived oxidants. Indeed, the dietary antioxidants glutathione (42), β-carotene (11), and o-tocopherol (75) have been shown to protect against this decline in immune function.

The Possible Role of Oxidants and Antioxidants in Metastases

Cancer cells assume increasingly aggressive phenotypes as tumor growth progresses to metastases (78). A metastasizing tumor is frequently associated with point mutations in the H-ras protooncogene and increased collagenase secretion, which is necessary for basement membrane breakdown and metastasis (43). Collagenase is released from the tumor cells in a latent form that is inhibited by tissue inhibitors of metalloproteinase (TIMPs)(22). Activation of collagenase or inactivation of TIMPs would be a critical step in progression of tumors to metastases (68).

Oxidative activation of collagenases either directly by oxidation of the critical metal complexed thiol or oxidative inactivation of TIMP would lead to the conversion of collagenase from its inactive to active form. The hypochlorous acid (HOC1) produced by PMNs autoactivate collagenases and lead to the breakdown of basement membranes (110). A similar effect by oxidants produced by tumors could facilitate metastases. Indeed, such oxidants have been shown to be released from tumor cells and tumoricidal macrophages. Cells isolated from various tumors, for example, can produce a copious flux of H_2O_2, comparable to that of phorbol ester-triggered PMNs (98).

Direct evidence for the production of oxidants from tumor cells *in vivo* has yet to be obtained. However, the dramatic increase in antioxidant enzyme activities that have been detected on the outer surface of a fibrosarcoma but not in the core of the tumor (16) suggests that the tumor cells may be inducing these enzymes at the surface to survive a high flux of self-generated oxidants. Indeed, an ovarian cell line has been reported to be resistant to the cytolytic effects of 10 mM H_2O_2 (79).

Human small cell lung cancer cells isolated at different stages of tumor progression exhibit increasing resistance to the oxidant-inducing chemotherapeutic drug adriamycin (32). Such adaptive responses may also play an important role in the radioresistance that is observed in many malignant cancers. In principle, antioxidants could inhibit metastasis by inhibiting the oxidant-induced activation of collagenase that catalyzes the breakdown of basement membranes.

THE ROLE OF MITOGENESIS IN
MUTAGENESIS AND CARCINOGENESIS

Henderson and co-workers (51,84) and others (7) have discussed the importance of chronic mitogenesis for many, if not most, of the known causes of human cancer, for example, hormones in breast cancer; hepatitis B (35) or C viruses, or alcohol in liver cancer; high salt or *Helicobacter pylori* infection in stomach cancer (14,81); papilloma virus in cervical cancer; asbestos or tobacco smoke in lung cancer; and excess animal fat and low calcium in colon cancer. For chemical carcinogens associated with occupational cancer, worker exposure has been primarily at high, near-toxic, doses that might be expected to induce mitogenesis. Permitted worker exposure levels for some rodent carcinogens are too close to the doses that induce tumors in test animals (45). For high occupational exposures, little extrapolation is required from the doses used in rodent bioassays, and, therefore, assumptions about extrapolation are less important.

Chronic inflammation is a risk factor for cancer (66,73,99,111). Oxidants produced by the inflammatory response induce both mutagenesis and mitogenesis, and antioxidants are important for decreasing this stimulus. The oxidants produced by phagocytic cells and activated neutrophils are signals for mitogenesis (to promote wound healing) (18,24,27,29,95) that lead to the expression of *fos* and *jun* protooncogenes (1). Activated neutrophils damage DNA in neighboring cells (92), and macrophages have been shown to be capable of inducing the formation of drug-resistant variants in mouse mammary tumor cells (117).

Geneticists have long known that cell division is critical for mutagenesis. If one accepts that mutagenesis is important for carcinogenesis, it follows that mitogenesis rates must be important. The inactivation of tumor suppressor genes is also known to be important in carcinogenesis, and recent evidence suggests that one of the functions of tumor suppressor genes is to inhibit mitogenesis (96). Once the first copy of a tumor suppressor gene is mutated, the inactivation of the second copy (loss of heterozygosity) is more likely to be caused by processes whose frequency is dependent on cell division (mitotic recombination, gene conversion, and non-disjunction) than by an independent second mutation (7,8). Therefore, loss of heterozygosity will be stimulated by increased mitogenesis. Thus, while the stimulation of mitogenesis increases the chance that a mutational step will occur, it is a much more important factor for tumor induction after the first mutation has occurred. This explains why mutagenesis and mitogenesis are synergistic (7,8) and why mitogenesis after the first mutation is more effective than before (Fig. 3).

Thinking of chemicals as "initiators" or "promoters" confuses mechanistic issues of carcinogenesis (59). The idea that "promoters" are not in themselves carcinogens is not credible on mechanistic grounds and is not correct on experimental grounds

(7,8,59). Every classical "promoter" that has been tested adequately, for example, phenobarbital, catechol, TPA, is a carcinogen (57,58). A "promoter" can be viewed more accurately as an agent that facilitates the development of neoplasia by inducing cell division and stimulating the rate at which key mutations needed to acquire a transformed phenotype are accumulated, rather than by selected growth of an initiated cell. Thus, the very word "promoter" is confusing, since mitogenesis may be caused by one dose of a chemical and not by a lower dose. Dominant oncogenes and their clonal expansion by mitogenesis can clearly be involved in carcinogenesis, adding complexity; however, these mechanisms are still consistent with the view that mitogenesis is an important factor in carcinogenesis. Nongenotoxic agents, for example, saccharin, can be carcinogens at high doses just by causing cell killing and inducing chronic mitogenesis and inflammation, and the dose response would be expected to show a threshold (7,19,26).

Chronic mitogenesis by itself can be a risk factor for cancer; theory predicts it and a large body of literature supports it (7,84). Genotoxic chemicals, because they hit DNA, are even more effective than nongenotoxic chemicals at causing cell killing and cell replacement at high doses. Since genotoxic chemicals also act as mutagens, they can produce a multiplicative interaction not found at low doses, leading to an upward curving dose response for carcinogenicity (7,19,26).

Mitogenesis is often an important and sometimes the dominant factor in chemical carcinogenesis at the high, nearly toxic doses used in rodent bioassays, even for mutagens. Mitogenesis can be caused by toxicity of chemicals at high doses (leading to cell killing and subsequent replacement), by interference with cell-cell communication at high doses (101-104), by substances that control cell division such as hormones binding to receptors (84), by oxidants (inducing the wound-healing response), by viruses, etc. (7). The important factor is not toxicity, but increased mitogenesis in those cells that are not discarded. Work on radiation has also supported the idea that both mutagenesis and mitogenesis are important in tumor induction (61,60,69,80).

Epigenetic factors are also involved in carcinogenesis. Epigenetic changes in DNA, such as in S-methylcytosine levels, appear to be important in turning off genes in differentiation and could play a role in both cancer (36,53,106,107) and aging (53,54). It has been observed that the 5-methylcytosine level decreases with age (115), and it is known that cells are dedifferentiating with age (48,49). Folate deficiency (13,28), mitogenesis (i.e., through mitotic recombination), and DNA damage (21) would all be expected to increase the rate of loss of 5-methylcytosine.

THE ROLE OF DIETARY IMBALANCES IN HUMAN CANCER

The high endogenous level of oxidative adducts reinforces evidence from epidemiology that deficiency of antioxidants (12,77) is likely to be an important risk factor for cancer.

Epidemiologists have been accumulating evidence that unbalanced diets are major contributors to heart disease and cancer and are likely to be as important as smoking. It has been estimated that approximately 30% of all cancers are related to diet and that the main culprit is a dietary imbalance of too few fruits and vegetables

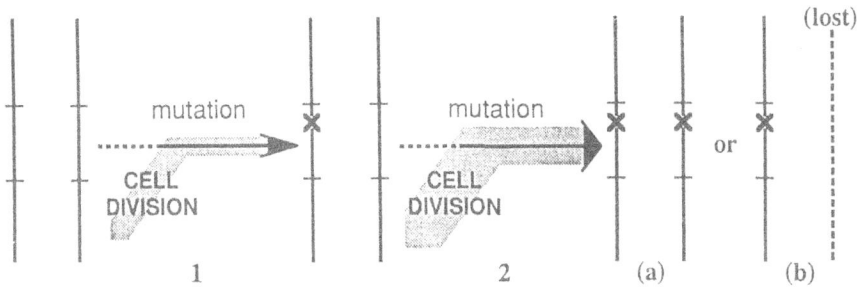

Figure 3. Mitogenesis (induced cell division) is a major multiplier of endogenous (or exogenous) DNA damage leading to mutation (7). The pathway to inactivating (X) both copies of a recessive tumor-suppressor gene is shown (two vertical lines represent the pair of chromosomes carrying the genes). Cell division increases mutagenesis due to the following: DNA adducts converted to mutations before they are repaired (1 and 2a); mutations due to DNA replication (1 and 2a); vulnerability of replicating DNA to damage (1 and 2a). Mitotic recombination (2a), gene conversion (2a), and nondisjunction (2b) are more frequent, and the first two give rise to the same mutation on both chromosomes. This diagram does not attempt to deal with the complex mutational pathway to tumors (36,63).

and too much fat. Particular micronutrients in fruits and vegetables that appear to be important in disease prevention are antioxidants (carotenoids, tocopherols, ascorbate) and folic acid, but many more vitamins and essential minerals may also be of interest (12,77,85). Micronutrients are components of the defenses against oxidants and other endogenous mutagens contributing to the degenerative diseases associated with aging, such as cancer, heart disease, cataracts, etc. Since endogenous oxidative DNA damage is enormous, there are good theoretical reasons for thinking that antioxidants should be as important as they are being found to be. Surveys have indicated that 91% of the U.S. population is not eating sufficient fruit and vegetables; for example, almost half of the population had eaten neither fruits nor vegetables on the day of the survey (82).

Work showing that folate deficiency in mice results in chromosome breakage (72) reinforces the large literature indicating that folate deficiency is an important cause of chromosome breaks, cancer, and birth defects (12,77). Again, a sizeable proportion of the population (30% or more) may not be ingesting sufficient folate. Hypomethylation of DNA that is associated with folate deficiency could also contribute to epigenetic effects, such as dedifferentiation of cells that occurs during tumorigenesis.

In the quest to delay aging and prevent cancer and heart disease, it is important to understand what level of each micronutrient is optimal for long-term effects. The U.S. Recommended Daily Allowance (RDA) for micronutrients is based on the level necessary to prevent an immediate pathological effect, but the optimal levels that may be needed to suppress tumor formation are likely to be much higher. The great genetic variability of the human species makes it likely that many people will require a higher than average optimal RDA for particular micronutrients. This will require the development of in vivo assays of short-term damage that can be measured readily in humans. The techniques for noninvasive measurement of DNA damage in humans are clearly relevant (31).

CALORIE RESTRICTION: EFFECTS ON MITOGENESIS

In rodents, a high-calorie diet appears to be carcinogenic (17,87,88). A calorie-restricted diet, compared to an *ad libitum* diet, significantly increases the lifespan of rats and mice and dramatically decreases the cancer rate. Though a causative mechanism to account for the beneficial effects of calorie restriction has not been established, several physiological indices change in rodents raised on this dietary regimen. Among the principal changes that occur is the activation of the pituitary adrenocorticotropic axis that leads to a decrease in the release of reproductive and mitogenic hormones. This change may be an adaptive response to the decreased availability of food resources. Indeed, it has been suggested that Darwinian fitness will be increased if reproductive function is delayed during periods of low food availability (52) and that the saved resources will be invested in maintenance of the body until food resources are available for successful reproduction (55).

If one accepts the concept of a trade-off between reproduction and maintenance as predicted by evolutionary biology (113), it becomes evident why calorie restriction is so effective in reducing all cancers. Decreases in mitogenic hormones such as insulin, thyroid-stimulating hormone, growth hormone, estrogen, and prolactin decrease the likelihood of hormone-induced cancers, as has been shown in various animal studies (89). The markedly lowered mitogenesis rates observed in a variety of tissues of calorie-restricted rodents (50, 70) is consistent with suppression of mitogenic hormones (9) and decreased protooncogene expression (76). The lowered incidence of mammary tumors observed in calorie-restricted rats has been attributed to reduced circulating levels of the mammotropic hormones estrogen and prolactin (97). Thus, the decrease in mitogenesis rates in calorie-restricted rats is likely to account for much of the decrease in cancer rates.

The suggestion that maintenance functions are enhanced in calorie-restricted rats is supported by the findings of more efficient DNA repair (112), better coupled mitochondrial respiration (109), and a delay in the age-dependent decline of antioxidant defenses (64). The overall effect of these enhanced maintenance activities would be lower oxidative damage to cellular macromolecules and decreased somatic mutations induced by endogenous and exogenous mutagens. The higher levels of antioxidant defenses in calorie-restricted rodents could in principle (as described above) account for the enhanced immune response (65) that is needed to effectively suppress tumor formation.

Protein restriction may prove to exhibit many of the beneficial effects of calorie restriction with regard to reducing cancer rates and increasing longevity. Rats that have been pretreated with aflatoxin B^1 and then shifted three weeks later to diets of varying protein concentrations (6,14, or 22% casein) exhibit liver cancer rates that are inversely correlated to protein intake (119). Spontaneous cancers, i.e, tumors arising at organ sites not known to be affected by aflatoxin, are also reduced dramatically in rats fed low-protein diets. While the life-extending properties of low-protein diets have been shown to be less impressive than those of calorie-restricted diets, the few studies that have investigated the relationship between protein restriction and lifespan indicate that an increase in maximal lifespan of approximately 20% can be achieved. A diet restricted in protein would be expected to be more easily accepted by humans than a diet restricted in calories by 40%.

CONCLUSIONS AND FUTURE DIRECTIONS

This review argues that oxidant-induced DNA damage and mitogenesis are two important causes of cancer. Oxidants that are produced by normal aerobic metabolism contribute to the extensive oxidative DNA damage that has been measured in rodents and humans. The oxidative damage rate in mammalian species such as rats, with a high metabolic rate, short lifespan, and high age-specific cancer rate, is much higher than the rate in species such as humans, with a lower metabolic rate, long lifespan, and a lower age-specific cancer rate. Ascorbate protects human sperm DNA from oxidative damage (40). Similar effects of dietary antioxidants in somatic tissues would be expected, and further studies designed to explain the relationship between risk of certain cancers and the consumption of antioxidants would be desirable (82).

Mitogenesis increases the probability of accumulating the mutations that are required for the development of malignant tumors. Risk factors for increased mitogenesis include chronic inflammatory diseases, which cause cell death and replacement (84), and ingestion of excessive calories, which results in increases of circulating mitogenic hormones, such as insulin, growth hormone, and estrogens. It remains to be seen whether the mitogenesis that accompanies chronic inflammatory diseases can be suppressed in humans by dietary antioxidants.

The production of oxidants by tumor cells deserves further attention. Such oxidants could facilitate the activation of latent collagenases that are necessary for metastases. Oxidant-producing tumor cells could also be viewed as possessing a mutator phenotype, that is, one that is capable of inducing in neighboring tumor cells genetic changes, some of which could lead to selective growth advantages. Further studies will need to be performed to delineate the role of oxidant-producing tumor cells in tumor progression and metastases and to determine the possible beneficial effects of dietary antioxidants in delaying this progression.

ACKNOWLEDGEMENTS

This work was supported by National Cancer Institute Outstanding Investigator Grant CA39910 and by National Institute of Environmental Health Sciences Center Grant ESO1896 to B.N. Ames. M.K. Shikenaga was supported by a National Institute of Aging Postdoctoral Fellowship AG 05489. This paper was adapted in part from B.N. Ames and L.S. Gold, Proc. Natl. Acad. Sci., USA (7), B.N. Ames and L.S. Gold, J. Nat. Cancer Inst. (in press), and B.N. Ames and L.S. Gold, Mutat. Res. (in press). We are indebted to M. Profet and T. Hagen for criticisms.

REFERENCES

1. Abate, C., L. Patel, F.J. Rauscher III, and T. Curran (1990) Redox regulation of *fos* and *jun* DNA-binding activity *in vitro*. Science 249:1157-1161.

2. Adelman, R., R.L. Saul, and B.N. Ames (1988) Oxidative damage to DNA: Relation to species metabolic rate and life span. Proc. Natl. Acad. Sci., USA 85:2706-2708.

3. Ames, B.N. (1983) Dietary carcinogens and anticarcinogens. Oxygen radicals and degenerative diseases. Science 221:1256-1264.

4. Ames, B.N. (1988) Measuring oxidative damage in humans: Relation to cancer and aging. In Methods for Detecting DNA Damaging Agents in Humans: Applications in Cancer Epidemiology and Prevention, H. Bartsch, K. Hemminki, and I.K. O'Neill, eds. IARC Scientific Publications, No. 89, International Agency for Research on Cancer, France, pp. 407-416.

5. Ames, B.N. (1989) Endogenous oxidative DNA damage, aging and cancer. Free Rad. Res. Comm. 7:121-128.

6. Ames, B.N., and R.L. Saul (1988) Cancer, aging, and oxidative DNA damage. In Theories of Carcinogenesis, O.H. Iversen, ed. Hemisphere Publishing Corp., Washington, D.C., pp. 203-220.

7. Ames, B.N., and L.S. Gold (1990) Chemical carcinogenesis: Too many rodent carcinogens. Proc. Natl. Acad. Sci., USA 87:7772-7776.

8. Ames, B.N., and L.S. Gold (1990) Too many rodent carcinogens: Mitogenesis increases mutagenesis. Science 249:970-971.

9. Armario, A., J.L. Montero, and T. Jolin (1987) Chronic food restriction and circadian rhythms of pituitary-adrenal hormones, growth hormone and thyroid-stimulating hormone. Ann. Nutr. Metab. 31:81-87.

10. Badwey, J.A., and M.L. Karnovsky (1980) Active oxygen species and the functions of phagocytic leukocytes. Ann. Rev. Biochem. 49:695-726.

11. Bendich, A. (1989) Carotenoids and the immune response. J. Nutr. 119:112-115.

12. Bendich, A., and C.E. Butterworth, Jr., eds. (1991) Micronutrients in Health and in Disease Prevention, Marcel Dekker, New York, (in press).

13. Bhave, M.R., M.J. Wilson, and L.A. Poirier (1988) c-H-*ras* and c-K-*ras* gene hypomethylation in the livers and hepatomas of rats fed methyl-deficient, amino acid-defined diets. Carcinogenesis 9:343-348.

14. Blaser, M.J. (1990) *Helicobacter pylori* and the pathogenesis of gastroduodenal inflammation. J. Infect. Dis. 161:626-633.

15. Block, G. (1991) Vitamin C and cancer prevention: The epidemiologic evidence. Am J. Clin. Nutr. 53:270S-282S.

16. Borunov, E.V., L.P. Smirnova, I.A. Shchepetkin, V.Z. Lankin, and N.V. Vasil'ev (1989) High activity of antioxidant enzymes in a tumor as a factor of "avoidance of control" in the immune system. Bull. Eksp. Biol. Med. 107:467-469.

17. Boutwell, R.K., and M.W. Pariza (1987) Historical perspective: Calories and energy expenditure in carcinogenesis. Am. J. Clin. Nutr. 45 (1 Suppl.):151-156.

18. Burdon, R.H., V. Gill, and C. Rice-Evans (1990) Oxidative stress and tumour cell proliferation. Free Rad. Res. Comm. 11:65-76.

19. Butterworth, B.E., and T. Slaga, eds. (1991) Chemically Induced Cell Proliferation: Implications for Risk Assessment, Alan R. Liss, New York, (in press).

20. Cadet, J., and M. Berger (1985) Radiation-induced decomposition of the purine bases within DNA and related model compounds. Int. J. Rad. Biol. 47:127-143.

21. Cannon, S.V., A. Cummings, and G.W. Teebor (1988) 5-Hydroxymethylcytosine DNA glycolases activity in mammalian tissue. Biochem. Biophys. Res. Comm. 151:1173-1179.

22. Carmichael, D.F., A. Sommer, R.C. Thompson, D.C. Anderson, C.G. Smith, H.G. Welgus, and G.P. Stricklin (1986) Primary structure and cDNA cloning of human fibroblast collagenase inhibitor. Proc. Natl. Acad. Sci.,USA 83:2407-2411.

23. Cathcart, R., E. Schwiers, R.L. Saul, and B.N. Ames (1984) Thymine glycol and thymidine glycol in human and rat urine: A possible assay for oxidative DNA damage. Proc. Natl. Acad. Sci., USA 81:5633-5637.

24. Chan, T.M., E. Chen, A. Tatoyan, N.S. Shargill, M. Pleta, and P. Hochstein (1986) Stimulation of tyrosine-specific protein phosphorylation in the rat liver plasma membrane by oxygen radicals. Biochem. Biophys. Res. Comm. 139:439-445.

25. Chung, M.H., H. Kasai, D.S. Jones, H. Inoue, H. Ishikawa, E. Ohtsuka, and S. Nishimura (1991) An endonuclease activity of Escherichia coli that specifically removes 8-hydroxyguanine residues from DNA. Mutat. Res. 254:1-12.

26. Cohen, S.M., and L. Ellwein (1990) Cell proliferation in carcinogenesis. Science 249:1007-1011.

27. Craven, P.A., J. Pfanstiel, and F.R. DeRubertis (1987) Role of activation of protein kinase C in the stimulation of colonic epithelial proliferation and reactive oxygen formation by bile acids. J. Clin. Invest. 79:532-541.

28. Cravo, M., J. Mason, R.N. Salomon, J. Ordovas, J. Osada, J. Selhub, and J.H. Rosenberg (1991) Folate deficiency in rats causes hypomethylation of DNA. FASEB 5:A914.

29. Crawford, D., and P. Cerutti (1989) Expression of oxidant stress-related genes in tumor promotion of mouse epidermal cells JB6. In Anticarcinogenesis and Radiation Protection, O. Nygaard and M.G. Simic, eds. Plenum Press, New York, pp. 183-190.

30. Cutler, R.G. (1984) Antioxidants, aging, and longevity. In Free Radicals in Biology, W.A. Pryor, ed. Academic Press, New York, pp. 371-428.

31. Degan, P., M.K. Shigenaga, E.-M. Park, P.E. Alperin, and B.N. Ames (1991) Immunoaffinity isolation of urinary 8-hydroxy-2'-deoxyguanosine and 8-hydroxyguanine and quantitation of 8-hydroxy-2'-deoxyguanosine in DNA by polyclonal antibodies. Carcinogenesis 12:865-871.

32. de Vries, E.G.E., C. Meijer, H. Timmer-Bosscha, H.H. Berendsen, L. de Leij, R.J. Scheper, and N.H. Mulder (1989) Resistance mechanisms in three human small cell lung cancer cell lines established from one patient during clinical follow-up. Cancer Res. 49:4175-4178.

33. Dizdaroglu, M. (1985) Formation of 8-hydroxyguanine moiety in deoxyribonucleic acid on gamma-irradiation in aqueous solution. Biochemistry 24:4476-4481.

34. Doll, R. (1971) The age distribution of cancer: Implications for models of carcinogenesis. J. Royal Stat. Soc. A134:133-166.

35. Dunsford, H.A., S. Sell, and F.V. Chisari (1990) Hepatocarcinogenesis due to chronic liver cell injury in hepatitis B virus transgenic mice. Cancer Res. 50:3400-3407.

36. Fearon, E.F., K.R. Cho, J.M. Nigro, S.E. Kern, J.W. Simons, J.M. Ruppert, S.R. Hamilton, A.C. Preisinger, G. Thomas, K.W. Kinzler, and B. Vogelstein (1990) Identification of a chromosome 18q gene that is altered in colorectal cancer. Science 247:49-56.

37. Fiala, E.S., C.C. Conaway, and J.E. Mathis (1989) Oxidative DNA and RNA damage in the livers of Sprague-Dawley rats treated with the hepatocarcinogen 2-nitropropane. Cancer Res. 49:5518-5522.

38. Floyd, R.A., J.J. Watson, P.K. Wong, D.H. Altmiller, and R.C. Richard (1986) Hydroxyl free radical adduct of deoxyguanosine: Sensitive detection and mechanisms of formation. Free Rad. Res. Comm. 1:163-172.

39. Fraga, C.G., M.K. Shigenaga, J.-W. Park, P. Degan, and B.N. Ames (1990) Oxidative damage to DNA during aging: 8-hydroxy-2' deoxyguanosine in rat organ DNA and urine. Proc. Natl. Acad. Sci., USA 87:4533-4537.

40. Fraga, C.G., P.A. Motchnik, M.K. Shigenaga, H.J. Helbock, R. Jacob, and B.N. Ames (1991) Ascorbate protects against endogenous oxidative damage in human sperm. Proc. Natl. Acad. Sci., USA (in press).

41. Frei, B., and B.N. Ames (1991) Small molecule antioxidant defenses in human extracellular fluids. In The Molecular Biology of Free Radical Scavenging Systems, J. Scandalios, ed. Cold Spring Harbor Laboratory Press, Cold Spring Harbor, New York (in press).

42. Furukawa, T., S.N. Meydani, and J.B. Blumberg (1987) Reversal of age-associated decline in immune responsiveness by dietary glutathione supplementation in mice. Mech. Ageing Dev. 38:107-117.

43. Garbisa, S., R. Pozzatti, R.J. Muschel, U. Saffiotti, M. Ballin, R.H. Goldfarb, G. Khoury, and L.A. Liotta (1987) Secretion of type IV collagenolytic protease and metastatic phenotype: Induction by transfection with c-Ha-*ras* but not c-Ha-*ras* plus Ad2-Ela. Cancer Res. 47:1523-1528.

44. Gensler, H.L., and H. Bernstein (1981) DNA damage as the primary cause of aging. Quart. Rev. Biol. 56:279-303.

45. Gold, L.S., G.M. Backman, N.K. Hooper, and R. Peto (1987) Ranking the potential carcinogenic hazards to workers from exposures to chemicals that are tumorigenic in rodents. Env. Health Perspect. 76:211-219.

46. Hagen, T.M., M.K. Shigenaga, J.W. Kitzler, and B.N. Ames (1990) Increased oxidative DNA damage following partial hepatectomy. Free Rad. Res. Comm. 9: Suppl. 1. 49.

47. Hartman, P. (1981) The aging process. Proc. Natl. Acad. Sci., USA 78:7124-7128.

48. Hartman, P.E. (1983) Mutagens: Some possible health impacts beyond carcinogenesis. Env. Mut. 5:139-152.

49. Hartman, P.E., and R.W. Morgan (1985) Mutagen-induced focal lesions as key factors in aging: A review. In Molecular Biology of Aging: Gene Stability and Gene Expression, R.S. Sohal, L.S. Birnbaum, and R.G. Cutler, eds. Raven Press, New York, pp. 93-136.

50. Heller, T.D., P.R. Holt, and A. Richardson (1990) Food restriction retards age-related histological changes in rat small intestine. Gastroenterology 98:387-391.

51. Henderson, B.E., R. Ross, and L. Bernstein (1988) Estrogens as a cause of human cancer: The Richard and Hinda Rosenthal Foundation Award Lecture. Cancer Res. 48:246-253.

52. Holehan, A.M., and B.J. Merry (1985) Modification of the oestrous cycle hormonal profile by dietary restriction. Mech. Ageing Dev. 32:63-76.

53. Holliday, R. (1987) DNA methylation and epigenetic defects in carcinogenesis. Mutat. Res. 181:215-217.

54. Holliday, R. (1987) The inheritance of epigenetic defects. Science 238:163-170.

55. Holliday, R. (1989) Food, reproduction and longevity: Is the extended lifespan of calorie-restricted animals an evolutionary adaptation? Bioessays 10:125-127.

56. Hollstein, M.C., P. Brooks, S. Linn, and B.N. Ames (1984) Hydroxymethyluracil DNA glycosylase in mammalian cells. Proc. Natl. Acad. Sci., USA 81:4003-4007.

57. IARC (1977) Phenobarbital and phenobarbital sodium. In IARC monographs on the evaluation of carcinogenic risk of chemicals to man. Some miscellaneous pharmaceutical substances, Vol 13. IARC, Lyon, France, pp. 157-181.

58. Ingram, A.J., and P. Grasso (1991) Evidence for and possible mechanisms of non-genotoxic carcinogenesis in mouse skin. Mutat. Res. 248:333-340.

59. Iversen, O.H. (1988) In Theories of Carcinogenesis, O.H. Iversen, ed. Hemisphere Publishing, Washington, D.C., pp. 119-126.

60. Jones, T.D. (1984) A unifying concept for carcinogenic risk assessments: Comparison with radiation-induced leukemia in mice and men. Health Phys. 4:533-558.

61. Jones, T.D., G.D. Griffin, and P.J. Walsh (1983) A unifying concept for carcinogenic risk assessments. J. Theor. Biol. 105:35-61.

62. Kasai, H., S. Nishimura, Y. Kurokawa, and Y. Hayashi (1987) Oral administration of the renal carcinogen, potassium bromate, specifically produces 8-hydroxydeoxyguanosine in rat target organ DNA. Carcinogenesis 8:1959-1961.

63. Kinzler, K.W., M.C. Nilbert, B. Vogelstein, T.M. Bryan, D.B. Levy, K.J. Smith, A.C. Preisinger, S.R. Hamilton, P. Hedge, A. Markham, M. Carlson, G. Joslyn, J. Groden, R. White, Y. Miki, Y. Miyoshi, I. Nishiso, and Y. Nakamura (1991) Identification of a gene located at chromosome 5q21 that is mutated in colorectal cancers. Science 251:1366-1370.

64. Koizumi, A., R. Weindruch, and R.L. Walford (1987) Influences of dietary restriction and age on liver enzyme activities and lipid peroxidation in mice. J. Nutr. 117:361-367.

65. Kubo, C., B.C. Johnson, N.K. Day, and R.A. Good (1984) Calorie source, calorie restriction, immunity and aging of (NZB/NZW)F1 mice. J. Nutr. 114:1884-1899.

66. Lewis, J.G., and D.O. Adams (1987) Inflammation, oxidative DNA damage, and carcinogenesis. Env. Health Persp. 76:19-27.

67. Lindahl, T. (1982) DNA repair enzymes. Ann. Rev. Biochem. 51:61-87.

68. Liotta, L.A., P.S. Steeg, and W.G. Stetler-Stevenson (1991) Cancer metastasis and angiogenesis: An imbalance of positive and negative regulation. Cell 64:327-336.

69. Little, J.B., A.R. Kennedy, and R.B. McGandy (1985) Effect of dose rate on the induction of experimental lung cancer in hamsters by radiation. Rad. Res. 103:293-299.

70. Lok, E., F.W. Scott, R. Mongeau, E.A. Nera, S. Malcolm, and D.B. Clayson (1990) Calorie restriction and cellular proliferation in various tissues of the female Swiss Webster mouse. Cancer Lett. 51:67-73.

71. Maccubbin, D.L., K.F. Mace, M.J. Ehrke, and E. Mihich (1989) Modification of host antitumor defense mechanisms in mice by progressively growing tumor. Cancer Res. 49:4216-4224.

72. MacGregor, J.T., R. Schlegel, C.M. Wehr, P. Alperin, and B.N. Ames (1990) Cytogenetic damage induced by folate deficiency in mice is enhanced by

caffeine. Proc. Natl. Acad. Sci., USA 87:9962-9965.

73. Madsen, C. (1989) Squamous-cell carcinoma and oral, pharyngeal and nasal lesions caused by foreign bodies found in food. Cases from a long-term study in rats. Lab. Animal 23:241-247.

74. Meydani, S.N., M. Meydani, and J.B. Blumberg (1990) Antioxidants and the aging immune response. Adv. Exp. Med. Biol. 262:57-68.

75. Meydani, S.N., M.P. Barklund, S. Liu, M. Meydani, R.A. Miller, J.G. Cannon, F.D. Morrow, R. Rocklin, and J.B. Blumberg (1990) Vitamin E supplementation enhances cell-mediated immunity in healthy elderly subjects. Am. J. Clin. Nutr. 52:557-563.

76. Nakamura, K.D., P.H. Duffy, M.H. Lu, A. Turturro, and R.W. Hart (1989) The effect of dietary restriction on myc protooncogene expression in mice: A preliminary study. Mech. Ageing Dev. 48:199-205.

77. National Research Council (1989) Diet and Health. Implications for Reducing Chronic Disease Risk. National Academy Press, Washington, D.C.

78. Nowell, P.C. (1986) Mechanisms of tumor progression. Cancer Res. 46:2203-2207.

79. O'Donnell-Tormey, J., C.J. DeBoer, and C.F. Nathan (1985) Resistance of human tumor cells in vitro to oxidative cytolysis. J. Clin. Invest. 76:80-86.

80. Ootsuyama, A., and H. Tanooka (1991) Threshold-like dose of local B-irradiation repeated throughout the life span of mice for induction of skin and bone tumors. Rad. Res. 125:98-101.

81. Parsonnet, J., D. Vandersteen, J. Goates, R.L. Sibley, J. Pritikin, and Y. Chang (1991) Helicobacter pylori infection in intestinal- and diffuse-type gastric adenocarcinomas. J. Nat. Cancer Inst. 83:640-643.

82. Patterson, B.H., and G. Block (1991) Fruit and vegetable consumption national survey data. In Micronutrients in Health and in Disease Prevention, A. Bendich and C.E Butterworth Jr., eds., Marcel Dekker, New York (in press).

83. Portier, C.J., J.C. Hedges, and D.G. Hoel (1986) Age-specific models of mortality and tumor onset for historical control animals in the National Toxicology Program's carcinogenicity experiments. Cancer Res. 46:4372-4378.

84. Preston-Martin, S., M.C. Pike, R.K. Ross, and P.A. Jones (1990) Increased cell division as a cause of human cancer. Cancer Res. 50:7415-7421.

85. Reddy, B.S., and L.A. Cohen, eds. (1986) In Diet. Nutrition. and Cancer: A Critical Evaluation, Vols. I and II. CRC Press, Florida, 175 pp. and 181 pp.

86. Richter, C., J.-W. Park, and B.N. Ames (1988) Normal oxidative damage to mitochondrial and nuclear DNA is extensive. Proc. Natl. Acad. Sci., USA 85:6465-6467.

87. Roe, F.J.C. (1989) Non-genotoxic carcinogenesis: Implications for testing and extrapolation to man. Mutagenesis 4:407-411.

88. Roe, F.J.C., P.N. Lee, G. Conybeare, G. Tobin, D. Kelly, D. Prentice, and B. Matter (1991) Risks of premature death and cancer predicted by body weight in early adult life. Human and Exp. Tox. 10:285-288.

89. Ruggeri, B. (1991) The effects of calorie restriction on neoplasia and age-related degenerative processes. In Cancer and Nutrition, Human Nutrition: A Comprehensive Treatise, Vol. 7, R.B. Alfin-Slater and D. Kritchevsky, eds. Plenum Press, New York, pp. 187-210.

90. Saul, R.L, and B.N. Ames (1986) Background levels of DNA damage in the population. In Mechanisms of DNA Damage and Repair: Implications for Carcinogenesis and Risk Assessment, M.G. Simic, L. Grossman, and A.C. Upton, eds. Plenum Press, New York, pp. 529-536.

91. Saul, R.L., P. Gee, and B.N. Ames (1987) Free radicals, DNA damage, and aging. In Modern Bioloical Theories of Aging, H.R. Warner, R.N. Butler, R.L. Sprott, and E.L. Schneider, eds. Raven Press, New York, pp. 113-129.

92. Shacter, E., E.J. Beecham, J.M. Covey, K.W. Kohn, and M. Potter (1988) Activated neutrophils induce prolonged DNA damage in neighboring cells. Carcinogenesis 9:2297-2304.

93. Shibutani, S., M. Takeshita, and A.P. Grollman (1991) Insertion of specific bases during DNA synthesis past the oxidation-damaged base 8-oxodG. Nature 349:431-434.

94. Shigenaga, M.K., C.J. Gimeno, and B.N. Ames (1989) Urinary 8-hydroxy-2'-deoxyguanosine as a biomarker of in vivo oxidative DNA damage. Proc. Natl. Acad. Sci., USA 86:9697-9701.

95. Sieweke, M.H., A.W. Stoker, and M.J. Bissell (1989) Evaluation of the cocarcinogenic effect of wounding in Rous Sarcoma virus tumorigenesis. Cancer Res. 49:6419-6424.

96. Stanbridge, E.J. (1990) Human tumor suppressor genes. Ann. Rev. Genet. 24:615-657.

97. Sylvester, P.W., C.F. Aylsworth, D.A. Van Vugt, and J. Meites (1982) Influence of underfeeding during the "critical period" or thereafter on carcinogen-induced mammary tumors in rats. Cancer Res. 42:4943-4947.

98. Szatrowski, T.P., and C.F. Nathan (1991) Production of large amounts of hydrogen peroxide by human tumor cells. Cancer Res. 51:794-798.

99. Templeton, A. (1980) Pre-existing, non-malignant disorders associated with increased cancer risk. J. Env. Path. Tox. 3:387-397.

100. Totter, J.R. (1980) Spontaneous cancer and its possible relationship to oxygen metabolism. Proc. Natl. Acad. Sci., USA 77:1763-1767.

101. Trosko, J.E. (1989) Towards understanding carcinogenic hazards: A crisis in paradigms. J. Am. Coll. Tox. 8:1121-1132.

102. Trosko, J.E., C.C. Chang, and A. Medcalf (1983) Mechanisms of tumor promotion: Potential role of intercellular communication. Cancer Invest. 1:511-526.

103. Trosko, J.E., C.C. Chang, and B.V. Madhukar (1990) In vitro analysis of modulators on intercellular communications: Implications for biologically based risk assessment models for chemical exposure. Tox. In Vitro. 4:635-643.

104. Trosko, J.E., C.C. Chang, B.V. Madhukar, and J.E. Klaunig (1990) Chemical, oncogene, and growth factor inhibition of gap junctional communication: An integrative hypothesis of carcinogenesis. Pathobiology 58:265-278.

105. von Sonntag, C., ed. (1987) The Chemical Basis of Radiation Biology. Taylor and Francis, England, 515 pp.

106. Vorce, R.L., and J.I. Goodman (1989) Hypomethylation of ras oncogenes in chemically induced and spontaneous B6C3F1 mouse liver tumors. Molec. Tox. 2:99-116.

107. Vorce, R.L., and J.I. Goodman (1989) Altered methylation of ras oncogenes in benzidine--induced B6C3F1 mouse liver tumors. Tox. Appl. Pharm. 100:398-410.

108. Weindruch, R.H., J.A. Kristie, K.E. Cheney, and R.L. Walford (1979) Influence of controlled dietary restriction on immunologic function and aging. Fed. Proc. 38:2007-2016.

109. Weindruch, R.H., M.K. Cheung, M.A. Verity, and R.L. Walford (1980) Modification of mitochondrial respiration by aging and dietary restriction. Mech. Ageing Dev. 12:375-392.

110. Weiss, S.J., G. Peppin, X. Ortiz, C. Ragsdale, and S.T. Test (1985) Oxidative autoactivation of latent collagenase by human neutrophils. Science 227:747-749.

111. Weitzman, S.A., and L.J. Gordon (1990) Inflammation and cancer: Role of phagocyte-generated oxidants in carcinogenesis. Blood 76:655-663.

112. Weraarchakul, N., R. Strong, W.G. Wood, and A. Richardson (1989) The effect of aging and dietary restriction on DNA repair. Exp. Cell Res. 181:197-204.

113. Williams, G.C. (1957) Pleiotropy, natural selection and the evolution of senescence. Evolution 11:398-411.

114. Williams, G.C., and R.M. Nesse (1991) The dawn of Darwinian medicine. Quart. Rev. Biol. 66:1-22.

115. Wilson, V.L., R.A. Smith, S. Ma, and R.G. Cutler (1987) Genomic 5-methyl-deoxycytidine decreases with age. J. Biol. Chem. 262:9948-9951.

116. Wood, M.L., M. Dizdaroglu, E. Gajewski, and J.M. Essigman (1990) Mechanistic studies of ionizing radiation and oxidative mutagenesis: Genetic effects of a single 8-hydroxyguanine (7-hydro-8-oxoguanine) residue inserted at a unique site in a viral genome. Biochemistry 29:7024-7032.

117. Yamashina, K., B.E. Miller, and G.H. Heppner (1986) Macrophage-mediated induction of drug-resistant variants in a mouse mammary tumor cell line. Cancer Res. 46:2396-2401.

118. Yarbrough, J., M. Cunningham, H. Yamanaka, R. Thurman, and M. Badr (1991) Carbohydrate and oxygen metabolism during hepatocellular proliferation: A study in perfused livers from mirex-treated rats. Hepatology 13:1229-1234.

119. Youngman, L.D. and T.C. Campbell (1991) Low protein intake inhibits the growth of hepatic foci in tumors: Evidence that altered foci indicate neoplastic potential. Cancer Res. (in press).

THE GROWTH OF PRENEOPLASTIC LESIONS BY

1,2-DIMETHYLHYDRAZINE IN RAT COLON IS

INHIBITED BY DIETARY STARCH

Piero Dolara, Giovanna Caderni, Franca Bianchini,
Cristina Luceri, Maria Teresa Spagnesi,[1] and
Giulio Testolin[2]

[1]Department of Pharmacology
University of Florence
Viale G.B. Morgagni, 65
50134 Firenze, ITALY

[2]Department of Food Technology and Microbiology
University of Milan
Milan, ITALY

ABSTRACT

The effect of dietary starch and sucrose on colon proliferation and the growth of foci of dysplastic crypts in the colon (FDC) were studied in female Sprague-Dawley rats, treated p.o. with 1,2-dimethylhydrazine (DMH). The animals were fed for 30 and 105 days with high fat (23% w/w corn oil) diets in which carbohydrates were represented by corn starch (starch diet) or sucrose (sucrose diet) (46% w/w). After 105 days of feeding, proliferation was markedly reduced in animals fed the starch diet. The number of FDC was not significantly affected by dietary treatments. However, after 30 and 105 days the percent of small FDC (formed by one-two dysplastic crypts) was higher in the animals fed the starch diet when compared to the sucrose diet.

In the cecum of the animals fed the starch diet the percent of butyrate, propionic, isovaleric, and valeric over total short-chain fatty acids (SCFA) was increased, whereas the percent of acetic acid was decreased.

Cecal pH was also decreased in the animals fed starch. The results suggest that starch diets have a protective role against DMH-colon carcinogenesis in the rat, mediated by a drop in cecal pH and an increased concentration of some SCFA.

Table 1. Composition of experimental diets

Component	g/100 g		
	Initial diet	Sucrose diet	Starch diet
Corn oil	23.1	23.1	23.1
Sucrose	23	46	/
Corn starch	23	/	46
Casein	23.1	23.1	23.1
Cellulose	2	2	2
Mineral mix 1	4	4	4
Vitamin mix 2	1.2	1.2	1.2
Choline bitartrate	0.2	0.2	0.2
DL-Methionine	0.3	0.3	0.3
Energy, kcal/g	4.86	4.81	4.90

1 AIN-76 mineral mix modified to provide 0.1% calcium in the diet. 2 AIN-76 vitamin mix .

INTRODUCTION

Dietary habits have been linked to the development of colon cancer in experimental and epidemiological studies (10,14). Diets containing high fat and low complex carbohydrates increase the risk of colon cancer (10,14). Epidemiological studies have also suggested that dietary starch might be a protective factor against this type of cancer (5,6).

To test this hypothesis, we first studied whether dietary starch alters proliferation in the colonic mucosa of rats. In fact, colon mucosal proliferation influences the process of colon carcinogenesis (1). We also studied whether high dietary starch affects the number and growth of 1,2-dimethylhydrazine (DMH)-induced foci of dysplastic crypts (FDC) in the colon of rats. These lesions are considered to be early events in the development of colon cancer (2). We also followed the variations of pH and short-chain fatty acids (SCFA) in the colon content, factors supposed to play a role in colon carcinogenesis (9,12).

MATERIALS AND METHODS

We used 8-week-old Sprague-Dawley rats. All the diets used in the experiments (Tab. 1) are based on the composition of AIN-76 semisynthetic diet, modified to contain a high level of fat (23% corn oil w/w). We followed two experimental protocols.

1) Dietary Treatment After DMH Administration

Rats were fed a high-fat diet containing the same amount of sucrose and starch; in Tab. 1, this diet is referred to as "Initial diet." One week after the beginning of the initial diet, rats were treated with saline or with DMH (Sigma Chemical, St. Louis, Mo.) (dose: 9 mg/rat, approximately 50 mg/Kg). Five days after

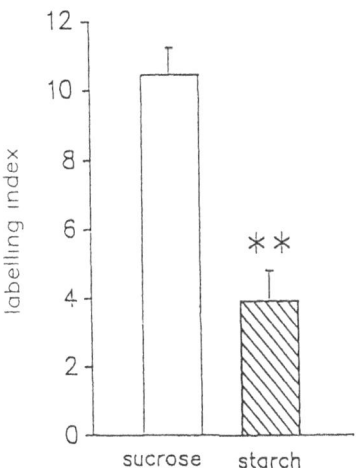

Figure 1. Proliferative activity in the colonic mucosa of rat fed diets containing sucrose (white bar) or starch (hatched bar) for 105 days. Means +SE. **=P<0.01 when compared to the sucrose diet.

the last treatment, the animals were assigned to diets containing as a source of carbohydrates either 46% sucrose or 46% starch (see Tab. 1); these diets were fed for 30 or 105 days.

2)   Dietary Treatment Before DMH Administration

Rats were fed for 105 days with the diets containing either starch or sucrose. After this period the animals were treated p.o. with DMH as in treatment 1, and after one month, in which the same diets were fed, the animals were sacrificed. Proliferative activity, number and dimension of dysplastic foci, cecal pH, and the concentration of cecal and fecal SCFA were determined as previously described (2,4,12).

RESULTS

Figure 1 shows that a diet containing starch induced a marked reduction in the proliferative activity of colonic mucosa after 105 days of feeding. To control possible effects of variations of the proliferative activity on the actions of colon carcinogens, we administered DMH both before and after the beginning of the dietary treatment with starch or sucrose.

Figure 2 shows that DMH administration induced a very high number of FDC after 30 days of feeding; this response was not affected by the variations of dietary regimens. By determining the mean number of dysplastic crypts/focus in the two dietary groups (Fig. 3), we found that the starch diet caused a significant decrease in the dimension of FDC (number of dysplastic crypts forming a focus).

In the experiments in which DMH was given before dietary treatments, we also analyzed the distribution of FDC according to their dimension. The results of this analysis (Fig. 4) indicated that after 30 days of feeding, the percent of small foci

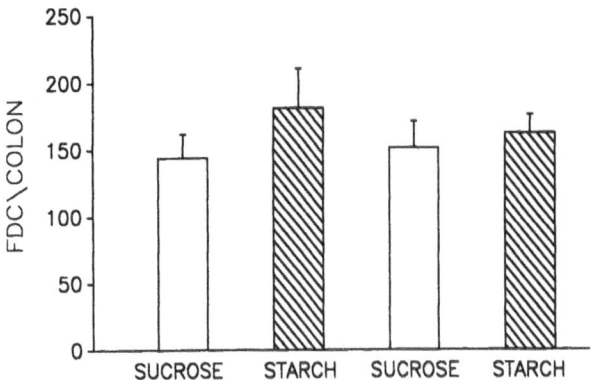

Figure 2. Number of foci of dysplastic crypts (FDC)/colon in animals treated with DMH before and after dietary treatment with diets containing sucrose (white bar) or starch (hatched bar) as a source of carbohydrates for 30 days. Means + SE.

(formed by one-two dysplastic crypts) was significantly higher in the animals fed the starch diet, while the percent of larger foci (formed by 3-4 dysplastic crypts) was significantly lower in this dietary group.

The results after 105 days of feeding (Fig. 5) indicated that in the animals fed the starch diet there was a significantly higher percent of foci formed by 1-2 crypts and a significantly lower percent of foci formed by 3-4 and by 5-6 dysplastic crypts.

At days 30 and 105, after the beginning of the feeding experiment, we observed a significant drop in pH at day 105 in the animals fed the starch diet when compared to the sucrose diet (Tab. 2).

The variations in cecal SCFA in the two experimental periods are shown in Fig 6. Total, propionic, butyric, isovaleric, and valeric acid concentrations were significantly higher in the animals fed starch for 105 days.

Table 3 shows the variations in the percentage of individual SCFA in the cecal content of animals fed the sucrose or starch diet for 105 days. The percent of acetic acid was lower, while butyric and valeric acids were significantly higher in the starch diet than in the sucrose diet. Table 3 also shows that the ratio between acetic and butyric acid was markedly lower in the animals fed a high starch diet. We then carried out a series of correlations between the values of proliferation and the various parameters pertaining to SCFA and pH in the cecum in each individual animal. We observed a negative correlation between the concentration of butyric acid and proliferation ($r = -0.65$; $P < 0.01$) and a positive correlation between cecal pH and proliferation ($r = 0.46$; $P < 0.01$).

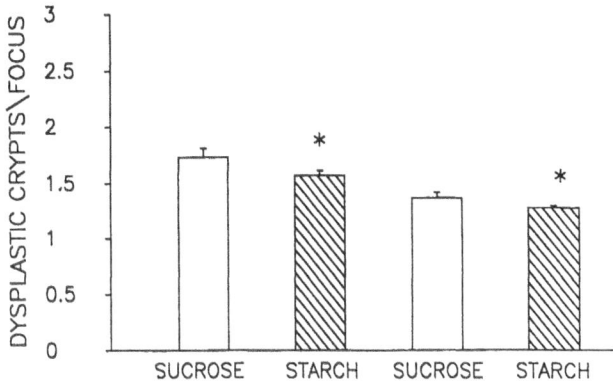

Figure 3. Dysplastic crypts/focus in animals treated with DMH before and after dietary treatment with diets containing sucrose (white bar) or starch (hatched bar) as a source of carbohydrates for 30 days. Means + SE; * = P <0.05 compared to the sucrose diet).

DISCUSSION

Dysplastic crypts have been observed and described in rodents treated with colon carcinogens, and are interpreted as early precancerous lesions (2). Dysplastic crypts have also been described in the colonic mucosa of patients with ulcerative colitis and familial polyposis, who are subject to a high risk of colon cancer (11). Recently it has been reported that foci of dysplastic crypts can be easily quantified in the unsectioned colon of experimental animals treated with carcinogens (2,8). It has also been reported that the growth of these dysplastic lesions is increased by certain diets (such as high-fat diets) that are believed to promote colon cancer (8).

The present results demonstrate that diets containing starch as a source of carbohydrates caused a decrease in the dimensions of DMH-induced dysplastic foci in the colon when compared to similar diets in which carbohydrates were supplied by sucrose. Moreover, we also found that the number of small foci (formed by 1-2 dysplastic crypts) was higher in the animals fed the diet containing starch, while the percent of larger foci was decreased.

These results suggest that starch has an inhibitory effect on the promotion of growth of dysplastic foci of the colon. Since starch markedly reduces mucosal proliferation, we were expecting a variation of the number of FDC when DMH was administered and while animals were fed with diets associated with high- and low-colonic proliferation (sucrose and starch diet respectively). On the contrary, no difference in the number of FDC following different protocols of DMH administration was observed. These last results suggest, therefore, no effect of starch on the initiation process by DMH.

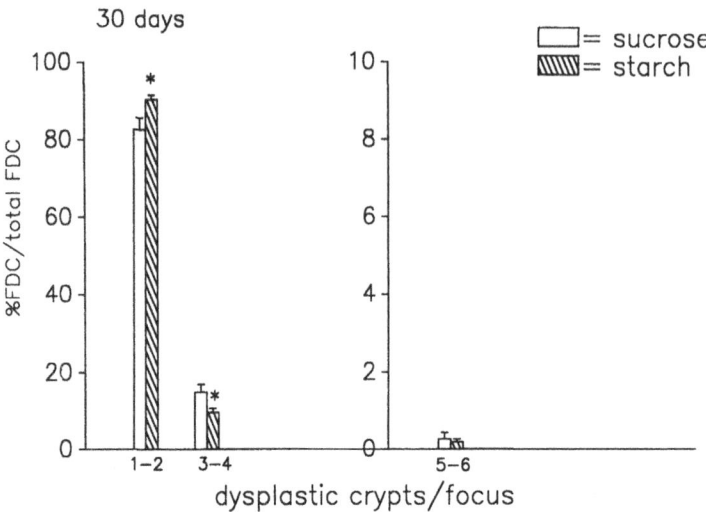

Figure 4. Size distribution of dysplastic foci, expressed as percent foci of dysplastic crypts of different size/total foci of dysplastic crypts (FDC) in animals treated with DMH and fed with diets containing sucrose (white bar) or starch (hatched bar) for 30 days. Means + SE; * = P <0.05 vs. animals fed the sucrose diet.

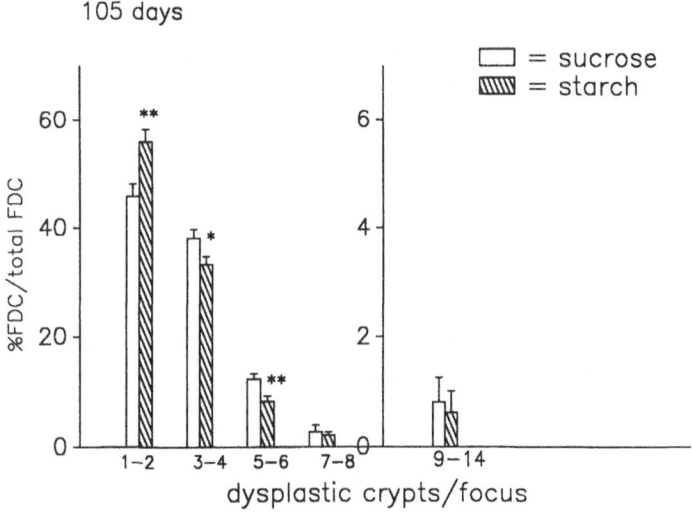

Figure 5. Size distribution of dysplastic foci in animals treated with DNH and fed with diets containing sucrose (white bar) or starch (hatched bar) for 105 days. Means + SE; * * = P <0.01 vs. animals fed the sucrose diet.

Table 2. Cecal pH in rats fed different diets for 30 days (n = 12) or 105 days (n = 15)

```
-----------------------------------------------------------------
                    30 days               105 days
            sucrose       starch       sucrose        starch
-----------------------------------------------------------------
Cecal pH  7.28+/- 0.08  7.19+/-0.07  7.37+/-0.04  7.16+/-0.03 **
-----------------------------------------------------------------
--Means +/-SE.
** P<0.01 compared to the sucrose diet.
```

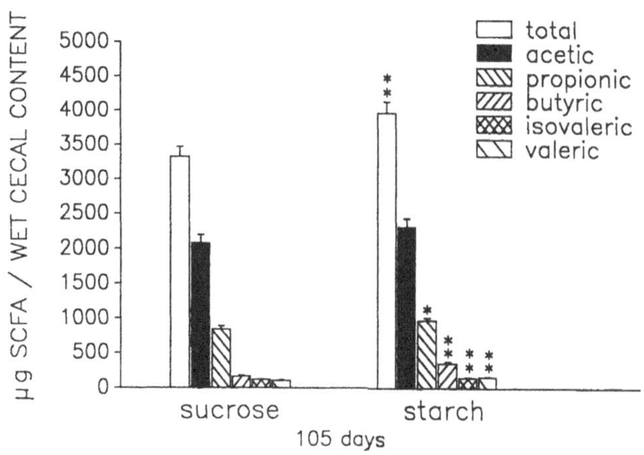

Figure 6. Concentration of SCFA (μg/g) in the wet cecal content of rats fed different diets for 105 days. Means + SE; ** P <0.01 compared to the sucrose diet.

Table 3. Percentage of individual/total SCFA in the cecal content of rats fed different diets for 105 days (n = 15)

```
-------------------------------------------------------------
                         sucrose       starch
-------------------------------------------------------------
Acetic                 62.8+/-1.0    58.4+/-0.8 **
Propionic              25.2+/-0.4    24.7+/-0.5
Butyric                 5.1+/-0.6     9.0+/-0.4 **
Isovaleric              3.7+/-0.3     3.9+/-0.2
Valeric                 3.0+/-0.3     4.1+/-0.2 **
Acetic/Butyric         14.7+/-1.7     6.8+/-0.4 **
-------------------------------------------------------------
Means +/- SE.
** P<0.01 compared to the sucrose diet.
```

The mechanisms underlying the inhibitory effect of starch on colonic mucosa proliferation and on the growth of dysplastic foci are not yet entirely defined. However, the consumption of starch in the diet has been suggested to alter the colonic environment (6,7). In the present set of experiments, starch diets increased the production of short-chain fatty acids and the percentage of butyrate in the cecum. Butyrate is known to induce differentiation of colon cancer lines, as well as other types of cancer cells in vitro (13). Moreover, we found that the animals fed the starch diets had a cecal pH significantly lower than the ones fed sucrose. A reduction in colonic pH has been suggested to be a protective factor against the development of colon cancer (3,9). We are proposing that the observed variations in short-chain fatty acids and in colonic pH are the causative agents explaining the inhibition of growth of dysplastic foci of the colon.

Since some epidemiological studies have suggested that starch is a protective factor in the development of colon cancer in humans (5,6), the present results, although obtained in the DMH-rat model, indicate a possible mechanism for explaining the correlation between a high starch consumption and a reduction of colon cancer incidence.

ACKNOWLEDGEMENTS

This work was supported by grants from Consiglio Nazionale delle Ricerche (Progetto Finalizzato Raisa, sottoprogetto 4) and partially by Ministero della Pubblica Istruzione, Italy.

REFERENCES

1. Barthold, S.W., and D. Beck (1980) Modification of early dimethylhydrazine carcinogenesis by colonic mucosa hyperplasia. Cancer Res. 40:4451-4455.
2. Bird, R.P. (1987) Observation and quantification of aberrant crypts in the murine colon treated with a colon carcinogen: Preliminary findings. Cancer Lett. 37:147-151.
3. Bruce, W.R. (1987) Recent hypotheses for the origin of colon cancer. Cancer Res. 47:4237-4242.
4. Caderni, G., F. Bianchini, P. Dolara, and D. Kriebel (1989) Proliferative activity in the colon of the mouse and its modulation by dietary starch, fat, and cellulose. Cancer Res. 49:1655-1659.
5. Decarli, A., and C. La Vecchia (1986) Environmental factors and cancer mortality in Italy: Correlational exercise. Oncology 43:116-126.
6. Jenkins, D.J.A, A.L. Jenkins, A.V. Rao, and L.U. Thompson (1986) Cancer risk: Possible protective role of high carbohydrate, high fiber diets. Am. J. Gastroenterol. 81:931-935.
7. Jenkins, D.J.A., A.L. Jenkins, T.M.S. Wolever, A.V. Rao, and L.U. Thompson (1986) Fiber and starchy foods: Gut functions and implications in disease. Am. J. Gastroenterol. 81:920-930.
8. McLellan, E.A., and R.P. Bird (1988) Aberrant crypts: Potential preneoplastic lesions in the murine colon. Cancer Res. 48:6187-6192.
9. Newmark, H.L., and J.R. Lupton, (1990) Determinants and consequences of colonic pH: Implications for colon cancer. Nutr. and Cancer 14:161-173.

10. Reddy, B.S. (1983) Tumor promotion in cancerogenesis. In <u>Mechanisms of Tumor Promotion,</u> T.S. Slaga, ed. CRC Press, Boca Raton, Fla., pp. 107-109.

11. Riddell, R.H. (1984) Dysplasia and cancer in ulcerative colitis: A soluble problem? <u>Scand. J. Gastroenterol.</u> 19 (Suppl.104):137-149.

12. Weaver, G.A., J.A. Krause, T.L. Miller, and M.J. Wolin (1988) Short chain fatty acid distribution of enema samples from sigmoidoscopy population: An association of high acetate and low butyrate ratio with adenomatous polyps and colon cancer. <u>Gut</u> 9:1539-1543.

13. Whitehead, R.H., G.P. Young, and P.S. Bhathal (1986) Effects of short chain fatty acids on a new human colon carcinoma cell line (LIM1215). <u>Gut</u> 27:1457-1463.

14. Willet, W. (1989) The search for the causes of breast and colon cancer. <u>Nature</u> 338:389-394.

PREVENTION OF GENOTOXIC EFFECTS BY DIETARY CONSTITUENTS

IN VARIOUS ORGANS OF MICE TREATED WITH NITROSAMINES

Siegfried Knasmüller, Wolfgang Huber, and Rolf Schulte-Hermann

Institute of Tumor Biology and Cancer Research
University of Vienna
Borschkegasse 8a, A-1090 Vienna, AUSTRIA

INTRODUCTION

In recent years, numerous compounds have been identified that exert anti-mutagenic effects toward nitrosamines *in vitro,* but it is unknown at present if they are also active under *in vivo* conditions (1). In addition, it has been shown that certain dietary constituents prevent nitrosamine-induced carcinogenesis in rodents, but it is not yet proven whether these properties are paralleled by prevention of DNA-damage (2,3). The aim of the present study was to investigate the inhibitory effects of four representatives of different groups of putative antimutagens or anticar-cinogens toward nitrosamine-induced genotoxicity in the living animal. The test system used was the differential DNA repair host mediated assay with *Escherichia coli* K-12 derivatives. This assay enables the detection of repairable DNA damage in various organs (liver, kidney, lungs, spleen, testes, and blood) of chemically treated mice. In addition, *in vitro* assays were carried out under comparable experimental conditions using the same indicator strains. Finally, biochemical experiments were conducted aimed at determining the influence of the various compounds on the α-hydroxylation of nitrosamines.

The test compounds were vitamin A (ROL), phenethyl isothiocyanate (contained in *Brassica* species, PEITC), oleic acid (OA) and its triglyceride (TO). All these compounds are contained in the human diet, and several studies give evidence that retinoids, isothiocyanates, and unsaturated fatty acids protect against the mutagenic and/or carcinogenic effects of various environmental compounds (for reviews see Refs. 4, 5, 6). The information available at present on antigenotoxic and anticarcinogenic effects of OA, TO, ROL, and PEITC is summarized in Tab. 1. For the former three compounds, evidence for antimutagenic effects toward nitrosamines is restricted to in *vitro* experiments. For PEITC, no data on antimutagenic effects are available, but in carcinogenicity studies it prevented tumor formation by tobacco-specific nitrosamines. The nitrosamines used to induce repairable DNA damage in the present experiments were: N-nitrosodimethylamine (NDMA), N-nitrosopyrolidine

Table 1. Antimutagenic and anticarcinogenic effects of the test compounds against nitrosamines

Compound	Antimutagenicity	Anticarcinogenicity	Remarks	Ref.
Oleic acid	effective against NDMA, NDEA, NDBA, NPYR, NMOR, NPIP in _E.coli_ WP 2 _in vitro_, no _in vivo_ data	?	inhibition of α-hydroxy-lation _in vitro_	6, 7, 8, 9
Triolein	effective against NDMA _in vitro_ in _E.coli_ WP 2, no _in vivo_ data	?	?	7
Retinol	effective against NDMA and NDEA in _S.typhimurium in vitro_, reduction of NDMA and NDEA induced SCEs in V-79 cells, no _in vivo_ data	retinoids are ineffective towards BOP-nitrosamine induced carcinogenesis in hamsters	?	9, 10
Phenethyliso-thiocanate	no data in combination with nitrosamines	prevention of carcinogene-sis induced by NNK and NBMA	inhibition of α-hydroxy-lation _in vivo_ in rats	2, 3, 11, 12

(NPYR), and N-nitrosodiethanolamine (NDELA). The former two are contained in substantial amounts in foodstuffs and tobacco smoke (13); the diethanol analogue is a contaminant of cosmetics and cutting fluids (14).

DIFFERENTIAL DNA-REPAIR ASSAYS WITH _ESCHERICHIA COLI_ K-12 DERIVATIVES

Test Principle

The test system was developed by G.R. Mohn (Univ. of Leiden). The differential survival of two _E. coli_ K-12 strains that differ vastly in their repair capacity (343/753 is _uvr_B/_rec_A; 343/765 is _uvr_+/_rec_+) serves as a measure for the induction of repairable DNA damage. Mixtures of the two indicator strains are exposed to the test compound either in liquid suspensions or in animal hosts and are subsequently plated on neutralred-streptomycin agar plates for determination of the individual strain survival. The two strains can be distinguished by the color of the colonies. 343/753 is _lac_$^+$ and forms red colonies on neutralred agar; 343/765 is _lac_$^-$ and forms white colonies. Both strains are streptomycin-dependent; thus infections of the selective plates from organ homogenates can be avoided. For a detailed description of the phenotypes, see Mohn (15).

The system has been validated with a variety of alkylating agents, N-nitroso compounds, cytostatic agents, and cooked food mutagens (16). We know from earlier experiments that the differential DNA-repair test is very sensitive to the genotoxic action of nitrosamines in contrast to the _Salmonella typhimurium_ assay (17). An important advantage of the system for antimutagenicity studies compared to microbial mutagenicity tests is that lag-division or bacteriocidal effects cannot mimic prevention of DNA damage.

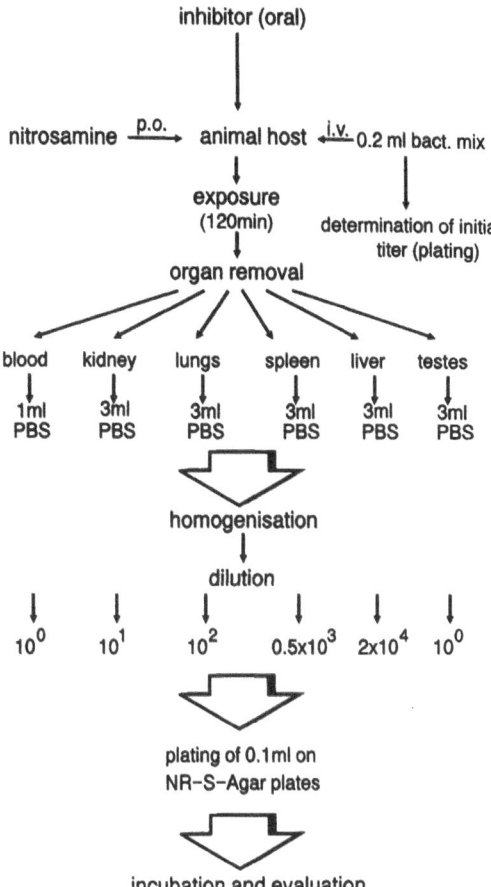

inhibitor (oral)

↓

nitrosamine ──p.o.──→ animal host ◄──i.v.── 0.2 ml bact. mix

exposure
(120min)

determination of initial
titer (plating)

organ removal

blood	kidney	lungs	spleen	liver	testes
1ml PBS	3ml PBS	3ml PBS	3ml PBS	3ml PBS	3ml PBS

homogenisation

dilution

10^0 10^1 10^2 0.5×10^3 2×10^4 10^0

plating of 0.1ml on
NR–S–Agar plates

incubation and evaluation

Figure 1. Schematic representation of animal-mediated differential DNA repair assays with *E. coli* K-12 343/753 (uvr^B/rec^B) and 343/765 (uvr^+/rec^+) for the detection of inhibitory effects on nitrosamine-induced genotoxicity.

Cultivation of the Strains

Overnight cultures of the two strains were grown in PEP-S bouillon. Immediately before the assays, stationary phase cultures of 343/753 and 343/765 were mixed 10:1, washed and resuspended in phosphate buffered saline (PBS) prior to use in animal mediated or liquid suspension assays. For a detailed description of the PEP-S bouillon, the PBS and the selective medium (NRS-agar) see Mohn (15).

In Vivo Assays

The test procedure used in the animal-mediated experiments can be seen in Fig. 1. In all assays, male Swiss Albino mice (25-30 g body weight) were used as host

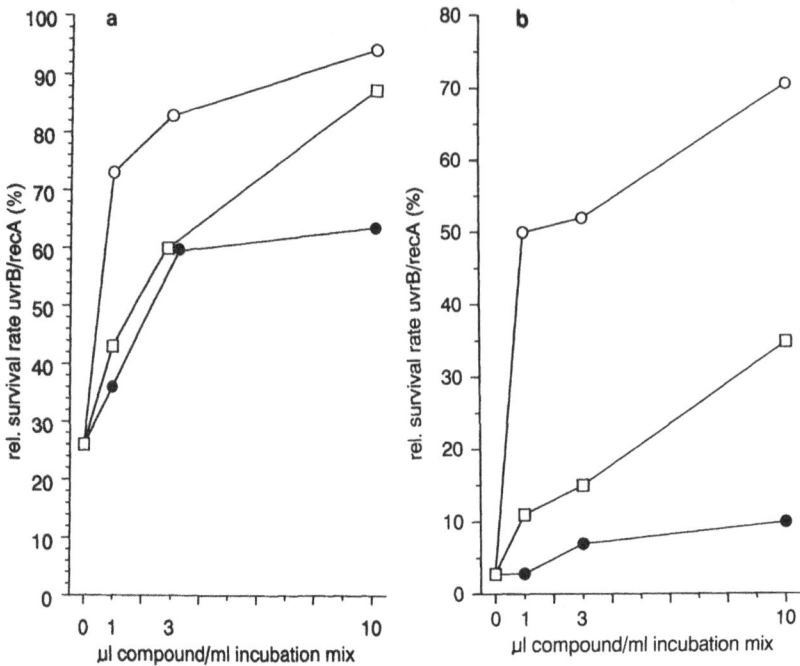

Figure 2. Inhibition of repairable DNA damage induced by NDMA (2a) or NPYR (2b) by addition of PEITC (O), TO (●), and OA (□) in liquid suspension assays with *E. coli* K-12 strains.

animals. Aqueous solutions of the nitrosamines (0,2 ml) were administered orally in all experiments simultaneously with a mixture of the two indicator strains (2-4 x 10^8 cells of each strain in 0,2 ml PBS) which was injected into the lateral vein. The inhibitors were given by gavage either 2 hrs before the administration of the indicator cells (acute treatment) or consecutively over 4 days (repeated treatment). The animals were deprived of feed 12 hrs before the experiments but received water ad libitum. ROL and PEITC were dissolved in corn oil; OA and TO were given, undiluted. In preliminary experiments it was ascertained that the solvent had no influence on the genotoxicity of the nitrosamines under the present experimental conditions. Two hours after administration of the bacteria, the animals were sacrificed and the indicator cells recovered and plated.

In Vitro Tests

Liquid suspension tests were performed under experimental conditions comparable to those in animal-mediated assays. Mixtures of the two strains (0,1 ml; 1-2x 10^8 cells of each strain) were incubated with 0,7 ml standard S-9 mix (prepared from untreated male Swiss Albino mice according to the protocol of Maron and Ames, Ref.18), 0,1 ml of aqueous solution of the nitrosamines, and various amounts of the inhibitors (1-10 µl) and PBS (ph 7,0, and 1 ml). Following an exposure time of 2 hrs, the suspensions were diluted 10^{-4} and 10^{-5} and 0,1 ml aliquots plated on NRS agar.

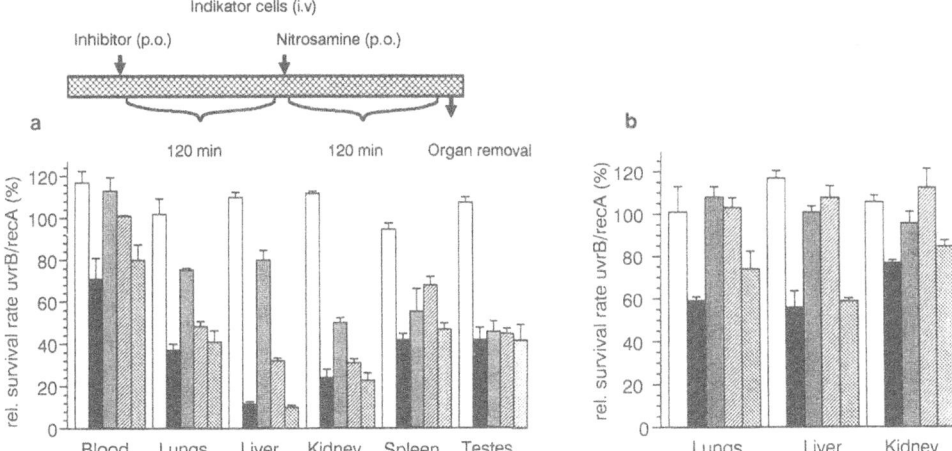

Figure 3. Inhibition of genotoxic effects induced in various organs of mice by NPYR (360 mg/kg; 3a) or NDELA (600 mg/kg, 3b) upon acute treatment of the animal hosts with PEITC (150 mg/kg, ▦), OA (2,000 mg/kg, ▩), and ROL (250 mg/kg, ▨). White bars represent the relative survival in untreated animals (as compared to the injection titer); black bars give relative survival rates in animals treated with the nitrosamines only.

Evaluation

Each experiment was repeated at least twice with 2 animals per experimental point in each experiment. From each organ homogenate and from the liquid suspension mixtures three replica plates were seeded. The plates were incubated for 24 hrs at 37°C in the dark and subsequently were stored for 24 hrs at room temperature prior to enumeration of the colonies. The relative survival rates of the repair-deficient strain (343/753) in animal-mediated assays were calculated on the basis of the ratio 343/753 vs. 343/765 in the injection titer.

DETERMINATION OF α-HYDROXYLATION

α-Hydroxylation was determined following the procedure that Schulte-Hermann used on other substrates (19). In brief, formaldehyde was formed from NDMA by the enzymatic activity of microsomal suspensions after addition of an NADPH-regenerating system. The microsomes were prepared from animals that had been pretreated with the putative modifiers either for 2 hrs (ROL-250 mg/kg, PEITC-150 mg/kg) or consecutively over 4 days (TO-4,000 mg/kg/day). The NDMA concentration in the incubation mixtures was 4 mM. Formaldehyde formed in the incubation mixtures after 20 min was determined spectrophotometrically through a reaction with acetylacetone at 412 nm.

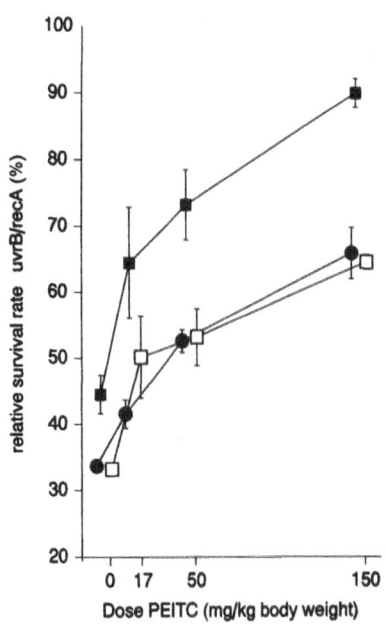

Figure 4. Dose-dependent inhibition of genotoxic effects of NDMA (8 mg/kg) in blood (■), lungs (●), and liver (□) of mice upon acute pretreatment with various doses of PEITC.

INHIBITION OF NITROSAMINE-GENOTOXICITY IN LIQUID SUSPENSION TESTS (IN VITRO)

The effects of various concentrations of OA, TO, and PEITC on the genotoxic effects of NDMA and NPYR are depicted in Figs. 2a and 2b. The y-axis indicates the relative survival of the repair-deficient strain (343/753) relative to that of the repair-proficient counterpart (343/765). In presence of 10 mM NDMA, the survival was about 25% compared to the controls; upon addition of PEITC, OA, and TO (Fig. 2a), the induction of repairable DNA damage is attenuated in a dose-dependent way. The most efficient inhibitor was PEITC, followed by OA and TO. A similar pattern of inhibition was observed in combination with NPYR (Fig. 2b). *In vitro*, NDELA induces only marginal effects at high concentrations; therefore a precise determination of inhibitory effects was not possible. For ROL, no reliable *in vitro* data are available at present, since suitable solvents such as DMSO or ethanol cause a potent reduction of the genotoxic activity of the nitrosamines which is due to inhibition of α-hydroxylation.

INHIBITION OF NITROSAMINE-INDUCED GENOTOXICITY IN ANIMAL-MEDIATED ASSAYS (IN VIVO)

Acute Treatment

The results obtained in a series of experiments with NPYR using the acute treatment protocol are depicted in Fig. 3a. NPYR induced pronounced genotoxic

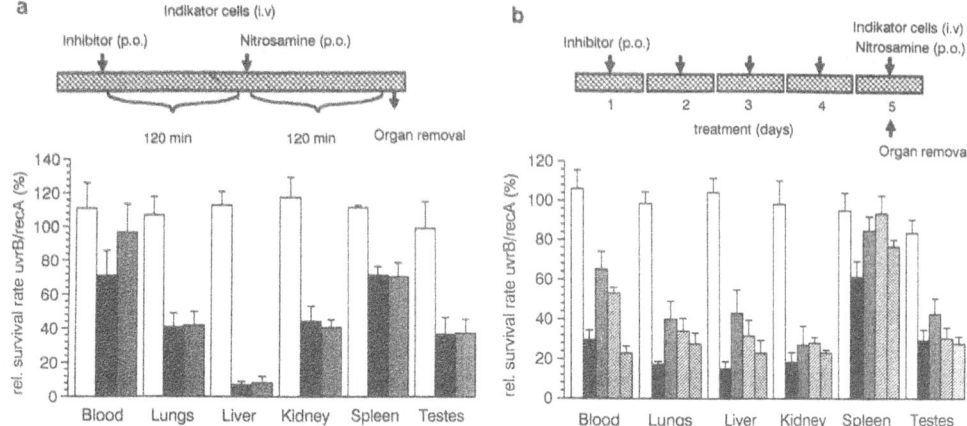

Figure 5. Influence of acute (5a) and repeated (5b) treatment with various doses of TO (⊞ -4,000 mg/kg/day; ▨ -2,000 mg/kg/day; ⊠ -1,000 mg/kg/day) on the genotoxic effects induced in mice with NDMA (80 mg/kg). White bars represent values in controls; black bars give values from hosts that received the nitrosamine only.

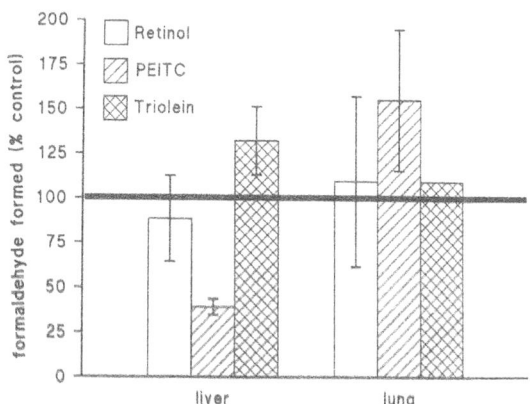

Figure 6. Summary of two experiments determining the influence of ROL, PEITC, and TO on the formation of formaldehyde from NDMA (α-hydroxylation) in vivo. Mice were exposed to ROL (250 mg/kg) and PEITC (150 mg/kg) for 2 hrs. TO (4,000 mg/kg/day) was given consecutively over 4 days (repeated treatment protocol). Mean +/- SD from 2-3 mice are shown.

effects in all organs (liver>kidney>lungs>spleen>testes>blood) in comparison to untreated controls. Administration of PEITC 2 hrs before nitrosamine treatment caused a pronounced reduction of genotoxic effects in most organs, except the testes. With ROL, a more moderate effect was measured in liver, kidneys, spleen, lungs, and in the blood, whereas with OA no antigenotoxic activity could be detected. Essentially similar effects were seen with NDMA (80 mg/kg; data not shown) and NDELA (Fig. 3b) under identical experimental conditions: PEITC and to a lesser extent ROL were active; OA was inactive. In all experiments, PEITC was the most effective inhibitor of DNA damage; therefore the dose dependency of its antigenotoxic efficiency was studied in further experiments with NDMA in liver, lungs, and in the blood. Already, at a dose of 17 mg/kg body weight, prevention of DNA damage could be observed (Fig. 4). TO was inactive in all acute treatment experiments. The lack of inhibition of the genotoxic effects of NDMA can be seen in Fig. 5a.

Repeated Treatment

In contrast to the negative results obtained with TO upon acute treatment, a dose-dependent inhibition of NDMA-induced genotoxicity was observed upon consecutive administration (over 4 days) in the range between 1-4 g/kg/day (Fig. 5b). All experiments with the other compounds failed to detect prevention of NDMA-induced genotoxic effects. The concentrations of ROL and OA were identical to those used in the acute treatment experiments (250 mg/kg/day for ROL; 2,000 mg/kg/day for OA; the PEITC dose was 50 mg/kg/day).

INFLUENCE OF ROL, PEITC, AND TO ON THE α-HYDROXYLATION OF NDMA

The results of preliminary experiments aimed at determining the influence of the various test compounds on the α-hydroxylation of NDMA are summarized in Fig. 6. PEITC and ROL were administered to the animals 2 hrs prior to preparation of microsomes; TO was administered consecutively over 4 days (according to the repeated treatment protocol used in the differential DNA repair experiments). The results show that PEITC clearly inhibits α-hydroxylation in the liver, whereas no such effects can be measured with ROL and TO.

CONCLUSIONS

The present experiments show that the results obtained with putative antimutagens under *in vitro* conditions reflect only partly the situation in the living animal. In agreement with earlier reports, our findings in liquid preincubation assays show that OA and TO prevent the genotoxic activity of nitrosamines (6-9). PEITC, whose antigenotoxic activities toward nitrosamines have not been examined earlier, was the most effective antigenotoxin *in vitro*.

In the living animal, a pronounced effect (at moderate dosages) is only observed with the isothiocyanate when the compound is given shortly before the nitrosamines, whereas no effect was detectable in the repeated dosage experiments. These findings and also the data from α-hydroxylation experiments are in good agreement with the results of earlier experiments in which the effects of PEITC on metabolic activation and on the carcinogenicity of NDMA, NNN, and NNK were investigated (2,3,11,12). It is notable that in the animal-mediated assays a reduction of NDMA-induced DNA damage by administration of PEITC was not only restricted to the liver, but also pronounced inhibitory effects were observed in lungs and in blood, whereas in the α-hydroxylation assays significant effects were restricted to the liver. This discrepancy is probably due to the fact that active metabolites formed in the liver do reach other tissues and are at least partly responsible for the induction of repairable DNA damage in other organs.

ROL exerted only a moderate effect *in vivo* which was restricted to acute treatment. It has to be noted that the doses of vitamin A required to cause inhibitory effects are substantially higher than those contained in the human diet. The results of Birt et al. (10) show that structural analogues of vitamin A have no effect on bis (2-oxo-propyl) nitrosamine-induced carcinogenesis.

In contrast to the pronounced antimutagenic effects of OA toward nitrosamines observed under *in vitro* conditions, no prevention of genotoxicity was observed in the animal-mediated assays. Preliminary biochemical experiments, not further described here, suggest that acute oral treatment of the animals with 2,000 mg/kg body weight does not result in an increased OA level in the liver. Due to acute inflammatory effects in the GI tract it is not possible to administer higher doses of the free acid. The results obtained with the triglyceride, which can be given at higher doses, indicate that repeated dosage results in a reduction of NDMA-induced genotoxicity. The prevention of repairable DNA damage by TO is apparently not associated with an inhibition of α-hydroxylation, although it has been shown that decreased metabolic activation is responsible for the antimutagenic effects of unsaturated fatty acids *in vitro* (7). On the other hand, it has been postulated that short-chain fatty acids inhibit cellular uptake of nitrosamine metabolites (7), and in a recent study it has been demonstrated in tests with *Salmonella* and *E. coli* WPs *in vitro* that these fatty acids also inhibit the mutagenic activity of α-hydroxylated nitrosamines (20). These findings suggest that, besides inhibition of α-hydroxylation, other mechanisms might be involved in the antimutagenic effects of these compounds.

In conclusion, the results of the present experiments show that differential DNA-repair assays are a convenient and rapid method to assess the prevention of nitrosamine-induced genotoxicity by dietary compounds in the living animal. On the basis of the present results it can be tentatively assumed that PEITC and TO might also be effective in the prevention of DNA damage caused by nitrosamines in humans.

ACKNOWLEDGEMENTS

The authors are thankful to Evelin Kainzbauer for competent technical assistance, and to Harald Kienzl for his help in the evaluation of the data. These studies were sponsored by an Otto Loewi grant of the "Fonds zur Förderung wissenschaftlicher Förschung" of Austria (to SK).

REFERENCES

1. Gichner, T., and J. Veleminski (1988) Mechanisms of inhibitors of N-nitroso compounds induced mutagenicity. Mutat. Res. 202:325-334.
2. Stone, G.D., D.T. Morissey, Y.-H. Heur, E.M. Daniel, A.J. Galati, and S.A. Wagner (1991) Inhibitory effect of phenethylisothiocyanate on N-nitroso-benzylmethylamine carcinogenesis in the rat oesophagus. Canc. Res. 51:2053-2068.
3. Morse, M.A., K.I. Elkind, S.S. Hecht, and F.L. Chung (1989) Effect of aromatic isothiocyanates on tumorigenicity, O^6-methylguanine formation and tumorigenicity of the tobacco specific nitrosamine 4-(methylnitrosamino)-1-(3-pyridil)-1-butanone in A/J mouse lung. Canc. Res. 49:2894-2987.
4. Hartman, P.E., and D.M. Shankel (1990) Antimutagens and anticarcinogens: A survey on putative interceptor molecules. Env. Molec. Mutagen. 15:145-182.
5. Wattenberg, L.W. (1985) Chemoprevention of cancer. Cancer Res. 45:1-8.
6. Hayatsu, H., S. Arimoto, and T. Negishi (1988) Dietary inhibitors of mutagenesis and carcinogenesis. Mutat. Res. 202:429-446.
7. Negishi, T., and H. Hayatsu (1984) Inhibitory effect of saturated fatty acids on the mutagenicity of N-nitrosodialkylamines. Mutat. Res. 135:87-96.
8. Negishi, T., Y. Ohara, and H. Hayatsu (1982) A sensitive assay for mutagenic activity of N-nitrosamines and its use for the detection of modulators of mutagenesis. In N-Nitroso Compounds: Occurrence and Biological Effects, H. Bartsch, et al., eds. IARC, Lyon, France, No. 41, pp. 685-693.
9. Huang, C.C. (1987) Retinol (vitamin A) inhibition of dimethylnitrosamine (DMN) and diethanolamine (DEN) induced sister chromatid exchanges in V-79 cells and mutations in Salmonella/microsome assays. Mutat. Res. 187:133-140.
10. Birt, D.F., M.H. Davies, P.M. Pour, and S. Salmas (1983) Lack of inhibition by retinoids of bis(2-oxopropyl) nitrosamine induced carcinogenesis in Syrian hamsters. Carcinogenesis 4:1215-1220.
11. Chung, F.-L., M. Wang, and S.S. Hecht (1985) Effects of dietary indoles on N-nitrosodimethylamine and 4-(methylnitrosamino)-1-(3-pyridil) 1-b butanone α-hydroxylation and DNA demethylation in rat liver. Carcinogenesis 6:539-543.
12. Chung, F.L., A. Juchaz, J. Vitarius, and S.S. Hecht (1984) Effects of dietary compounds on α-hydroxylation of N-nitrosopyrolidine and N-nitrosonornicotine in rat target tissues. Canc. Res. 44:2924-2928.
13. Eisenbrand, G. (1981) N-Nitrosoverbindungen in Nahrung und Umwelt. WVG, Stuttgart, pp. 51-89.
14. Lijinski, W., M.D. Reuber, and W.B. Manning (1980) Potent carcinogenicity of N-nitrosodiethanolamine in rats. Nature 288:589-590.
15. Mohn, G.R. (1984) The DNA repair host mediated assay as a rapid and sensitive procedure for the detection of genotoxic factors present in various organs of mice, preliminary results with mitomycin C. Arch. Toxikol. 55:268-271.
16. Knasmüller, S., A. Szakmary, and F. Wottawa (1986) Microbial host mediated assays: A convenient method to investigate the bio-transformation of xenobiotics to genotoxins in vivo. Bull. Pol. Acad. Sci. 36:225-234.
17. Knasmüller, S., A. Szakmary, and M. Kehrer (1990) Use of differential DNA repair host mediated assays to investigate the biotransformation of xenobiotics

in *Drosophila melanogaster.* I. Genotoxic effects of nitrosamines. <u>Chem. Biol. Interact.</u> 75:17-29.

18. Maron, D.M., and B.N. Ames (1984) Revised methods for the *Salmonella*/ microsome assay. In <u>Handbook of Mutagenicity Testing Procedures,</u> B.J. Kilbey et al., eds. Elsevier, Amsterdam-N.Y.-Oxford, pp. 93-141.

19. Schulte-Hermann, R., and W. Parzefall (1980) Adaptive responses to the gestagen and anti-androgen cypoterone acetate and other inducers. I. Induction of drug metabolizing enzymes. <u>Chem. Biol. Interact.</u> 31:279-286.

20. Takeda, K., S. Ukawa, and M. Mochizuki (1991) Inhibition of fatty acids of direct mutagenicity of nitrosamines. In <u>Relevance to Human Cancer of N-Nitroso Compounds, Tobacco Smoke and Mycotoxins,</u> J.K. O'Neill et al., eds. IARC, Lyon, France, No. 105, pp. 558-563.

EFFECT OF ANTIOXIDANT SUPPLEMENTATION

IN AN ELDERLY POPULATION

R.J Šrám[1], B. Binková[1], J. Topinka[1], F. Kotěšovec[2],
I. Fojtíková[3], I. Haneľ[3], J. Klaschka[3], J. Kočišová[3],
M. Prošek[2], and J. Machálek[4]

[1]Institute of Experimental Medicine
Czechoslovak Academy of Sciences
120 00 Prague 2, CZECHOSLOVAKIA

[2]District Institute of Hygiene
416 65 Teplice, CZECHOSLOVAKIA

[3]Psychiatric Center Prague
181 03 Prague 8, CZECHOSLOVAKIA

[4]Regional Institute of Hygiene
700 00 Ostrava, CZECHOSLOVAKIA

INTRODUCTION

According to some contemporary theories, the aging process in humans may be the result of the accumulation of unsuccessfully repaired genetic damage. The critical changes determining cell senescence are most probably the changes that take place in macromolecules such as DNA and RNA, enzymes, and structural parts of biomembranes--lipids and proteins.

It is supposed that long-term human exposure to various factors inducing mutations that continually trigger DNA repair mechanisms may decrease the ability of man to repair induced damage. The reduced repair ability results in a modification of genes coding the enzymes taking part in the process of DNA repair. If the human organism is simultaneously affected by environmental mutagens, the impact of genetic damage is still more pronounced.

Several products of normal metabolism may act as endogenous alkylating agents or can induce reactive oxygen radicals that act as mutagens, promoters, and carcinogens. Oxygen radicals, especially reactive hydroxy radicals, that cannot be eliminated from cells by a protective enzymatic system (e.g., superoxide dismutase,

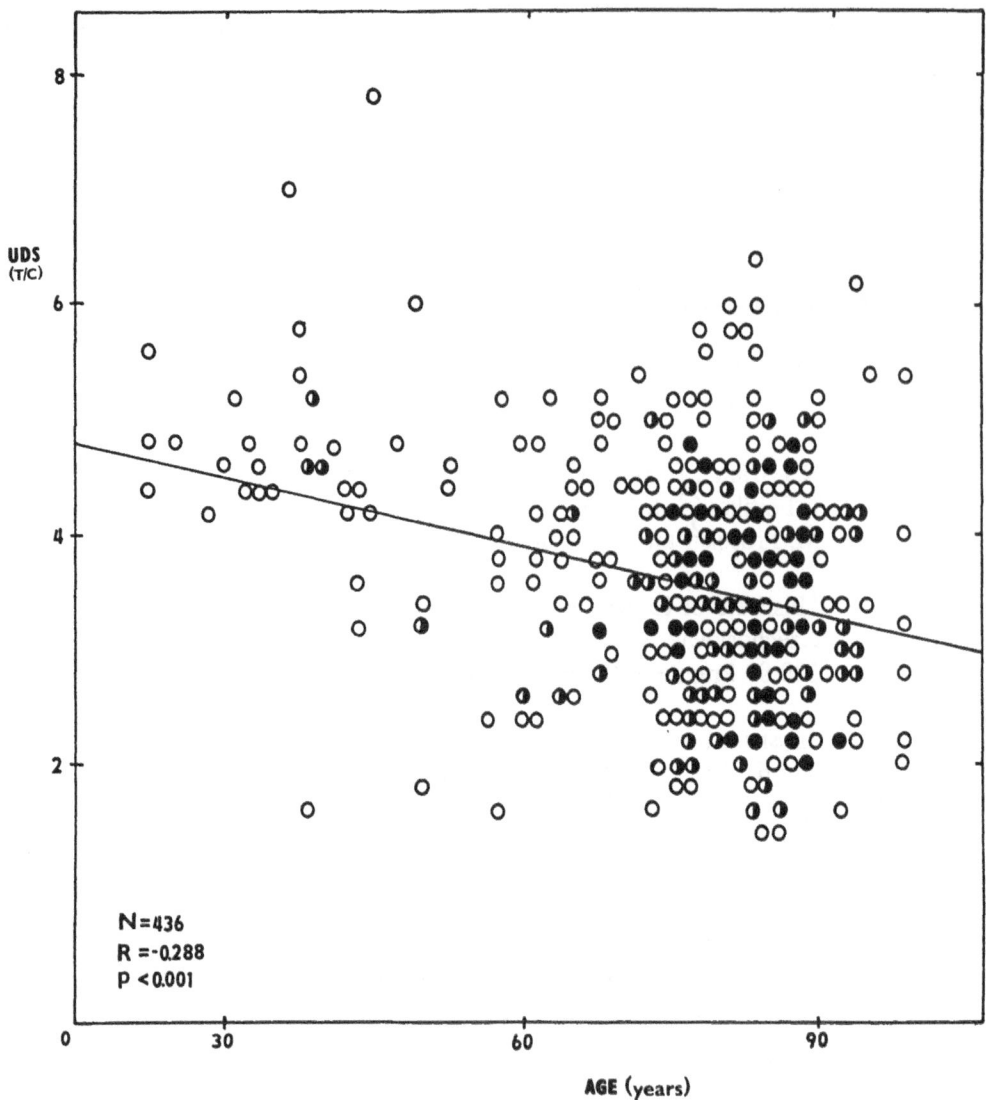

Figure 1. Effects of UDS supplementation according to age.

catalase, glutathione peroxidase) as superoxide radicals or hydrogen peroxide initiate the chain reaction of lipid peroxidation (LPO) resulting in the structural degradation of biomembranes, as well as oxidative damage to proteins, RNA, and DNA. The activation of these oxidative processes (by the increased concentration of oxygen radicals or the decreased activity of protecting enzymes or by the deficit of natural antioxidants) may induce gradual cellular degeneration and damage (1).

Present knowledge indicates the significance of oxidative damage to humans, and the connection between the processes of lipid peroxidation and DNA repair

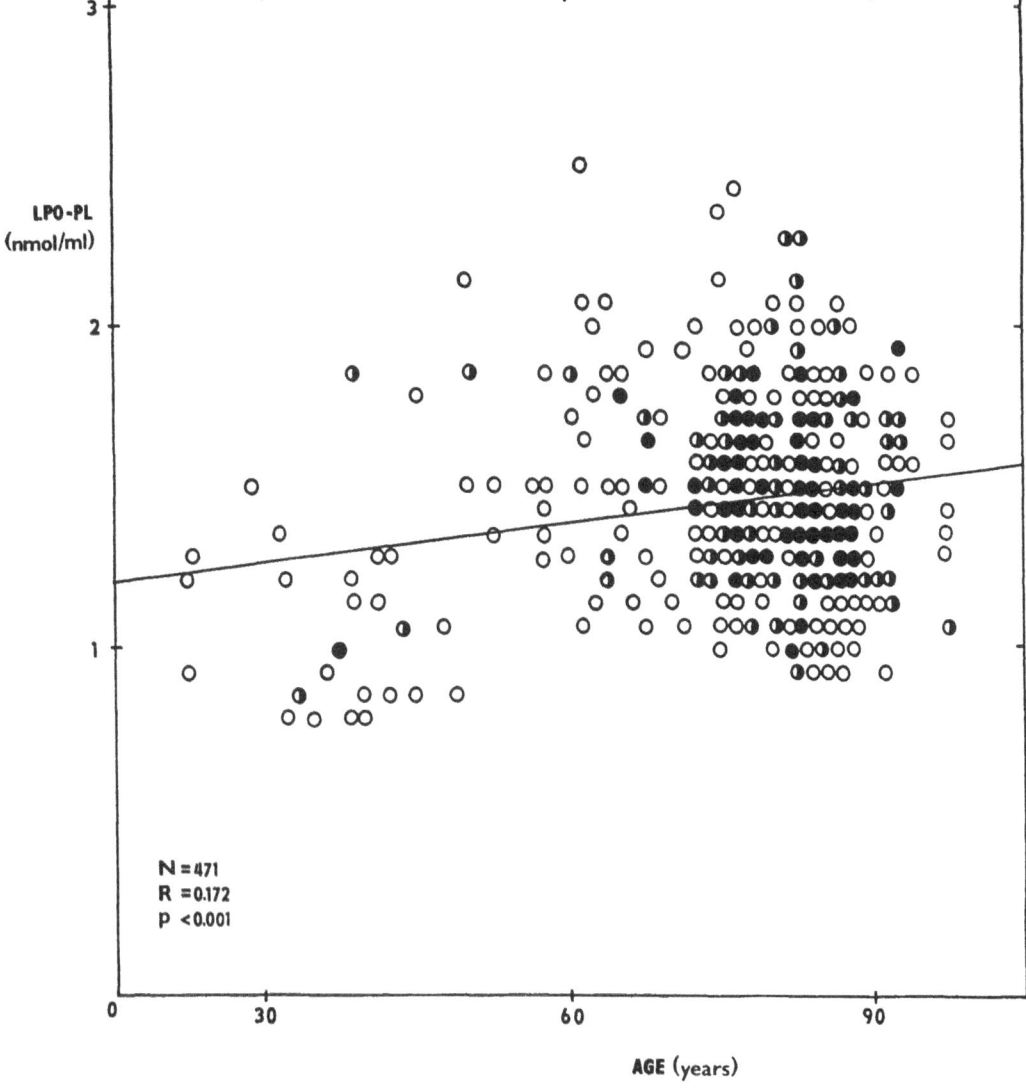

Figure 2. Effects of LPO-PL supplementation according to age.

mechanisms as processes that may speed up the aging process, as well as the course of several diseases (e.g., cancer, atherosclerosis, hypertension, Alzheimer's disease) (2).

The impact of free oxygen radicals on the aging process should be related to the age-dependent increase of LPO and the decrease of DNA repair. It has been repeatedly postulated that oxidative damage may be suppressed by antioxidants acting as free radical scavengers (3,4,5,). Therefore, we have studied the effects of long-term supplementation by α-tocopherol and ascorbic acid in elderly males and females.

Table 1. Cytogenetic analysis of peripheral lymphocytes in elderly population--effect of 12 months of antioxidant (AO) supplementation

AO	N	ANALYSED CELLS	% AB.C.	B'	B''	E'	E''	B/C
–	80	7 461	1.70	115	15	3	1	0.018
				(1.54)	(0.20)	(0.04)	(0.01)	
+	70	6 695	1.61	93	13	2	0	0.016
				(1.39)	(0.20)	(0.03)		

%AB.C.-%aberrant cells, B'—chromatid breaks, B''—chromosome breaks, E'— chromatid exchanges, E''—chromosome exchanges, B/C—breaks per cell; number in the second line indicate the number of aberrations per 100 cells

MATERIALS AND METHODS

Subjects

The subjects for the study were chosen from males and females in two homes for old people. All were between 60 and 90 years of age. They were placed in these homes because of an incapacity to take care of their personal requirements due to age and various diseases. They were interviewed and voluntarily agreed to blood sampling (20 ml). All inhabitants in the home for old people A were under antioxidant therapy; all inhabitants in the home for old people B were matching controls. In home A, they were supplemented for 24 mos with antioxidants in the form of Celaskon effervescens Spofa (L-ascorbic acid) at 1,000 mg per day, and with vitamin E Spofa (α-tocopherol) at 300 mg per day.

From each home, 80 residents who passed the first clinical interview were selected for psychological tests. All of them were mentally normal.

Scheme of Study

The effect of antioxidant therapy was followed after 0, 3, 6, 12, 18, and 24 mos. Specific results analyzed were: unscheduled DNA synthesis, lipid peroxidation, and vitamin level. In the same period, psychological tests were used.

Table 2. Effect of 12-month antioxidant supplementation on UDS, LPO, and vitamin levels in elderly populations

SEX	MALES		FEMALES	
ANTIOXIDANTS	+	−	+	−
N	32	21	26	54
AGE	74.3	73.4	78.6	80.2
s.d.	10.2	9.7	8.6	8.2
UDS (T/C)	4.11	3.91	3.92	3.74
s.d.	0.64	0.85	0.85	0.68
LPO PLASMA				
(nmol MDA/ml)	1.35^{**}	1.55	1.37^{**}	1.54
s.d.	0.21	0.22	0.21	0.26
LPO LYMPHOCYTES				
(nmol MDA/mg)	1.45^{***}	1.68	1.42	1.61
s.d.	0.21	0.21	0.24	0.32
ASCORBIC ACID				
(mg/l)	17.8^{***}	2.2	18.1^{***}	3.6
s.d.	4.6	1.9	5.2	2.4
α−TOCOPHEROL				
(mg/l)	13.4^{*}	7.6	18.1^{*}	9.5
s.d.	4.9	3.2	8.7	3.4
RETINOL				
(mg/l)	0.59	0.47	0.63	0.59
s.d.	0.18	0.28	0.19	0.17
VITAMIN B 12				
(ng/l)	187	207	199	169
s.d.	266	184	218	146

* $p < 0.05$

** $p < 0.01$

*** $p < 0.005$

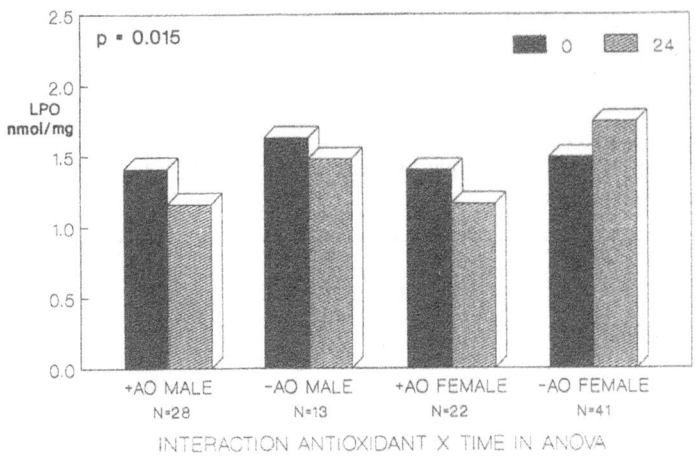

Figure 3. Twenty-four months' AO-supplementation LPO lymphocytes.

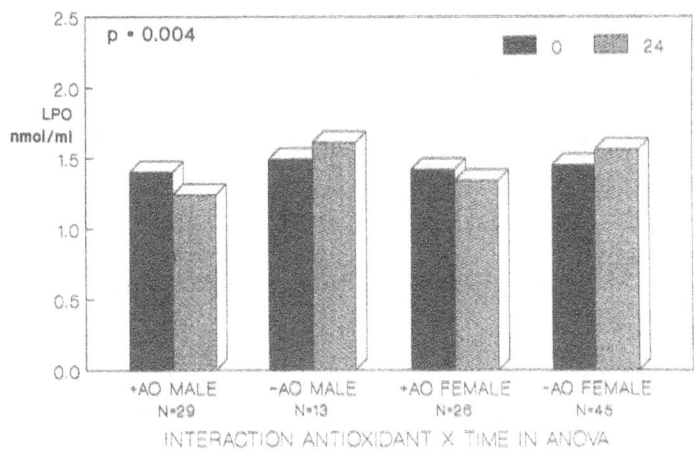

Figure 4. Twenty-four months' AO-supplementation LPO plasma.

Cytogenetic Analysis

The cytogenetic studies were done on short-term lymphocyte cultures as described by Hungerford (6). The scheme of cultivation and the evaluation of chromosome aberrations were as reported by Šrám et al. (7).

Unscheduled DNA Synthesis

The level of the oxidative damage of DNA was estimated by the scintillometric measurement of in vitro unscheduled DNA synthesis (UDS) induced by alkylating agent 1-methyl-3-nitro-1-nitrosoguanidine (MNNG). DNA from cells was isolated as described by Martin et al. (8). The incorporated radioactivity was measured on a Beckman LS 5801 liquid scintillation counter. The concentration of DNA was

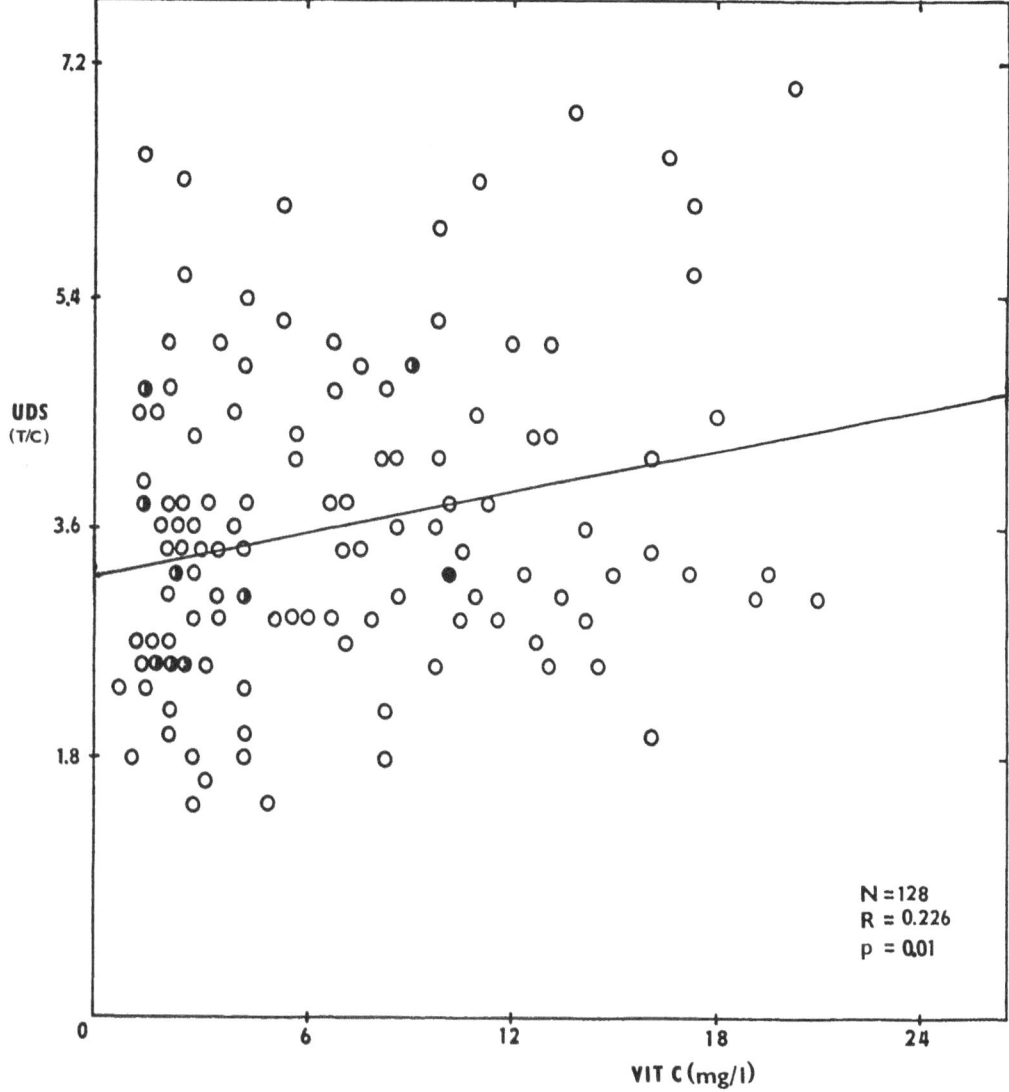

Figure 5. Effect of UDS supplementation in the presence of vitamin C.

estimated by the diphenylamine method. Specific activity of samples was expressed as CPM/μg DNA and calculated for all treated (T) and control (C) samples. The ratio T/C gives information about the increased incorporation of the radiolabeled nucleoside (methyl- H)-thymidine as the consequence of the DNA damage (the excision repair).

Lipid Peroxidation

The LPO level in lymphocytes and plasma was determined by the modified thiobarbituric acid (TBA) assays (9,10). The obtained values were expressed as μmol of MDA/mg proteins or μmol MDA/ml plasma, using 1,1,3,3,-tetraethoxypropane as a standard. The content of the proteins in the suspension of lymphocytes was determined according to Bradford (11).

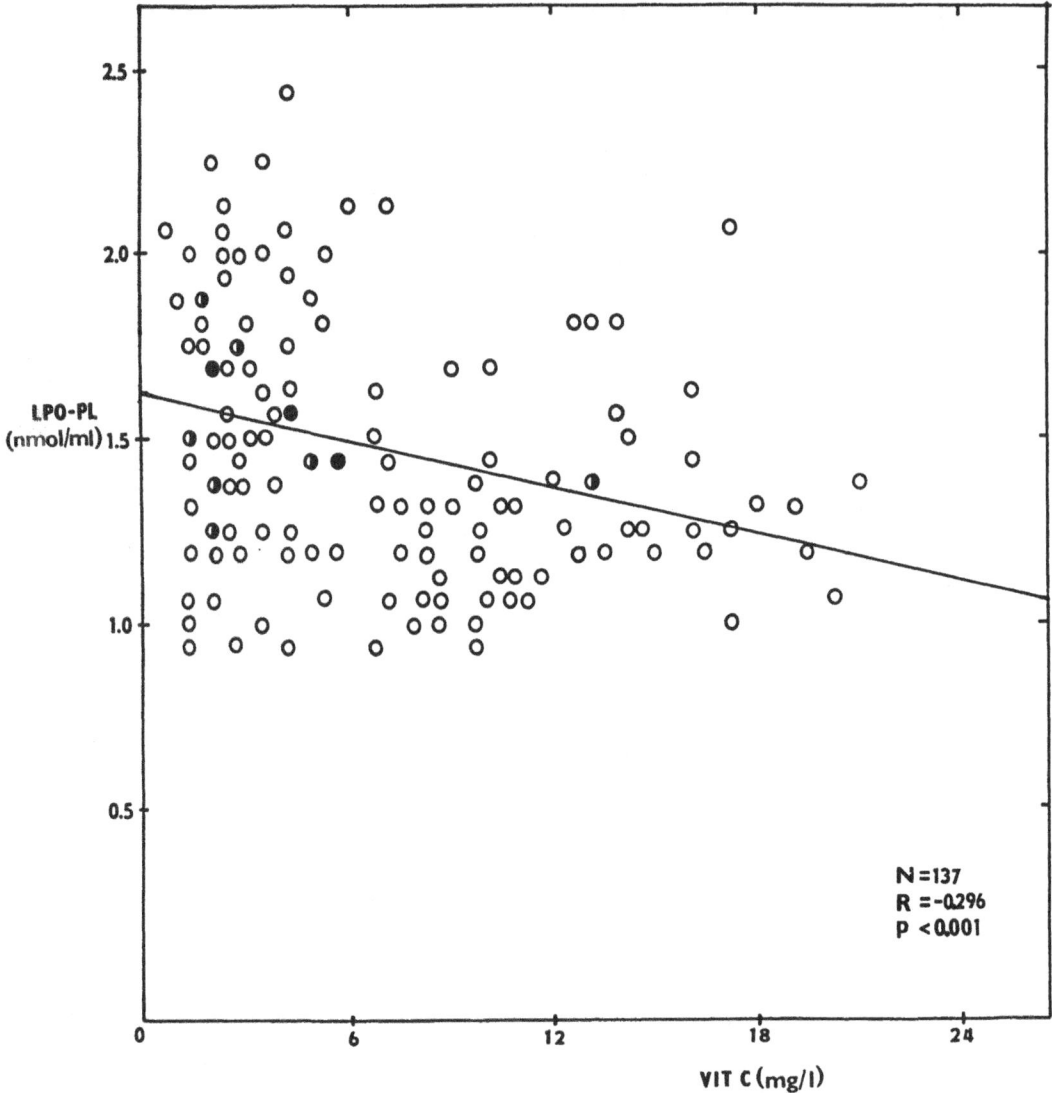

Figure 6. Effects of LPO-PL in the presence of vitamin C.

Vitamin Analysis

Ascorbic acid was estimated by a colorimetric method at 500 nm using its reaction with 2,4-dinitro-phenylhydrazine. α-Tocopherol (E-vit) and retinol (A-vit) in plasma were determined after extraction into n-heptan by an HPLC method (12). Plasma level of vitamin B-12 was estimated according to the competitive protein-binding analysis with RIA-set (Laborchemie, GDR).

Humoral Immunity

The concentrations of IgA, IgM, IgG, α_2-macroglobulin, arosomucoid, and prealbumin were determined by the single radial immunodiffusion method (13). The concentration of C-reactive proteins (C3) was determined by the semiquantitative capillary precipitation reaction (14).

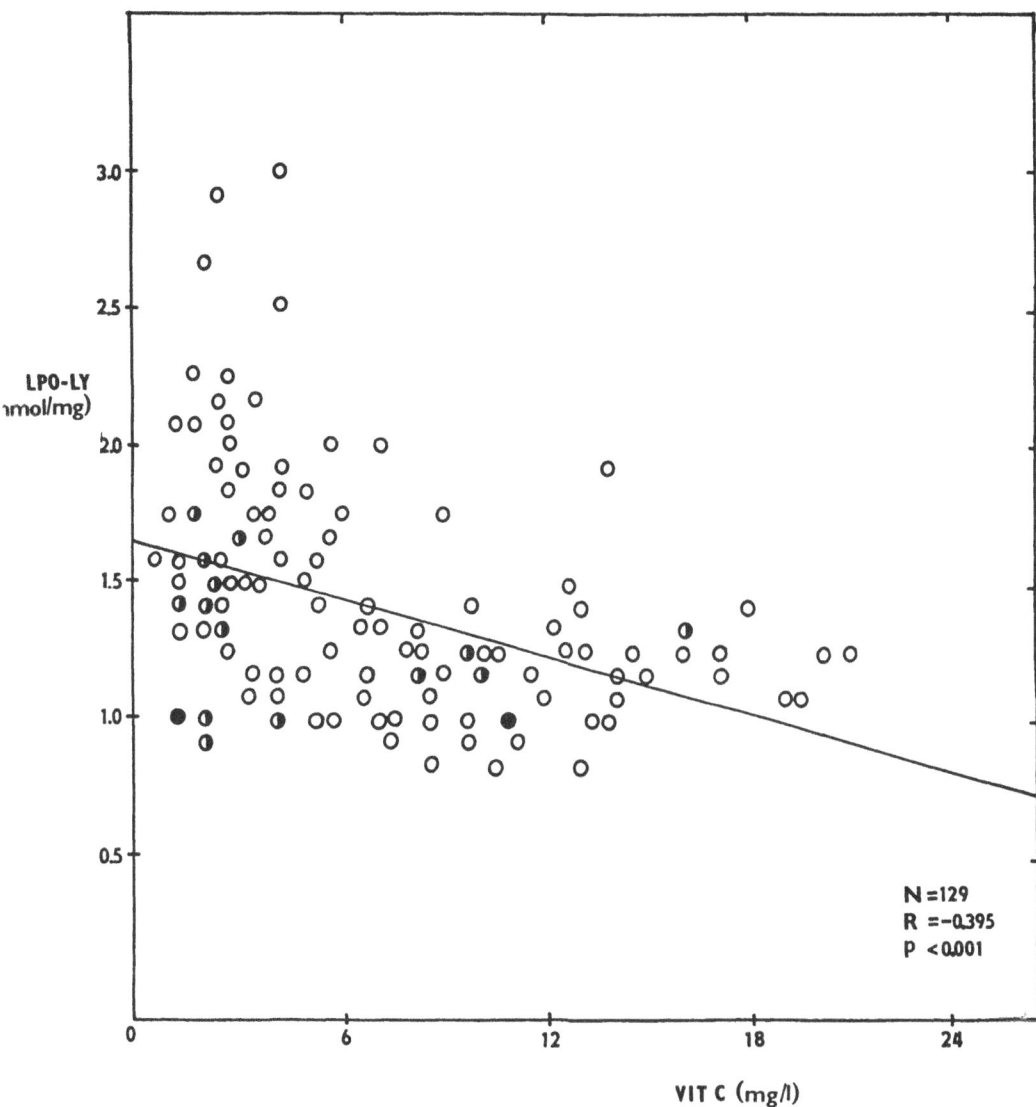

Figure 7. Effects of LPO-LY in the presence of vitamin C.

Psychological Tests

We analyzed digit span memory, verbal memory, drawing test, and subjective symptoms in those subjects who passed the first clinical interview. All residents were examined by the Crichton geriatric rating scale:

a) Digit Span - subtest from Wechsler-Bellevue test, version WAIS was used: test I, repeating the digits forward and test II, repeating the digits backward.

b) Verbal Memory - subtest from Wechsler-Bellevue test, version WAIS was used. It is a short story and its reproduction.

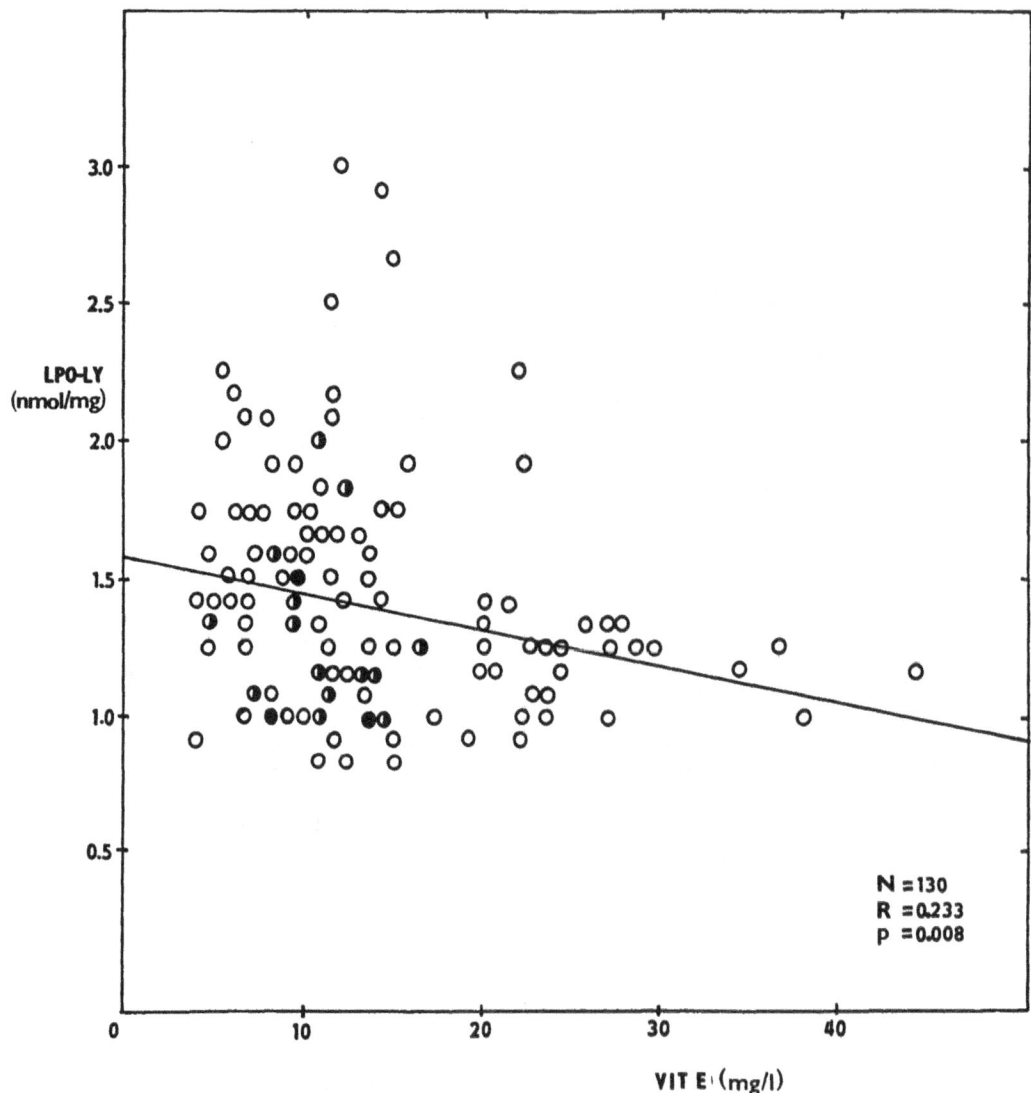

Figure 8. Effects of LPO-LY in the presence of vitamin E.

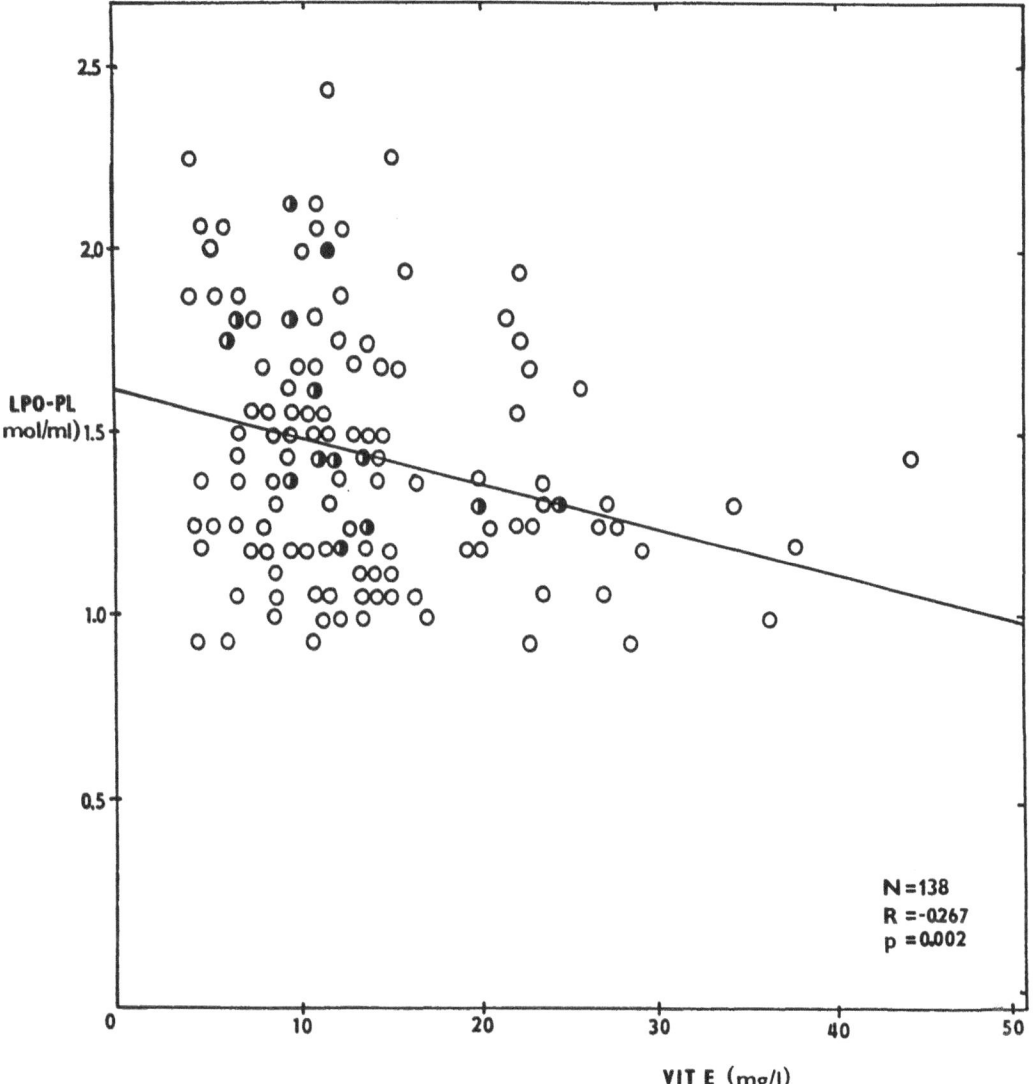

Figure 9. Effects of LPO-PL in the presence of vitamin E.

Table 3. Effect of 12-month antioxidant supplementation on psychological tests in elderly populations

SEX		MALES		FEMALES	
ANTIOXIDANTS		+	−	+	−
Digit	I	4.79**	3.80	4.51*	3.89
Span	s.d.	1.15	0.63	1.00	1.25
	N	34	10	33	27
	II	2.47**	1.22	2.29**	1.22
	s.d.	1.11	1.20	1.16	1.34
	N	34	9	31	27
Verbal		6.73	5.70	7.77***	4.55
Memory	s.d.	2.15	2.75	2.26	2.46
	N	33	10	31	22
Drawing		3.95***	1.15	4.03***	1.03
	s.d.	2.36	1.81	2.42	2.36
	N	38	20	39	33
Subjective		39.1	34.0	50.2	55.3
Symptoms	s.d.	20.2	23.1	19.1	19.3
	N	35	10	30	23
Crichton		30.6**	34.7	32.3**	35.2
	s.d.	8.0	10.4	9.7	9.8
	N	122	61	161	184
Depression		6.15**	7.15	6.31***	7.07
Scorew					
	s.d.	1.43	2.12	1.64	1.96
	N	122	61	161	184

* $p < 0.05$

** $p < 0.01$

*** $p < 0.001$

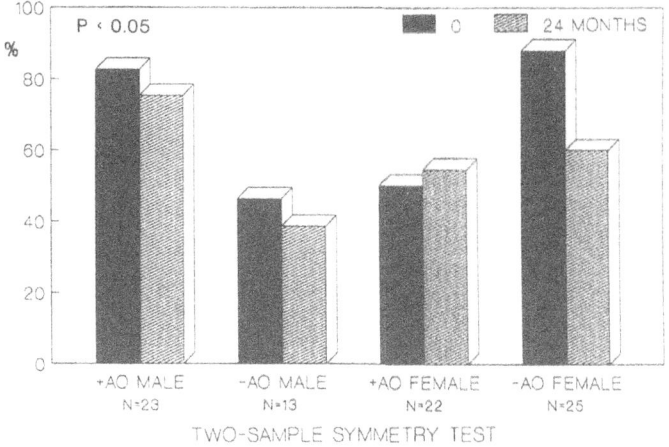

Figure 10. Subjects' ability to psychological test numerical memory I.

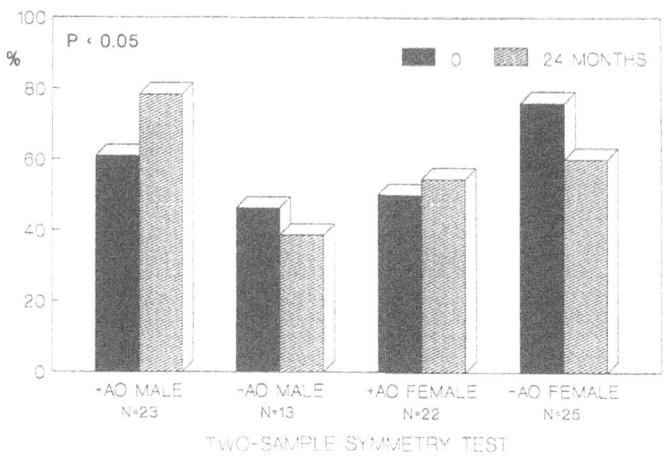

Figure 11. Subjects' ability to psychological test numerical memory II.

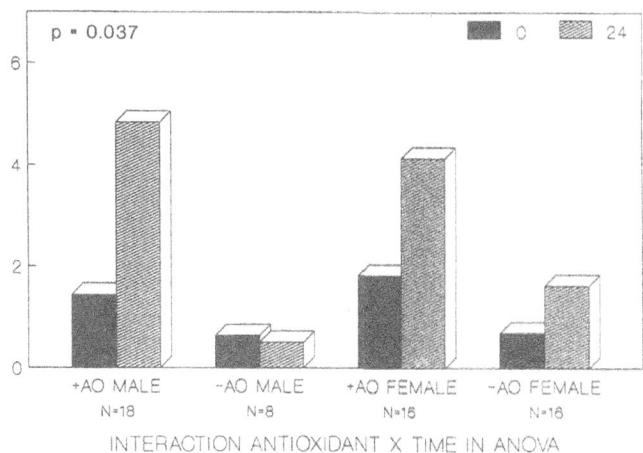

Figure 12. Twenty-four months' AO-supplementation psychological test-drawing.

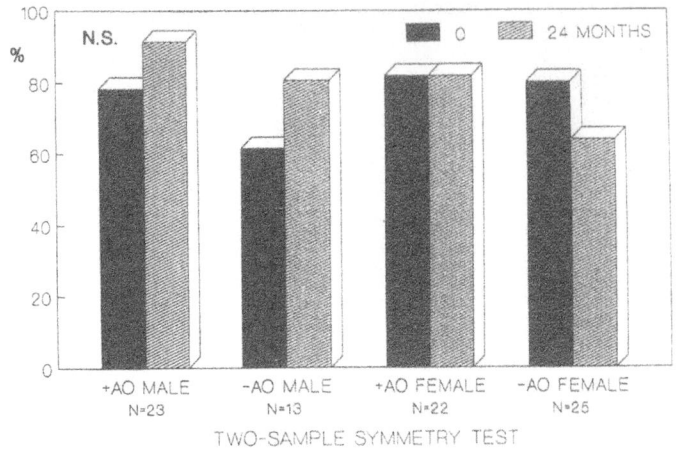

Figure 13. Subjects' ability to psychological test drawing.

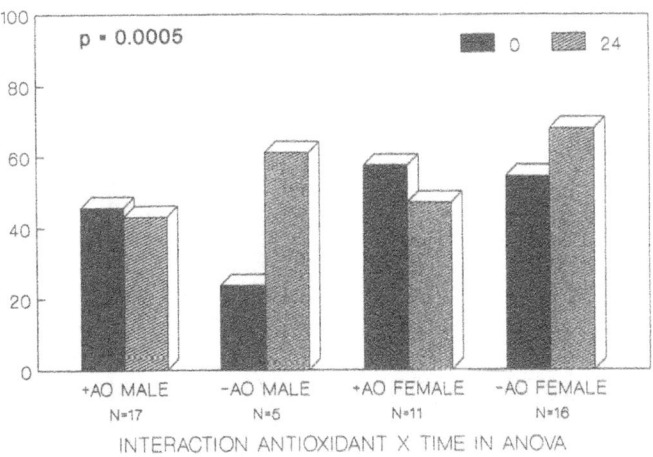

Figure 14. Twenty-four months' AO-supplementation psychological test-subjective feeling.

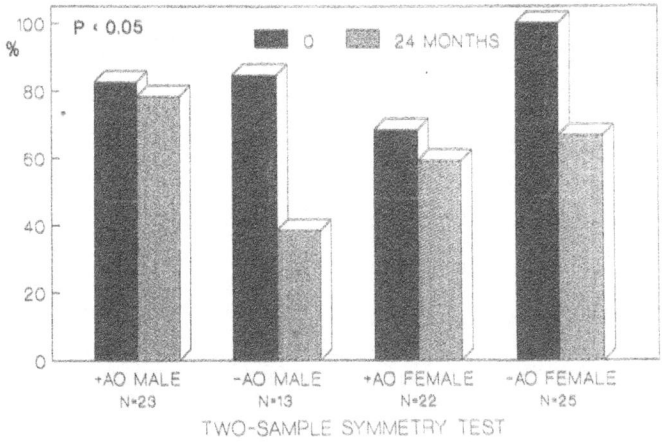

Figure 15. Subjects' ability to psychological test subjective feeling.

Table 4. Effect of 24-month antioxidant supplementation on death rate in elderly populations

A G E	A O	N	D E A T H (%)
< 60	+	14	7.1
	–	8	37.5
60 – 69	+	34	23.5
	–	32	34.4
70 –79	+	87	28.7
	–	64	48.4
> 80	+	107	42.1
	–	134	43.3

A O + administration of α-tocopherol and ascorbic acid
A O – without antioxidants' administration

c) <u>Psychomotor Performance</u> was evaluated according to figure drawing. The test uses a set of three pictures.

d) <u>Subjective Symptoms Registration</u> (subjective feeling). The test is a rating scale system recording pain, fear, mood, eating difficulties, sleep disorders, anger, fatigue, motor difficulties, and memory difficulties.

e) <u>Crichton Geriatric Rating Scale</u> was performed on all subjects in homes A and B. Depression score was calculated as the sum of items 10 and 11 (activity and satisfaction).

Mortality

The total death rate was analyzed according to each month and age group.

Statistical Methods

The intergroup difference were evaluated statistically using Student's paired t-test (Program BMDP - 3-D-87, Dixon et al., 15) and by the two-sample symmetry test (16).

Table 5. Effect of 24-month antioxidant supplementation on death rate in elderly populations according to sex and the year of death

A G E	SEX	A O	N	DEATH (%) 1988	DEATH (%) 1989
< 60	M	+	10	10.0	0
		−	3	66.7	0
	F	+	4	0	0
		−	5	20.0	0
60 – 69	M	+	24	12.5	20.8
		−	18	16.7	11.1
	F	+	10	0	0
		−	14	28.6	14.3
70 – 79	M	+	45	15.6	13.3
		−	19	26.3	21.1
	F	+	42	14.3	14.3
		−	45	17.8	31.1
80 +	M	+	40	37.5	12.5
		−	19	26.3	15.8
	F	+	67	19.4	17.9
		−	115	21.7	21.7

A O + administration of α-tocopherol and ascorbic
        acid
A O − without antioxidants' administration

RESULTS AND DISCUSSION

The relationships between the age and the UDS as well as LPO were analyzed in a group of elderly males and compared to a control group. UDS was decreased in the senescent population by 30% (Fig. 1), LPO in plasma was increased by 25% (Fig. 2), and LPO in lymphocytes by 30%. Regression equations were calculated for all given characteristics (UDS = -0.356 x age + 5.95; LPO in plasma = 0.0061 x age + 0.90; LPO in lymphocytes = 0.0066 x age + 0.94).

LPO and UDS as well as other parameters were followed after 3, 6, 12, 18, and 24 mos of antioxidant supplementation, while cytogenetic analysis was done only after 12 mos. No changes in the frequency of AB.C. were observed (Tab. 1). UDS

Table 6. Effect of 24-month antioxidant supplementation on immunological data in elderly populations

MONTHS	A O +		A O −	
	0	24	0	24
I g A (g/l)	3.44 1.44	3.88 1.75	3.72 1.74	4.19 2.70
I g M (g/l)	1.43 0.71	1.08 0.84	1.68 0.85	1.56 1.55
I g G (g/l)	13.09 4.20	14.72 5.21	14.19 4.43	14.93 6.03
C 3 (g/l)	0.77 0.24	0.97 0.21	0.79 0.21	0.81 0.15
α 2 − M A C (g/l)	2.85 0.99	3.22 1.06	2.79 0.89	3.37 0.77
OROSOMUCOID (g/l)	0.92 0.32	0.93 0.32	0.89 0.31	0.84 0.33
PREALBUMIN (g/l)	0.25 0.08	0.26 0.06	0.35 0.16	0.28 0.09

A O + administration of α-tocopherol and ascorbic acid
A O − without antioxidants' administration

was increased after 3 mos. vitaminization, but there were no significant changes between the vitaminized and nonvitaminized groups after 12 mos (Tab. 2), as well as after 24 mos using ANOVA. The most significant changes were observed in LPO levels in plasma and lymphocytes (Tab. 2, Fig. 3, and Fig. 4). This effect can be observed only after 6 mos. of vitaminization. There were also continual increases in the levels of ascorbic acid and α-tocopherol. Expected decrease in B12 vitamin level due to ascorbic acid in the vitaminized group was not confirmed (Tab. 2). No changes were observed in the cholesterol or triglyceride levels or in hematological data (leu, lymph, ery, Hb, Hct, MCV).

UDS was related to the level of vitamin C in serum (Fig. 5), LPO in plasma and lymphocytes to the concentration of vitamin C (Figs. 6 and 7), as well as of vitamin E (Figs. 8 and 9).

The improvement in vitamin C and E levels was related to improvement in psychological tests. The antioxidant supplementation for 12 mos. significantly improved the short-term memory (digit span) in males and females, verbal memory in females, as well as drawing test results, indicating sensomotor abilities. Also the score on the Crichton scale was improved. Particularly interesting may be the improvement of the depression score (Tab. 3).

When these scales are compared in the same subjects before and after supplementation for 24 mos, antioxidants did not affect the short-term memory as much as verbal memory, but improved subjects' ability to pass these tests (Figs. 10 and 11). Vitamin supplementation significantly improved results of drawing tests (Figs. 12 and 13), indicating sensomotor abilities, and of the test of subjective feeling (Figs. 14 and 15). The ability of vitamin supplementation to improve cognitive and emotional functions was observed if vitaminization lasted for a period longer than 6 mos. This supports the idea that shorter supplementation cannot prevent an increase in either cognitive impairment or behavioral disturbance (17).

During 24 mos. in home A, of the subjects examined by the psychological test, 22 inhabitants died and 8 cases were evaluated as in the advanced stages of dementia. In home B, 30 inhabitants died and 12 cases were evaluated as in the advanced stages of dementia.

The death rate was significantly lower in the group aged 60 to 79 years (Tabs. 4 and 5). Comparing mortality by month, there was a notable difference during March and April 1989, which was related to the influenza epidemic. The data indicate that antioxidants could unspecifically improve the patients' immunological defense to viruses. However, no differences in the immunological data were observed (Tab. 6).

CONCLUSION

Long-lasting antioxidant supplementation by vitamin C (1,000 mg per day) and vitamin E (300 mg per day) was followed in two homes for old people for 24 mos. In home A, 279 subjects were supplemented. In home B, 243 subjects were used as matching controls. Eighty subjects were selected from each home for psychological tests. Approximately 70 blood samples were collected from each group for biochemical analysis.

Our results indicate that long-term antioxidant supplementation, lasting for more than 6 mos, may decrease the LPO process in aging organisms and thus decrease the DNA damage induced by free radicals. It is speculated that it may also improve cognitive and emotional functions in elderly populations.

It may also be expected that antioxidant supplementation could be useful in the prevention of free radical damage in younger organisms when induced changes are not irreversible.

REFERENCES

1.   Simic, M.G., D.S. Bergtold, and L.R. Karam (1989) Generation of oxyradicals in biosystems. Mutat. Res. 214:3-12.

2. Ames B.N. (1983) Dietary carcinogens and anticarcinogens (oxygen radicals and degenerative diseases). Science 221:1256-1264.

3. Šrám, R.J., M. Černá, and N. Holá (1986) Effect of ascorbic acid prophylaxis in groups occupationally exposed to mutagens. In Genetic Effects and Applied Mutagenesis, C. Ramel, B. Lambert, and J. Magnusson, eds. Alan R. Liss, New York, pp. 327-335.

4. Ames, B.N. (1989) Endogenous DNA damage as related to cancer and aging. Mutat. Res. 214:41-46.

5. Dowson, J.H. (1989) Drug treatment for cognitive impairment due to aging and disease: Current and future strategies. Int. J. Ger. Psychiatry 4:345-353.

6. Hungerford, D.A. (1965) Leucocytes cultured from small inocula of whole blood and the preparation of metaphase chromosomes by treatment with hypotonic KCl. Stain Technol. 40:333-338.

7. Šrám, R. J., L. Dobiáš, A. Pastorková, P. Rossner, and L. Janča (1983) Effect of ascorbic acid prophylaxis on the frequency of chromosome aberrations in the peripheral lymphocytes of coal tar workers. Mutat. Res. 120:181-186.

8. Martin, C.N., A.C. McDermid, and R.C. Garner (1978) Testing of known carcinogens for their ability to induce unscheduled DNA synthesis in HeLa cells. Cancer. Res. 38:2621-2627.

9. Yagi, K. (1976) A simple fluorometric assay for lipoperoxide in blood plasma. Biochem. Med. 15:212-216.

10. Ohkawa, H., N. Ohishi, and K. Yagi (1979) Assay for lipid peroxides in animal tissues by thiobarbituric acid reaction. Analyt. Biochem. 95:351-358.

11. Bradford, M.M. (1976) A rapid sensitive method for quantitation of microgram quantities of protein utilizing the principle of protein-dye binding. Analyt. Biochem. 72:248-254.

12. Driskell, J.W. (1982) Measurement of vitamin A and vitamin E in human serum by high-performance liquid chromatography. J. Chromatogr. 231:439-444.

13. Manciny, C., A.O. Carbonara, and J.F. Heremans (1965) Immunochemical quantitation of antigens by single radial immunodiffusion. Immunochemistry 2:235-254.

14. Kwapinski, J.B.G. (1972) Methodology of Immunochemistry and Immunological Research. Wiley, New York, pp. 324-325.

15. Dixon, W.J. (1985) BMDP Statistical Software, 1985 Printing, University of California, Berkeley, CA, 733 pp.

16. Havranek, T., and G.A. Lienart (1986) Remission Control of Pre-Post Treatment Comparisons by Two-Sample Symmetry Testing. Inf. Med. 25:116-122.

17. Burns, A., A. Marsch, and D.A. Bender (1989) A trial of vitamin supplementation in senile dementia. Int. J. Ger. Psychiatry 4:333-338.

SPONTANEOUS MUTATIONS AND FIDELOGENS

R.C. von Borstel and Ursula G.G. Hennig

Department of Genetics
University of Alberta
Edmonton, Canada T6G 2E9

TERMINOLOGY

The word "antimutagenesis" was first used by Novick and Szilard (51) to describe the lowering of the spontaneous mutation rate. Since then, Kada (26) has made antimutagenesis into a generic word to describe two types of events: the inactivation of mutagens or carcinogens before they can reach the DNA, and the events which can restore the DNA after lesions have occurred. J.W. Drake suggested the word "desmutagens" to Kada for mutagen inactivators, and P.J. Hastings since has suggested "countermutagens" as a synonym, and perhaps a more appropriate term. "Inhibition" of the spontaneous mutation rate, which Novick and Szilard were talking about, has plodded along without a species name, although Kada et al. (27) used the term "bioantimutagenesis" for the types of reactions which act at the level of the DNA, including both inhibition of inducibility of mutagenic repair processes and the lowering of the spontaneous mutation rate. The lowering of the spontaneous mutation rate deserves a name of its own, perhaps reflecting the essential quality of improving the fidelity with which DNA normally is replicated and repaired. Steinberg has recommended the term "fidelogen" as an alternative word for "bioantimutagen." Fidelogens cover three classes, the "rate droppers," which lower the spontaneous mutation rate, the "gash healers," which reverse the lesions inducted by mutagens and carcinogens, and the "SOS stoppers," which eliminate inducibility of mutagenic DNA repair. The perfect fidelogen is one that will carry out at least the first two of these three functions.

Endogenous processes that reverse lesions by lowering the induced mutation frequency fit the general category of DNA repair. For survival it is known as "liquid-holding recovery" (LHR).

DNA repair with respect to mutations instead of survival was first recognized as "mutation frequency decline" (MFD), most notably seen when split doses of mutagens were used, and the time between the first and second doses permitted recovery so that the split dose did not add up to the acute dose. Also, mutation frequency decline can be demonstrated by holding cells in culture, after a single treatment with a mutagen, for different lengths of time before plating. As the holding time is increased after exposure to a mutagen, a lowering of induced mutation frequency is observed when compared with those cells plated immediately. For chromosome breaks, the split-dose effect was known as chromosome rejoining, which inevitably could lead to chromosomal aberrations of all types, excluding only aneuploidy and polyploidy. So, DNA repair could have been deduced at the time the first chromosomal translocation was discovered in Drosophila by Alfred Sturtevant back in the legendary days (believed to have been in 1922).

DNA repair became subdivided into error-prone DNA repair and error-proof (= error-free) DNA repair by Witkin (73). Error-prone DNA repair is the process which turns lesions in the DNA into mutations, and error-free DNA repair is the process which brings about mutation frequency decline. More general and more descriptive are the words mutagenic (= error-prone) DNA repair and nonmutagenic (= error-free) DAN repair, first suggested by Hastings et al. (22).

The inducibility of noninducibility of any one DNA repair system is species specific, and not a general phenomenon. For example, SOS repair of DNA was named by Radman as the inducible response to the mutagenic DNA repair system in *Escherichia coli*. Inducibility of recombination (= double-strand break) repair of DNA in yeast was demonstrated by the elegant experiment by Fabre and Roman (16) who used the reversion of two conditional mutants as the assay for recombination. Excision repair (of a short polynucleotide) and recombination repair had already been shown to be the principal nonmutagenic repair systems in yeast by Game and Cox (19) and Brendel and Haynes (5). Mutagenic DNA repair in yeast is mainly constitutive, in contradistinction to the inducible mutagenic DNA repair in *E. coli*. On the whole, the excision-repair system seems not to be inducible in either yeast or *E. coli*, even though some of the genes in yeast have inducible mRNA elements (see, for example, Ref. 18). It came as somewhat of a surprise when Samson and Cairns (59) described another inducible DNA repair system in *E. coli*, the so-called mismatch repair system, another kind of nonmutagenic excision DNA repair pathway where one base is excised rather than having a short polynucleotide removed. They called this "adaptive repair," because of its inducibility, but it is now known more formally as mismatch repair, and is one of two types of excision repair. Mismatch repair is difficult to demonstrate in yeast, taking place at the first postmeiotic division for certain (29), and, possibly, by glycosylation in vegetative cells at any time (7).

In contrast to yeast, recombination repair is weak or absent in *E. coli* (that is why a double-strand DNA break is usually lethal in bacteria and bacteriophages). If recombination repair is present in a prokaryote, it generally is not inducible.

THE ORIGIN OF SPONTANEOUS MUTATIONS

Likely Origins

Historically, the most likely origins of spontaneous mutations were established during the decade of the 1960s. These are the three Rs of DNA metabolism, namely, recombination, replication, and repair. The likely origins of spontaneous mutations are listed in Table 1.

Recombinations. When it was found that spontaneous mutations took place at a higher rate during meiosis than during mitosis (41), it was only natural to suspect that crossing over itself was the factor in mutation rate enhancement. This was verified by the demonstration that each reversion of a homozygous marker exhibited an exchange of outside markers (39). Moreover, it was soon found that reversions of base substitution mutations did not exhibit the "meiotic effect" (40,69), implying that frameshift mutations were caused by the meiotic effect. All mutants showing the strongest meiotic effect were then assumed to be frameshift mutations, and this has turned out to be true for the one mutant which has been studied thoroughly (71).

Replication. Mutants of the gene encoding DNA polymerase in bacteriophage T4 were shown to have an enhanced spontaneous mutation rate (61). Moreover, it also was shown that some of the mutant alleles of the gene encoding this same DNA polymerase exhibited a decreased spontaneous mutation rate (13). Loeb (34) and Loeb and Cheng (35) have provided an argument for DNA replication errors being a cause of cancer.

DNA Repair. A number of investigators working with a variety of organisms demonstrated independently, and almost at the same time, that some repair-deficient mutants were mutators in *Proteus mirabilis* (4), *Neisseria meningitides* (25), *E. coli* (47), and *S. cerevisiae* (70,74). It is worth noting that the mutator activity of repair-deficient mutants had been predicted by Hanawalt and Haynes (21).

Deamination of Hydroxymethyl Cytosine. One of the likely sources of spontaneous mutations is the deamination of 5-methylcytosine in those species where this methylated base is present in the DNA. 5-Methylcytosine is present in the DNA of *E. coli* and mammalian cells, but it is not present in yeast in detectable quantities. When 5-methylcytosine is deaminated, it will pair with adenine, and thus the location of 5-methylcytosine can be a hotspot for mutations (11). Baltz et al. (3) have shown that mutations occur in phage T4 when the phage is stored in the cold for extended periods of time. This is caused by spontaneous deamination of 5-hydroxymethylcytosine, but there is no clear understanding of what brings about that

Table 1. Some likely origins of spontaneous mutation and cancer initiation

Likely Origin	References
DNA Recombination	41
DNA Replication	30, 61
DNA Repair	4
Deamination of methylcytosine	11, 14, 72
Intracellular oxidative reactions and peroxidations	1, 2, 10, 17, 28, 42, 46, 62
Depurination	33, 36, 57

Table 2. Some suggested but unlikely origins of spontaneous mutation and cancer initiation

Possible Origin	Likelihood* Yes	No	References
Imbalance of deoxynucleoside and deoxynucleotide pools	x		23, 38
Intracellular alkylations		x	37
Intracellular nitrosations		x	44
Intracellular aldehyde formation		x	20
Estrogens	x		32
Reducing sugars (e.g., glucose and glucose 6-phosphate)		x	31
Endogenous and exogenous radiation		x	50, 65

*Likelihood is suggested because it is known that imbalance of deoxynucleoside pools is either recombinagenic or mutagenic, and it is known that an oxidized estrogen analog, diethylstilbestrol, is highly mutagenic (68). Lack of likelihood is defined as being the state where a mutant that results in known mutator activity has yet to be found, or, in the case of background radiation, is too low to ever be detectable biologically.

"spontaneous" action, although Wink et al. (72) have made a good case for endogenous nitrous oxide.

Free Radicals, Oxidation, and Peroxidation Reactions. It has long been known that oxygen can induce chromosome breakage (10). Kasai and Nishimura (28) demonstrated that one of the principal oxidatively damaged forms of liver DNA contains 8-hydroxyguanine, but since the compound is oxidized, it is better called 8-oxo-7,8-dihydro-2'-deoxyriboguanine (8-OxodG) (64). Deoxyadenosine tends to pair with template 8-OxodG in vitro forming G-C → T+A transversions (60), and Michaels et al. (46) have shown that the lack of the MutM protein of *E. coli* formamido-pyrimidine-DNA glycosylase, results in mutator activity. Moreover, with respect to oxidative radicals, Storz et al. (62) have demonstrated that strains of *Salmonella typhimurium* with a deletion of a positive regulator for reductases and superoxide dismutases have an enhanced spontaneous mutation rate. Consequently, one can conclude that oxidative damage to DNA is likely to be a source of some spontaneous mutations.

Apurinic (AP) Lesions. DNA containing apurinic lesions was first described by Tamm and Chargaff (63). These are removed by exonucleases and endonucleases. The search for mutator activity in strains of *E. coli* deficient for exonuclease III was reported not to have been successful (36), yet exonuclease III (the major AP endonuclease) is believed to be responsible for eliminating 90% of the apurinic (AP) lesions. Nevertheless, Cunningham and Weiss (12) report one experiment that indicates that the mutant exonuclease III strain is indeed a mutator, and still contains other AP endonuclease activities. The MutY protein of *E. coli* is an adenine glycosylase which is active on G-A mispairs, and it is quite homologous in sequence to endonuclease III which acts at AP sites (45). In yeast, strains deficient in the Apr1 AP endonuclease/3'-repair diesterase exhibit mutator activity (57). Consequently, it can be concluded that AP lesions are likely to be a source of spontaneous mutations.

Possible Origins

Other metabolic events have been suggested as sources of spontaneous mutations (Table 2): Estrogen (32), nucleotide imbalance (38), metabolic alkylating agents (14), and metabolic aldehydes (15,20). When radiation was found to be mutagenic (49), numerous individuals immediately believed that background radiation might be the source of spontaneous mutations.

The estrogen analog diethylstilbestrol is a special case; it is a powerful mutagen in yeast only when oxidized (43). Since mammals are highly oxygenated creatures, the oxidation of at least a few molecules of diethylstilbestrol is inevitable. These molecules will go directly to cells with estrogen receptors. It seems reasonable to assume that the genotoxic action of the oxidized diethylstilbestrol on these cells will initiate cancer.

Nucleotide imbalances, which can be caused by drugs of the sulfonamide family, will bring about increased spontaneous mutation rates. Normally, the nucleotide pools are so well regulated that spontaneous mutations would be infrequent.

Metabolic alkylating agents are aldehydes, as well as other conceivable metabolic mutagens are at the state of understanding where a case is still to be made for their acting as sources for spontaneous mutations.

Background radiation was once thought to be a primary source of spontaneous mutations, but Muller and Mott-Smith (50) demonstrated that background radiation would not suffice, and von Borstel (65) pointed out a way to quantitate the frequency of spontaneous mutations induced by background radiation. It was shown that background radiation induces no more than one mutation per 10^4 or 10^5 of all spontaneous mutations occurring in Drosophila. For normal mammalian cells in culture, it would be at least 10-fold less than that--and that does not even take into account the efficiency of DNA repair.

FIDELOGENESIS AND ITS TRAPS

Antimutator genes have been described in the yeast *S. cerevisiae* (55) and in *E. coli* (56), and antimutator alleles have been described for the gene encoding DNA polymerase in bacteriophage T4 (13). One of the antimutator mutants described in yeast is *rev3* (55). The wild-type locus, *RAV3*, is known to encode a DNA polymerase which is believed to be the DNA polymerase that engages in mutagenic DNA repair (48). It follows that when it is mutant (*rev3⁻*), lesions which would ordinarily be repaired by the mutagenic DNA polymerase either die or are channeled into a nonmutagenic DNA repair system (22). In either case, the spontaneous mutation rate would be lowered.

Seeking perfect fidelogens is literally seeking the Fountain of Youth, if there is any truth to the vast numbers of papers and books which categorically state that aging is a consequence of somatic spontaneous mutations. It is possible that some somatic mutations may be prevented by antioxidants, blocking agents, and P450 cytochromes, but none of these agents appear to lower the spontaneous mutation rate below the control level.

Clarke and Shankel (8,9) and Inoue et al. (24) have listed a number of quite unrelated compounds which can lower the spontaneous mutation rate in *E. coli*. Other compounds have been studied in yeast. (see, for example, Ref. 53 and 54).

The lowering of the spontaneous mutation rate with chemicals is fraught with difficulties, the most important being that of accurate measurement of small effects (see Ref. 66) because most such fidelogenic effects, at least in eukaryotes, are small. Another difficulty is that an agent which reduces the mutation rate may be producing a phenocopy. For example, the salt-induced antimutator effects (52) can perhaps, in part, be ascribed to nonsurvival of some classes of revertants. Moreover, inhibition of mutagenic DNA repair polymerases may cause an otherwise repairable spontaneous lesion to simply kill that cell; the rarity of mutations/cell would make cellular killing by a lesion difficult to distinguish from a spontaneous lesion in the DNA which had been channeled into a nonmutagenic pathway. Further, Rutten and Gocke (58) have made the observation that simple growth inhibition by cinnamaldehyde may account for its "antimutagenic" action against 4-nitroquinoline-*N*-oxide. Also, Cairns (6) has warned against taking mutation frequency decline lightly because strains with the capacity for mutation frequency decline continue DNA repair under holding conditions. This is why we elected two decades ago to use strain XV185-14C for mutagen testing; it does not exhibit mutation frequency decline (67) and thus can be used to detect extremely weak mutagens.

Likewise, the effects observed with antimutator genes (55,56) may be that the cells die because mutagenic DNA repair is missing, and not because the lesion is channeled into nonmutagenic DNA repair pathways (22).

Therefore, great care must be taken in order to distinguish "apparent antimutagenicity" from fidelogen action, and to interpret accurately any possible DNA repair which decreases the level of the normal spontaneous mutation rate.

ACKNOWLEDGEMENTS

This project was supported by operating and strategic grants from the Natural Sciences and Engineering Research Council of Canada and the ILSI Risk Science Institute.

REFERENCES

1. Ames, B.N. (1983) Dietary carcinogens and anticarcinogens. *Science* 221:1256-1264.
2. Ames, B.N., and L.S. Gold (1991) Endogenous mutagens and the causes of aging and cancer. *Mutat. Res.* 250:3-16.

3. Baltz, R.H., P.M. Bingham, and J.W. Drake (1976) Heat mutagenesis in bacteriophage T4: The transition pathway. *Proc. Natl. Acad. Sci., USA* 73:1269-1273.

4. Böhme, H. (1967) Genetic instability of an ultraviolet-sensitive mutant of *Proteus mirabilis. Biochem. Biophys. Res. Comm.* 28:191-196.

5. Brendel, M., and R.H. Haynes (1973) Interactions among genes controlling sensitivity to radiation and alkylation in yeast. *Molec. Gen. Genet.* 125:197-216.

6. Cairns, J. (1986) A summary: And a look ahead. In *Antimutagenesis and Anticarcinogenesis Mechanisms*, D.M. Shankel, P.E. Hartman, T. Kada, and A. Hollaender, eds. Plenum Press, New York, pp. 531-535.

7. Chen, J., B. Derfler, A. Maskati, and L. Samson (1989) Cloning of eukaryotic DNA glycosylase repair gene by the suppression of a DNA repair defect in *Escherichia coli. Proc. Natl. Acad. Sci., USA* 86:7961-7965.

8. Clarke, C.H., and D.M. Shankel (1975) Antimutagenesis in microbial systems. *Bacteriol. Rev.* 39:33-53.

9. Clarke, C.H., and D.M. Shankel (1988) Antimutagens against spontaneous and induced reversion of a *lacZ* frameshift mutation in *E. coli* K-12 strain ND-160. *Mutat. Res.* 202:19-23.

10. Conger, A.D., and L.M. Fairchild (1952) Breakage of chromosomes by oxygen. *Proc. Natl. Acad. Sci., USA* 38:289-299.

11. Coulondre, C., J.H. Miller, P.J. Farabaugh, and W. Gilbert (1978) Molecular basis of base substitution hotspots in *E. coli. Nature* 274:775-780.

12. Cunningham, R.P., and B. Weiss (1985) Endonuclease III (*nth*) mutants of *Escherichia coli. Proc. Natl. Acad. Sci., USA* 82:4740478.

13. Drake, J.W., and E.F. Allen (1968) Antimutagenic DNA polymerases of bacteriophage T4. *Cold Spring Harbor Symp. Quant. Biol.* 33:339-344.

14. Ehrlich, M., X.-Y. Zhang, and M.N. Inamdar (1990) Spontaneous deamination of cytosine and 5-methylcytosine residues in DNA and replacement of 5-methylcytosine residues with cytosine residues. *Mutat. Res.* 238:277-286.

15. Esterbauer, H., P. Eckl, and A. Ortner (1990) Possible mutagens derived from lipids and lip precursors. *Mutat. Res.* 238:223-233.

16. Fabre, F., and H. Roman (1977) Genetic evidence for inducibility of recombination competence in yeast. *Proc. Natl. Acad. Sci., USA* 74:1667-1671.

17. Fridovich, I. (1978) The biology of oxygen radicals. *Science* 201:875-880.

18. Friedberg, E.C. (1984) *DNA Repair*. W.H. Freeman and Company, New York.

19. Game, J.C., and B.S. Cox (1972) Epistatic interactions between four *rad* loci in yeast. *Mutat. Res.* 16:353-362.

20. Grafström, R.C. (1990) In vitro studies of aldehyde effects related to human respiratory carcinogenesis. *Mutat. Res.* 238:175-184.

21. Hanawalt, P.C., and R.H. Haynes (1965) Repair replication of DNA in bacteria: Irrelevance of chemical nature of base defect. *Biochem. Biophys. Res. Comm.* 19:462-467.

22. Hastings, P.J., S.-K. Quah, and R.C. von Borstel (1976) Spontaneous mutation by mutagenic repair of spontaneous lesions in DNA. *Nature* 264:719-722.

23. Haynes, R.H., and B.A. Kunz (1985) A possible role for deoxyribonucleotide pools in carcinogenesis. In *Basic and Applied Mutagenesis*, A. Muhammed and R.C. von Borstel, eds. Plenum Press, New York, pp. 147-156.

24. Inoue, T., T. Ohta, Y. Sadaie, and T. Kada (1981) Effect of cobaltous chloride on spontaneous mutation induction in a *Bacillus subtilis* mutator strain. *Mutat. Res.* 91:41-45.

25. Jyssum, K. (1968) Mutator factor in *Neisseria meningitides* associated with increased sensitivity to ultraviolet light and defective transformation. *J. Bact.* 96:41-45.

26. Kada, T. (1982) Mechanisms and genetic implications of environmental antimutagens. In *Environmental Mutagens and Carcinogens*, T. Sugimura, S. Kondo, and H. Takebe, eds. University of Tokyo Press, Tokyo, pp. 355-359.

27. Kada, T., T. Inoue, T. Ohta, and Y. Shirasu (1986) Antimutagens and their mode of action. In *Antimutagenesis and Anticarcinogenesis Mechanisms*, D.M. Shankel, P.E. Hartman, T. Kada, and A. Hollaender, eds. Plenum Press, New York, pp. 181-196.

28. Kasai, H., and S. Nishimura (1984) Hydroxylation of deoxyguanosine at the C-8 position by ascorbic acid and other reducing agents. *Nucleic Acids Res.* 12:2137-2145.

29. Kramer, W., B. Kramer, M.S. Williamson, and S. Fogel (1989) Cloning and nucleotide sequence of DNA mismatch repair gene PMS1 from *Saccharomyces cerevisiae*: Homology of PMS1 to procaryotic MutL and HexB. *J. Bact.* 171:5339-5346.

30. Kunkel, T.A., and L.A. Loeb (1981) Fidelity of mammalian DNA polymerases. *Science* 213:765-767.

31. Lee, A.T., and A. Cerami (1990) In vitro and in vivo reactions of nucleic acids with reducing sugars. *Mutat. Res.* 238:185-191.

32. Liehr, J.G. (1990) Genotoxic effects of estrogens. *Mutat. Res.* 238:269-276.

33. Lindahl, T. (1990) Repair of intrinsic DNA lesions. *Mutat. Res.* 238:305-311.

34. Loeb, L.A. (1989) Endogenous carcinogenesis: Molecular oncology into the twenty-first century. *Cancer Res.* 49:5489-5496.

35. Loeb, L.A., and K.C. Cheng (1990) Errors in DNA synthesis: A source of spontaneous mutations. *Mutat. Res.* 238:297.304.

36. Loeb, L.A., and B.D. Preston (1986) Mutagenesis by apurinic/apyrimidinic sites. *Ann. Rev. Genet.* 20:201-230.

37. Lutz, W.K. (199) Endogenous genotoxic agents and processes as a basis of spontaneous carcinogenesis. *Mutat. Res.* 238:287-295.

38. MacPhee, D.G., R.H. Haynes, B.A. Kunz, and D. Anderson, eds. (1988) Genetic aspects of deoxyribonucleotide metabolism.. *Mutat. Res.* 200 (special issue):1-256.

39. Magni, G.E. (1963) The origin of spontaneous mutations during meiosis. *Proc. Natl. Acad. Sci., USA* 50:975-980.

40. Magni, G.E. (1964) Origin and nature of spontaneous mutations in meiotic organisms. *J. Cell. Comp. Physiol.* 64 (Suppl. 1): 165-172.

41. Magni, G.E., and R.C. von Borstel (1962) Different rates of spontaneous mutations during mitosis and meiosis in yeast. *Genetics* 47:1097-1108.

42. McBride, T.J., B.D. Preston, and L.A. Loeb (1991) Mutagenic spectrum resulting from DNA damage by oxygen radicals. *Biochemistry* 30:207-213.

43. Mehta, R.D., and R.C. von Borstel (1982) Genetic activity of diethylstilbestrol in *Saccharomyces cerevisiae*: Enhancement of mutagenicity by oxidizing agents. *Mutat. Res.* 92:49-61.

44. Meier, I., S.E. Shepard, and W.K. Lutz (1990) Nitrosation of aspartic acid, aspartame, and glycine ethylester. Alkylation of 4-(*p*-nitrobenzylpyridine) (NBP) in vitro and binding to DNA in the rat. *Mutat. Res.* 238:193-201.

45. Michaels, M.L., L. Pham, Y. Nghiem, C. Cruz, and J.H. Miller (1990) MutY, an adenine glycosylase active on G-A mispairs, has homology to endonuclease III. *Nucl. Acids Res.* 18:3841-3845.

46. Michael, M.L., L. Pham, C. Cruz, and J.H. Miller (1991) MutM, a protein that prevents G.C →T.A transversions, is a formamidopyrimidine-DNA glycosylase. *Nucl. Acids Res.* 19:3629-3632.

47. Mohn, G (1968) Korrelation zwischen verminderter Reparatur-fähigkeit für UV-Läsionen und hoher Spontanmutabilität eines Mutatorstammes von *E. coli* K-12. *Molec. Gen. Genet.* 101:43-50.

48. Morrison, A., R.B. Christensen, J. Alley, A.K. Beck, E.G. Bernstine, J.F. Lemontt, and C.W. Lawrence (1989) *REV3*, a *Saccharomyces cerevisiae* gene whose function is required for induced mutagenesis, is predicted to encode a nonessential DNA polymerase. *J. Bacteriol.* 171:5659-5667.

49. Muller, H.J. (1927) Artificial transmutation of the gene. *Science* 66:85-87.

50. Muller, J.H., and L.M. Mott-Smith (1930) Evidence that natural radioactivity is inadequate to explain the frequency of "natural" mutations. *Proc. Natl. Acad. Sci., USA* 16:277-285.

51. Novick, A., and L. Szilard (1952) Anti-mutagens. *Nature* 170:926-927.

52. Parker, K.R., and R.C. von Borstel (1990) Antimutagenesis in yeast by sodium chloride, potassium chloride, and sodium saccharin. In *Antimutagenesis and Anticarcinogenesis Mechanisms II*, Y. Kuroda, D.M. Shankel, and M.D. Waters, eds. Plenum Press, New York, pp. 367-371.

53. Puglisi, P.O. (1966) Mutagenic and antimutagenic effects of acridine salts in yeast. *Genetics* 54:315-322.

54. Puglisi, P.P. (1967) Mutagenic and antimutagenic effects of acridine salts in yeast. *Mutat. Res.* 4:289-294.

55. Quah, S.-K., R.C. von Borstel, and P.J. Hastings (1980) Spontaneous mutations in yeast. *Genetics* 96:819-839.

56. Quinones, A., and R. Piechocki (1985) Isolation and characterization of *Escherichia coli* antimutators. *Molec. Gen. Genet.* 201:315-322.

57. Ramatar, D., and B. Demple (1991) Biological role of yeast Apn1 AP endonuclease/3'-repair diesterase and functional substitution in yeast by *E. coli* endonuclease IV. Book of abstracts, Symposium on Cellular Responses to Environmental DNA Damage, Banff, 1-6 December 1991, Abstract A-20.

58. Rutten, B., and E. Gocke (1988) The "antimutagenic" effect of cinnamaldehyde is due to a transient growth inhibition. *Mutat. Res.* 201:97-105.

59. Samson, L., and J. Cairns (1977) A new pathway for DNA repair in *Escherichia coli.* *Nature* 267:281-283.

60. Shibutani, S., M. Takeshita, and A.P. Grollman (1991) Insertion of specific bases during DNA synthesis past the oxidation-damaged base 8-oxodg. *Nature* 349:431-434.

61. Speyer, J.F., J.D. Daram, and A.B. Lenny (1966) On the role of DNA polymerase in base selection. *Cold Spring Harbor Symp. Quant. Biol.* 31:693-697.

62. Storz, G., M.F. Christman, H. Sies, and B.N. Ames (1987) Spontaneous mutagenesis and oxidative damage to DNA in *Salmonella typhimurium. Proc. Natl. Acad. Sci., USA* 84:8917-8921.

63. Tamm, C., and Erwin Chargaff (1953) Physical and chemical properties of the apurinic acid of calf thymus. *J. Biol. Chem.* 203:689-694.

64. Tchou J., H. Kasai, S. Shibutani, M.H. Chung, J. Laval, A.P. Grollman, and S. Nishimura (1991) 8-Oxoguanine (8-hydroxyguanine) DNA glycosylase and its substrate specificity. *Proc. Natl. Acad. Sci., USA* 88:4690-4694.

65. von Borstel, R.C. (1969) On the origin of spontaneous mutations. *Japan J. Genet.* 4(Suppl.):102-105.

66. von Borstel, R.C. (1978) Measuring spontaneous mutation rates in yeast. In *Methods in Cell Biology*, D.M. Prescott, ed. Academic Press, New York. Vol 20, pp. 1-24.

67. von Borstel, R.C. (1986) The relation of activation and inactivation to antimutagenic processes. In *Antimutagenesis and Anticarcinogenesis Mechanisms*, D.M. Shankel, P.E. Hartman, T. Kada, and A. Hollaender, eds. Plenum Press, New York, pp. 39-43.

68. von Borstel, R.C., and R.D. Mehta (1982) Nonmutagenic carcinogens. In *Mutagens in Our Environment*, Marja Sora and H. Vainio, eds. Alan R. Liss, Inc., New York, pp. 47-47.

69. von Borstel, R.C., M.J. Bond, and C.M. Steinberg (1964) Spontaneous reversion rates of a super-suppressible mutant during mitosis and meiosis. *Genetics* 50:293 (abstract).

70. von Borstel, R.C., D.E. Graham, K.J. La Brot, and M.A. Resnick (1968) Mutator activity of an X-radiation-sensitive yeast. *Genetics* 60:233 (abstract).

71. Wang, W., U.G.G. Hennig, R.G. Ritzel, E.A. Savage, and R.C. von Borstel (1990) Double-stranded base-sequencing confirms the genetic evidence that the *hom3-10* allele of *Saccharomyces cerevisiae* is a frameshift mutant. yeast 6 (Supplement): S76 (Abstract 02-10A).

72. Wink, D.A., K.S. Kasprzak, C.M. Maragos, R.K. Elespuru, M. Misra, T.M. Dunams, T.A. Cebula, W.H. Koch, A.W. Andrews, J.S. Allen, L.K. Keefer (1991) DNA deaminating ability and genotoxicity of nitric oxide and its progenitors. *Science* 254:1001-1003.

73. Witkin, E.M. (1967) Mutation-proof and mutation-prone modes of survival in derivatives of *Escherichia coli* B differing in sensitivity to ultraviolet light. Brookhaven Symp. Biol. 20:17-55.

74. Zakharov, I.A., T.N. Kozina, and I.V. Federova (1968) Increase of spontaneous mutability in ultraviolet-sensitive yeast mutants. *Dokl. Akad. Sci., SSSR* 181:470-472.

LIST OF PARTICIPANTS AND HOME COUNTRIES*

Abbondandolo, Angelo, *Italy*
Aikawa, Katsuhiro, *Japan*
Albini, Adriana, *Italy*
Anderson, Diana, *England*
Anwar, Wagida, *Egypt*
Arimoto, Sakae, *Japan*
Barrai, Italo, *Italy*
Bagewadikar, R., *Germany*
Bagnasco, Maria, *Italy*
Balanski, Roumen, *Italy*
Balbinder, Elias, *USA*
Banerjee, Samarendra, *India*
Barale, Roberto, *Italy*
Bartsch, Helmut, *France*
Belisario, Maria, *Italy*
Benova, Donka, *Bulgaria*
Bhide, S.V., *India*
Bianchi, Livia, *Italy*
Bichet, Nicole, *France*
Binkova, Blanka, *Czechoslovakia*
Bonatti, Stefania, *Italy*
Boone, Charles, *USA*
Boothman, David, *USA*
Borek, Carmia, *USA*
Brambilla, Giovanni, *Italy*
Bronzetti, Giorgio, *Italy*
Cantelli-Forti, Giorgio, *Italy*
Cerutti, Peter, *Switzerland*
Cesarone, Frederico, *Italy*
Conforti-Froes, Nivea, *Brazil*
Corneanu, Gabriel, *Romania*
Corneanu, Michaela, *Romania*
Cruces-Martinez, Martha, *Mexico*
D'Agostini, Francesco, *Italy*
De Flora, Silvio, *Italy*
Degan, Paolo, *Italy*
Del Carratore, Renata, *Italy*
Della Croce, Clara, *Italy*
Della Morte, Rosella, *Italy*

Della Porta, Giuseppe, *Italy*
De Luca, Luigi, *USA*
De Meo, Michel, *France*
Di Lorenzo, Ann Marie, *USA*
Di Paolo, Joseph, *USA*
Dolara, Piero, *Italy*
Dumenil, G., *France*
Ebata, Junko, *Japan*
Ember, Istvan, *Hungary*
Fabiani, Leila, *Italy*
Feo, Francesco, *Italy*
Fiorio, Roberto, *Italy*
Frank, Norbert, *Germany*
Fratta, Duccio, *Italy*
Furukawa, Hideyuki, *Japan*
Galli, Alvaro, *Italy*
Gandolfi, Ivana, *Italy*
Gaspar, Jorge, *Portugal*
Gebhart, E., *Germany*
Gelormini, Alfonso, *Italy*
Giannoni, Paolo, *Italy*
Harris, Curtis, *USA*
Hartman, Philip, *USA*
Hartwig, Andrea, *Germany*
Hayatsu, Hikoya, *Japan*
Hirose, Masao, *Japan*
Hisamatsu, Yoshiharu, *Japan*
Hrelia, Patrizia, *Italy*
Ioannides, C., *England*
Itsukaichi, Osamu, *Japan*
Izzotti, Alberto, *Italy*
Kappas, Andreas, *Greece*
Kasai, Hiroshi, *Japan*
Kelloff, Gary, *USA*
Kennedy, Ann, *USA*
Kensler, Thomas, *USA*
Kikugawa, Kiyomi, *Japan*
Knasmüller, Siegfried, *Austria*
Kopsidas, George, *Australia*

Krishna, Gopal, *USA*
Kuroda, Yukiaki, *Japan*
Krutovskikh, Vladimir, *France*
Laxminarayana, D., *USA*
Laget, Michele, *France*
Le Bon, Anne-Marie, *France*
Leoni, Valerio, *Italy*
Maltoni, Cesare, *Italy*
Martelli, Antonietta, *Italy*
Matula, Tibor, *Canada*
Maturo, Maria, *Italy*
Mazué, Guy, *Italy*
Melchiori, Antonella, *Italy*
Mendelsohn, Mortimer, *USA*
Meyer, David, *England*
Mitscher, Lester, *USA*
Mohammed, Abdulmalik, *Sweden*
Moldéus, Peter, *Sweden*
Moreno, Fernando, *Brazil*
Moretti, Massimo, *Italy*
Mori, Hideki, *Japan*
Morichetti, Elena, *Italy*
Morishita, Y., *Japan*
Nagao, Minako, *Japan*
Nagel, Rosa, *Argentina*
Nandan, Sadhu, *USA*
Narbonne, Jean-F., *France*
Negishi, Tomoe, *Japan*
Niro, Antonio, *Italy*
Nistor, C., *Romania*
Nohmi, Takehiko, *Japan*
Partrineli, Alexandra, *Greece*
Perticone, Paolo, *Italy*
Pluygers, Eric, *Belgium*
Pool-Zobel, Beatrice, *Germany*
Rainaldi, Giuseppe, *Italy*
Ramel, Claes, *Sweden*
Ribeiro, L.R., *Brazil*
Rodrigues, A.S., *Portugal*

Rosellini, Dino, *Italy*
Rosin, Miriam, *Canada*
Salvadori, Claudio, *Italy*
Salvadori, Daisy, *Brazil*
Scarabelli, Linda, *Italy*
Schulte-Frohlinde, D., *Germany*
Semba, Mari, *Japan*
Sen, Soumitra, *India*
Shankel, Delbert, *USA*
Shigenaga, Mark, *USA*
Simic, Draga, *Yugoslavia*
Simic, Michael, *USA*
Sobels, Fritz, *Netherlands*
Sram, R.J., *Czechoslovakia*
Staiano, Norma, *Italy*
Stavenuiter, Johannes, *Netherlands*
Suganuma, Masami, *Japan*
Suzuki, Kunio, *Japan*
Takahashi, Kazuhiko, *Japan*
Toma, Salvatore, *Italy*
Ujica, Mihaela, *Romania*
Vellosi, Rita, *Italy*
Villani, Guglielmo, *Italy*
Von Borstel, R.C., *Canada*
Wall, Monroe, *USA*
Barbour, Warren, *USA*
Waters, Michael D., *USA*
Weisburger, John, *USA*
Willems, Marianne, *Netherlands*
Yamazoe, Ysushi, *Japan*
Yoshikawa, Kentaro, *Japan*

* This list contains the names of some participants who preregistered for the conference but were unable to attend due to international tensions or funding problems.

INDEX

The manufacturer's authorised representative in the EU is Springer
Nature Customer Service Centre GmbH, Europaplatz 3, 69115 Heidelberg,
Germany. If you have any concerns regarding our products, please
contact ProductSafety@springernature.com

Printed and bound by CPI Group (UK) Ltd, Croydon, CR0 4YY
23/04/2026
02095607-0019